AF249078

Series on Quality, Reliability and Engineering Statistics Vol. 7

Mathematical and Statistical Methods in Reliability

SERIES IN QUALITY, RELIABILITY & ENGINEERING STATISTICS

Series Editors: M. Xie (National University of Singapore)
T. Bendell (Nottingham Polytechnic)
A. P. Basu (University of Missouri)

Published

Series on Quality, Reliability and Engineering Statistics Vol. 7

Mathematical and Statistical Methods in Reliability

editors

Bo H Lindqvist

Norwegian University of Science and Technology,
Trondheim, Norway

Kjell A Doksum

University of Wisconsin, Madison, USA

Published by

World Scientific Publishing Co. Pte. Ltd.

5 Toh Tuck Link, Singapore 596224

USA office: Suite 202, 1060 Main Street, River Edge, NJ 07661

UK office: 57 Shelton Street, Covent Garden, London WC2H 9HE

British Library Cataloguing-in-Publication Data
A catalogue record for this book is available from the British Library.

MATHEMATICAL AND STATISTICAL METHODS IN RELIABILITY
Series on Quality, Reliability and Engineering Statistics — Vol. 7

ISBN 981-238-321-2

Printed in Singapore by Multiprint Services

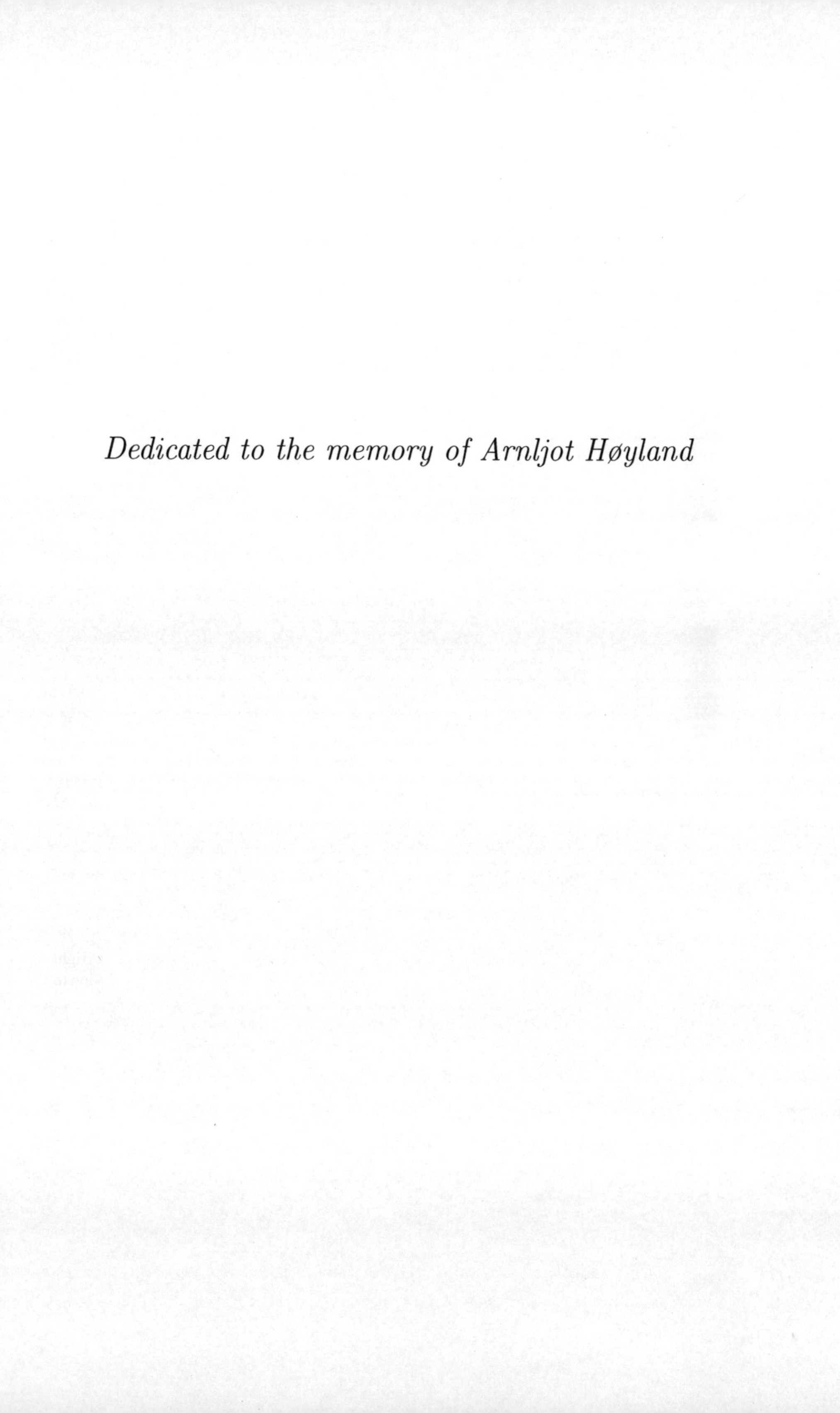

Dedicated to the memory of Arnljot Høyland

CONTENTS

PREFACE

Reliability usually refers to the ability of some piece of equipment or component to satisfactorily perform the task for which it was designed or intended. Originally, reliability was considered a qualitative property. For example, the desirability of having more than one engine on an airplane was well recognized without relying on any kind of performance or failure data. In the same spirit, an experienced low reliability of a system was often compensated by an unsystematic attempt to improve its individual components.

Today reliability is almost always used as a quantitative concept. Reliability in this sense is commonly characterized as the probability of an item fulfilling its function along its life cycle. This is of course where mathematical and statistical methods enter the scene. A natural question is therefore: "What is mathematical reliability theory?" In the preface of their classical book, "Mathematical Theory of Reliability", the authors Richard E. Barlow and Frank Proschan start by asking exactly this question. They give the following comprehensive answer which, although written in 1964, we still find appropriate and well suited as an introduction to the contents of the present volume:

"Generally speaking, it is a body of ideas, mathematical models, and methods directed toward the solution of problems in predicting, estimating, or optimizing the probability of survival, mean life, or, more generally, life distribution of components or systems; other problems considered in reliability theory are those involving the probability of proper functioning of the system at either a specified or an arbitrary time, or the proportion of time the system is functioning properly. In a large class of reliability situations, maintenance, such as replacement, repair, or inspection, may be performed, so that the solution of the reliability problem may influence decisions concerning maintenance polices to be followed".

Although many important building blocks of reliability theory thus were settled already by the mid sixties, a truly significant development of the field has taken place since then. The main driving force has been the needs of the industry and society to improve on reliability and safety of equipment, products and processes. Much of todays research is directed towards improving quality, productivity, and reliability of manufactured products. Further, during the last decades, reliability theory has been considerably developed through its important applications in the construction and operation of nuclear power plants, in connection with the offshore oil industry and in the design and operation of modern aircrafts. There has also throughout the last years been an increasing and fruitful cooperation between researchers in reliability theory and researchers in biostatistics working on survival analysis.

The present volume is published in connection with the Third International Conference on Mathematical Methods in Reliability, which was organized in Trondheim, Norway, June 17-20, 2002. The volume is the successor of the books "Statistical and Probabilistic Models in Reliability" (Birkhäuser Boston, 1999) and "Recent Advances in Reliability Theory" (Birkhäuser Boston, 2000), which were published on the occasions of the two first conferences in the MMR series (Bucharest, 1997, Bordeaux, 2000, respectively). The MMR conferences serve as a forum for discussing fundamental issues on mathematical and statistical methods in reliability theory and its applications, assembling researchers from universities and research institutions all over the world.

The volume contains extended versions of 34 carefully selected papers presented at the conference. The intention of the book is to give an overview of current research activities in reliability theory and survival analysis. The chapters present original and important research and are all refereed. To facilitate the use of the book, we have divided the individual contributions into nine parts, listed below. This division also reflects the wide range of topics being covered:

Reliability theory in the past and present centuries
General aspects of reliability modelling
Reliability of networks and systems
Stochastic modelling and optimization in reliability
Modelling in survival and reliability analysis
Statistical methods for degradation data
Statistical methods for maintained systems
Statistical inference in survival analysis
Software reliability methods

The intended audience is academics, professionals and students in probability and statistics, reliability analysis, survival analysis, industrial engineering, software engineering, operations research and applied mathematics.

The editors would like to thank the authors for their contributions, as well as the members of the scientific committee of MMR 2002 for their advice and help in the preparation of the volume. In particular we are grateful to Min Xie for his help in the beginning of the project and for his continuous support throughout the process. We are also grateful for the careful and thorough work by those who acted as referees of the individual chapters. Indeed, many of the authors commented on the high quality and importance of the reviews.

Many thanks to Randi Håpnes at NTNU for her assistance in the project. We would finally like to express our sincere thanks to Chelsea Chin, the Editor, whose patience has been outstanding and who has always been there to help us.

Bo H. Lindqvist,
Trondheim,
Norway

Kjell A. Doksum,
Madison, Wisconsin,
USA

CONTRIBUTORS

Odd O. Aalen, University of Oslo, Norway

D. L. Antzoulakos, University of Piraeus, Greece

Elja Arjas, University of Helsinki, Finland

Terje Aven, Stavanger University College, Norway

V. Bagdonavičius, University of Vilnius, Lithuania

N. Balakrishnan, McMaster University, Hamilton, Canada

Richard E. Barlow, University of California, Berkeley, USA

Prasanta Basak, Penn State Altoona, Pennsylvania, USA

Bruno Bassan, Università "La Sapienza", Roma, Italy

S. Bersimis, University of Piraeus, Greece

Madhuchhanda Bhattacharjee, University of Helsinki, Finland

A. Bikelis, University of Vilnius, Lithuania

Philip J. Boland, National University of Ireland, Dublin, Ireland

Cornel Bunea, Delft University of Technology, The Netherlands

Adrian Constantinescu, "POLITEHNICA" University, Bucharest, Romania

Roger Cooke, Delft University of Technology, The Netherlands

Richard Dagg, City University, London, UK

Kjell Doksum, University of Wisconsin, Madison, USA

Laurent Doyen, Institut National Polytechnique de Grenoble, France

Luis A. Escobar, Louisiana State University, Baton Rouge, USA

Olivier Gaudoin, Institut National Polytechnique de Grenoble, France

Håkon K. Gjessing, Norwegian Institute of Public Health, Oslo, Norway

Tina Herberts, University of Ulm, Germany

Li-Fei Huang, University of Wisconsin, Madison, USA

Arne Bang Huseby, University of Oslo, Norway

Michael Ingleby, University of Huddersfield, UK

Dumitru Cezar Ionescu, "POLITEHNICA" University, Bucharest, Romania

Uwe Jensen, University of Ulm, Germany

Jiancheng Jiang, Peking University, Beijing, China

Richard Johnson, University of Wisconsin, Madison, USA

Waltraud Kahle, University of Applied Sciences Magdeburg-Stendal, Magdeburg, Germany

V. Kazakevičius, University of Vilnius, Lithuania

Sallie A. Keller-McNulty, Los Alamos National Laboratory, New Mexico, USA

Chungwai Kong, Singapore Airlines, Singapore

M. V. Koutras, University of Piraeus, Greece

Laura L. Kramer, Hewlett-Packard, Corvallis, Oregon, USA

Danny L. Kugler, Hewlett-Packard, Corvallis, Oregon, USA

Jan Terje Kvaløy, Stavanger University College, Norway

C. D. Lai, Massey University, Palmerston North, New Zealand

Helge Langseth, Norwegian University of Science and Technology, Trondheim, Norway

James Ledoux, Centre de Mathmatiques INSA & IRMAR, Rennes, France

Gregory Levitin, The Israel Electric Corporation, Haifa, Israel

Nikolaos Limnios, Université de Technologie de Compiègne, France

Bo Henry Lindqvist, Norwegian University of Science and Technology, Trondheim, Norway

Anatoly Lisnianski, The Israel Electric Corporation, Haifa, Israel

Charles E. Love, Simon Fraser University, Burnaby, British Columbia, Canada

William Q. Meeker, Iowa State University, Ames, Iowa, USA

D. N. P. Murthy, University of Queensland, Brisbane, Australia

Arvid Naess, Norwegian University of Science and Technology, Trondheim, Norway

Morten Naustdal, University of Oslo, Norway

Martin Newby, City University, London, UK

M. Nikulin, Victor Segalen University Bordeaux 2, France

Brahim Ouhbi, Ecole Nationale Supérieure d'Arts et Métiers de Meknès, Morocco

Edsel A. Peña, University of South Carolina, Columbia, USA

David F. Percy, University of Salford, UK

Urho Pulkkinen, VTT Industrial Systems, Finland

Ioan Rotaru, National Company "NuclearElectrica Inc.", Bucharest, Romania

F. J. Samaniego, University of California, Davis, California, USA

Santos Faundez Sekirkin, National University of Ireland, Dublin, Ireland

Carlo Sempi, Università di Lecce, Italy

Harshinder Singh, West Virginia University, Morgantown, USA

Nozer Singpurwalla, The George Washington University, Washington, DC, USA

Fabio Spizzichino, Università "La Sapienza", Roma, Italy

Andrew Swift, Worcester Polytechnic Institut, Massachusetts, USA
Paul Ulmeanu, "POLITEHNICA" University, Bucharest, Romania
Igor Ushakov, San Diego, USA
E. M. Vestrup, DePaul University, Chicago, Illinois, USA
Margaret West, University of Huddersfield, UK
Alyson G. Wilson, Los Alamos National Laboratory, New Mexico, USA
M. Xie, National University of Singapore, Singapore
Shelemyahu Zacks, Binghamton University, New York, USA

Part I

RELIABILITY THEORY IN THE PAST AND PRESENT CENTURIES

1

MATHEMATICAL RELIABILITY THEORY: FROM THE BEGINNING TO THE PRESENT TIME

Richard E. Barlow

College of Engineering, University of California, Berkeley, CA 94720, USA
E-mail: barlow@newton.berkeley.edu

It is argued that the mathematical theory of reliability as a separate discipline began in 1961 with the publication of "Multi-component systems and their structures and their reliability" by Birnbaum, Esary and Saunders[8]. Prior to this time, mathematicians were just applying standard mathematical techniques such as queueing theory, statistics and probability to engineering reliability problems. We will describe how the 1965 book[5] "Mathematical Theory of Reliability" came to be written. Some personal historical perspectives will follow on probabilistic concepts of aging. Finally, we will discuss more recent work on Schur functions and Bayesian implications for reliability research.

1. Coherent Systems

Reliability became a subject of great engineering interest in the 1950's due to the failure of American rockets as well as the failure of the first commercial jet aircraft; the British de Havilland comet. Life testing was part of this engineering interest. Epstein and Sobel's[10] 1953 paper studying the exponential distribution was a landmark contribution. However it was not until 1961 with the publication of Birnbaum, Esary and Saunders[8] paper on coherent structures that reliability theory began to be treated as a separate subject. The emphasis in this paper is on *theory*.

The Boeing 707 was under development at the time the de Havilland comets were crashing. It was partly for this reason that the Boeing Scientific Research Laboratories in Seattle began to emphasize reliability theory in their mathematics division. Z. W. Birnbaum from the University of Washington was a consultant to this group. Z. W. had a strong mathematical background. He studied under Steinhaus and Banach among others. He had

3

  R. E. Barlow

Fig. 1.   Z. W. Birnbaum, 1903-2000

a special talent for getting quickly to the nub of a problem, especially in his consulting. He was adept at formulating appropriate theoretical models to capture the essential aspects of applied problems. In a 1974 conference dedicated to Birnbaum, Sam Saunders introduced him as follows: "I must remind you that Birnbaum is not a man to whom everyone looks up. In fact I estimate that 85% of all the men and 50% of all the women working in Reliability Theory today look down on him. The sad fact is, he is not a giant among the workers in his field." (Birnbaum was 5' 4" tall.)

A coherent system can be defined in terms of a binary function, $\phi(x)$; a system structure function which is non-decreasing in each vector argument and such that each component is relevant. (A component is *irrelevant* if it doesn't matter whether or not it is working.) Such systems are called coherent. Not all systems of interest are coherent but the class is sufficiently large to be of considerable significance. Figure 2 is an example of a coherent system. The system is operational if there is a working path from source to terminal. Arcs are assumed to fail independently but may have different failure probabilities.

Perhaps the most beautiful result concerning coherent systems is the IFRA closure theorem. (IFRA stands for increasing failure rate on the average.) The theorem says that the class of IFRA life distributions is the smallest class 1) containing the exponentials, 2) that is closed under the

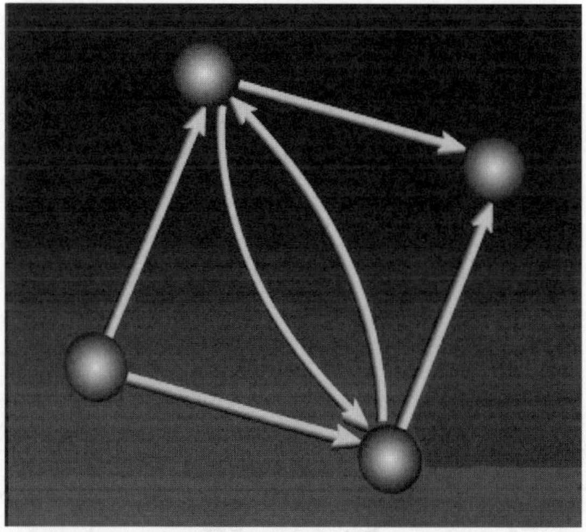

Fig. 2.   Directed Graph

formation of coherent systems and 3) is closed under limits in distribution. Although life times are assumed independent, life distributions may differ. This theorem, due to Birnbaum, Esary and Marshall[7], was published in 1966. Sheldon Ross[14] in 1972 provided a simplified proof of the theorem.

Since coherent structures may be very complex, there has been a great deal of interest in their efficient probability calculation. A. Satyanarayana[15] in 1978 introduced the idea of *domination* into the reliability literature. Starting with the minimal path sets, the *signed domination* is the number of odd formations of a coherent system minus the number of even formations. A *formation* is a set of minimal path sets whose union is the set of all components (or arcs in the case of networks). The *domination* is the absolute value of the signed domination. Using these ideas, Satyanarayana and Chang[16] proved in 1983 that the *Factoring Algorithm* is the most efficient algorithm for undirected networks based on series-parallel probability reductions and pivoting when 1) arcs fail independently and 2) nodes are deemed perfect. Figure 3 illustrates the factoring algorithm by a binary computational tree. The two leaves at the bottom of the tree are series-parallel reducible. The domination coincides with the number of leaves at the bottom of the binary computational tree. The domination is 2 in this case. Arne Bang Huseby[12] in 1984 provided a more abstract unified theory of domination and signed domination with application to exact reliability calculation.

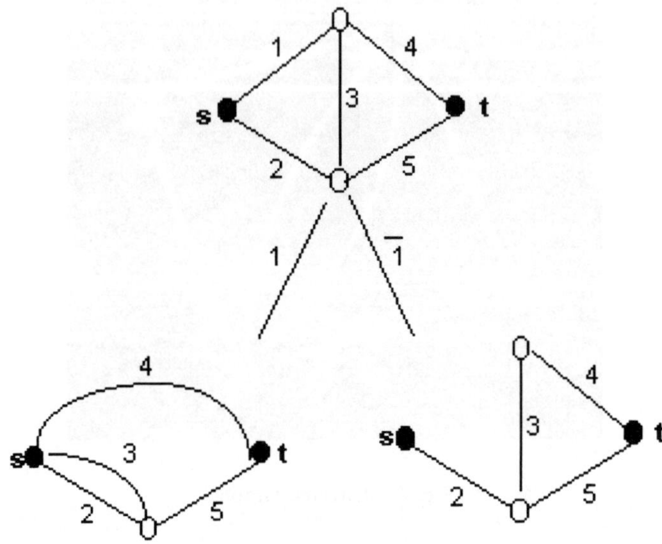

Fig. 3.   The Binary Computational Tree

2. Mathematical Theory of Reliability Published in 1965

The research monograph, "Mathematical Theory of Reliability"[5], was in part a product of the Cold War. I first met Frank Proschan in 1958 at the Electronic Counter measures Laboratory (EDL), a Cold War quick reaction facility in Mountain View, California. We were both Ph.D. students in Statistics at Stanford. Frank finished first in 1959. Frank had a poker face when telling the most outrageous stories. He was very witty even when delivering technical papers. I remember once when I was visiting Frank in Tallahassee sitting around a swimming pool in his apartment complex. A woman who lived in the complex began telling how wonderful it was since her recent divorce. After about 5 minutes of this, Frank said: "Your divorce must have been made in heaven." Often, however, his humorous comments were about himself rather than other people.

In the summer of 1960 Rudy Drenick, representing SIAM, came to EDL and proposed that we write a research monograph in reliability theory. He had first approached Z. W. Birnbaum who had declined but suggested

Fig. 4. Frank Proschan, Jim Esary and Al Marshall

Frank's name. We thought that this was a great opportunity and quickly signed a book contract. Originally Larry Hunter was to join us but had to drop out.

We submitted a monograph proposal to John Wiley & Sons, the publisher of the SIAM series. A reviewer of the proposal suggested that we include material about life distributions. This was missing in the proposal. Much of the literature on the failure rate function or the hazard rate was in the insurance literature. This inspired our interest in working on IFR (increasing failure rate) distributions. In 1960-61 I began working with Albert Marshall at Princeton while Frank went to the Boeing Scientific Research Laboratories. Together we began the study of IFR distributions.

Frank's Ph.D. thesis was in part concerned with optimal redundancy. At this time, spare parts allocation was of great interest to the military. The problem was to achieve an optimal allocation of redundancy; that is, maximize system reliability for the cost, weight, or volume, etc., allowed. Frank developed an algorithm for determining an undominated family of optimal solutions; that is each member of the family has the property that any allocation achieving higher reliability must be costlier, heavier or bulkier. His algorithm depended on the log concavity of the survival distribution corresponding to the convolution of n iid random variables. (The survival distribution of a single item is log concave iif it is IFR.) For this reason we spent a great deal of time trying to prove the IFR convolution theorem. Using total positivity, Frank provided the neatest though not the first proof. The convolution result and other IFR results were published in a 1963 IMS paper. In a 1975 textbook[2] with Frank we conjectured that IFRA distributions were also closed under convolution. In a 1976 paper, Block and Savits[9] provided an elegant proof of this result.

In the summer of 1961 I joined the GTE Research Laboratories in Menlo Park (now defunct). I discovered the crossing property for IFR survival distributions with respect to exponential survival distributions with the same mean. Continuing to work with Al Marshall, we published several papers on inequalities and bounds for IFR distributions starting in 1964[1].

During the Cold War, the Russians were also very interested in reliability. In 1965, Gnedenko, Belyayev and Solovyev published "Mathematical Methods of Reliability Theory" in Russian. An English version[11] appeared in 1969. They applied queueing theory ideas to reliability problems especially involving maintenance and replacement problems. In 1967 Frank visited Gnedenko in Moscow.

3. Reliability and Fault Tree Analysis

A conference on Reliability and Fault Tree Analysis occurred at Berkeley in 1974. In retrospect, perhaps one of the most influential papers in this volume leading to reliability research in the 1990's was the elegant paper by Proschan[13], "Applications of Majorization and Schur Functions in Reliability". In this expository paper, Frank introduces the ideas of majorization and Schur functions from inequality theory and uses them to obtain bounds, comparisons, and inequalities in reliability and life testing. Much later, Max Mendel and myself used these ideas to define multivariate IFR distributions for exchangeable random quantities. Subsequently, Fabio Spizzichino[19] in 2001 published an excellent monograph further developing these ideas.

Engineers who studied engineering systems in great detail, with little or no contribution by mathematicians, developed fault tree analysis. A possible explanation for this comes from the fact that the construction of the fault tree, a basic step in fault tree analysis, requires an intimate knowledge of the manner in which a system is designed and operated. A key feature of this approach is that it is failure oriented rather than success oriented. Often for this reason it can be very useful in improving the reliability of systems.

4. Concepts of Aging and Schur Concavity

Although a great deal of research dealt with IFR distributions, there were theoretical problems.

4.1. *First difficulty with the IFR idea*

Consider the case of two components in parallel as in Figure 5.

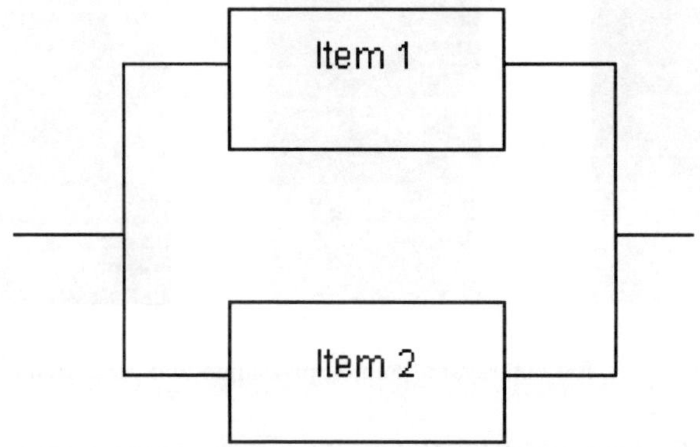

Fig. 5.   Two items in parallel

Jim Esary at the Boeing Scientific Research Laboratories noticed that if each had an exponential life distribution, i.e. constant failure rate, but the failure rates were different, then the life distribution of the parallel system was not IFR. However the failure rate is increasing on the average; i.e. is IFRA. This observation motivated research leading to the IFRA closure theorem.

4.2. *Second difficulty with IFR distributions*

Suppose you index a distribution, say the exponential using the mean life, θ, e.g.

$$F(x|\theta) = 1 - e^{-x/\theta}, \quad \text{for } x \geq 0$$

Usually, in applications, you do not know the mean life. Suppose, however, that you have prior knowledge that can be encapsulated in a probability distribution. In this case your unconditional distribution for lifetime is

$$\int_0^\infty F(x|\theta)p(\theta)d\theta = \int_0^\infty [1 - e^{-x/\theta}]p(\theta)d\theta.$$

The unconditional or predictive life distribution has a decreasing failure rate function. Likewise IFR distributions are not closed under mixing. This again suggests that something is wrong with the univariate IFR characterization of aging.

                                R. E. Barlow

Fig. 6.   A. Satyanarayana, Fabio Spizzichino and Max Mendel

4.3. *Aging Represented by Schur Concave Joint Survival Distributions*

New ideas are what make good research both possible and interesting. The new idea relative to answering the second difficulty was to consider not just a single lifetime but instead a collection of exchangeable lifetimes. Fabio Spizzichino[18] said it best. In words, his mathematical result says that among any two items from n similar items (i.e. exchangeable items) that have survived a life test, the "younger" is the "better", if and only if the joint survival function is *Schur concave*. Mathematically, he proved that $\bar{F}(x_1, x_2, \ldots, x_n)$ is Schur concave, if and only if, for any $t > 0$ and $x_i < x_j$,

$$P(X_i > x_i + t | X_1 > x_1, \ldots, X_n > x_n) \geq$$
$$P(X_j > x_j + t | X_1 > x_1, \ldots, X_n > x_n).$$

This is an intuitive restatement of the IFR idea only now for *conditional joint* survival distributions. If the conditional joint survival distribution $\bar{F}(x_1, x_2, \ldots, x_n | \theta)$ is Schur concave, then it is still Schur concave unconditionally since Schur concavity is defined in terms of an inequality on $\bar{F}$. Using Schur concavity of the joint survival probability as our new definition of aging, the second difficulty with the univariate IFR definition is overcome. In the case of the exponential distribution, Schur constancy is preserved under mixing.

The ideas in Spizzichino's[18] 1992 paper were extended to a more general analysis of Bayesian multivariate aging using different stochastic comparisons of residual lifetimes for units having different ages. Bassan and

Spizzichino[6] define a notion of multivariate IFRA but not in terms of Schur concavity. The IFRA notion of aging seems to be very different from the IFR notion of aging or of its generalization in terms of Schur concavity.

5. Physical Foundations for Probability Distributions

Since reliability is concerned with uncertainty questions about engineering systems, it would make sense to "derive" appropriate probability distributions based on the engineering physics of a given problem. This would seem

to make more sense than picking a mathematically convenient probability distribution more or less at random. It seems, however, that to do this we need to adopt a Bayesian approach to probability. That is, judgments such as indifference relative to certain basic random quantities must be made. For example, suppose we are interested in the stress level relative to yielding in a given material. Suppose furthermore that we believe Hooke's Law is valid in this case. Starting with an indifference assumption regarding vectors of distortion energies with the same mean, we are led to the Weibull distribution for stress level at yielding with shape parameter equal to 2, Barlow and Mendel[3].

A controversial figure, Max Mendel, appeared on the reliability scene in 1989. His MIT Ph.D. thesis in Mechanical Engineering concerned probability derivations based on engineering principles. Beginning in 1994, Mendel began exploring the use of differential geometry for the purpose of deriving probability distributions. This eventually led to the conclusion that lifetime spaces are *not* physical Euclidean spaces. The use of the hazard gradient, for example, to model multivariate hazard rates is therefore incorrect since it relies on the Euclidean metric.

Shortle and Mendel[17] argue as follows: Let L^N be the space of possible lifetimes for N items. Euclidean space is not a good representation for L^N for two reasons:

(1) L^N has a preferred orientation for its axes.
(2) L^N has no natural notion of distance.

Observe that Euclidean space is invariant under rotations, since rotations preserve the value of the inner product; i.e. there is no preferred orientation for the axes. We can characterize the physical structure of a space by the transformations that leave the space invariant. For Euclidean space, these are translations and rotations. For L^N these are changes of units of the individual items. This is because physical properties about lifetimes should not depend on the units used to measure lifetimes. In the language of differential geometry, the correct representation for the space of lifetimes is a collection of fiber bundles.

The new ideas in Shortle and Mendel[17] suggest a surprising and exciting new line of research in mathematical reliability theory.

References

1. R.E. Barlow and A. W. Marshall, Bounds for distributions with monotone hazard rate, I and II. *Annals of Mathematical Statistics*, **35**, 1234-1274 (1964).
2. R.E. Barlow and F. Proschan, *Statistical Theory of Reliability and Life Testing*, (Holt, Rinehart and Winston, New York, 1975).
3. R.E. Barlow and M. B. Mendel, The operational Bayesian approach, in *Aspects of Uncertainty*, Eds. P. R. Freeman and A. F. M. Smith (Wiley, Chichester, 1994), pp. 19–28.
4. R. E. Barlow, A. W. Marshall and F. Proschan, Properties of probability distributions with monotone hazard rate, *Annals of Mathematical Statistics*, **34**, 375-389 (1963).
5. R. E. Barlow and F. Proschan, *Mathematical Theory of Reliability*, (Wiley & Sons, New York, 1965), (Reprinted SIAM, Philadelphia, PA, 1996).
6. B. Bassan and F. Spizzichino, Dependence and multivariate aging: the role of level sets of the survival function, in *System and Bayesian reliability*, Eds. Y. Hayakawa, T. Irony and M. Xie, (World Scientific, Singapore, 2001) pp. 229-242.
7. Z. W. Birnbaum, J. D. Esary and A. W. Marshall, Stochastic characterization of wearout for components and systems, *Annals of Mathematical Statistics*, **37**, 816-825 (1966).
8. Z. W. Birnbaum, J. D. Esary and S. C. Saunders, Multi-component systems and structures and their reliability, *Technometrics*, **3**, 55-77 (1961).
9. H. Block and T. H. Savits, The IFRA closure problem, *Annals of Probability*, **4**, 1030-1032 (1976).
10. B. Epstein and M. Sobel, Life testing, *Journal of American Statistical Association*, **48**, 486-502 (1953).
11. B. V. Gnedenko, Yu. Belyayev and A. D. Solovyev, *Mathematical Methods of Reliability Theory* (Academic Press, New York, 1969).
12. A. B. Huseby, Domination theory and the Crapo beta-invariant, *Networks*, **19**, 135-149 (1989).
13. F. Proschan, Applications of majorization and Schur functions in reliability and life testing, in *Reliability and Fault Tree Analysis*, Eds. R. E. Barlow, J. B. Fussell and N. D. Singpurwalla, (Society for Industrial and Applied Mathematics, Philadelphia, 1975) pp. 237-258.
14. S. M. Ross, *Introduction to Probability Models with Optimization Applications* (Academic Press, New York, 1972).
15. A. Satyanarayana and A. Prabhakar, New topological formula and rapid

algorithm for reliability analysis of complex networks, *IEEE Transactions on Reliability*, **R-27**, 82-100 (1978).

16. Satyanarayana, A., and M. K. Chang (1983). Network reliability and the factoring theorem. Networks 13, 107-120.

17. J. F. Shortle and M. B. Mendel, Physical foundations for lifetime distributions, in *System and Bayesian reliability*, Eds. Y. Hayakawa, T. Irony and M. Xie, (World Scientific, Singapore, 2001) pp. 257-266.

18. F. Spizzichino, Reliability decision problems under conditions of ageing, in *Bayesian Statistics 4*, Eds. J. M. Bernardo, J. O. Berger, A. P. Dawid and A. F. M. Smith, (Clarendon Press, Oxford, 1992).

19. F. Spizzichino, it Subjective Probability Models for Lifetimes, (Chapman & Hall/CRC, Boca Raton, 2001).

2

RELIABILITY FOR THE 21ST CENTURY

Sallie A. Keller-McNulty and Alyson G. Wilson

Statistical Sciences Group
Los Alamos National Laboratory
P. O. Box 1663, D-1, MS F600
Los Alamos, NM 87545 USA
E-mail: sallie@lanl.gov

The sophistication of science and technology is growing almost exponentially. Government and industry are relying more and more on science's advanced methods to assess reliability coupled with performance, safety, surety, cost, schedule, etc. Unfortunately, policy, cost, schedule, and other constraints imposed by the real world inhibit the ability of researchers to calculate these metrics efficiently and accurately using traditional methods. Because of such constraints, reliability must undergo an evolutionary change. The first step in this evolution is to reinterpret the concepts and responsibilities of scientists responsible for reliability calculations to meet the new century's needs. The next step is to mount a multidisciplinary approach to the quantification of reliability and its associated metrics using both empirical methods and auxiliary data sources, such as expert knowledge, corporate memory, and mathematical modeling and simulation.

1. Introduction

By definition, *reliability* is the probability a *system* will perform its intended function for at least a given period of time when operated under some specified conditions. The 20th century solution to this problem[a] has been to define a reliability function as

$$R(t) = P(T > t) = \int_t^\infty f(x)dx = 1 - F(t) \tag{1}$$

---

[a] An excellent review of reliability theory and corresponding references can be found in Martz.[11]

and to use the function as the basis of definition for other important concepts, such as failure rate and mean time between failures. Powerful parametric (e.g., binomial, Poisson, exponential, Weibull) and nonparametric statistical models have been developed to estimate reliability and its associated properties. These traditional reliability methods were developed for industrial, mass-produced products such as electronics and consumer goods. Everything works quite nicely provided we have coherent system representations and clean, typically single, sources of quantitative data about the system. Problems today, however, are much more complex and include systems such as nuclear weapons, infrastructure networks, supercomputer codes, jumbo jets, etc. These systems demand more of reliability than our current methodology allows. In many instances it is not possible to mount vast numbers of full system tests, and frequently none are available.[2]

System assessment is complicated by the need to consider more than what has been traditionally considered as reliability, because a system's ability to perform is intertwined with other concepts such as its age, safety, and surety. In addition, our ability to do reliability assessments may be severely constrained by policy, cost, and schedule, particularly in problems dealing with the inherent reliability of an existing system. Therefore, we must expand our definition of the system to include all aspects that affect its performance and all constraints (e.g., test schedule) that affect the confidence we have in the assessment. The end result should be a reliability assessment that is an expression of our complete state of knowledge about the system.

Statisticians are frequently the scientists responsible for driving the reliability assessment process. Due to the demands stated above, their roles in this process must correspondingly (and significantly) broaden. This chapter provides a broad overview of some of the concepts and research that need to be brought together to address the reliability challenges of the 21st century. Section 2 motivates the fact that these are decision, not simply analysis or modeling, problems. Section 3 outlines the diverse research areas needed, Section 4 gives motivating examples, and Section 5 provides conclusions.

2. Decision Context

Today's reliability problems are driven by the need to support decision-making at any point in the life of the system under study, using the broader definition of system given above. This requires continuous, over the life of the system, integration of all information and knowledge into the decision-making process. Traditional decision analysis is based on objective functions and constraints that are well-defined and static. This is a serious flaw and may be the reason why formal decision-analytic projects frequently do not succeed. In addition, the incorporation of scientific knowledge, such as that derived from reliability assessments, into the decision-making process is weak. This is due to the inability to model evolving scientific discovery quickly and clearly and communicate the results in useful ways to decision-makers. The development of rigorous statistical and mathematical methods for this integration and evolution holds the promise for bridging the gap between narrowly defined reliability analyses, based on Equation 1, and these dynamic decision problems.

Once it is understood that the true *system* problem is a decision problem, this introduces the realization that diverse information sources, not simply test data, must be clearly understood and modeled. This is because diverse sources of information are what is used to guide prudent decision-making. More specifically, the diversity of scientific information that is used to support decision-making arises from:

- the sources of information, including theoretical models, test data, observational data, computer simulations, and expertise from scientists, field personnel, and decision-makers;
- the content of the information, including information about the system structure and behavior, decision-maker constraints, options, and preferences; and
- the multiple communities of practice that are the stakeholders in the decision process.

Figure 1 is a graphical view of the integrated reliability assessment that is needed to capture the full context of the decision problem, including the information sources listed above. Techniques from various disciplines (statistics, probability, mathematics, computer science, decision theory, graph theory, expert knowledge elicitation and representation, and simulation) must be merged to develop formal methods to integrate the multiple information sources relative to the content of the system evaluation (e.g., simultaneous assessment of performance, reliability, sustainability, dependability, safety, etc.). These assessments must support the needs of the stakeholders, such as complex resource allocation (e.g., mount a full-system test, build a new experimental facility, develop more scientific capability, implement a higher fidelity computer model) and continuous evaluation (i.e., supporting decisions that need to be made at, frequently unanticipated, multiple points in time). The methods must explicitly incorporate uncertainty and allow for dynamic changes in the evolution of knowledge and processes about the system being studied, thus enabling a flow of new information to support continuous decision-making. The solution and development of such methods is the 21st century challenge for reliability.

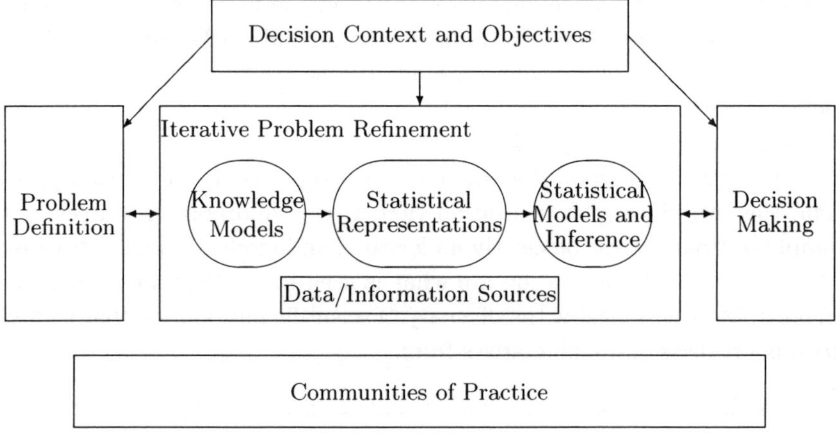

Fig. 1.   Schematic view of integrated reliability assessment.

3. Overview of Research Areas

This section provides a nontechnical overview of several, possibly nonobvious, research areas that must be further developed and integrated if the 21st century reliability challenges are to be realized. Section 4 demonstrates the need for these research areas and points to gaps in what is well understood vis-à-vis specific examples.

3.1. *Knowledge Representation*

If the goal is to support continuous decision-making for highly complex and innovative systems, statisticians will be frequently faced with the dilemma of minimal data for the system under scrutiny. Complicating such situations is the increasing ubiquity of multidisciplinary and multinational research teams. Statisticians often find themselves asked to contribute to complex, emergent projects that challenge their ability to build predictive models capable of integrating multiple types of data, information, and knowledge from a wide range of sources. Therefore, there is a need to develop a multidisciplinary approach to knowledge elicitation, representation, and transformation. This approach must mesh techniques from cultural anthropology, computer science, and statistics to address the complexities of multidisciplinary research.

Specifically, elicitation techniques derived from cultural anthropology can be used to elicit tacit problem-solving structures from the "natives"— generally, the scientists and engineers collaborating on difficult system problems. The elicited information, in turn, can be used to develop ontologies that represent the problem space in the "native language" of the research team, but which are more mathematically tractable to the computer science and statistical communities. Iterative cycles of representational refinement and quantification will lead to the emergence of predictive statistical models that make intuitive sense to all parties: the scientists, engineers, elicitation experts, knowledge modelers, and statisticians.[10] Important methodological challenges include advancing research on conceptual graphs, statistical graphical models, and the translation from qualitative to quantitative representations.

3.2. *Statistical Methods*

The need for aggressive research in statistical methods is clear. The research must address the treatment of heterogeneous and diverse information sources. Because pure, full-system testing approaches are frequently infeasible, methods that can parse the problem into subsystems and constituently fold together subsystem analyses to build full-system predictions must be developed. Mathematically, this is a complex, ill-posed problem. Solutions may easily take the statistician outside the comfort zone of traditional probability theory and into areas of fuzzy measurement, belief functions, and possibility theory. Methodological focuses will clearly include Bayesian inference, Bayesian hierarchical models, and computational methods for reliability and lifetime estimation; system reliability and lifetime analysis; computer model evaluation; methods for accelerated life testing; models for degradation data; and demonstration testing.[15,16]

Special challenges arise when the analyses must include expert judgment or the output from computer models. There is a body of work that looks at various aspects of expert judgment, from the construction of priors for statistical models, to the development of utility functions and decision analyses within economics, to the anthropological and psychological work on elicitation strategies.[4,14] Research is needed on the development of a conceptual framework for utilizing "statistical" information elicited from experts. When presented with a dataset containing experimental data, statisticians have a "toolkit" of methods and a set of canonical examples to use when deciding how to analyze the data. This toolkit is still absent from the realm of expert judgment.

Another important source of data that must be leveraged into 21st century system reliability problems is that derived from computer simulations. These computer models are often highly complex themselves, and there are many research challenges associated with the calibration and validation of these models under the conditions of limited real world data. Simulation data are often limited as well, because the models require many hours, days, or even months to run. Another challenge to the evaluation of these simulation data is the high dimensionality of the output. New research in the areas of complex computer model evaluation, applying techniques from sensitivity analysis, Bayesian interpolation methods, and extensions of techniques from spatial statistics is beginning to formalize the use and limitations of such data.[1,3,6,9,12,13]

Traditional experimental design is concerned with allocating trials within a single experiment. Suppose, however, that data is available from many diverse kinds of experiments including different types of physical experiments (e.g., destructive or nondestructive tests) and runs of a computer code. The newly coined area of *hybrid experimental design*[5] considers the allocation of test resources across different types of experiments by trading off the costs of performing any particular trial with the information gained. Significant research on the extension and optimization of such methods is needed.

3.3. *Knowledge Management*

The way in which information is organized has a major influence on how that information is used. With the wide range of information (and knowledge) needed to solve the problems for continuous decision-making and evaluation, formal organization of information is critical. This will require a variety of tools to capture information, organize it, and make the results available for subsequent analysis to the distributed communities working on a problem. Significant research is needed in knowledge management with a focus on the development of these tools both for the statistical and knowledge modeling researchers and for the stakeholders.

4. Motivating Examples

This section contains three current problems that illustrate the complexity of "system assessment" and the need for rigorous mathematical solutions. All three examples are defense related, which is fitting if one considers the role that defense agencies around the world have played in the development and adoption of reliability methods.

4.1. *Science-Based Stockpile Stewardship*

An example of the complexity facing scientists responsible for reliability assessment is Science Based Stockpile Stewardship (SBSS) at Los Alamos National Laboratory (LANL) and the history that has brought about this problem. From its earliest days, LANL has had a prominent role in the development and evaluation of the United States nuclear weapons stockpile, but the end of the Cold War brought significant changes to how this mission could be carried out. There have been significant reductions in the number of weapons, leading to a smaller, "enduring" stockpile. The United States is no longer manufacturing new-design weapons, and it is consolidating facilities across the nuclear weapons complex. In 1992, the United States declared a moratorium on underground nuclear testing; in 1995, the moratorium was extended, and President Clinton decided to pursue a "zero yield" Comprehensive Test Ban Treaty. However, the basic mission of LANL remains unchanged: LANL must evaluate the weapons in the aging nuclear stockpile and certify their safety, reliability, and performance even though the live test data that have traditionally been used for this evaluation can no longer be collected.

To complete this mission, a two-pronged approach of experiments and computational modeling was adopted. The experimental approach is exemplified by the Dual-Axis Radiography for Hydrotesting (DARHT) facility, which enables experimenters to better understand the nature of explosions. The computational modeling effort is exemplified by the Accelerated Strategic Computing Initiative (ASCI), which uses supercomputers to model the types of complex nuclear experiments that are no longer performed. At a fundamental level, though, the new experimental and computer technologies have not been developed to address SBSS; rather a "zero yield" policy could be negotiated and implemented because advances in computer technology made it seem feasible that the sophisticated modeling could be done to realize SBSS. In short, the promise of the technology drove the policy. It created an expectation that certain tough questions could be answered with adequate justification.

Alongside the efforts at experimentation and modeling, statisticians and knowledge modelers have been working to integrate historical data and to quantify the vast resources of expertise at LANL in such a way as to facilitate their inclusion through Bayesian statistical methods. The challenge is to integrate experimental data, computational models, past tests, subsystem tests, and the expert judgment of subject-matter experts to provide a rigorous, quantitative assessment, with associated uncertainties, of the safety, reliability, and performance of the stockpile.

Without careful attention to the whole picture, or purpose of the system assessment, the accomplishments of individual scientists can become lost and detached. Figure 2 is a notional representation of several elements of the SBSS problem. Within parts of the figure, traditional methodology works well for various questions. For example, event-tree methods can be used to define the critical paths for successful completion on the physical experiments and the risks involved that could affect the schedule. But, what happens if an experiment that is needed to help resolve some of the equation-of-state parameterizations for the computational experiments cannot be done? The uncertainty that results must be propagated through the computational models and accounted for in our statements about confidence in our assessments. This in turn will affect the design of other computational experiments. This is not a standard problem addressed through traditional reliability analysis.

The engineering portion of certification depicted in Figure 2 can be thought of as a traditional engineering reliability problem based on coherent system representations. However, there is rarely direct data available on all parts of the system. Therefore, we must develop methodology that can integrate other, related information and be able to propagate information up and down throughout the system representation.[7,15,16] A major challenge is to then integrate the engineering reliability information with the physics performance assessment, material degradation models, etc. In contrast to the discrete nature of the engineering component condition representations of coherent systems, the physics is represented as continuous, time-dependent, integrated processes. It is these two elements, engineering and physics, in combination, that are needed to understand the condition of the enduring stockpile. Once again our traditional reliability representations and treatments of problems do not address this integrated assessment.

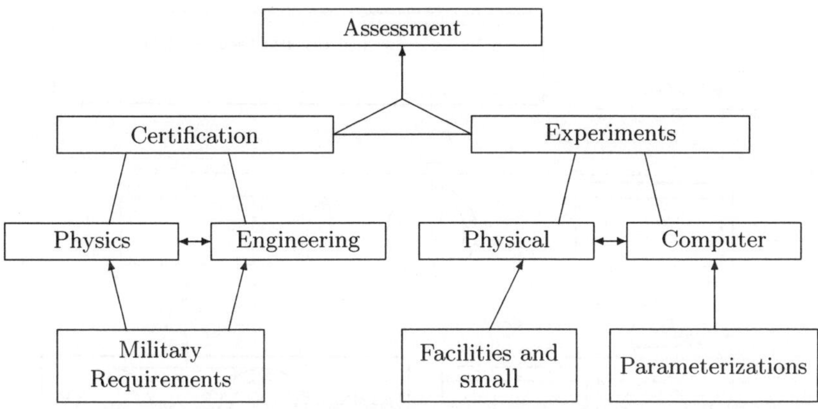

Fig. 2. Notional representation of an integrated assessment process for SBSS.

4.2. *Conventional Munitions Stockpile Surveillance*

An extension of the SBSS problem is the development of methodology
for assessing the reliability of aging stockpiles of conventional munitions.
These problems are similar to the assessment of nuclear weapons stockpiles,
but there is usually some full-system testing. Stockpile surveillance gathers
many kinds of data, including full-system tests, component and subsystem
information, and nondestructive evaluation. In addition, there are computer
codes that can predict properties of materials aging. The challenge is the
combination of information to support decisions about stockpile life extension programs. Figure 3 gives a notional representation of a system, its
subsystems, and the type of information available at a snapshot in time for
this system.

For this problem, a Bayesian hierarchal model has been developed,[7]
where the hierarchy reflects the similar component behavior in Figure 3.
The hierarchy was developed using knowledge modeling techniques and
formal elicitation. With the model, information can be leveraged across the
hierarchy to build reliability distributions for components with no test or
computer simulation data, and improve the reliability distributions for components with data. As more data become available at the subsystem and
component levels, the hierarchical model is used to do a downward propagation. This downward propagation results in a belief function representation
of the subsystem and component reliabilities. Figure 4 shows the estimated

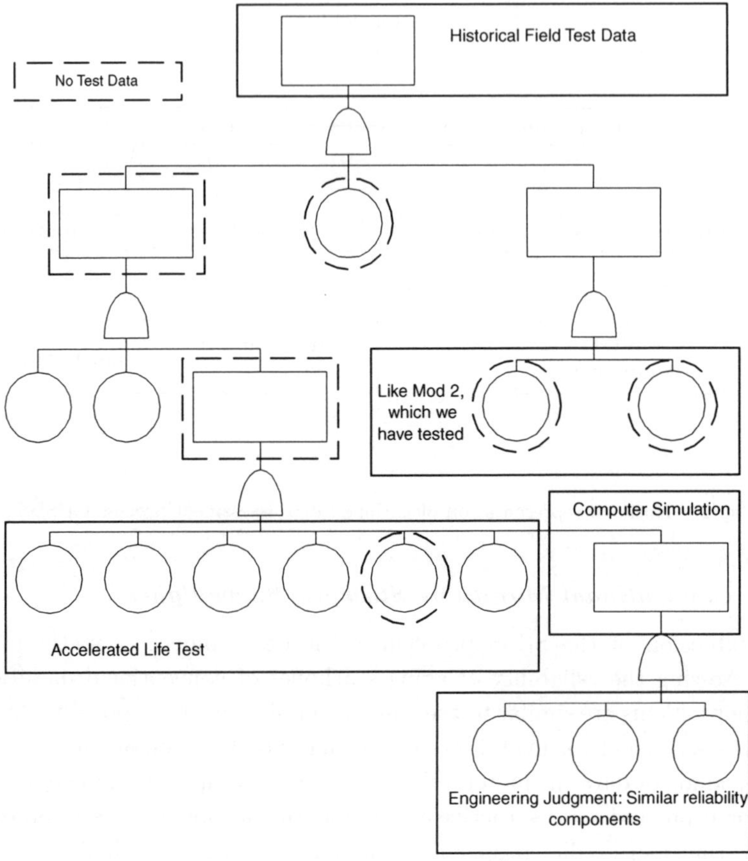

Fig. 3. Structural display of information sources for anti-aircraft missile problem.

reliability distribution for the full system and for two subsystems. For one of the subsystems, no test data existed. In all three cases, reliability distributions that included information for the full-system flight tests are displayed and estimated distributions without that information are displayed. As one would expect, more uncertainty exists when full-system test data are not used (available). The model is being extended to take into account time-dependent changes in the system architecture and material degradation. The model is also being extended to account for covariates, e.g., storage conditions. The ability to solve the covariate and time-dependent problem will be what helps support the dynamic and continuous decisions that must be made regarding this missile system. Mathematical details can be found in Johnson et al.[7]

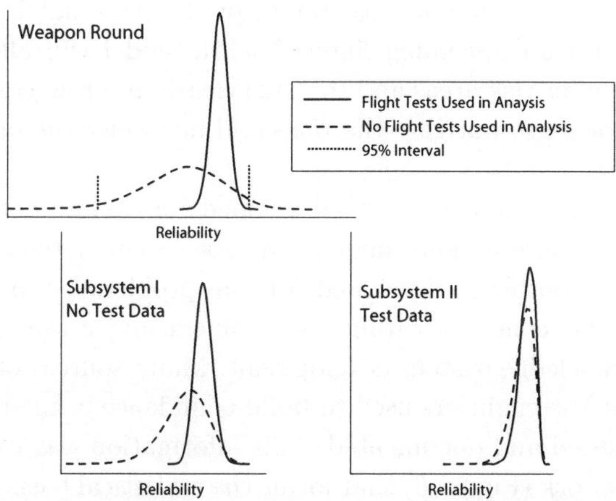

Fig. 4.   Reliability estimation for anti-aircraft missile stockpile surveillance.

4.3. *Mission Success for a Ballistic Missile*

As a third example, consider a research and development program that is charged with developing, flying, and collecting data about ballistic missile systems that emulate the flight characteristics of "threat" missiles. These flights are expensive and politically visible, and technical risk mitigation is an important element of the program. These flight tests are also very difficult, because they are "one-of-a-kind" events with many complex factors. One organization is in charge of project management, cost controls, and scheduling; another is in charge of building the missile booster to send the rocket into the upper atmosphere; a third is in charge of building the missile payload.

The program managers approached LANL with a specific problem: how does one develop a predictive reliability model for an engineering system that is still in the design stages? Multiple concerns drove this question: the rocket development program is extremely expensive. Only one or two of the prototypes are built and flown and are usually destroyed in the process; rarely are the engineers able to salvage subsystems for reuse in further iterations of the program. Because each system flown is unique, there is little direct performance or reliability data available for parts or subsystems on the test rocket. Hence the program managers had little idea how to make predictions or assess risk areas for the flights.

The goal is to develop an integrated, predictive reliability and performance model for an upcoming flight. Such a model will allow the stakeholders to pinpoint risk areas and to make clearly informed decisions about resource allocation to mitigate the risks and maximize the opportunity for mission success.

In developing the model, LANL developed a model framework that captured the critical interactions among the rocket's subsystems during flight. Figure 5 is the ontology developed for the problem—it maps the basic relationships and concepts within the problem and serves as a basis for subsequent knowledge system development. Many sources of data and information that the engineers used to build confidence in their rocket before flight were elicited and documented. This information was used to develop a Bayesian network (Figure 6) that forms the statistical basis for combining multiple sources of information in a rigorous, quantitative framework.

5. Conclusions

Traditional statistical science approaches to reliability based strictly on the reliability function given in Section 1 are no longer sufficient to address the reliability assessment process for multifaceted 21st century problems.[8] The complexities of big science problems such as SBSS, and the other examples in Section 4, demonstrate the impossibility of static system solutions. Today the overall assessment process is more about "decision-making" than "modeling." Many problems, such as these, are politically and economically charged. Therefore, even the best data collection design and corresponding statistical models for the problem at hand may not be feasible, or even allowed. Our 21st century reliability challenge is to be able to structure and overlay statistical models on integrated assessment processes, such as that represented in Figure 1. These models will need to be robust enough to support decision-making at various resolutions (e.g., about a specific experiment, engineering component design, or facility resource allocation to support the overall assessment process). States of knowledge about the system will be a collection of heterogeneous and diverse sources of information. These sources of information will need to be integrated via tractable mathematical models. The information will be coming from very different disciplines (e.g., physics, materials, chemistry, and engineering). Therefore, uncertainty quantification inherent in the statistical models will need to be

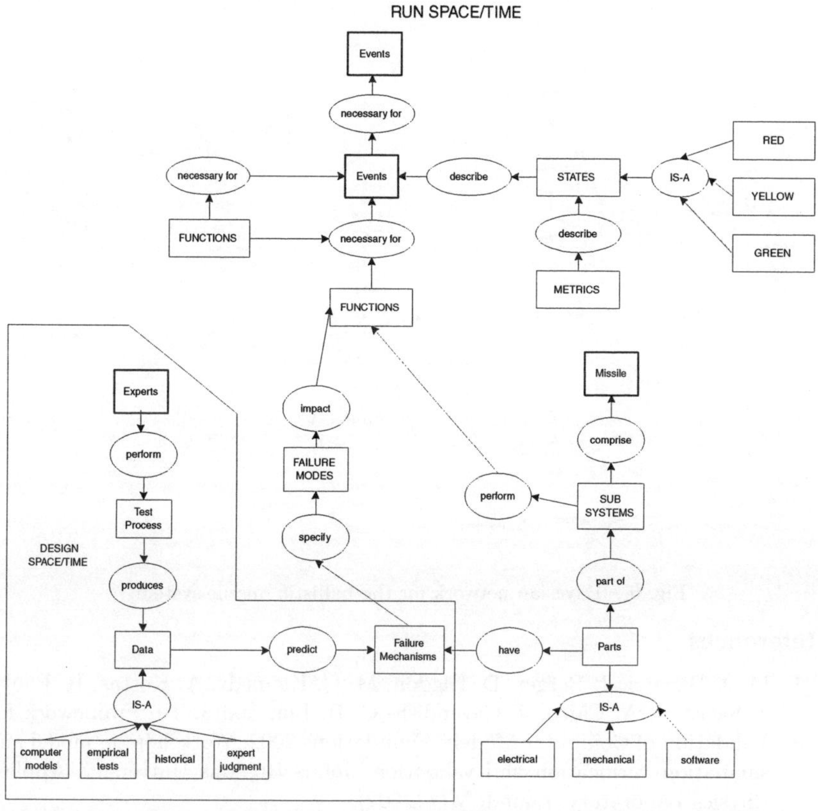

Fig. 5. Ballistic missile system ontology.

flexible to account for natural ways to represent the information (e.g., probability, fuzzy measures, belief functions, possibility theory, etc). With these challenges come wonderful opportunities for the advancement of reliability analysis and the significant advancement of science.

Acknowledgments

The authors acknowledge several colleagues involved in the problems described in Section 4. These include Art Dempster, Todd Graves, Michael Hamada, Nicolas Hengartner, David Higdon, Valen Johnson, Deborah Leishman, Laura McNamara, Mark McNulty, Jerome Morzinski, Shane Reese, and Nozer Singpurwalla. Los Alamos National Laboratory report LA-UR-02-7283.

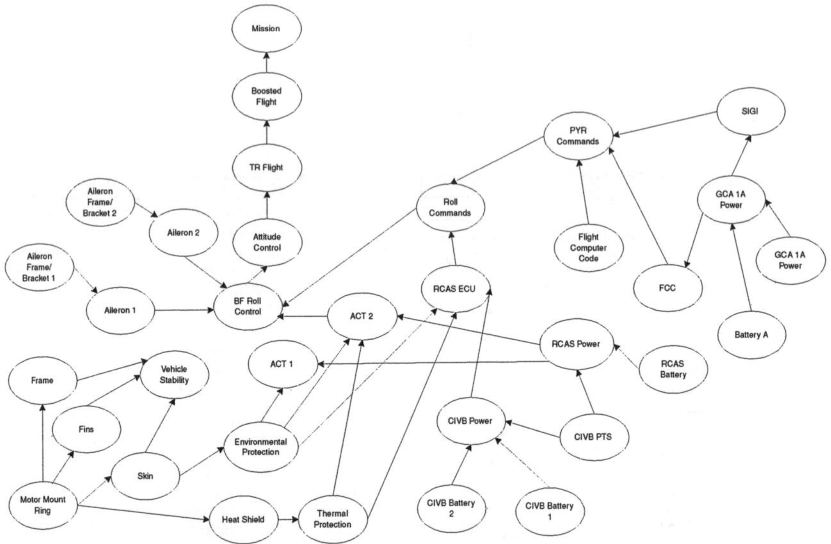

Fig. 6.   Bayesian network for the ballistic missile system.

References

1. M. J. Bayarri, J. Berger, D. Higdon, M. C. Kennedy, A. Kottas, R. Paulo, J. Sacks, J. A. Cafeo, J. Cavendish, C. H. Lin, and J. Tu, Framework for Validation of Computer Models, Foundations 2002: Workshop on model and simulation verification and validation, Johns Hopkins University, Applied Physics Laboratory, Laurel, MD (2002).

2. T. Bement, J. Booker, S. Keller-McNulty, and N. Singpurwalla, Testing the Untestable: Reliability in the 21st Century, *IEEE Transactions on Software Reliability* (to appear).

3. R. Berk, P. Bickel, K. Campbell, R. Fovell, S. Keller-NcNulty, E. Kelly, R. Linn, B. Park, A. Perelson, N. Roupail, J. Sacks, and F. Schoenberg, Workshop on Statistical Approaches for the Evaluation of Complex Computer Models, *Statistical Science* **17**, 173-192 (2002).

4. J. Booker and L. McNamara, Expertise and Expert Judgment in Reliability Characterization: A Rigorous Approach to Eliciting, Documenting, and Analyzing Expert Knowledge, in *Engineering Design Reliability Handbook*, Eds. D. Ghiocel and S. Nikolaidis (CRC Press, 2002).

5. M. Hamada, H. Martz, C. S. Reese, and A. Wilson, Finding Near-Optimal Bayesian Experimental Designs via Genetic Algorithms, *The American Statistician* **55** (2001).

6. D. M. Higdon, H. Lee, and C. Holloman, Markov Chain Monte Carlo-based Approaches for Inference in Computationally Intensive Inverse Problems, in *Bayesian Statistics 7*, Eds. J. M. Bernardo, M. J. Bayarri, J. O. Berger, A. P. Dawid, D. Heckerman, A. F. M. Smith, M. West (Oxford University Press, 2002).

7. V. Johnson, T. Graves, M. Hamada, and C. S. Reese, A Hierarchical Model for Estimating the Reliability of Complex Systems, in *Bayesian Statistics 7*, Eds. J. M. Bernardo, M. J. Bayarri, J. O. Berger, A. P. Dawid, D. Heckerman, A. F. M. Smith, M. West (Oxford University Press, 2002).

8. S. Keller-McNulty, G. Wilson, and A. Wilson, Integrating Scientific Information: Reconsidering the Scientific Method as a Path to Better Policy Making, Los Alamos National Laboratory Report LA-UR-01-5739 (2001).

9. H. Lee, D. M. Higdon, Z. Bi, M. Ferreira, and M. West, Markov Random Field Models for High-Dimensional Parameters in Simulations of Fluid Flow in Porous Media, *Technometrics* **44**, 230-241 (2002).

10. D. Leishman and L. McNamara, Interlopers, Translators, Scribes, and Seers: Anthropology, Knowledge Representation and Bayesian Statistics for Predictive Modeling in Multidisciplinary Science and Engineering Projects, Conference on Visual Representations and Interpretations 2002, Liverpool, UK (2002).

11. H. Martz, Reliability Theory, in *Encyclopedia of Physical Science and Technology Volume 14*, Ed. R.A. Meyers, (Academic Press, San Diego, 2002) pp. 143-159.

12. M. D. McKay, Nonparametric Variance-based Methods of Assessing Uncertainty Importance, *Reliability Engineering and System Safety* **57**, 267-279 (1997).

13. M. D. McKay, J. D. Morrison, and S. C. Upton, Evaluating Prediction Uncertainty in Simulation Models, *Computer Physics Communications* **117**, 44-51 (1999).

14. M. Meyer and J. Booker, Eliciting and Analyzing Expert Judgment: A Practical Guide, 2nd Edition. American Statistical Association and the Society for Industrial and Applied Mathematics (2001).

15. C. S. Reese, A. Wilson, M. Hamada, H. Martz, and K. Ryan, Integrated Analysis of Computer and Physical Experiments, Los Alamos National Laboratory report LA-UR-00-2915, submitted to Technometrics (2001).

16. A. Wilson, C. S. Reese, M. Hamada, and H. Martz, Integrated Analysis of Computational and Physical Experimental Lifetime Data, in *Mathematical Reliability: An Expository Perspective*, Eds. R. Soyer, N. Singpurwalla, T Mazzuchi (Kluwer Press, 2001).

Part II

GENERAL ASPECTS OF RELIABILITY MODELLING

3

ON THE USE OF MODELS AND PARAMETERS IN A BAYESIAN SETTING

Terje Aven

Stavanger University College
P.O. Box 8002, N-4068 Stavanger, Norway
E-mail: terje.aven@tn.his.no

The description of uncertainty about the world through probability is the defining idea of the Bayesian thinking. Probability models and parameters are just tools for expressing these uncertainties. Nonetheless, the Bayesian literature is mainly focused on probability models and the problem of assessing uncertainties of the parameters of these models. In this chapter we argue for the use of a stronger predictive emphasis in a Bayesian analysis, where the key elements are observable quantities to be predicted, models linking observable quantities on different level of detail, and probability used to express uncertainty of these observable quantities. And for some of the observable quantities, the assignments of the probabilities can be done without introducing parameters and adopting the somewhat sophisticated procedure of specifying prior distributions of parameters. Furthermore, if parameters are introduced they should have a physical meaning, they should represent states of the world. In this setting, model uncertainty has no meaning. What we look for is the "goodness" of the model, to represent the world and to perform its function as a tool in the analysis.

1. Introduction

The purpose of engineering reliability and risk analyses is to provide decision support for design and operation. In this presentation we are particularly interested in the planning of complex man-machine systems. We focus on risk analysis, but the discussion also applies to reliability analysis.

We would like to establish some guidelines on how to think when performing risk analyses of such systems, using a Bayesian approach. Our starting point is the Bayesian approach as it is presented in the literature, see *e.g.* Barlow[1], Bernardo and Smith[2], Lindley[3], Singpurwalla[4] and

Singpurwalla and Wilson[5].

However, we find it somewhat difficult to proceed. The literature is mainly addressing the analysis within chosen probability models, but our main concerns relate to the phase before that; how we should think when approaching the problem, regarding

- what type of performance measures we should look at,
- how we should model the world (system),
- whether or not to introduce a probability model, and
- the interpretation of parameters of the probability models, if we decide to introduce such models.

We use a simple risk analysis example to illustrate ideas.

2. An Illustrative Example

We consider an offshore installation. As a part of a risk analysis of the installation, a separate study is to be carried out of the risk associated with the operation of the control room that is placed in a compressor module. Two persons operate the control room. The purpose of the study is to assess risk to the operators as a result of possible fires and explosions in the module and to evaluate the effect of implementing risk reducing measures. Based on the study a decision will be made on whether to move the control room out of the module or to implement some risk reducing measures. The risk is currently considered to be too high, but management is not sure what is the overall best arrangement taking into account both safety and economy.

Management decides to conduct a risk analysis to support decision making. To simplify, suppose the analysis is based on one event tree as shown in Figure 1 below. The tree models the possible occurrence of gas leakages in the compression module during a period of time, say one year. The number of gas leakages, referred to as the initiating events, is denoted X. If an initiating event I occurs, it leads to Y number of fatalities, where $Y = 2$ if the events A and B occur, $Y = 1$ if the events A and not B occur, and $Y = 0$ if the event not A occurs. We may think of the event A as representing ignition of the gas and B as explosion.

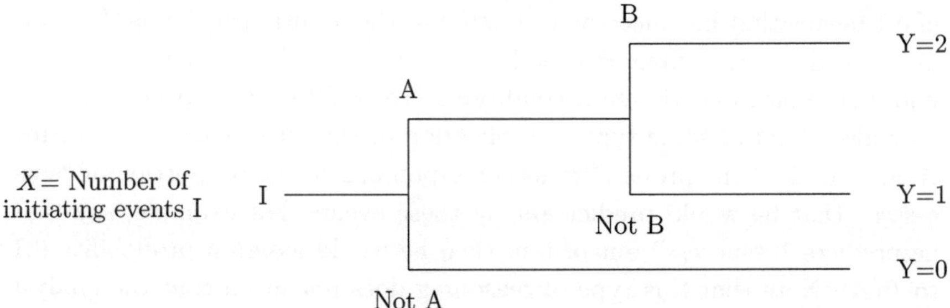

Fig. 1.   Event tree example

How should the analysis be carried out, and how should we use the results?

- What should be the objectives of the analysis? What is the focus?
- How should we define and express risk?
- How is uncertainty understood and addressed?
- What is the meaning of models and how do we use them?
- How should we understand and use (parametric) probability models like the Poisson model?

First, let us consider the problem of specifying the probability of ignition, $P(A)$, and the distribution of the number of leakages X occurring in a one year period.

2.1. *Specifying the Probability of Ignition*

To assess the uncertainty of the ignition event we would use the approach "Analyst judgement using all sources of information". This is a method commonly adopted when data are absent or are only partially relevant to the assessment endpoint. A number of uncertain exposure and risk assessment situations are in this category. The responsibility for summarising the state of knowledge, producing the written rationale, and specifying the probability distribution rests with the analyst.

Now, how does the analyst derive one particular probability? The starting point is that the analyst is experienced in assigning probabilities expressing uncertainty, so that he has a number of references points—he has a feeling for what 0.5 means in contrast to 0.1, for example. A probability of 0.1 means that his uncertainty related to the occurrence of A is the same as when drawing a favorable ball from an urn with 10% favorable balls under standard experimental conditions. To facilitate the specification he may also think of some type of replication of similar events as generating A, and think of the probability as corresponding to the proportion of "successes" that he would predict among these events. For example, say that he predicts 1 "success" out of ten, then he would assign a probability 0.1 to $P(A)$. Note that this type of reasoning does not mean that the analyst presume the existence of a true probability, it is just a tool for simplifying the specification of the probability.

The basis for the assignment is available data and the use of expert opinions.

This approach is in line with the Bayesian thinking. But the Bayesian literature prescribes a better procedure; to use probability models. How should the analysis then be conducted and what are the possible benefits of adopting such a procedure compared to the more direct approach?

Adopting a full Bayesian analysis, the first step would be to introduce a parameter. In this case this would be p, interpreted as the proportion of times ignition would occur when considering an infinite or very large number of similar situations to the one analysed. If we were to know p, we would assign a probability of A equal to p, i.e., $P(A|p) = p$. Hence we obtain

$$P(A) = \int p\,dH(p), \tag{1}$$

where H is the prior distribution of p. Now how should we interpret this formula?

The standard Bayesian framework and interpretation is as follows. To specify the probabilities related to A, a direct assignment could be used, based on everything we know. Since this knowledge is often complex, of high dimension, and much in the background information may be irrelevant to A, this approach is often replaced by the use of "probability models", which is a way of abridging the background information so that it is manageable. Such probability models play a key role in the Bayesian approach. In this case the probability model is simply $P(A|p) = p$, where p is the parameter of the probability model. The parameter p is also referred to as a chance—it is an objective property of the constructed sequence or population of situations—it is not a probability for the assessor, though were p known to the assessor it would be the assessor's probability of A, or any event of the sequence. The parameter p is unknown, and our uncertainty related to its value is specified through a prior distribution $H(p)$. According to the law of total probability, the equation (1) follows, showing that the unconditional distribution of A is simply given by the mean in the prior distribution of p. Note that both $P(A)$ and H are specified given the background information. Thus the uncertainty distribution of A is expressed via two probability distributions, p and H. The two distributions reflect what is commonly referred to as aleatory (stochastic) uncertainty and epistemic (state of knowledge) uncertainty, respectively.

This framework is based on the idea that there exists, or can be constructed through a thought experiment, a sequence of events A_is related to "similar" situations to the one analysed. The precise mathematical term used to define what is "similar" is "exchangeability". Random quantities $X_1, X_2, ..., X_n$ are judged exchangeable if their joint probability distribution is invariant under permutations of coordinates, i.e.,

$$F(x_1, x_2, ..., x_n) = F(x_{r_1}, x_{r_2}, ..., x_{r_n}),$$

where F is a generic joint cumulative distribution function for $X_1, X_2, ..., X_n$ and equality holds for all permutation vectors $(r_1, r_2, ..., r_n)$, obtained by switching (permuting) the indices $\{1, 2, ..., n\}$. Exchangeability means a judgement about indifference between the random quantities. It is a weaker requirement than independence, because, in general, exchangeable random quantities are dependent.

In our case we may view the random quantities as binary, i.e., they take either the value 0 or 1, and if we consider an infinite number of such quantities, judged exchangeable, then it is a well-known result from Bayesian theory that the probability that k out of n are 1 is necessarily of the form

$$P(\sum_{i=1}^{n} X_i = k) = \binom{n}{k} \int_0^1 p^k (1-p)^{n-k} dH(p), \tag{2}$$

for some distribution H. This is a famous result and is referred to as de Finetti's representation theorem, cf. e.g. Bernardo and Smith[2]. Thus, we can think of the uncertainties (beliefs) about observable quantities as being constructed from a parametric model, where the random quantities can be viewed as independent, given the parameter, together with a prior distribution for the parameter. The parameter p has interpretation as the long run frequency of 1's. Note that it is the assessor that judges the sequence to be exchangeable, and only when that is done does the frequency limit exist for the assessor.

Bayesian statistics is mainly concerned about inference about parameters of the probability models. Starting from the prior distribution H, this distribution is updated to a posterior distribution using Bayes theorem.

We see that that the Bayesian approach as presented above allows for fictional parameters, based on thought experiments. Such parameters are introduced and uncertainty of these assessed.

2.2. *Specifying the Probability Distribution for the Number of Leakages*

Now, let us examine a somewhat more complex case; assessing the uncertainty of X, the number of leakages occurring in one year.

Suppose we have observations $x_1, x_2, ..., x_n$ related to previous years, and let us assume that these data are considered relevant for the year studied. We would like to predict X. How should we do this? The data allow a prediction simply by using the mean $\bar{x}$ of the observations $x_1, x_2, ..., x_n$. But what about uncertainties? How should we express this uncertainty? Suppose the observations $x_1, x_2, ..., x_n$ are 4,2,6,3,5, so that $n = 5$ and the observed mean is equal to 4. In this case we have rather strong background information, and we suggest to use the Poisson distribution with mean 4 as our uncertainty distribution of X. For an applied risk analyst, this would be the natural choice as the Poisson distribution is commonly used for event type analysis and the historical mean is 4. Now, how can this uncertainty distribution be "justified"? Well, if this distribution reflects our uncertainty about X, it *is* justified, and there is nothing more to say. This is a subjective probability distribution and there is no need for further justification. But is a Poisson distribution with mean 4 "reasonable", given the background information? We note that this distribution has a variance not larger than 4. By using this distribution, 99% of the mass is on values less than 10.

Adopting the standard Bayesian thinking, as outlined above, using the Poisson distribution with mean 4, means that we have no uncertainty about the parameter λ, which is interpreted as the long run average number of failures when considering an infinite number of exchangeable random quantities, representing similar systems as the one being analyzed. According to the Bayesian theory, ignoring the uncertainty about λ, gives misleading overprecise inference statements about X, cf. e.g. Bernardo and Smith [2], p. 483. This reasoning is of course valid if we work within a setting where we are considering an infinite number of exchangeable random quantities. In our case, however, we just have one X, so what do we gain by making a reference to limiting quantities of a sequence of similar hypothetical Xs? The point is that given the observations $x_1, x_2, ..., x_5$, the choice of the Poisson distribution with mean 4, is in fact "reasonable", under certain conditions on the uncertainty assessments. Consider the following argumentation. Suppose that we divide the year $[0, T]$ into time periods of length T/k, where k is for example 1000. Then we may ignore the possibility of having two events occurring in one time period, and we assign an event probability of $4/k$ for the first time period, as we predict 4 events in the whole interval $[0, T]$. Suppose that we have observations related to $i-1$ time periods. Then for the next time period we should take these observations into account— using independence means ignoring available information. A natural way of balancing the prior information and the observations is to assign an event probability of $(d_i + 4n)/((i-1) + nk)$, where d_i is equal to the total number of events that occurred in $[0, T(i-1)/k]$; i.e., we assign a probability which is equal to the total number of events occurred per unit of time. It turns out that this assignment process gives an approximate Poisson distribution for X. This can be shown for example by using Monte Carlo simulation. The Poisson distribution is justified as long as the background information dominates the uncertainty assessment of the number of events occurring in a time period. Thus from a practical point of view, there is no problem in using the Poisson distribution with mean 4. The above reasoning provides a "justification" of the Poisson distribution, even with not more than one or two years of observations.

Now consider a case with no historical data. Then we will probably find the direct use of the Poisson distribution as described above to have too small variance. The natural approach is then to implement a full parametric Bayesian procedure. But how should we interpret the various elements of the set-up? We suggest the following interpretation.

The Poisson probability distribution $p(x|\lambda)$ is a candidate for our subjective probability for the event $X = x$, and $H(\lambda)$ is a confidence measure, reflecting for a given value of λ, the confidence we have in $p(x|\lambda)$ for being able to predict X. If we have several X_is, similar to X, and λ is our choice, we believe that about $p(x|\lambda) \cdot 100\%$ of the X_is will take a value equal to x, and $H(\lambda)$ reflects for a given value of λ, the confidence we have in $p(x|\lambda)$

for being able to predict the number of X_is taking the value x. We refer to this as the confidence interpretation.

Following this interpretation, we avoid the reference to an hypothetical infinite sequence of exchangeable random quantities. We do not refer to $H(\lambda)$ as an uncertainty distribution as λ is not an observable quantity.

If a suitable infinite (or large) population of "similar units" can be defined, in which X and the X_is belong, then the above standard Bayesian framework applies as the parameter λ represents a state of the world, an observable quantity. Then $H(\lambda)$ is a measure of uncertainty and $p(x|\lambda)$ truly is a model—a representation of the proportion of units in the population having the property that the number of failures is equal to x. We may refer to the variation in this population, modeled by $p(x|\lambda)$, as aleatory uncertainty, but still the uncertainty related to the values of the X_is is seen as a result of lack of knowledge, i.e., the uncertainty is epistemic. This nomenclature is in line with the basic thinking of e.g. Winkler[6], but not with that commonly used in the standard Bayesian framework.

The same type of thinking can be used for the uncertainty assessment of the ignition event A. The confidence interpretation would in this case be as follows: Our starting point is that we consider alternative values p for expressing our uncertainty about A. The confidence we have in p for being able to predict A, is reflected by the confidence distribution H. If we have several A_is, similar to A, and p is our choice, we believe that about $p \cdot 100\%$ of the A_is would occur, and $H(p)$ reflects for a given value of p, the confidence we have in p for being able to predict the number of A_is occurring.

2.3. *Models and Model Uncertainty*

The above analysis provides a tool for predicting the observable quantities and assessing associated uncertainties. When we have little data available, modeling is required to get insights and hopefully reduce our uncertainties. The modeling also makes it possible to see the effects of changes in the system and to identify risk contributors.

In the risk analysis example, the event tree is a model, a representation of the world. It is not relevant to talk about uncertainty of such a model. What is interesting to address is the goodness or appropriateness of a specific model to be used in a specific risk analysis and decision context. Clearly, a model can be more or less good in describing the world. No model reflect all aspects of the world, per definition—it is model—but key features should be reflected. In our setting, a model is a purely deterministic representation of factors judged essential by the analyst. It provides a framework for mapping uncertainty about the observable quantity of interest from expressions of epistemic uncertainty related to the observable quantities on a more detailed system level, and does not in itself introduce additional uncertainty. In this setting the model is merely a tool judged useful for expressing knowledge of the system. The model is a part of the background information of the probability distribution specified for Y. If we change the model, we change the background information. We refer to Nilsen and Aven[7] for a further discussion on models and modeling uncertainty in a risk analysis context.

3. Conclusions

We see the need for a rethinking how to present the Bayesian paradigm in a practical setting. The aim of the discussion in this chapter has been to give a basis for such a thinking.

The presentation of the Bayesian paradigm should in our view have a clear focus and an understanding of what can be considered as technicalities. A possible way of structuring the various elements of the analysis is shown in Fig. 2, which highlights the way the reliability (risk) analyst uses the model and probability calculus. The figure is read as follows: A risk (reliability) analyst (or an analyst team) conducts a risk (reliability) analysis. Focus is on the world, and in particular some future observable quantities reflecting the world; Y and $\mathbf{X} = (X_1, X_2, ..., X_n)$. Based on the analyst's understanding of the world the analyst develops a model (several models), that relates the overall system performance measure Y to $\mathbf{X}$, which is a vector of quantities on a more detailed level. The analyst assesses uncertainties of $\mathbf{X}$, and that could mean the need for simplifications in the assessments, for example using independence between the quantities X_i. Using probability calculus, the uncertainty assessments of $\mathbf{X}$ together with the model g, gives the results of the analysis, *i.e.*, the probability distribution of Y and a prediction of Y. The uncertainties are a result of lack of knowledge, *i.e.*, it is epistemic.

This way of presenting the Bayesian approach to reliability and risk analysis we refer to as a predictive, Bayesian approach. It is also sometimes referred to as the predictive, epistemic approach, cf. Aven[8,9,10] and Apeland et al. [11]. The essential steps of the analysis can be summarized as follows:

(1) Identify the overall system performance measures (observable quantities

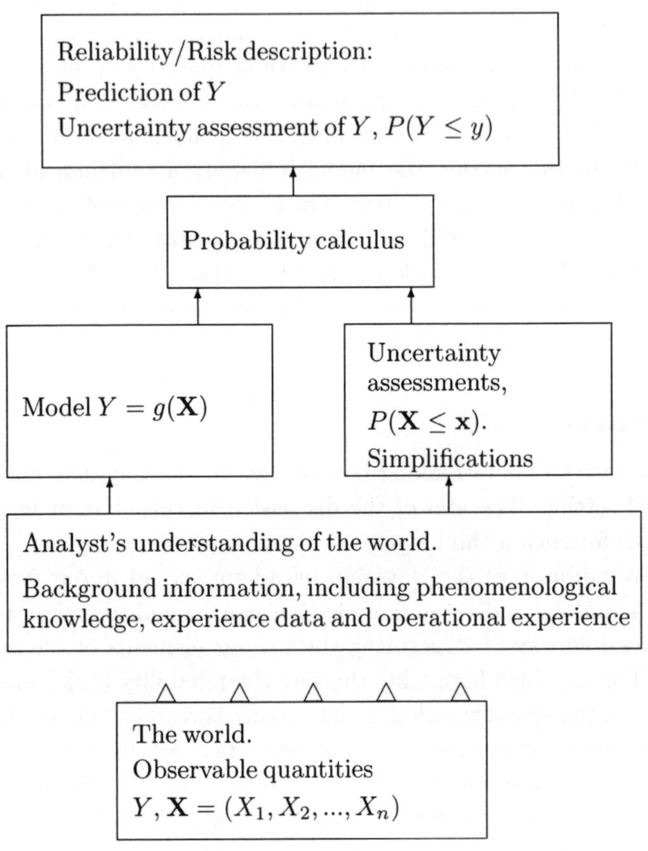

Fig. 2.   Basic elements of a reliability or risk analysis

on a high level). These are typically associated with the objectives of the system performance.

(2) Develop a deterministic model of the system, linking the system performance measures and observable quantities on a more detailed level.

(3) Collect and systematize information about these low level observable quantities.

(4) Use probabilities to express uncertainty of these observable quantities.

(5) Calculate the uncertainty distributions of the performance measures and determine suitable predictions from these distributions.

Sometimes a model is not developed as the analysis is just a transformation from historical data to a probability distribution and predictions related to a performance measure. Often the predictions are derived directly from the historical data without using the probability distributions.

The reliability or risk description needs then to be evaluated and related to costs, and other aspects, to support decision making. A utility-based analysis could be carried out as a tool for identifying a "good" decision alternative. We refer to Aven[10] for a discussion on when to use such an analysis.

The above way of thinking, emphasizing observable quantities and using the reliability or risk analysis as a tool for prediction, is in line with the modern, predictive Bayesian theory, as described in *e.g.* Bernardo and Smith[2], Barlow[1] and Barlow and Clarotti[11], cf. also Singpurwalla[13]. The objective of this presentation, which is partly based on Aven[10], has been to add some perspectives to this thinking, in order to strengthen the practical applicability of the Bayesian paradigm.

Acknowledgments

The author is grateful to J. T. Kvaløy and a referee for valuable comments and suggestions, and the Norwegian Research Council for financial support.

References

1. R.E. Barlow, *Engineering Reliability* (SIAM, Philadephia 1998).
2. J.M. Bernardo, and A. Smith, *Bayesian Theory* (Wiley & Sons., New York, 1994).
3. D.V. Lindley, The philosophy of statistics, *The Statistician*, **49**, 293-337 (2000).
4. N.D. Singpurwalla, Foundational issues in reliability and risk analysis, *SIAM Review*, **30**, 264-281 (1988).
5. N.D. Singpurwalla and S.P. Wilson, *Statistical Methods in Software Engineering* (Springer Verlag, New York, 1999).
6. R.L. Winkler, Uncertainty in probabilistic risk assessment, *Reliability Engineering and System Safety* **54**, 127-132 (1996).
7. T. Nilsen and T. Aven, Models and modeling uncertainty in a risk analysis context, submitted to *Reliability Engineering and System Safety* (2002).
8. T. Aven, Risk analysis – a tool for expressing and communicating uncertainty, In *Proceedings of European Safety and Reliability Conference (ES-REL)*, Eds. Cottam, M.P. et al. (Balkema, Rotterdam, 2000) pp. 21-28.
9. T. Aven, Reliability analysis as a tool for expressing and communicating uncertainty, in *Recent Advances in Reliability Theory: Methodology, Practice*

and Inference, Eds. N. Liminios and M. Nikulin (Birkhauser, Boston, 2000) pp. 23-28.

10. T. Aven, *How to Approach Risk and Uncertainty to Support Decision Making* (Wiley, New York, 2003) to appear.

11. S. Apeland, T. Aven and T. Nilsen, Quantifying uncertainty under a predictive epistemic approach to risk analysis, *Reliability Engineering and System Safety* **75**, 93-102 (2002).

12. R.E. Barlow and C.A. Clarotti, *Reliability and Decision Making*, Preface (Chapman & Hill, London, 1993).

13. N.D. Singpurwalla, Some cracks in the empire of chance (flaws in the foundations of reliability), *International Statistical Review*, **70**, 53-78 (2002).

4

CAUSAL INFLUENCE COEFFICIENTS: A LOCALISED MAXIMUM ENTROPY APPROACH TO BAYESIAN INFERENCE

Michael Ingleby and Margaret West

School of Computing and Engineering
University of Huddersfield
HD1 3DH, UK
E-mail: M.Ingleby@hud.ac.uk, M.M.West@hud.ac.uk

We consider the problem of incomplete conditional probability tables in Bayesian nets, noting that marginal probabilities for an effect, given a single cause are usually easy to elicit and can serve as constraints on the full conditional probability table (CPT) for occurrence of an effect given all possible conditions of its causes. A form of maximum entropy principle, local to an effect node is developed and contrasted with existing global methods. Exact maximum-entropy CPTs are computed and a conjecture about the exact solution for effects with a general number N of causes is examined.

1. Introduction

1.1. *Background and Related Work*

The use of directed acyclic graphs (DAGs) to represent both complex multivariate probability distributions and the dependency relations that reflect causality is very widespread. The recent review material in[10,4] gives the breadth and tenor of current practice. The usage encompasses Bayesian nets, and nets enhanced with non-random decision or action nodes. Also, effect nodes with utilities attached are admitted, and the usual causal propagation procedures extended to derive distributions of utilities. With such enhancements, Bayesian nets are applicable whenever causal or diagnostic inference is needed in expert systems or other intelligent agents. Good examples of these Bayesian decision nets can be found in the review volume of Cowell, Dawid, Lauritzen and Spiegelhalter[3].

A DAG, fully furnished with prior probability tables (PPTs) for all base

45

nodes and conditional probability tables (CPTs) for all higher effect nodes, determines the joint probability density (JPD) of all its node variates. The point is made with many illustrative examples in the classic text of Pearl[9] and its more recent sequel[10]. In systems that reason automatically about uncertainty, it is not usually efficient to build the full JPD. Two kinds of inference, based only on PPTs and CPTs are operated: *causal* inference propagates *forwards* from priors along causal links characterised by $P(E \mid C)$ to compute probabilities at all effect nodes; *diagnostic* inference operates *backwards* to estimate marginals $P(C \mid E)$.

The interest in reasoning from limited knowledge in PPTs and CPTs is mainly practical. The amount of sampling data needed to populate these tables with frequentist estimates is much less than that needed to populate JPDs directly. Nevertheless, if a DAG has effect nodes linked to many causes the amount of training data needed to populate the CPTs of such nodes may be a source of practical difficulty. It is the main barrier to practical use of DAGs in some applications. For example, a Boolean effect E with n Boolean causes $C_1, C_2, \ldots, C_n$ has a CPT with 2^n entries $P(E \mid C_1, C_2, \ldots C_n), P(E \mid \neg\, C_1, C_2, \ldots, C_n)$, etc. If the causes and effects are represented on a scale of 5, as is common in human factors work, the table has 4×5^n entries. Obtaining frequency data sufficient to estimate such large numbers of CPT entries is costly, in some cases impossibly so.

Some of the best known examples of automated Bayesian inference in DAD frameworks are expert systems. Broadly, they deal with a small range of causes (such as diseases) each of which produce many effects (symptoms). In this chapter, we are responding to the *dual* of this situation: a hazardous effect that can result from unfortunate combinations of many causes. Such combinations of multiple causes are not always anticipated by the designers of complex systems. They figure in the *post hoc* analysis of catastrophic system failures where human factors and multiple component degradation 'conspire' to produce the unexpected. Rail and air traffic accidents usually have this multiple causality.

Although multi-causal CPTs may be too costly to complete, in many environments experts willingly assess the marginal influence of a single causal factor on a multiply caused effect. They use weights or influence coefficients justified by limited data. These are interpretable in a Bayesian framework as conditional probabilities, reinforcing the effect $E = e$ with weight $p(e \mid c_i)$ where $c_i \in \mathcal{C}_i$ (set $\mathcal{C}_i$ containing all the possible values for cause node C_i). In the case of boolean E and C_i, these reinforcement factors are simply the two influence coefficients $P(E \mid C_i)$ and $P(E \mid \neg\, C_i)$.

Such marginals do not in general determine the full CPT for a multi-cause effect node, and hence do not provide a basis for the causal and diagnostic reasoning that make Bayesian nets such attractive processors of uncertain knowledge. We therefore suggest a maximal entropy (ME) principle that leads to a unique CPT by minimising a Shannon information measure. The principle amounts to minimising that part of the Shannon information in a CPT not already present in the marginal influence coefficients. It effectively views a node as a communication channel: its output is E determined by inputs $C_1 \ldots C_n$, but constrained by influence coefficients. More generic aspects of determining a JPD under constraint is covered by a theorem of Jirousek[8]. The placing of Jirousek's very general theorem in the context of expert systems – dual to our multiple cause context – has been detailed in the review volume[5].

Our ME approach contrasts with others in the literature, notably Dempster-Schafer theory[11,9], Huffman coding (as e.g. outlined by Heckermann[6]) and the theory of vines developed by Bedford and Cooke in a series of papers exemplified by[1]. Dempster-Schafer theory and the theory of vines operate outside the Bayesian net framework, but nevertheless support a type of belief revision similar to causal propagation. The focus of Dempster and Schafer is not on completing a JPD but on propagating only what can be proved from evidence and on combining different evidence forms. The belief functions that result from their evidential revision do not amount to full knowledge of the JPD. In the theory of vines, the focus is on a minimal set of marginal probabilities that is large enough to generate the whole JPD but more general than Pearl's PPTs and CPTs. It makes none of the Bayesian net suppositions that relate the causal linkage of a DAG to notions of conditional independence.

Huffman coding takes place within the Bayesian framework, but does not keep the causal arcs fixed; it allows extra links to be inserted experimentally after experts have indicated which are the most vital links. The Huffman code of the extended net is the length of binary string needed to represent the available training data and the DAG connectivity. This is minimised, globally.

More elementary approaches to global information minimisation were also described by Cheeseman[2] and used for expert system building. The idea was essentially practical: complete the JPD with parameters in place of missing frequencies, fixing the parameter values to maximise the Shannon entropy of the whole JPD. Deeper studies than Cheeseman's have also focused on optimising the information in a JPD. Spiegelhalter and Lauritzen[3], have used JPDs in many studies of conditional independence and of learning from sample data. They use global entropies related to likelihood functions, and engage in a form of maximum likelihood estimation. They do this in a Bayesian learning context that starts with prior parametric distributions assumed on all nodes. Training data is used to estimate the parameters from a likelihood function defined for all the parametric distributions employed.

All these approaches differ from ours in their main concern with the information residing in the whole DAG. We are concerned with the local CPT of each effect node. We are especially interested in multi-cause nodes and the situation of insufficient training data for the CPTs of these. We regard such a node as a communication channel down which information flows during causal or diagnostic propagation. The channel parameters are set by maximising the Shannon entropy.

1.2. *Notation*

We illustrate our local ME approach by working out in detail the maximum entropy CPTs for 3-cause Boolean nodes. The former arises in simple hardware systems with one failure mode arising from joint action of three causal factors. For nodes of arity n, we need a ladder representation based on binary nodes. The general idea and notation is now described.

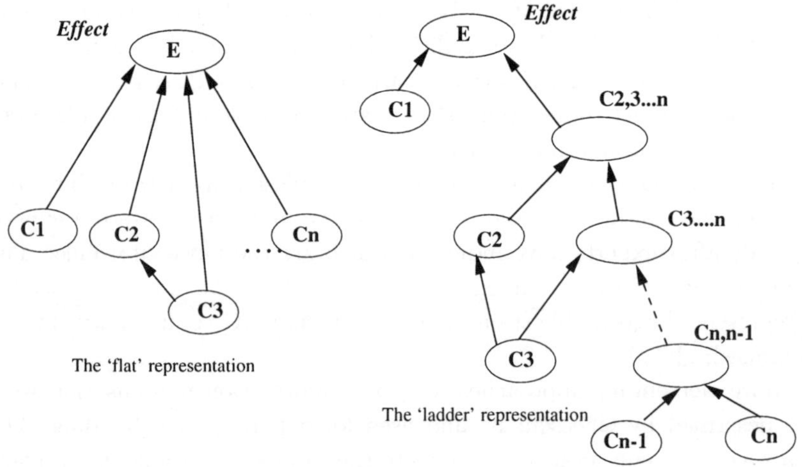

Fig. 1.   Example of an effect node with n causes

Consider Figure 1 which shows an effect node E taking values $e \in \mathcal{E}$

of a CPT with parents $C_1, C_2, \ldots, C_n$. Nodes $C_1, C_2, \ldots, C_n$ take values $c_1, c_2, \ldots, c_n$, with $c_i \in C_i$. No assumptions are made about independence of the C_i and the figure emphasises this by showing a causal link from C_3 to C_2. *Two* representations of the node are shown, the 'flat' representation and the 'ladder'. The ladder graph shows the decomposition of an n-ary node as a ladder of successive ternary nodes involving joint causes. It allows us to use ternary node solutions to build up the CPT for the n-ary case. The ternary nodes in the ladder representation are characterised by a trivial CPT $P(C_i \wedge C_j \mid C_i \wedge C_j)$ which has value 1.

The associated probability distribution functions (pdfs) of a multi-cause node are denoted:

$$p(c_1, \ldots, c_n) \quad \equiv P((C_1 = c_1) \wedge \ldots \wedge (C_n = c_n))$$
$$\textit{joint distribution of causes;}$$

$$p(e \mid c_1, \ldots, c_n) \equiv P(E = e \mid (C_1 = c_1) \wedge \ldots \wedge (C_n = c_n))$$
$$\textit{distribution of effects,}$$
$$\textit{conditional on all causes;}$$

$$p(c_1, \ldots, c_n \mid e) \equiv P((C_1 = c_1) \wedge \ldots \wedge (C_n = c_n) \mid E = e)$$
$$\textit{diagnostic distribution of}$$
$$\textit{causes, conditional on effects;}$$

$$p(c_1, \ldots, c_n; \; e) \equiv P((C_1 = c_1) \wedge \ldots \wedge (C_n = c_n) \wedge E = e)$$
$$\textit{joint distribution of causes and effects.}$$

The definition of conditional probability implies that:

$$p(c_1, c_2, \ldots, c_n; \; e) = p(c_1, c_2, \ldots, c_n) p(e \mid c_1, c_2, \ldots, c_n)$$

and Bayes' rule takes the form:

$$p(e) p(c_1, c_2, \ldots, c_n \mid e) = p(c_1, c_2, \ldots, c_n) p(e \mid c_1, c_2, \ldots, c_n).$$

If there are no direct links between the C_i nodes pointing into E, then by conditional independence, the diagnostic conditional above reduces to a product:

$$p(c_1, c_2, \ldots, c_n \mid e) = \prod_i p(c_i \mid e).$$

The absence of direct links is plausibly true in the case of a diagnostic expert system with many distinct effects of few causes. In the multi-cause context underlying this chapter, however, we do *not* assume this, as is made explicit in the example Figure 1.

1.3. *Chapter Structure*

The chapter solves the problem of finding the maximum entropy CPT for a ternary effect node in Section 2 - using a variant Lagrange technique to take into account the constraints arising from influence coefficients.

In Section 3 an inductive approach to nodes of higher arity is outlined, and including remarks on direction of future research are collected in Section 4.

2. Nodes of Arity 3

We develop the maximum entropy calculation for nodes of small arity, commencing with arity 3. For simplicity, we take the nodes to be boolean: each of the values c_1, c_2, c_3 belong to the set $\{0, 1\}$. Table 1 shows the values of the joint distribution function in 8 rows. The values of the joint distribution

Table 1.   Definition of u_i.

c_1	c_2	c_3	$p(c_1, c_2, c_3; \quad e)$
1	1	1	u_1
1	1	0	u_2
1	0	1	u_3
1	0	0	u_4
0	1	1	u_5
0	1	0	u_6
0	0	1	u_7
0	0	0	u_8

are variables u_i determined from the following linear equations:

$$u_1 + u_2 + u_3 + u_4 = P(C_1 \wedge E) = P(C_1)P(E \mid C_1) = v_1;$$

$$u_5 + u_6 + u_7 + u_8 = P(\neg C_1 \wedge E) = v_2;$$

$$u_1 + u_2 + u_5 + u_6 = P(C_2 \wedge E) = P(C_2)P(E \mid C_2) = v_3;$$

$$u_3 + u_4 + u_7 + u_8 = P(\neg C_2 \wedge E) = v_4;$$

$$u_1 + u_3 + u_5 + u_7 = P(C_3 \wedge E) = v_5;$$

$$u_2 + u_4 + u_6 + u_8 = P(\neg C_3 \wedge E) = v_6.$$

The $v_1 \ldots v_6$ in this system of equations are constants determined only by constraints: the influence coefficients and the $P(C_i)$s. The latter are fixed by causal propagation from the PPTs through effect nodes preceding the current node in a given DAG. We envisage that ME estimation is to be carried out *step-wise*, starting with the effect nodes immediately forward of the PPT nodes and proceeding forwards layer by layer.

The corresponding system of equations for a binary node take the form:

$$u_1 + u_2 = v_1;$$
$$u_3 + u_4 = v_2;$$
$$u_1 + u_3 = v_3;$$
$$u_2 + u_4 = v_4,$$

which have a solution set:

$$\{(u_1, u_2, u_3, u_4) \mid (x, v_1 - x, v_3 - x, v_2 - v_3 + x) \text{ and } 0 \leq x \leq 1\}$$

with a continuously varying parameter x. The existence of many solutions prompts us to search for an entropy principle that supplies an objective function of parameter x. The value of x that maximises the entropy objective determines the full CPT.

Like the simplest binary node case, the system of linear equations for nodes of arity 3 is under-determined, and this remains the case for higher arity nodes. For arity 3 there are 6 equations in 8 unknowns. Inspection of these reveals the fact that row 1 plus row 2 equals row 3 plus row 4 which in turn equals row 5 plus row 6. We thus have 4 independent equations in 8 unknowns, and therefore expect that the solution will require 4 parameters $x_1 \ldots x_4$.

2.1. *Solution of the linear equations*

To solve in a systematic way, we write the equations in terms of an augmented matrix and use row manipulation to transform the augmented matrix to its echelon form. The most straightforward way is to use our knowledge about the dependencies: thus R3 becomes R3 -R2 + R4 - R1, a null row, and similarly for R6. The augmented matrix and its re-arranged form is

$$
\begin{pmatrix}
1 & 1 & 1 & 1 & 0 & 0 & 0 & 0 & v_1 \\
0 & 0 & 0 & 0 & 1 & 1 & 1 & 1 & v_2 \\
1 & 1 & 0 & 0 & 1 & 1 & 0 & 0 & v_3 \\
0 & 0 & 1 & 1 & 0 & 0 & 1 & 1 & v_4 \\
1 & 0 & 1 & 0 & 1 & 0 & 1 & 0 & v_5 \\
0 & 1 & 0 & 1 & 0 & 1 & 0 & 1 & v_6
\end{pmatrix}
\rightarrow
\begin{pmatrix}
1 & 1 & 1 & 1 & 0 & 0 & 0 & 0 & v_1 \\
0 & 1 & 0 & 1 & 0 & 1 & 0 & 1 & v_6 \\
0 & 0 & 1 & 1 & 0 & 0 & 1 & 1 & v_4 \\
0 & 0 & 0 & 0 & 1 & 1 & 1 & 1 & v_2 \\
0 & 0 & 0 & 0 & 0 & 0 & 0 & 0 & 0 \\
0 & 0 & 0 & 0 & 0 & 0 & 0 & 0 & 0
\end{pmatrix}
$$

As expected, the matrix indicates 4 linearly independent equations in 8 unknowns, so the solution set involves parameters. We fix these by letting:

$u_4 = x_1$, $u_6 = x_2$, $u_7 = x_3$, $u_8 = x_4$. Substituting these values we thus obtain solutions

$$u_1 = -v_4 - v_2 + v_5 + x_1 + x_2 + x_3 + 2x_4;$$

$$u_2 = v_6 - x_1 - x_2 - x_4;$$

$$u_3 = v_4 - x_1 - x_3 - x_4;$$

$$u_5 = -x_2 - x_3 - x_4 + v_2.$$

2.2. *Maximum Entropy Calculation*

The maximum entropy principle was originally developed by Jaynes[7]. We apply it here to chose from the above solution set the value that maximises an appropriate entropy function of the node. The entropy or *uncertainty* of the above pdf is $H(x_1, x_2, x_3, x_4)$ where $H(x_1, x_2, x_3, x_4) = -\Sigma_i u_i \log u_i$.

Let

$$H_j = \frac{\partial}{\partial x_j} H(x_1, x_2, x_3, x_4) \qquad (j = 1 \ldots 4).$$

By maximising the uncertainty H we *minimise* the Shannon information and this requires $H_j = 0$ $(j = 1 \ldots 4)$. The simplest way of calculating H_j is to use the chain rule:

$$\partial(u_i \log u_i)/\partial x_j = (1 + \log u_i)\partial u_i/\partial x_j.$$

Calculating the $\partial u_i/\partial x_j$ for the direct solutions in Section 2.1, the condition $H_j = 0$ $(j = 1 \ldots 4)$ simplifies to yield:

$$u_3/u_1 = u_8/u_6; \qquad u_4/u_2 = u_8/u_6; \qquad u_4/u_2 = u_7/u_5$$

To solve we require a multiplier λ:

$$u_4/u_2 = u_3/u_1 = u_8/u_6 = u_7/u_5 = \lambda.$$

Substitution into the equations in u_i yields $\lambda = v_4/v_3$ and hence u_i in terms of v_i, from which the full CPT can be expressed in terms of the v_i. The solution obtained has the expected symmetry:

$$P(C_1 \wedge C_2 \wedge C_3 \wedge E) = \frac{P(C_1 \wedge E)P(C_2 \wedge E)P(C_3 \wedge E)}{P(E)^2};$$

$$P(C_1 \wedge C_2 \wedge \neg\, C_3 \wedge E) = \frac{P(C_1 \wedge E)P(C_2 \wedge E)P(\neg\, C_3 \wedge E)}{P(E)^2};$$

$$\ldots$$

$$P(\neg\, C_1 \wedge \neg\, C_2 \wedge \neg\, C_3 \wedge E) = \frac{P(\neg\, C_1 \wedge E)P(\neg\, C_2 \wedge E)P(\neg\, C_3 \wedge E)}{P(E)^2}.$$

We observe that the estimated CPT obtained from ME considerations is the CPT that we would have obtained by assuming $C_1 \wedge E, C_2 \wedge E, C_3 \wedge E$ to be independent events. The (simpler) derivation for nodes of arity 2 and more complex derivation for nodes of arity 4 also yield a result consistent with independence of $C_i \wedge E$. The next section attempts to generalise, to nodes of arity n.

3. Extension to nodes of arity n

Consider $C_1, C_2, \ldots, C_n$ which take boolean values $c_1, c_2, \ldots, c_n$. There are $2n$ equations in $M = 2^n$ unknowns, $u_1, u_2, \ldots, u_M$. Of these, $n + 1$ are independent. To solve these equations requires $L = 2^n - (n + 1)$ arbitrary parameters $x_1 \ldots x_L$ and to solve in a systematic way, we again use row manipulation to transform the augmented matrix to its echelon form. An attempt can be made to obtain a solution, but the exact pattern of the solution is difficult to discern. The transformation of the matrix to its echelon form and subsequent calculation of maximum entropy leads us to deduce a result requiring multipliers $\lambda_0, \lambda_1, \lambda_2, \ldots$ where for example:

$$u_1/u_2 = u_3/u_4 \ldots = \lambda_0; \quad u_1/u_3 = u_5/u_7 \ldots = \lambda_1;$$
$$u_1/u_5 = u_9/u_{13} \ldots = \lambda_2 \ldots$$

We can evaluate some of the multipliers; for example the first and second linear constraints are:

$$(1 + \lambda_0)(u_2 + \ldots + u_{M/2}) = v_1;$$
$$(1 + \lambda_0)(u_{2+M/2} + \ldots + u_M) = v_2$$

and we have $(1 + \lambda_0)(u_2 + \ldots + u_M) = v_1 + v_2$. The last constraint is: $u_2 + u_4 + \ldots + u_M = v_{2n}$ so that, assuming the u factor is non-zero:

$$(1 + \lambda_0) = (v_1 + v_2)/v_{2n}.$$

Proceeding in a similar manner to the 3-node case, we conjecture that for all values of $c_i \in \{0, 1\}$:

$$p(c_1, c_2, \ldots, c_n; \ e) = (\ p(c_1; \ e)p(c_2; \ e) \ldots p(c_n; \ e)\ )/p(e)^n.$$

The conjecture is consistent with our results for small n, that the calculation of maximum entropy for a system with a factor node E with parents $C_1, C_2, \ldots, C_n$ yields the result obtainable assuming independence of $C_1 \wedge E, C_2 \wedge E, \ldots, C_n \wedge E$. The following inductive argument is inspired by the ladder decomposition, and may be called the 'ladder method'.

3.1. *An Inductive Approach*

To reason by induction, we assume that the maximum entropy principle is associative in the following sense: to extend from a 2-node to a 3-node one can replace C_2 by $C_{2,3}$ for the '2 node':

$$P(C_1 \wedge C_{2,3} \wedge E) = \frac{P(C_1 \wedge E)P(C_{2,3} \wedge E)}{P(E)};$$

$$P(C_1 \wedge \neg\, C_{2,3} \wedge E) = \frac{P(C_1 \wedge E)P(\neg\, C_{2,3} \wedge E)}{P(E)};$$

$$P(\neg\, C_1 \wedge C_{2,3} \wedge E) = \frac{P(\neg\, C_1 \wedge E)P(C_{2,3} \wedge E)}{P(E)};$$

$$P(\neg\, C_1 \wedge \neg\, C_{2,3} \wedge E) = \frac{P(\neg\, C_1 \wedge E)P(\neg\, C_{2,3} \wedge E)}{P(E)}.$$

Since $C_{2,3} = C_2 \wedge C_3$, we can rewrite as follows, using the initial definitions of u_i:

$$P(C_1 \wedge C_2 \wedge C_3 \wedge E) = \frac{P(C_1 \wedge E)P(C_2 \wedge C_3 \wedge E)}{P(E)};$$

$$P(C_1 \wedge \neg\, (C_2 \wedge C_3) \wedge E) = \frac{P(C_1 \wedge E)P(\neg\, (C_2 \wedge C_3) \wedge E)}{P(E)};$$

$$P(\neg\, C_1 \wedge C_2 \wedge C_3 \wedge E) = \frac{P(\neg\, C_1 \wedge E)P(C_2 \wedge C_3 \wedge E)}{P(E)};$$

$$P(\neg\, C_1 \wedge \neg\, (C_2 \wedge C_3) \wedge E) = \frac{P(\neg\, C_1 \wedge E)P(\neg\, (C_2 \wedge C_3) \wedge E)}{P(E)}.$$

If we use the results for arity 2 (assuming we can) then

$$u_1 = \frac{P(C_1 \wedge E)P(C_2 \wedge C_3 \wedge E)}{P(E)} = \frac{P(C_1 \wedge E)}{P(E)} \frac{P(C_2 \wedge E)P(C_3 \wedge E)}{P(E)}$$

To obtain the rest of the pdf, use:

$$P(C_1 \wedge \neg\, C_2 \wedge C_3 \wedge E) + P(C_1 \wedge C_2 \wedge C_3 \wedge E) = P(C_1 \wedge C_3 \wedge E).$$

Thus

$$u_1 = P(C_1 \wedge C_3 \wedge E) - u_3$$

(from definitions) and so

$$u_3 = P(C_1 \wedge C_3 \wedge E) - \frac{P(C_1 \wedge E)P(C_2 \wedge E)P(C_3 \wedge E)}{P(E)P(E)}$$

$$= \frac{P(C_1 \wedge E)P(C_3 \wedge E)P(\neg\, C_2 \wedge E)}{P(E)^2}.$$

The expressions for $u_2, u_4, u_5, u_6, u_7, u_8$ can also be determined and also yield the same result. Thus, assuming that maximum entropy is associative we can proceed straightaway to induction. The rule is

"Assuming that the maximum entropy criteria for a factor node E of a CPT with parents $C_1, C_2, C_3, \ldots, C_n$ for $n \geq 2$ yields, for each value of C_i:

$$P(C_1 \wedge C_2 \wedge \ldots \wedge C_n \wedge E) = \frac{P(C_1 \wedge E)P(C_2 \wedge E)\ldots P(C_n \wedge E)}{P(E)^{n-1}}$$

then for a CPT with parents $C_1, C_2, \ldots, C_{n+1}$,

$$P(C_1 \wedge C_2 \wedge \ldots \wedge C_{n+1} \wedge E) = \frac{P(C_1 \wedge E)P(C_2 \wedge E)\ldots P(C_{n+1} \wedge E)}{P(E)^{n}}$$

where the base case is $n = 2$."

From this type of associativity assumption, the result follows easily using the summation rule:

$$P(C_1 \wedge \ldots \wedge C_i \wedge \ldots \wedge C_{n+1} \wedge E)$$
$$+ P(C_1 \wedge \ldots \wedge \neg C_i \wedge \ldots \wedge C_{n+1} \wedge E) =$$
$$P(C_1 \wedge \ldots \wedge C_{i-1} \wedge C_{i+1} \wedge \ldots C_{n+1} \wedge E).$$

Then, as before:

$$p(c_1, c_2, \ldots, c_n;\ e) = (\ p(c_1;\ e)p(c_2;\ e)\ldots p(c_n;\ e)\ )/p(e)^n.$$

4. Conclusions and Further Work

The chapter shows that knowledge of single-cause influence coefficients at multi-cause nodes is sufficient to enable Bayesian belief propagation under our local maximum entropy assumptions.

There remain, however, some open questions to be answered by further research. Firstly, our associativity assumption appears to relate to notions of conditional independence and this needs closer investigation. At present the connection has not been established in a general way. Secondly, it is not fully established that the ladder decomposition method produces exact solutions for nodes of all arities. We can only confirm this via explicit solutions up to arity 4. The possibility of obtaining the result more rigorously from Jirouseks general results remains open. Finally, it is not clear whether the local ME approach is exportable to the other forms of belief propagation such as the relatively rigorous Dempster-Schafer theory of Evidence and the various semi-empirical ways of propagating influence coefficients.

References

1. T. Bedford and R. M. Cooke, Vines – a new graphical model for dependent random variables, *Annals of Statistics*, **30**, 1031–1068 (2002).
2. P. Cheeseman, A method of computing generalised Bayesian probability values for expert systems, IJCAI- 83 Proceedings, 198–202 (1983).

3. R. G. Cowell, A. P. Dawid, S. L. Lauritzen, and D. J. Spiegelhalter, *Probabilistic Networks and Expert Systems* (Springer, New York, 1999).
4. A. P. Dawid, Influence diagrams for causal modelling and inference, *International Statistical Review*, **70**, 161–189 (2002).
5. P. Hajek and T. Havranek R. Jirousek, *Uncertain Information Processing in Expert Systems* (CRC Press, 1992).
6. D. Heckermann, A tutorial on learning with Bayesian networks, Technical Report MSR-TR-95-06, Microsoft Research, Redmond, WA, USA (1996).
7. E. T. Jaynes, Information theory and statistical mechanics, *Physical Review*, **106**, 620–630 (1957).
8. R. Jirousek, Solution of the marginal problem and decomposable distributions, *Kibernetica*, **27**, 403–412 (1991).
9. J. Pearl, *Probabilistic Reasoning in Intelligent Systems* (Morgan Kauffman, CA, USA, 1988).
10. J. Pearl, *Causality* (Cambridge University Press, UK, 2000).
11. G. Schafer, *A Mathematical Theory of Evidence* (Princeton University Press, USA, 1979).

5

SUBJECTIVE RELIABILITY ANALYSIS USING PREDICTIVE ELICITATION

David F. Percy

Centre for Operational Research and Applied Statistics,
University of Salford, Greater Manchester, M5 4WT, United Kingdom
E-mail: d.f.percy@salford.ac.uk

This chapter tackles the difficulties of specifying subjective prior distributions for parameters that arise in reliability modelling. We review strategies for selecting families of priors, and propose some simplifications and enhancements. Sampling distributions of particular interest here are binomial, negative binomial, Poisson, exponential, normal, lognormal, gamma and Weibull. We then consider the formulation of suitable priors for generalized linear models and stochastic processes.

Our research also investigates methods of predictive elicitation to determine values for hyperparameters encountered in priors for subjective reliability analysis, and illustrates the philosophical beauty and practical benefits of this approach. We briefly discuss numerical algorithms needed to resolve the computational difficulties incurred, as an aid to decision making. Finally, we demonstrate these procedures by applying them to a problem of quality control in the electronics industry.

1. Background

Reliability applications are notorious for their lack of data, leading to poor parameter estimates and inaccurate decisions about replacement intervals, preventive maintenance, action thresholds, warranty schemes and so on. Subjective Bayesian analysis can resolve this problem, by retaining mathematical models and using expert knowledge to enhance empirical observations. This approach provides more information than frequentist alternatives and often leads to better inference.[8]

We consider stochastic decision problems involving one or more random variables X with probability density (mass) function $f_{X|\theta}(x|\theta)$ or just $f(x|\theta)$, ignoring the subscript where it is obvious from the function argument. This distribution depends on one or more unknown parameters θ, existing knowledge of which can be expressed by a prior density $g(\theta)$. Any data $\mathcal{D}$ that become available can be represented by a likelihood function $L(\theta; \mathcal{D})$. For a random sample of observations $\mathcal{D} = \{x_1, \ldots, x_n\}$ which are typically lifetimes, the likelihood function is given by

$$L(\theta; \mathcal{D}) \propto \prod_{i=1}^{n} p_i(x_i|\theta) \tag{1}$$

where

$$p_i(x_i|\theta) = \begin{cases} f(x_i|\theta); \ x_i \text{ observed} \\ R(x_i|\theta); \ x_i \text{ right censored} \\ F(x_i|\theta); \ x_i \text{ left censored} \end{cases} \tag{2}$$

in terms of the probability density (mass) function, survivor function and cumulative distribution function respectively.

1.1. *Bayesian Analysis*

The Bayesian approach to inference then evaluates the posterior density

$$g(\theta|\mathcal{D}) \propto L(\theta; \mathcal{D})g(\theta) \tag{3}$$

for direct inference about θ from the model. We can also use the posterior density to determine the posterior predictive density (mass) function

$$f(x|\mathcal{D}) = \int_{-\infty}^{\infty} f(x|\theta)g(\theta|\mathcal{D})d\theta \tag{4}$$

to make direct inference about X from the model. This posterior predictive distribution can now be combined with a suitable cost (utility) function $c(X)$ to determine the posterior expected cost (utility)

$$E\{c(X)|\mathcal{D}\} = \int_{-\infty}^{\infty} c(x)f(x|\mathcal{D})dx, \tag{5}$$

which is encountered in many stochastic decision problems, where the recommended strategy is that which minimizes (maximizes) this expectation.

1.2. *Other Approaches*

The frequentist approach dominated statistics for much of the twentieth century but the hypothesis tests and confidence intervals it generates have limited use and are often misinterpreted. In this context, such analysis typically involves evaluating the approximations

$$f(x|\mathcal{D}) \approx f(x|\hat{\theta}) \tag{6}$$

and

$$E\{c(X)|\mathcal{D}\} \approx \int_{-\infty}^{\infty} c(x)f(x|\hat{\theta})dx \tag{7}$$

where $\hat{\theta}$ is a point estimate for θ, usually the maximum likelihood estimate defined by

$$L(\hat{\theta}; \mathcal{D}) \geq L(\theta; \mathcal{D}) \forall \theta \neq \hat{\theta}. \tag{8}$$

A fully subjective approach involves making inference about X directly, using expert knowledge and observed data to determine the posterior predictive density (mass) function and posterior expected cost without reference to any parameters. Ignoring model structure in this way reduces the predictive accuracy of the analysis, though the method is useful for complex systems and unique applications, where parametric models can be difficult to establish.

2. Prior Specification

Many authors, including Bernardo and Smith[1], discuss the relative merits of various types of subjective prior. A common approach is to adopt natural conjugate priors for sampling distributions in the exponential family and location-scale priors for sampling distributions that are not. We start by investigating the choice of subjective priors for the most common models in reliability analysis, building on the ideas of Percy[10].

2.1. *Binomial and Negative-binomial*

The binomial distribution is common in reliability and risk analysis. It is the appropriate model for a sequence of Bernoulli trials, such as the total number of failures X out of n components that operate independently with the same probability of failure. Its probability mass function is given by

$$f(x|\theta) = \binom{n}{x} \theta^x (1 - \theta)^{n-x}; x = 0, 1, \ldots, n \tag{9}$$

and the special case where $n = 1$ is known as the Bernoulli distribution. A close relative is the negative-binomial distribution, which models the number of Bernoulli trials X until r events have occurred, such as the number of spare parts that must be tested to find two that work. Its probability mass function is given by

$$f(x|\theta) = \binom{x-1}{r-1} \theta^r (1-\theta)^{x-r}; x = r, r+1, \ldots \tag{10}$$

and the special case with $r = 1$ corresponds to the geometric distribution.

Both binomial and negative-binomial models belong to the exponential family and the natural conjugate prior in each case is a beta density

$$g(\theta) = \frac{\theta^{a-1}(1-\theta)^{b-1}}{\mathrm{B}(a,b)}; 0 < \theta < 1. \tag{11}$$

This depends upon two hyperparameters a and b whose values are set by the analyst so that $g(\theta)$ fairly represents his or her prior knowledge and that of available experts. We consider this aspect in Section 3.

2.2. *Poisson and Exponential*

The Poisson distribution with probability mass function

$$f(x|\mu) = \frac{\mu^x \exp(-\mu)}{x!}; x = 0, 1, \ldots \tag{12}$$

is also common in reliability analysis. It is used to model numbers X of defects in manufactured items, cracks in pipes or structures, and errors in computer software. It is a member of the exponential family and its natural conjugate prior is a gamma density

$$g(\mu) = \frac{b^a \mu^{a-1} \exp(-b\mu)}{\Gamma(a)}; \mu > 0. \tag{13}$$

The exponential distribution with probability density function

$$f(x|\lambda) = \lambda \exp(-\lambda x); x > 0 \tag{14}$$

is a common model for lifetimes X of parts when ageing does not occur, which is typical for many electrical components. It is also a member of the exponential family and its natural conjugate prior is again a gamma density

$$g(\lambda) = \frac{b^a \lambda^{a-1} \exp(-b\lambda)}{\Gamma(a)}; \lambda > 0. \tag{15}$$

There are two hyperparameters a and b for both of these models.

2.3. *Normal and Lognormal*

The normal distribution with probability density function

$$f(x|\mu, \psi) = \sqrt{\frac{\psi}{2\pi}} \exp\left\{-\frac{\psi}{2}(x-\mu)^2\right\}; x \in \mathbb{R} \tag{16}$$

is used to model many measurable quantities. In this expression, the unknown parameters μ and ψ represent the mean and precision (reciprocal variance) respectively. A lognormal random variable Y with density

$$f(y|\mu, \psi) = \sqrt{\frac{\psi}{2\pi}} \frac{1}{y} \exp\left\{-\frac{\psi}{2}(\log y - \mu)^2\right\}; y > 0 \tag{17}$$

is best dealt with by considering the transformed random variable $X = \log Y$ which has the normal distribution above.

The normal model is in the two-parameter exponential family and its natural conjugate prior is normal-gamma with density

$$g(\mu, \psi) = \sqrt{\frac{b\psi}{2\pi}} \exp\left\{-\frac{b\psi}{2}(\mu-a)^2\right\} \frac{c^{\frac{b+1}{2}} \psi^{\frac{b-1}{2}} \exp(-c\psi)}{\Gamma\left(\frac{b+1}{2}\right)}; \mu \in \mathbb{R}, \psi > 0 \tag{18}$$

which depends on three hyperparameters: a, b and c. Some authors relax the constraint of natural conjugacy, to yield a normal-gamma conjugate prior with four hyperparameters. In the above expression, the hyperparameter b is allowed to differ in the normal and gamma components. This relaxation introduces greater flexibility at the expense of more effort in eliciting hyperparameters. Further relaxation is possible by considering the parameters μ and ψ to be independent *a priori* and using the conditional conjugate prior[8] whereby $\mu \sim No(a, b)$ and $\psi \sim Ga(c, d)$.

2.4. *Gamma and Weibull*

The gamma distribution with probability density function

$$f(x|\alpha, \lambda) = \frac{\lambda^\alpha x^{\alpha-1} \exp(-\lambda x)}{\Gamma(\alpha)}; x > 0 \tag{19}$$

is commonly used as a model for reliability analysis, as it corresponds to a sum of independent and identically distributed exponential lifetimes. It is a member of the two-parameter exponential family and its natural conjugate prior takes the form

$$g(\alpha, \lambda) \propto \left\{\frac{\lambda^\alpha}{\Gamma(\alpha)}\right\}^c \exp(a\alpha + b\lambda); \alpha > 0, \lambda > 0. \tag{20}$$

The constant of proportionality cannot be determined analytically, though the conditional distribution of λ given α clearly has a gamma form. This bivariate prior depends on three hyperparameters a, b and c. Again, a conjugate prior could be obtained by relaxing the constraint of natural conjugacy, so introducing an extra hyperparameter.

The Weibull distribution with density

$$f(x|\alpha, \lambda) = \alpha \lambda x^{\alpha-1} \exp(-\lambda x^{\alpha}); x > 0 \tag{21}$$

is very common in reliability applications, because it is justifiable as an extreme value model for the first time to failure of many components. However, it is not in the exponential family and so does not have a conjugate prior. Soland[12], Canavos and Tsokos[3], and Singpurwalla[11] suggested alternatives with restricted ranges for one of the two parameters. A better choice, without restriction, might be the joint probability density function

$$g(\alpha, \lambda) = \sqrt{\frac{2b}{\pi}} \frac{c^{\frac{b+1}{2}} \alpha^b}{\Gamma\left(\frac{b+1}{2}\right) \lambda} \exp\left\{ -c\alpha^2 - \frac{b}{2}(a\alpha + \log \lambda)^2 \right\}; \alpha > 0, \lambda > 0 \tag{22}$$

which is a location-scale prior for $\log X$; see Bury[2] and Percy[10]. Unlike the natural conjugate prior for the gamma distribution, this joint density is determined explicitly. It also depends on three or four hyperparameters and the conditional distribution of λ given α clearly has a lognormal form. However, this bivariate prior's non-conjugacy renders much subsequent calculation analytically impossible and therefore heavily dependent upon numerical computation—this continues to become less of an issue as computing power and numerical algorithms improve. Note that no location-scale prior of the form in Equation (22) exists for the gamma distribution.

2.5. *Other Reliability Models*

The generic phrase *generalized linear models*[7] encompasses multiple linear regression, analysis of variance and covariance, logistic regression, log-linear and survival regression, proportional odds and hazards models and rejuvenation models. These models also assume a form for the probability distribution of one or more random variables, now usually labelled Y, conditional upon one or more unknown parameters θ. However, some (link) function of θ is equated with a linear predictor $x^{\mathrm{T}}\beta$ which is the product of a vector of observable predictor variables x and a vector of unknown regression coefficients β .

As these regression coefficients are real parameters, a multivariate normal distribution is appropriate to express their joint prior density. By transforming predictor variables if necessary, it is often reasonable to assume prior sphericity (independence and equal variance) to reduce the numbers of hyperparameters to be determined. Where the response distribution has a nuisance scale parameter, then independent multivariate normal and gamma priors can be adopted. However, a multivariate-normal-gamma distribution[1] is conjugate for the general linear model with a normal response variable

$$Y|\boldsymbol{x}, \boldsymbol{\beta}, \psi \sim No(\boldsymbol{x}^{\mathrm{T}}\boldsymbol{\beta}, \psi) \tag{23}$$

and so is preferable to avoid unnecessary algebraic complexity in this extremely common situation, where ψ again represents the unknown precision.

Although time series analysis is less common in reliability applications, *stochastic processes*[5] feature prominently. Perhaps the two most common types in reliability analysis are renewal processes and Poisson processes. The former are often applied to model successive lifetimes of components that are replaced upon failure, such as light bulbs in a socket (renewal). The latter are typically used to model complex repairable systems, in which the failures of single components cause system failures and replacing the components restores the systems to the condition immediately before failure (minimal repair). Renewal processes are easily dealt with by formulating priors for the probability distributions discussed earlier in Section 2 and constructing likelihood functions as described in Section 1. Analysis of a Poisson process is not so easy.

Popular choices of intensity function for a Poisson process, as functions of system age t, are the power-law intensity

$$\iota(t) = \alpha\beta t^{\beta-1} \tag{24}$$

and the log-linear intensity

$$\iota(t) = \exp(\alpha + \beta t), \tag{25}$$

yet the possibilities are infinite. Although both of these contain two unknown parameters, $\alpha, \beta \in \mathbb{R}^+$ for the former whereas $\alpha, \beta \in \mathbb{R}$ for the latter. In these cases, the first times to failure are Weibull and Gumbel respectively, so we can specify the prior density by considering these lifetime distributions as before. To proceed with the analysis, consider a Poisson process with intensity function $\iota(t)$ which depends on one or more unknown parameters θ. Suppose that we observe all failure times $t_1, \ldots, t_n$ during the interval $(0, T]$. The likelihood function is then given by

$$L(\theta; \mathcal{H}_T) \propto \left\{ \prod_{i=1}^{n} \iota(t_i) \right\} \exp \left\{ - \int_0^T \iota(t)dt \right\} \tag{26}$$

where $\mathcal{H}_T$ is the history of all failures up to time T. The prior-posterior analysis can now proceed as in Section 1.

2.6. *Discussion*

Section 2 presents recent developments relating to the specification of subjective prior distributions for common reliability models. Although natural conjugate priors are most appealing, they only exist for members of the exponential family. Location-scale priors are equally unable to act as a panacea for problems of specifying subjective priors. Our investigation enables us to propose the following simple, though robust, strategy.

Noting that beta, gamma and normal distributions are both natural conjugate and location priors for the simplest of models (respectively binomial, Poisson and normal with known variance), we recommend using these distributions for corresponding parameters in other models. If the parameter range is finite, a beta prior is appropriate. If semi-infinite, we use a gamma prior. If the parameter range is infinite, then a normal prior suffices. For the beta and gamma priors, linear scaling might first be required to match the density to the relevant parameter range. For multi-parameter situations, we advocate that prior independence is often a reasonable assumption.

More complex priors will be required in some applications, though the strategy suggested here works well for most analyses, especially those involving lifetime distributions and stochastic processes. A simple extension of this recommended strategy applies to generalized linear models, in which case a multivariate-normal distribution can be adopted for the vector of regression coefficients, possibly assuming sphericity as suggested earlier to concur with the use of independent priors for multi-parameter situations. In any case, nuisance scale parameters are modelled independently using gamma priors. Naturally, the subsequent algebra is often intractable and the analysis requires numerical computation.

The only difficulty with our proposed approach is that natural conjugacy is occasionally ignored, thereby introducing an extra hyperparameter and losing some computational advantage. This effect is most prominent for a normal sampling distribution, for which a natural conjugate prior (the normal-gamma distribution) exists and simplifies the resulting integrals. However, our assumption of prior independence recommends the product of a normal and a gamma density, which prevents us from obtaining closed-form expressions for predictive distributions. Nevertheless, this prior is conditionally conjugate for the normal model and ensures consistency with other multi-parameter analyses. Iterative numerical, Markov chain Monte Carlo methods can be used to overcome the computational difficulties incurred.

Our recommended simple approach also has one ambiguity, related to the non-invariance of subjective priors. Suppose we have a positive real parameter, such as λ when our model is exponential. The proposed strategy recommends imposing a gamma prior on this parameter. However, one could reparameterize the model in terms of a transformed parameter, such as $\mu = 1/\lambda$. We would then propose a gamma prior for μ, which does not correspond with a gamma prior for λ in general. Fortunately, this does not present a problem because both of these priors are sufficiently flexible to model available knowledge adequately well for practical purposes. This flexibility is universal, as our strategy ensures that each unknown parameter has an associated prior containing two hyperparameters.

3. Hyperparameter Elicitation

Hyperparameter elicitation from the prior $g(\theta)$ directly is conceptually difficult. The consensus of opinion amongst researchers is now to elicit expert knowledge about hyperparameters from observable quantities only; see Kadane and Wolfson[6] and O'Hagan[9]. This superior approach is achievable by specifying summary features of the prior predictive density (mass) function

$$f(x) = \int_{-\infty}^{\infty} f(x|\theta)g(\theta)d\theta, \tag{27}$$

which describes the probability distribution of the random variables X without conditioning on the parameters θ, yet is still a function of the unknown hyperparameters.

The moments (mean, variance, ...) are unreasonable summary features of $f(x)$, as they are based on the non-trivial concept of mathematical expectation. The mode (most likely value) is perhaps the obvious summary feature, though ambiguity arises if the maximum is at an endpoint. Furthermore, the mode's extensions to relative likelihoods are not usually amenable for analysis. Perhaps the best summary features are quantiles or cumulative probabilities. In principle, both tertiles would suffice when there are two hyperparameters to be determined, whereas the quartiles would be needed to determine three hyperparameters. Modifications of this basic strategy are usually desirable though, as constraints that cannot always be judged in advance are often imposed on the quantiles.

Consequently, we sometimes prefer to leave one or more hyperparameters free for fine-tuning. With two hyperparameters, the median alone can now be specified and one parameter is free for interactive adjustment. With three hyperparameters, the median or tertiles allow for further adjustments. This approach has the added advantage that there are fewer simultaneous nonlinear equations to be solved numerically, so the computational problem is better controlled. Currently, the author is developing a suite of programs using the mathematical software package Mathcad, for its interactive facility and impressive graphics. However, a more robust version is planned using Fortran95 with the NAG library. We now consider the feasibility of hyperparameter specification for the reliability models discussed in Section 2, thereby extending the work of Percy[10].

3.1. *Binomial and Negative-binomial*

For both binomial and negative-binomial models, which include Bernoulli and geometric models as special cases, the recommended prior is the beta density given in Equation (11). For these distributions only, the hyperparameters a and b can be determined by eliciting tertiles from the prior density directly. This is because the unknown parameter θ here is potentially an observable quantity: the long-run proportion of successes. The beta cumulative distribution function is equivalent to the incomplete beta function

$$G(\theta) = \int_0^\theta \frac{\theta^{a-1}(1-\theta)^{b-1}}{\mathrm{B}(a,b)}\, d\theta = I_\theta(a,b) \tag{28}$$

which is not expressible in closed form and must be evaluated numerically for each iteration in the numerical solution of two nonlinear simultaneous equations. Subroutines are available for these purposes in Mathcad and Fortran95 via the NAG library and we present an example of this in Section 4. In less common situations where experts have no information relating to the repetition of trials, such as when contemplating the reliability of a new system, we can do no better than judge the probability of failure completely subjectively.

3.2. *Poisson and Exponential*

For the Poisson distribution with gamma prior as in Equation (13), the prior predictive mass function is

$$f(x) = \frac{b^a \Gamma(x+a)}{\Gamma(a)x!(b+1)^{x+a}}; x = 0, 1, \ldots \tag{29}$$

and is referred to as Poisson-gamma[1]. The prior predictive distribution corresponding to the exponential model and prior of Equation (15) is gamma-gamma, with probability density function

$$f(x) = \frac{ab^a}{(x+b)^{a+1}}; x > 0. \tag{30}$$

In both cases, we determine the hyperparameters a and b by eliciting tertiles from the corresponding prior predictive distribution.

For the Poisson model, this involves evaluating cumulative probabilities from the prior predictive probability mass function by finite summation

$$F(x) = \sum_{y=0}^{\lfloor x \rfloor} \frac{b^a \Gamma(y+a)}{\Gamma(a)y!(b+1)^{y+a}}; x \geq 0 \tag{31}$$

and allowing for its stepwise increasing nature in specifying the quantiles. Here, y is simply a dummy variable that represents realized values of the random variable X. For the exponential model, the prior predictive cumulative distribution function is available in closed form as

$$F(x) = \int_0^x \frac{ab^a}{(y+b)^{a+1}} dy = 1 - \left(\frac{b}{x+b}\right)^a ; x > 0. \tag{32}$$

Hence, for both models we need only solve two simultaneous nonlinear equations numerically to determine hyperparameter values from expert knowledge. Indeed, these are readily condensed into a single nonlinear equation for the exponential model.

3.3. *Normal and Lognormal*

For a normal distribution with natural conjugate prior as in Equation (18), the prior predictive distribution is Student-t[1] with probability density function

$$f(x) = \sqrt{\frac{b}{2(b+1)c\pi}} \frac{\Gamma\left(\frac{b+2}{2}\right)}{\Gamma\left(\frac{b+1}{2}\right)} \left\{1 + \frac{b(x-a)^2}{2(b+1)c}\right\}^{-\frac{b+2}{2}} ; x \in \mathbb{R}. \tag{33}$$

However, in Subsection 2.6 we elected instead to work with independent normal and gamma priors, $\mu \sim No(a,b)$ and $\psi \sim Ga(c,d)$, for consistency with other multi-parameter analyses. Partial integration gives a prior predictive probability density function of the form

$$f(x) = \int_{-\infty}^{\infty} \int_{0}^{\infty} f(x|\mu, \psi)g(\mu)g(\psi)d\psi d\mu$$

$$= \frac{1}{2\mathrm{B}(c, \frac{1}{2})}\sqrt{\frac{b}{\pi d}} \int_{-\infty}^{\infty} \frac{\exp\left\{-\frac{b}{2}(\mu - a)^2\right\}}{\left\{1 + \frac{(x-\mu)^2}{2d}\right\}^{c+\frac{1}{2}}} d\mu; x \in \mathbb{R} \qquad (34)$$

and we cannot proceed analytically, so numerical computation is required.

One might suppose that we should elicit the expert's four pentiles from this distribution numerically, to determine suitable values for the hyperparameters a, b, c and d. However, it is a symmetric distribution about $x = a$ and so the pentiles are equal in pairs. Consequently, we choose to elicit the largest four octiles corresponding to cumulative probabilities of 1/2 (median), 5/8, 3/4 (upper quartile) and 7/8 respectively. Due to the symmetry identified above, the median immediately equates with the hyperparameter a and so we only have three nonlinear simultaneous equations to solve numerically for b, c and d.

3.4. *Gamma and Weibull*

The gamma distribution described earlier, with independent gamma priors $\alpha \sim Ga(a, b)$ and $\lambda \sim Ga(c, d)$ on its two parameters, has prior predictive density function given by

$$f(x) = \int_{0}^{\infty} \int_{0}^{\infty} f(x|\alpha, \lambda)g(\alpha)g(\lambda)d\lambda d\alpha$$

$$= \frac{b^a d^c}{\Gamma(a)} \int_{0}^{\infty} \frac{x^{\alpha-1}\alpha^{a-1}\exp(-b\alpha)}{\mathrm{B}(\alpha, c)(x + d)^{\alpha+c}} d\alpha \qquad (35)$$

for $x > 0$. Regrettably, this integral cannot be resolved analytically and numerical quadrature is required. Similarly, the Weibull distribution described earlier, with independent gamma priors on its two parameters as above, has prior predictive density function given by

$$f(x) = \frac{cb^a d^c}{\Gamma(a)} \int_{0}^{\infty} \frac{x^{\alpha-1}\alpha^a \exp(-b\alpha)}{(x^\alpha + d)^{c+1}} d\alpha \qquad (36)$$

for $x > 0$. Again, this integral is intractable and a closed form expression for the prior predictive density is not available.

In both the gamma and Weibull cases, numerical computation is needed for these prior predictive probability density functions in order to evaluate cumulative probabilities corresponding to the experts' stated quantiles. As there are four hyperparameters in each model, these take the form of pentiles and extra investigations are in progress to improve the efficiency of this strategy for these models.

3.5. *Other Reliability Models*

Consider first the general linear model with normal response variable Y and vector of predictor variables x, as given by Relation (23). With a conjugate multivariate-normal-gamma prior on the vector of regression coefficients β and precision ψ, the conditional prior predictive distribution for Y given x is Student-t, similar to the form in Relation (33). However, for consistent treatment of multi-parameter prior modelling, we again choose to work with independent multivariate-normal and gamma priors. For many generalized linear models, including logistic and log-linear regression, only the multivariate-normal prior is needed.

The corresponding prior predictive distributions for these models are desirable for hyperparameter elicitation. They must be determined conditionally, given some specified values for the observable predictor variables. The problem then reduces to considering the sampling distributions discussed earlier, for which the predictive distributions cannot be evaluated analytically in general. Efficient experimental strategies are used to select suitable values for the predictor variables, based on the theory of fractional factorial and response surface designs. Which design to choose for a given application depends on how many hyperparameters must be quantified to reflect elicited expert knowledge.

For any particular problem, we now have the choice of eliciting just one quantile (the median) at each of many design points, or several quantiles at each of few design points. The former strategy is more robust against errors in experts' probability assessments and provides better coverage of the sample space for predictor variables. Furthermore, the symmetrical nature of the predictive density identified earlier for the normal linear model implies that the median equates with the linear predictor each time, avoiding the need for extensive numerical quadrature in this case.

As for elicitation of hyperparameters in stochastic processes, we earlier noted that the intensity of a nonhomogeneous Poisson process corresponds to the hazard of the first time to failure. Consequently, one could easily formulate the prior predictive cumulative distribution function and hence elicit expert knowledge based on this first time to failure. However, a better strategy for elicitation might be to request the expert to envisage a system of any stated age t (such as that of one in current operation) and consider the time to next failure τ, whose conditional density is

$$f(\tau|t,\theta) = \iota(t+\tau)\exp\left(-\int_t^{t+\tau}\iota(t)dt\right), \tag{37}$$

from which the conditional prior predictive density can be determined as

$$f(\tau|t) = \int_{-\infty}^{\infty} f(\tau|t,\theta)g(\theta)d\theta. \tag{38}$$

4. Application

To illustrate some of the preceding methods, we now consider a reliability application investigated by Christer[4]. This involves the automatic checking of printed circuit boards in a high-speed, high-volume production line. There are essentially three unknown parameters, each representing a Bernoulli probability, which we denote as:

- $\alpha \approx 0$, the probability that a board is defective;
- $\beta \approx 1$, the probability that a defective board fails the test;
- $\gamma \approx 1$, the probability that a functional board passes the test.

Some data $\mathcal{D}$ are available, from which the maximum likelihood estimates are found to be $\hat{\alpha} = 0.025$, $\hat{\beta} = 1.000$ and $\hat{\gamma} = 0.980$. Note that $\hat{\beta}$ is at an endpoint of the feasible range for β and suggests that no defective boards pass the test, which is known to be untrue.

However, we would first elicit an expert's judgement about the long-run proportion of defective boards. The expert might indicate that this proportion is equally likely to be under 1%, between 1% and 3%, and over 3%. We translate this as $G_\alpha(0.01) = \frac{1}{3}$ and $G_\alpha(0.03) = \frac{2}{3}$, corresponding to the beta prior $\alpha \sim Be(0.9, 30.0)$. We would then elicit similar priors for the other parameters using similar logic, leading perhaps to $\beta \sim Be(30.0, 0.9)$ and $\gamma \sim Be(30.0, 0.9)$. The joint posterior density is then given by

$$g(\alpha,\beta,\gamma|\mathcal{D}) \propto L(\alpha,\beta,\gamma;\mathcal{D})g(\alpha,\beta,\gamma) \tag{39}$$

which can be used to compute the Bayes estimates $\hat{\alpha} = 0.027$, $\hat{\beta} = 0.982$ and $\hat{\gamma} = 0.982$. Not only have we included the observed data in this analysis, but also knowledge based upon experience. This avoids the problem that we previously encountered when estimating β. Better still, the posterior can be incorporated directly in a decision analysis as follows.

Define r to be the net gain from passing a functional board and s to be the net loss from passing a defective board. The former is typically very small, as individual boards are inexpensive, whereas the latter is typically very large, as the company incurs replacement costs and loss of goodwill. The expected profit per board is then

$$u(\alpha, \beta, \gamma) = r(1 - \alpha)\gamma - s\alpha(1 - \beta). \tag{40}$$

Arbitrarily setting $r = 1$ and $s = 100$ gives an estimated utility of $u(\hat{\alpha}, \hat{\beta}, \hat{\gamma}) = 0.96$ if the maximum likelihood estimates are used. The Bayesian approach, which we claim to be more accurate, gives an expected profit per board of

$$E\{u(\alpha, \beta, \gamma)|\mathcal{D}\} = \int_0^1 \int_0^1 \int_0^1 u(\alpha, \beta, \gamma)g(\alpha, \beta, \gamma|\mathcal{D})d\alpha d\beta d\gamma = 0.90 \tag{41}$$

which suggests an error of 5.8% when relying upon the frequentist approach.

5. Conclusion

A simple strategy for selecting subjective priors is proposed. The method of eliciting quantiles directly from prior distributions (Bernoulli models only) or from prior predictive distributions is shown to be feasible and promising when applied to a wide variety of reliability models. The procedure requires some numerical computation including quadrature, random sampling and solution of nonlinear simultaneous equations. Further investigations and practical experience are needed before this approach can be adopted routinely in practice.

References

1. J. M. Bernardo and A. F. M. Smith, *Bayesian Theory* (Wiley, Chichester, 1993).
2. K. V. Bury, Bayesian decision analysis of the hazard rate for a two-parameter Weibull process, *IEEE Transactions on Reliability*, **21**, 159-169 (1972).
3. G. C. Canavos and C. P. Tsokos, Bayesian estimation of life parameters in the Weibull distribution, *Operations Research*, **21**, 755-763 (1973).
4. A. H. Christer, Modelling the quality of automatic quality checks, *Journal of the Operational Research Society*, **45**, 806-816 (1994).
5. G. R. Grimmett and D. R. Stirzaker, *Probability and Random Processes* (Clarendon Press, Oxford, 1992).
6. J. B. Kadane and L. J. Wolfson, Experiences in elicitation, *The Statistician*, **47**, 3-19 (1998).
7. P. McCullagh and J. A. Nelder, *Generalized Linear Models* (Chapman and Hall, London, 1989).

8. A. O'Hagan, *Kendall's Advanced Theory of Statistics, Volume 2B: Bayesian Inference* (Arnold, London, 1994).
9. A. O'Hagan, Eliciting expert beliefs in substantial practical applications, *The Statistician*, **47**, 21-35 (1998).
10. D. F. Percy, Bayesian enhanced strategic decision making for reliability, *European Journal of Operational Research*, **139**, 133-145 (2002).
11. N. D. Singpurwalla, An interactive PC-based procedure for reliability assessment incorporating expert opinion and survival data, *Journal of the American Statistical Association*, **83**, 43-51 (1988).
12. R. M. Soland, Bayesian analysis of the Weibull process with unknown scale and shape parameters, *IEEE Transactions on Reliability*, **18**, 181-184 (1969).

6

COPULÆ AND THEIR USES

Carlo Sempi

Dipartimento di Matematica "Ennio De Giorgi"
Università di Lecce
I–73100 Lecce, Italy
E-mail: carlo.sempi@unile.it

This survey of copulas reviews some aspects of copulas, their properties, tries to stress their relevance for statistics and their connection with Markov processes and conditional expectations.

1. What is a Copula?

Copulæ were introduced in 1959 by Sklar[25]. Nowadays the literature on copulas is very large. The reader is referred in the first place to the authoritative books and surveys by Schweizer, Sklar and Nelsen[26,21,20,27,15]. Also the books by Joe[13] and by Hutchinson and Lai[12] contain much useful information; so do also a wealth of papers, some of which will be cited when the need arises. The present paper aims at introducing some of the properties and of the uses of copulas, even of those not (yet) of immediate relevance for the field of reliability; of course, it cannot pretend to have either the same breadth or the same depth of the works just cited.

A copula is a function $C : [0,1] \times [0,1] \to [0,1]$ that satisfies the following properties:

- for all t in $[0,1]$, $C(t,0) = C(0,t) = 0$;
- for all t in $[0,1]$, $C(t,1) = C(1,t) = t$;
- if x, x', y, y' are in $[0,1]$ with $x \leq x'$ and $y \leq y'$, then

$$C(x',y') - C(x,y') - C(x',y) + C(x,y) \geq 0. \tag{1}$$

As a consequence of these properties it follows that

(a) C satisfies the Lipschitz condition

$$|C(x,y) - C(x',y')| \le |x - x'| + |y - y'|,$$

(b) it is non–decreasing in each variable,

(c) it is absolutely continuous.

In other words, a copula is (the restriction of) an absolutely continuous bivariate distribution function that concentrates all the probability mass on the unit square $[0,1] \times [0,1]$ and which has uniform marginals. We consider random variables that may take the values $-\infty$ and/or $+\infty$ with probability different from zero. As a consequence, their distribution functions are defined on $\overline{\mathbf{R}} := [-\infty, +\infty]$, in the univariate case, and on $\overline{\mathbf{R}}^2 = \overline{\mathbf{R}} \times \overline{\mathbf{R}}$, in the bivariate case; also they may have a jump at points with at least an infinite coordinate.

The importance of the concept of copula stems from the following

Theorem 1: (Sklar) *Let X and Y be two random variables on the probability space $(\Omega, \mathcal{F}, P)$ having H as their joint distribution function and let F and G be the marginals of H,*

$$F(x) = H(x, +\infty), \qquad G(y) = H(+\infty, y).$$

Then, there exists (at least) a copula C such that

$$H(x,y) = C\left(F(x), G(y)\right). \tag{2}$$

If both F and G are continuous, then the copula C is uniquely determined; otherwise, C is uniquely determined on Ran $F \times$ Ran G, *where* Ran F *is the image of $\overline{\mathbf{R}}$ under F,* Ran $F := F\left(\overline{\mathbf{R}}\right)$. *Conversely, if C is a copula and F and G are univariate distribution functions, then the function H defined by (2) is a joint distribution function with marginals given by F and G.*

If either F or G, or both, is not continuous, then more than one copula may satisfy (2); all of these coincide on Ran $F \times$ Ran G. Notice that the discrete case is important for the applications because empirical versions of copulas are obtained by substituting empirical distributions for F and G. However, by a method of bilinear interpolation, it is always possible to choose a single copula that satisfies (2), see Section 2.3.5 in the book by Nelsen[15]. This method is essential for some of the developments presented in the sequel. More specifically, let C' be any of the copulas that satisfy (2) and which are uniquely determined on Ran $F \times$ Ran G; extend it by continuity to the closure of the set Ran $F \times$ Ran G. Now let (s,t) be any point in $[0,1] \times [0,1]$ and let s_1 and s_2 be the greatest and the least element, respectively, in $\overline{\text{Ran } F}$—the closure of Ran F—such that $s_1 \leq s \leq s_2$. Similarly, let t_1 and t_2 be the greatest and the least element, respectively, in $\overline{\text{Ran } G}$ such that $t_1 \leq t \leq t_2$. If s belongs to $\overline{\text{Ran } F}$, then $s_1 = s = s_2$; and, if t belongs to $\overline{\text{Ran } G}$, then $t_1 = t = t_2$. Then define

$$\lambda_1 := \begin{cases} \dfrac{s - s_1}{s_2 - s_1}, & \text{if } s_1 < s_2, \\ 1, & \text{if } s_1 = s_2; \end{cases}$$

$$\lambda_2 := \begin{cases} \dfrac{t - t_1}{t_2 - t_1}, & \text{if } t_1 < t_2, \\ 1, & \text{if } t_1 = t_2. \end{cases}$$

It is a long, but not hard, task to check that the function defined on $[0, 1] \times [0, 1]$ by

$$\begin{aligned} C(s, t) := {} & (1 - \lambda_1)\,(1 - \lambda_2)\,C'(s_1, t_1) + (1 - \lambda_1)\,\lambda_2\,C'(s_1, t_2) \\ & + \lambda_1\,(1 - \lambda_2)\,C'(s_2, t_1) + \lambda_1\,\lambda_2\,C'(s_2, t_2) \end{aligned}$$

is a copula and that it satisfies (2).

Among the copulas, three are particularly important; they are denoted by W, Π and M and are defined by

$$W(s, t) := \max\{0, s + t - 1\},$$
$$\Pi(s, t) := s\,t,$$
$$M(s, t) := \min\{s, t\}.$$

Two continuous random variables X and Y are independent if, and only if, their copula C_{XY} is equal to Π; two continuous random variables X and Y have W as their copula if, and only if, one of them is a strictly decreasing function of the other one, while they have M as their copula if, and only if, one of them is a strictly increasing function of the other one.

Let X and Y be two continuous random variables and let C_{XY} be their copula. If φ and ψ are strictly increasing functions defined on Ran X and Ran Y respectively, then the copula $C_{\varphi \circ X, \psi \circ Y}$ of the random variables $\varphi \circ X$ and $\psi \circ Y$ satisfies

$$C_{\varphi \circ X, \psi \circ Y} = C_{XY}.$$

Thus the copula C_{XY} is invariant under strictly increasing transformations of X and Y. This property has an important consequence: it makes copulas well suited to express the "scale invariant" properties and measures of association for random variables. This aspect will be exploited in section 3.

Moreover, if C is any copula, the following inequalities (the Fréchet bounds[14]) hold for all s and t in $[0,1]$,

$$W(s,t) \leq C(s,t) \leq M(s,t).$$

The properties of the partial derivatives of a copula C are important for what follows. For every $t \in [0,1]$, the function $s \mapsto C(s,t)$ is non-decreasing and, hence, differentiable for almost every t; where the derivative exists, one has

$$0 \leq D_1 C(s,t) := \frac{\partial C(s,t)}{\partial s} \leq 1 \qquad \text{a.e..}$$

Similarly, for every $s \in [0,1]$, the function $t \mapsto C(s,t)$ is non-decreasing and differentiable almost everywhere in I; where the derivative exists, one has

$$0 \leq D_2 C(s,t) := \frac{\partial C(s,t)}{\partial t} \leq 1 \qquad \text{a.e..}$$

2. Special Classes of Copulas

In the literature, one can find several methods of constructing copulas; see, for this, Chapter 3 in the book by Nelsen[15]. The most widely studied class of copulas is probably that of *Archimedean* copulas. The reason for which these copulæ have been so extensively studied is twofold: on one hand, they are symmetric and, therefore, they lend themselves admirably to the study of pairs of exchangeable random variables, and, on the other hand, many of the families presented in the literature depend on one or more parameters, which allows the usual statistical procedures of best estimation and goodness of fit in concrete case studies; see, for instance De Michele and Salvadori[7].

In order to define Archimedean copulas, let $\varphi : [0,1] \to [0,+\infty]$ be continuous, convex, strictly decreasing and such that $\varphi(1) = 0$. The *pseudo-inverse* of φ is the function $\varphi^{[-1]}$ from $[0,+\infty]$ into $[0,1]$ defined by

$$\varphi^{[-1]}(t) := \begin{cases} \varphi^{-1}(t), & t \in [0,\varphi(0)], \\ 0, & t \in [\varphi(0),+\infty]. \end{cases}$$

For every $t \in [0,1]$, one has

$$\varphi^{[-1]}\left(\varphi(t)\right) = t,$$

while, for every $x \in [0,+\infty]$, one has

$$\varphi\left(\varphi^{[-1]}(x)\right) = \min\{x, \varphi(0)\}.$$

Then, an Archimedean copula C is defined via

$$C(s,t) := \varphi^{[-1]}\left(\varphi(s) + \varphi(t)\right).\tag{3}$$

The function φ is called the *additive generator* of C.

In the work of Averous and Dortet–Bernadet[2] a nice relationship is established between the dependence properties of the copula (3) and the aging properties of the distribution function

$$F_\varphi(t) := 1 - \varphi^{[-1]}(t).$$

In the class of Archimedean copulas the most widely studied is the family of Frank's copulas[9]; if θ is a real parameter different from zero, then the Frank's family of copulas is defined through

$$C_\theta(s,t) := -\frac{1}{\theta}\ln\left(1 + \frac{\left(e^{-\theta s} - 1\right)\left(e^{-\theta t} - 1\right)}{e^{-\theta} - 1}\right).$$

The additive generator of C_θ is

$$\varphi_\theta(t) := -\ln\frac{e^{-\theta t} - 1}{e^{-\theta} - 1}.$$

It is important to note the following limiting values

$$\lim_{\theta\to-\infty} C_\theta = W, \qquad \lim_{\theta\to 0} C_\theta = \Pi, \qquad \lim_{\theta\to+\infty} C_\theta = M.$$

3. Statistical Properties

Copulæ are widely used in non–parametric statistics, especially in the study of dependence of random variables. For the notion of dependence see, for instance, the books by Szekli[28] and by Joe[13]. Many interesting known measures of association between random variables may be expressed in terms of copulas. If X and Y are continuous random variables and C is their copula, then Kendall's tau is

$$\tau_{X,Y} = 4\int_{[0,1]\times[0,1]} C(s,t)\,dC(s,t) - 1;$$

Spearman's rho is

$$\rho_{X,Y} = 12\int_{[0,1]\times[0,1]} st\,dC(s,t) - 3 = 12\int_{[0,1]\times[0,1]} C(s,t)\,ds\,dt - 3;$$

and Gini's measure of association is

$$\gamma_C = 4\int_0^1 C(t, 1-t)\,dt - 4\int_0^1 (1 - C(t,t))\,dt.$$

A measure of dependence was introduced by Schweizer and Wolff[22] in terms of the copula of the two random variables involved; its L^1–version is given by

$$\sigma_1(X,Y) := 12 \int_{[0,1]\times[0,1]} |C(s,t) - st| \, ds \, dt.$$

The L^p–version, with $p \in \,]1, +\infty[$, is given by

$$\sigma_p(X,Y) := \left\{ k_p \int_{[0,1]\times[0,1]} |C(s,t) - st|^p \, ds \, dt \right\}^{1/p};$$

here k_p is a suitable normalization factor, necessary for the inequality

$$0 \leq \sigma_p(X,Y) \leq 1.$$

For instance, in the case $p = 2$, one has $k_2 = 90$. There is also a L^∞–version given by

$$\sigma_\infty := 4 \sup \{|C(s,t) - st| : s, t \in [0,1]\}.$$

All these measures of dependence meet a slight modification of Rényi's list of requirements for such a measure[18]; among other properties, Rényi requested that a measure of dependence $R(X,Y)$ of two random variables X and Y defined on a common probability space should satisfy $R(X,Y) = 1$ if either $X = f \circ X$ or $Y = g \circ X$ for some Borel–measurable functions f and g, and $R(f \circ X, g \circ Y) = R(X,Y)$ when f and g are one–to–one Borel-measurable functions. Instead the measures σ_p defined above are such that $\sigma_p(X,Y) = 1$ and $\sigma_p(f \circ X, g \circ Y) = \sigma_p(X,Y)$, if, and only if, f and g are strictly monotone.

Beside satisfying Rényi's requirements in the modified form just mentioned, the measures of dependence σ_p are such that if $\{(X_n, Y_n)\}$ converges weakly to (X,Y), then $\lim_{n \to +\infty} \sigma_p(X_n, Y_n) = \sigma(X,Y)$.

One can define *empirical copulas*. If $\{(X_j, Y_j) : j = 1, 2, \ldots, n\}$ is a sample of size n from a bivariate distribution and if the order statistics from the same sample are denoted by $X_{(j)}$ and by $Y_{(j)}$ then the empirical copula is defined via

$$C_n\left(\frac{j}{n}, \frac{k}{n}\right) := \frac{\text{number of pairs } (X,Y) \text{ with } X \leq X_{(j)} \text{ and } Y \leq Y_{(k)}}{n}.$$

Empirical copulæ were first used by Deheuvels[6] in order to construct tests of independence.

Recently copulæ have been used in Finance, see, for instance[10,8].

4. Copulæ and Markov Processes

Darsow, Nguyen and Olsen[5] established a connection between copulas and Markov processes through an operation on the set $\mathcal{C}$ of all copulas, which will now be described. Let A and B be copulæ; if x and y are in $[0,1]$ an operation on $\mathcal{C}$ is defined via

$$(A * B)(x,y) := \int_0^1 D_2 A(x,t)\, D_1 B(t,y)\, dt. \tag{4}$$

Then $A * B$ is a copula, the operation $*$ is associative, and, if $\{A_n\}$ converges to A, then one has both

$$A_n * B \xrightarrow[n \to +\infty]{} A * B$$

and

$$B * A_n \xrightarrow[n \to +\infty]{} B * A;$$

however, the operation $*$ is not jointly continuous. The operation $*$ has both a zero and an identity: they are, respectively, the copulæ Π and M; in fact, for every copula C, one has

$$\Pi * C = C * \Pi = \Pi,$$
$$M * C = C * M = C.$$

It follows from these relationships, and this remark will be important in the sequel, that both Π and M are idempotent, in the sense that

$$\Pi * \Pi = \Pi,$$
$$M * M = M.$$

The importance of the operation $*$ stems from the following

Theorem 2: (Darsow, Nguyen, Olsen) *Let $\{X_t : t \in T\}$ be a real–valued stochastic process and let C_{st} be the copula of the random variables X_s and X_t. If, for s and t in T with $s < t$ and for a Borel set A, one sets*

$$P(s,x,t,A) := P\left(X_t \in A \mid X_s = x\right),$$

then the following are equivalent:

(a) *The Chapman–Kolmogorov equations*

$$P(s, x, t, A) = \int_{\mathbf{R}} P(u, \xi, t, A)\, P(s, x, u, d\xi) \quad (u \in \,]s, t[\,\cap T)$$

hold for almost all $x \in \mathbf{R}$;

(b) $C_{st} = C_{su} * C_{ut}$.

The crucial step in the proof of Theorem 2 is the following equality

$$P(s, t, x,]{-}\infty, a]) = D_1 C_{st}\left(F_s(x), F_t(a)\right) \qquad \text{a.e..}$$

Of course $\{X_t\}$ may satisfy the Chapman–Kolmogorov equations without being a Markov process. A necessary and sufficient condition in terms of the copulas C_{st} is known. In order to state it, it is necessary to have recourse to the notion of n–dimensional copula[21,15], or, briefly n–copula, i.e. the function $C : [0,1]^n \to [0,1]$ that expresses an n–dimensional distribution function H in terms of its one–dimensional marginals F_1, $F_1, \ldots F_n$,

$$H(x_1, x_2, \ldots, x_n) = C\left(F_1(x_1), F_2(x_2), \ldots, F_n(x_n)\right).$$

Let A be an m–copula and let B be an n–copula; then define $A \star B :$ $[0,1]^{m+n-1} \to [0,1]$ via

$$(A \star B)(x_1, x_2, \ldots, x_{m+n-1})$$
$$:= \int_0^{x_m} D_m A(x_1, \ldots, x_{m-1}, t)\, D_1 B(t, x_{m+1}, \ldots, x_{m+n-1})\, dt.$$

Here again D_m denotes the partial derivative with respect to the m–th variable.

Then $A \star B$ is an $(m + n - 1)$–copula and the operation $\star$ is associative, viz. $(A \star B) \star C = A \star (B \star C)$. The proclaimed necessary and sufficient condition is contained in the following theorem[5].

Theorem 3: *For a real–valued stochastic process* $\{X_t : t \in T\}$*, the following conditions are equivalent*

(a) $\{X_t : t \in T\}$ *is a Markov process;*

(b) *for every natural number* n *and for every choice of* n *elements* t_1, t_2, $\ldots$, t_n *in* T*, with* $t_1 < t_2 < \cdots < t_n$*, one has*

$$C_{t_1, t_2, \ldots, t_n} = C_{t_1, t_2} \star C_{t_2, t_3} \star \cdots \star C_{t_{n-1}, t_n},$$

where $C_{t_1, t_2, \ldots, t_n}$ *is the copula of the random vector* $\left(X_{t_1}, X_{t_2}, \ldots, X_{t_n}\right)$ *and* C_{t_{k-1}, t_k} *is the copula of the vector* $\left(X_{t_{k-1}}, X_{t_k}\right)$*, with* $k = 2, 3, \ldots, n$.

5. Copulæ, Markov Operators and Conditional Expectations

While proving Theorem 2, Darsow, Nguyen and Olsen[5] proved that if the random variables X and Y have copula C, then one has almost surely

$$E\left(1_{\{X<x\}} \mid Y\right)(\omega) = D_2 C\left(F_X(x), F_Y(Y(\omega))\right),$$
$$E\left(1_{\{Y<y\}} \mid x\right)(\omega) = D_1 C\left(F_X(X(\omega)), F_Y(y)\right).$$

These latter relationships point at a connection between Conditional Expectations(=CE's) and copulas. This connection is best established through Markov operators. Given a probability space $(\Omega, \mathcal{F}, P)$, a *Markov operator* is a linear operator $T : L^\infty \to L^\infty$ such that

(a) T is positive, $f \geq 0 \implies Tf \geq 0$;
(b) $T1 = 1$;
(c) for every function f in L^∞, one has $E(Tf) = E(f)$.

Above, L^∞ may be replaced by L^1. Notice that if $\mathcal{G}$ is a sub–σ–field of $\mathcal{F}$, then the CE $E_{\mathcal{G}} := E(\cdot \mid \mathcal{G})$ is a Markov operator. It is then natural to ask which Markov operators are also CE's. The answer to this question is obtained by exploiting the expectation invariance of a Markov operator, the property expressed by (c) above and the characterization of CE's given by Pfanzagl[17] (but see also the previous works of Bahadur[3] and Šidák[24]); it is then possible to identify those Markov operators that are also CE's, when the probability space under consideration is $([0,1], \mathcal{B}, \lambda)$.

Theorem 4: *A Markov operator* $T : L^\infty([0,1]) \to L^\infty([0,1])$ *is the restriction to* $L^\infty([0,1])$ *of a CE if, and only if, it is idempotent, viz.* $T^2 := T \circ T = T$. *When this latter condition is satisfied, then* $T = E_{\mathcal{G}}$, *where* $\mathcal{G} := \{A \in \mathcal{B} : T 1_A = 1_A\}$.

On the other hand, an explicit one–to–one correspondence between Markov operators and copulas can be established. It was shown[16] that, for every copula C, the operator T_C defined by

$$(T_C f)(x) := \frac{d}{dx} \int_0^1 D_2 C(x,t) f(t)\, dt \tag{5}$$

is a Markov operator on $L^\infty([0,1])$ and that, conversely, if T is a Markov operator on $L^\infty([0,1])$, then the function $C_T : [0,1] \times [0,1] \to [0,1]$ defined by

$$C_T(x,y) := \int_0^x \left(T 1_{[0,y]}\right)(s)\, ds \tag{6}$$

is a copula. Thus, to a copula C there corresponds a unique Markov operator T_C and, conversely, to a Markov operator T there corresponds a unique copula C_T. Moreover, the composition of the two Markov operators T_A and T_B that correspond to the two copulas A and B is connected to the $*$ operation by means of the relationship

$$T_{A*B} = T_A \circ T_B.$$

Therefore the Markov operator T_C corresponding to a copula C is idempotent, and, hence a CE, if, and only if, the copula C is itself idempotent with respect to the operation $*$ introduced above, namely if, and only if, $C = C * C$. Thus, there exists a one–to–one correspondence between idempotent copulas and CE's in the probability space $([0,1], \mathcal{B}, \lambda)$; for this, see the author's paper[23].

The correspondence between Markov operators and copulas allows to establish a correspondence between copulas and measure preserving transformations on the unit interval[16,30]. We recall that a function $f : [0,1] \to [0,1]$ is said to be a *measure preserving transformation* if, for every Borel subset B of $\mathcal{B}$, $f^{-1}(B)$ is measurable, namely it belongs to $\mathcal{B}$, and one has

$$\lambda\left(f^{-1}(B)\right) = \lambda(B).$$

If f and g are measure preserving transformations, the function $C_{f,g} : [0,1]^2 \to [0,1]$ defined by

$$C_{f,g}(s,t) := \lambda\left(f^{-1}([0,s]) \cap g^{-1}([0,t])\right),$$

is a copula. Conversely, for every copula C there exists a pair of measure preserving transformations f and g, such that

$$C = C_{f,g}.$$

6. Generalizations of Copulas

In spite of their many uses in probability and statistics, uses which we have tried to sketch in the previous sections, it has been necessary to generalize the notion of copula in order to deal with certain problems. The first such generalization was introduced by Alsina, Nelsen and Schweizer[1] in order to characterize a class of operations on distribution functions that derive from corresponding operations on random variables defined on the same probability space. The concept of *track* is needed: a track is a subset B of the unit square $[0,1] \times [0,1]$ that can be written in the form

$$B = \{(F(t), G(t)) : t \in [0,1]\},$$

for some continuous distribution functions F and G such that $F(0) = G(0) = 0$ and $F(1) = G(1) = 1$. Then, a *quasi-copula* is a function $Q : [0,1] \times [0,1] \to [0,1]$ such that, for every track B, there exists a copula C_B that coincides with Q on the points of B: for all $(s,t) \in B$

$$Q(s,t) = C_B(s,t).$$

Later it was proved[11] that Q is a quasi–copula if, and only if, it meets the following requirements

(a) $\forall t \in [0,1] \qquad Q(0,t) = Q(t,0) = 0, \quad \text{and} \quad Q(1,t) = Q(t,1) = t$;
(b) the function $(s,t) \mapsto Q(s,t)$ is non-decreasing in each of its arguments;
(c) Q satisfies Lipschitz's condition, viz., for all s, s', t and t' in $[0,1]$,

$$|Q(s',t') - Q(s,t)| \le |s' - s| + |t' - t|.$$

The properties of quasi–copulas have been extensively studied in Úbeda Flores's dissertation[29].

Bassan and Spizzichino[4] have introduced another generalization of the concept of copula in their investigation of bivariate aging. Let X and Y be two positive and exchangeable random variables and let $\overline{F}$ be their joint survival function

$$\overline{F}(s,t) := P(X > s, Y > t) \qquad s,t > 0.$$

Finally, let $\overline{G}$ be the univariate marginal survival function of $\overline{F}$, $\overline{G}(x) := \overline{F}(x,0) = \overline{F}(0,x)$. Then they introduce the *bivariate aging function* $B : [0,1]^2 \to [0,1]$ defined by

$$B(s,t) := \exp\left\{-\overline{G}^{[-1]}\left(\overline{F}\left(-\ln s, -\ln t\right)\right)\right\}.$$

This is an increasing function of each of its arguments, but is not necessarily a quasi–copula (nor, *a fortiori*, a copula).

A generalization in a different, more abstract, direction, has been introduced by Scarsini[19]; here, we shall briefly touch only on the case $n = 2$. Let $(\Omega_1, \mathcal{B}_1, P_1)$ and $(\Omega_2, \mathcal{B}_2, P_2)$ be probability spaces, where Ω_1 and Ω_2 are Polish (*i.e.* complete metrizable) spaces, each of them endowed with its Borel σ–field $\mathcal{B}_i$ ($i = 1, 2$). Let $\mathcal{A}_i$ be an increasing class of sets belonging to $\mathcal{B}_i$ ($i = 1, 2$) (if the sets A and B are in $\mathcal{A}_i$, then either $A \subset B$ or $B \subset A$, or $A = B$) and let $(\Omega_1 \times \Omega_2, \mathcal{B}_1 \otimes \mathcal{B}_2, \mu)$ be a probability space such that, for all $B_1 \in \mathcal{B}_1$ and $B_2 \in \mathcal{B}_2$, one has

$$\mu(B_1 \times \Omega_2) = P_1(B_1), \qquad \text{and} \qquad \mu(\Omega_1 \times B_2) = P_2(B_2).$$

Then, if one sets

$$\mathcal{A} := \{A_1 \times A_2 : A_1 \in \mathcal{A}_1, A_2 \in \mathcal{A}_2\},$$

there exists a copula $C_\mu^{\mathcal{A}}$ such that, for every choice of $A_1 \times A_2$ in $\mathcal{A}$, one has

$$\mu\,(A_1 \times A_2) = C_\mu^{\mathcal{A}}\,(P_1(A_1), P_2(A_2)).$$

References

1. C. Alsina, R.B. Nelsen, B. Schweizer, On the characterization of a class of binary operations on distribution functions, *Statistics & Probability Letters*, **17**, 85–89 (1993).
2. J. Averous, J.-L. Dortet–Bernadet, *Dependence for archimedean copulas and aging properties of their generating functions*, preprint, 2002, to appear.
3. R.R. Bahadur, Measurable subspaces and subalgebras, *Proceedings of the American Mathematical Society*, **6**, 565–570 (1955).
4. B. Bassan, F. Spizzichino, Dependence and multivariate aging: the role of level sets of the survival function, *Series on Quality and Reliability in Engineering and Statistics*, **5**, 229–242 (2001).
5. W.F. Darsow, B. Nguyen, E.T. Olsen, Copulas and Markov processes, *Illinois Journal of Mathematics*, **36**, 600-642 (1992).
6. P. Deheuvels, La fonction de dépendence empirique et ses propriétés. Un test non paramétrique d'indépendence, *Académie Royale de Belgique. Bulletin de la Classe de Sceinces*, (5) **65**, 274–292 (1979).
7. C. De Michele, G. Salvadori, A generalized Pareto intensity–duration model of storm rainfall exploiting 2–copulas, to appear
8. P. Embrechts, A. McNeil, D. Straudermann, Correlation and Dependence in Risk Management: Properties and Pitfalls, in *Risk Management: Value at Risk and Beyond*, Ed. M. Dempster (Cambridge University Press, 2002) pp. 176–223.
9. M.J. Frank, On the simoultaneous associativity of $F(x, y)$ and $x + y - F(x, y)$, *Aequationes Mathematicae*, **19**, 194–226 (1979).
10. E.W. Frees E. Valdez, Understanding Relationships using Copulas, *North American Actuarial Journal*, **2**, 1–25 (1998).
11. C. Genest, J.J. Quesada Molina, J.A. Rodríguez Lallena, C. Sempi, A characterization of quasi–copulas, *Journal of Multivariate Analysis*, **69**, 193–205 (1999).
12. T.P. Hutchinson, C.D. Lai, *Continuous bivariate distributions. Emphasising applications* (Rumsby Scientific Publishing, Adelaide, 1990).
13. H. Joe, *Multivariate models and dependence concepts* (Chapman & Hall, London, 1997).
14. P. Mikusiński, H. Sherwood, M.D. Taylor, The Fréchet bounds revisited, *Real Analysis Exchange*, **17**, 759–764 (1991-92).
15. R.B. Nelsen, *An Introduction to Copulas* (Springer, New York, 1999).

16. E.T. Olsen, W.F. Darsow, B. Nguyen, Copulas and Markov operators, in *Distributions with fixed marginals and related topics*, Eds. L. Rüschendorf, B. Schweizer, M.D. Taylor, (Institute of Mathematical Statistics (Lecture Notes — Monograph Series Volume 28), Hayward CA, 1996) pp. 244–259.

17. J. Pfanzagl, Characterizations of conditional expectations, *Annals of Mathematical Statistics*, **28**, 415–421 (1967).

18. A. Rényi, On measures of dependence, *Acta Mathematica Academiae Scientiarum Hungaricae*, **10**, 441–451 (1959); also in *Selected papers of Alfréd Rényi*, Ed. P. Turán (Akadémiai Kiadó, Budapest, 1976, Vol. 2) pp. 402–412.

19. M. Scarsini, Copulæ of probability measures on product spaces, *Journal of Multivariate Analysis*, **31**, 201–213 (1989).

20. B. Schweizer, Thirty years of copulas, in *Probability distributions with given marginals*, Eds. G. Dall'Aglio, S. Kotz and G. Salinetti (Kluwer, Dordrecht, 1991) pp. 13–50.

21. B. Schweizer, A. Sklar, *Probabilistic Metric Spaces* (North–Holland, New York, 1983).

22. B. Schweizer, E.F. Wolff, On nonparametric measures of dependence for random variables, *Annals of Statistics*, **9**, 870–885 (1981).

23. C. Sempi, Conditional Expectations and Idempotent Copulæ, in *Distributions with fixed marginals and statistical modeling*, Eds. C. Cuadras, J. Fortiana, J.A. Rodríguez Lallena (Kluwer, Dordrecht, 2002) pp. 223–228.

24. Z. Šidák, On relations between strict–sense and wide–sense conditional expectations, *Theory of Probability and Applications*, **2**, 267–272 (1957).

25. A. Sklar, *Fonctions de répartition à n dimensions et leurs marges*, Publ. Inst. Statist. Univ. Paris, **8** (1959), 229–231.

26. A. Sklar, Random variables, joint distribution functions and copulas, *Kybernetika*, **9**, 449–460 (1973).

27. A. Sklar, Random variables, distribution functions, and copulas — A personal look backward and forward, in *Distributions with fixed marginals and related topics*, Eds. L. Rüschendorf, B. Schweizer, M.D. Taylor, (Institute of Mathematical Statistics (Lecture Notes — Monograph Series Volume 28), Hayward CA, 1996) pp. 1–14.

28. R. Szekli, *Stochastic orderings and dependence in applied probability* (Springer, New York, Lecture Notes in Statistics 97, 1995).

29. M. Úbeda Flores, *Cópulas y quasicópulas: interrelaciones y nuevas propiedades. Aplicaciones* (Ph. D. Dissertation, Universidad de Almería, Servicio de Publicaciones de la Universidad de Almería, Spain, 2002).

30. R.A. Vitale, Parametrizing doubly stochastic measures, in *Distributions with fixed marginals and related topics*, Eds. L. Rüschendorf, B. Schweizer, M.D. Taylor, (Institute of Mathematical Statistics (Lecture Notes — Monograph Series Volume 28), Hayward CA, 1996) pp. 358–364.

Part III

RELIABILITY OF NETWORKS AND SYSTEMS

7

LINKING DOMINATIONS AND SIGNATURES IN NETWORK RELIABILITY THEORY

P. J. Boland

Department of Statistics
National University of Ireland - Dublin, Belfield - Dublin 4, Ireland

F. J. Samaniego

Department of Statistics,
University of California, One Shields Avenue, Davis, CA 95616

E. M. Vestrup

Department of the Mathematical Sciences,
DePaul University, 2230 N. Kenmore, Chicago, IL 60615.

"Domination theory" and the notion of the "signature" of a network, and their respective roles in calculating the reliability of a network, are briefly reviewed. The computational advantages of the former, and the interpretive richness of the latter, raise the question: how are the two related? The exact functional relationship between the signature vector and the vector of signed dominations is obtained.

1. Introduction

In this paper, networks will be viewed as undirected graphs with a fixed number of vertices (or nodes) and a fixed number of edges, each edge connecting two vertices. A network G with v vertices and n edges is typically denoted by the symbol $G(v, n)$. We follow the usual convention that postulates that nodes cannot fail, but that edges can be in either a functioning or a failed state. In communication networks, as in many other types of networks, the primary "quality" characteristic of interest is connectivity. A two-terminal network is connected if there is at least one set of functioning edges providing a path from one terminal to the other. One can also

89

consider k-terminal connectivity and, in its fullest extension, all-terminal connectivity, which occurs when every vertex is connected, via some collection of functioning edges, with every other vertex. We will restrict attention to networks that are "coherent", that is, to networks with the property that every edge is relevant and that all supersets of path sets are also path sets.

The network characteristics upon which we will be focusing are defined in terms of edges whose lifetimes are treated as independent and identically distributed random variables. We will be concerned with the distribution of T, the failure time of the network (i.e., the time at which it becomes disconnected in any of the senses above) and, in particular, with the probability that it is connected at a given time t_0. In the latter instance, we'll treat the states of edges (i.e. working or failed states) as independent Bernoulli variables. In most of what follows, we will be interested in the special case in which these variables are also identically distributed, that is, are i.i.d. Bernoulli variables with common "success" probability p at time t_0. For the most part, we will in the sequel be primarily interested in two-terminal reliability, that is, in the probability that two particular nodes are connected, given that the edges function independently with probability p. But the ideas and expressions that follow are equally applicable to other reliability measures.

It is known, of course, that the reliability of the network $G(v, n)$ in i.i.d. edges can be expressed as a polynomial $h(p)$ of order n, that is, as

$$h(p) = \sum_{r=1}^{n} d_r p^r, \tag{1}$$

where p is the common success probability for the edges. Satyarananaya and Prabhakar[10] showed that the coefficients in (1) could be obtained as the "signed dominations" associated with the network. Indeed, domination theory, described in 1984 as a breakthrough by Agrawal and Barlow[1] among computational tools in network reliability, continues to be a widely used algorithmic vehicle for calculating the reliability polynomial. We review the concept of dominations in Section 2. We note that, as useful as domination theory has proven to be in simplifying the computation of the reliability of a network, it has not been found particularly useful in comparing one network design with another as is required, for example, in searching for universally optimal networks of a given size (v, n).

A quite different tool was introduced by Samaniego[9] for studying the performance properties of coherent systems. The concept of "signature" applies equally well to network reliability. The signature of a network is a probability vector s whose components are simply the respective probabilities that the first, second, $\cdots$, and nth edge failures (ordered by time of occurence) are fatal to the network. Assuming, again, i.i.d. edge states at a fixed time t_0, the reliability polynomial of a network can be expressed in terms of the network's signature vector. Unlike the domination vector, the properties of the signature vector are readily interpretable and have a close relationship to the failure time T of the network itself. We review the notion of "signature", and some of the problems to which it has been applied, in Section 3.

The main goal of this paper is to identify the exact relationship between the vector of signed dominations d and the signature vector s. This is accomplished in Section 4. Because dominations are central to the computation of the reliability of a network, and signatures are rich in interpretation regarding the relative performance of competing networks, the exact linkage of the two through the functional relationship $s = f(d)$ established here enables one to exploit the benefits of both. Our closing example demonstrates the utility of this linkage.

2. A Brief Look at Domination Theory

The notion of dominations was discovered in the process of seeking a reduction in the complexity of the well-known "inclusion-exclusion" formula[6] for calculating the probability that all edges are functioning in at least one of a given network's minimal path sets. The inclusion-exclusion rule applies to the union of any m sets, and may be written as

$$P\left(\bigcup_{i=1}^{m} A_i\right) = \sum_1 P(A_i) - \sum_2 P(A_i \cap A_j) + \sum_3 P(A_i \cap A_j \cap A_k)$$

$$- \cdots + (-1)^{m+1} P\left(\bigcap_{i=1}^{m} A_i\right), \tag{2}$$

where $\sum_i$ represents a sum over all i-fold intersections. If A_i in (2) represents the event that all edges in the ith minimal path set are working, and there are m minimal path sets in all, then the formula in (2) provides the probability that the network will function. That probability is, of course, the network's reliability. One may recognize the result as the reliability polynomial by noting that the probability is p^r that a particular collection of r edges function simultaneously. Since each of the intersections appearing in the inclusion-exclusion formula is precisely the collection of edges appearing in one or more sets of the intersection in question, each term in the expansion (2) contributes elements of the form p^k or $-p^k$ to the reliability polynomial. Summing, and grouping elements of the same order, one would obtain the expression (1).

While the inclusion-exclusion formula will itself provide an explicit expression for the reliability polynomial, it falls short of the desired solution because of its inherent computational complexity. Since the generation of the m minimal path sets of a given network requires exponential time, and since the number of intersections in (2) is also exponential in m, the result is a doubly exponential algorithm for network reliability. This unpleasant fact motivated research that ultimately led to the simplifying notion of "dominations".

Suppose that one has a list of minimal path sets of a given network in n i.i.d. edges. A "formation" is defined as a union of minimal path sets. Note that the "intersection" of any k events in (2) - each event representing a particular set of working edges - may be thought of as the event that each edge in any of the original events works. As a consequence, the probability of an intersection of events representing functioning minimal path sets can be seen to be equivalent to the probability of the set consisting of the union of all these working edges, that is, a union of all these path sets. A *formation* is thus the union of the edges in a fixed collection of minimal path sets. Finally, an *i-formation* is a union of the components in a set of i minimal path sets. For example, the union $\{1, 2, 3, 4\}$ of the minimal path sets $\{1, 2\}$, $\{2, 3\}$, and $\{3, 4\}$ would be an example of a formation that is both a 2-formation and a 3-formation. We will refer to a particular formation as "even" if it is the union of an even number of minimal path sets and as "odd" if it is the union of an odd number of minimal path sets. It can, of course, be both at the same time.

Keeping track of formations is simply an accounting mechanism that helps one to track of what is happening in the inclusion-exclusion formula. An even formation occurs with each k-fold intersection in (2) when k is even, and an odd formation results from every intersection when k is odd. The signs associated with formations are drawn directly from (2), since the intersection of k events contributes a negative term to the sum if and only if k is even. The total number of formations for a network having m minimal path sets is precisely $2^m - 1$. We illustrate below the cataloguing of formations for the "Wheatstone bridge" network $G(4, 5)$ shown in Figure 2.1 below. We will calculate the probability that terminals A and D are connected given that each edge functions independently with probability p.

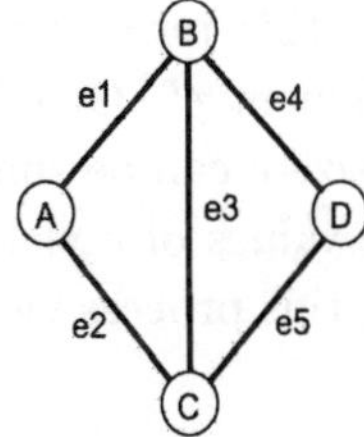

Figure 2.1: The Wheatstone bridge network

The minimal path sets of this network are the sets of edges $\{1,4\}$, $\{2,5\}$, $\{1,3,5\}$, and $\{2,3,4\}$. The *signed domination* of a given union of minimal path sets is simply the difference between the number of even dominations and the number of odd dominations for that union. The relevant accounting is tabulated below.

Min Path Set Unions	# Odd Formations	# Even Formations	Signed Domination
$\{1, 4\}$	1	0	1
$\{2, 5\}$	1	0	1
$\{1, 3, 5\}$	1	0	1
$\{2, 3, 4\}$	1	0	1
$\{1, 2, 3, 4\}$	0	1	-1
$\{1, 2, 3, 5\}$	0	1	-1
$\{1, 2, 4, 5\}$	0	1	-1
$\{1, 3, 4, 5\}$	0	1	-1
$\{2, 3, 4, 5\}$	0	1	-1
$\{1, 2, 3, 4, 5\}$	4	2	2

Table 2.1: Signed dominations for the Wheatstone bridge network

For example, denoting the minimal path sets in the bridge network above as A_1, A_2, A_3, and A_4, the formations associated with the set $\{1,2,3,4,5\}$ are the four odd ones $A_1 \cup A_2 \cup A_3$, $A_1 \cup A_2 \cup A_4$, $A_1 \cup A_3 \cup A_4$, and $A_2 \cup A_3 \cup A_4$ and the two even ones $A_3 \cup A_4$ and $A_1 \cup A_2 \cup A_3 \cup A_4$.

If A is a union of minimal path sets consisting of exactly r components, then the marginal probability that this set works is p^r. The signed domination of a particular set A consisting of r working edges is equal to the sum of the coefficients (1's and -1's) in the terms of the inclusion-exclusion formula corresponding to the occurrences of the probability $P(A)$ in the expansion

(2). The total contribution of this term is the signed domination of the set A times the probability element p^r. Given this type of accounting, the signed dominations of sets of size r can be summed and then multiplied by p^r. Summing over all possible values of r yields the reliability polynomial. For the bridge network above, this process yields the reliability polynomial $h(p)$ given by

$$h(p) = 2p^2 + 2p^3 - 5p^4 + 2p^5. \tag{3}$$

In the notation of (1), we have identified the coefficient vector $\boldsymbol{d}^t$ of the reliability polynomial as $(0, 2, 2, -5, 2)$ for the bridge network in Figure 2.1. We will henceforth refer to $\boldsymbol{d}$ as the vector of (signed) dominations, subsuming mention of the fact that each element of this vector is actually the sum of signed dominations for terms of the same order. It is worth noting that the general process described above applies equally well to the computation of the reliability function for networks with independent but non-identical edges. This generalization will not, however, be required in the developments here.

3. A Brief Look at Signatures

The signature of a network of order n (that is, having n edges) is defined as the probability distribution $\boldsymbol{s}$ on the integers $\{1, 2, \cdots, n\}$ for which

$$s_i = P(T = X_{(i)}), \quad i = 1, 2, \cdots, n, \tag{4}$$

where $X_{(1)} < X_{(2)} < \cdots < X_{(n)}$ are the order statistics from a random (i.i.d.) sample drawn from the (arbitrary) continuous lifetime distribution F, and T is the lifetime of the network.

The fact that the signature $\boldsymbol{s}$ depends only on the network design, and not on the distribution F, is a consequence of the fact that each of the $n!$ orderings of the failure times $X_1, X_2, \cdots, X_n$ of the n edges is equally likely to occur under the i.i.d. assumption. Thus, the probability that the ith edge failure is fatal to the network is solely dependent on the likelihood that the last working edge in some minimal cut set is the ith edge to fail overall. In other words, calculating s_i is simply a matter of examining minimal cut sets and counting how many among the equally likely permutations of $X_1, X_2, \cdots, X_n$ coincide precisely with a particular minimal cut set failing, before any other, upon the occurrence of $X_{(i)}$, the (ordered) ith edge failure time. We illustrate these ideas below in the computation of the signature of the Wheatstone bridge network in Figure 2.1.

As noted above, the Wheatstone bridge has min cut sets $\{1,2\}$, $\{4,5\}$, $\{1,3,5\}$, and $\{2,3,4\}$. It is clear that the first edge failure cannot cause the network to fail, so that $s_1 = 0$. For the second component failure to cause network failure, the first two failures would both have to fall into one of the min path sets $\{1,2\}$ or $\{4,5\}$, accounting for four possible arrangements of the first two failures, followed by 3! possible orderings of the remaining edges. There are thus 4! of the 5! permutations of the 5 failure times that result in network failure upon the second edge failure, leading to the probability $s_2 = 1/5$. Proceeding in this fashion, we obtain $s_3 = 3/5$, $s_4 = 1/5$, and $s_5 = 0$. Of these probabilities, the latter two are especially simple calculations, since the bridge network cannot function with just one working edge and the network fails upon the fourth edge failure if and only if the final two failures are either 1 and 4 or 2 and 5. There are 4! orderings, out of a possible 5!, that satisfy this latter condition. To recap, the signature of the 5-component bridge network above is $s^t = (0, 1/5, 3/5, 1/5, 0)$.

As shown in Samaniego[9] (see also Kochar, Mukerjee and Samaniego[7]), the survival function of a network's lifetime T can be written as a simple function of s and F. When focusing on the reliability of the network at a fixed time t_0, where $P(X_j > t_0) = p$ for all j, this representation reduces to the reliability polynomial in "pq-form", that is, in the form

$$h(p) = \sum_{j=1}^{n} \left(\sum_{i=n-j+1}^{n} s_i \right) \binom{n}{j} p^j q^{n-j}. \tag{5}$$

The "tail probabilities" of the signature vector s have an interpretation through the concept of path set. This connection was noted by Boland (2001) and exploited in his study of indirect majority systems. As is apparent from (5), the coefficient of $p^j q^{n-j}$ in the reliability polynomial in pq-form can be interpreted as the number of path sets of order j, as it is precisely those sets, among the collection of $\binom{n}{j}$ sets with exactly j working components, that each contribute the positive probability $p^j q^{n-j}$ to the reliability polynomial. If we let a_j stand for the proportion of path sets among the $\binom{n}{j}$ sets of j working components (with the complementary components non-working), then we see that the reliability polynomial can be written as

$$h(p) = \sum_{j=1}^{n} a_j \binom{n}{j} p^j q^{n-j}. \tag{6}$$

It follows that the vector $\boldsymbol{a}$, which is fundamentally related to path sets, and the vector $\boldsymbol{s}$, which is fundamentally related to cut sets, are related to each other through the system of equations

$$a_j = \sum_{i=n-j+1}^{n} s_i, \quad j = 1, \cdots, n, \tag{7}$$

or equivalently,

$$s_j = a_{n-j+1} - a_{n-j}, \quad j = 1, \cdots, n, \tag{8}$$

where $a_0 \equiv 0$. For future reference, we note that $\boldsymbol{a} = \boldsymbol{P}\boldsymbol{s}$, and $\boldsymbol{s} = \boldsymbol{P}^{-1}\boldsymbol{a}$, where $\boldsymbol{P}$ and $\boldsymbol{P}^{-1}$ are the $n \times n$ matrices

$$\boldsymbol{P} = \begin{pmatrix} 0\,0\,0 \cdots 0\,0\,0\,1 \\ 0\,0\,0 \cdots 0\,0\,1\,1 \\ 0\,0\,0 \cdots 0\,1\,1\,1 \\ 0\,0\,0 \cdots 1\,1\,1\,1 \\ \vdots\,\vdots\,\vdots\,\vdots\,\,\vdots\,\vdots\,\vdots\,\vdots \\ 0\,1\,1 \cdots 1\,1\,1\,1 \\ 1\,1\,1 \cdots 1\,1\,1\,1 \end{pmatrix} \tag{9}$$

and

$$\boldsymbol{P}^{-1} = \begin{pmatrix} 0 & 0 & 0\,0 \cdots & 0 & 0 & -1\,1 \\ 0 & 0 & 0\,0 \cdots & 0 & -1 & 1\,0 \\ 0 & 0 & 0\,0 \cdots & -1 & 1 & 0\,0 \\ \vdots & \vdots & \vdots\,\vdots & \vdots & \vdots & \vdots\,\vdots \\ 0 & -1 & 1\,0 \cdots & 0 & 0 & 0\,0 \\ -1 & 1 & 0\,0 \cdots & 0 & 0 & 0\,0 \\ 1 & 0 & 0\,0 \cdots & 0 & 0 & 0\,0 \end{pmatrix}. \tag{10}$$

In the introduction, we alluded to the fact that signatures are rich in interpretation and are particularly useful in the comparison of competing networks. We summarize here a collection of results that support this remark. These results demonstrate that properties of signatures have immediate and interesting applicability to comparing the reliability of competing network designs. First, we briefly explain some relevant terminology regarding stochastic ordering. More detail on these and related ideas can be found in Shaked and Shanthikumar[11].

The random variables X_1 and X_2, discrete or continuous, are stochastically ordered (i.e., $X_1 \leq_{st} X_2$) if the survival functions $S_i(x) = P(X_i > x)$ are suitably ordered, that is, if $S_1(x) \leq S_2(x)$ for all x. We say that X_1 is smaller than X_2 in the hazard rate (or uniform stochastic) ordering if the ratio of survival functions $S_2(x)/S_1(x)$ is nondecreasing in x. This ordering will be denoted by $X_1 \leq_{hr} X_2$. Finally, X_1 is said to be smaller than X_2 in the likelihood ratio ordering ($X_1 \leq_{lr} X_2$) if the ratio $f_2(x)/f_1(x)$ is nondecreasing in x, where f_i represents the density or probability mass function of X_i. As is well known, stochastic order is the weakest of these three relations; indeed, it is easy to verify that $lr \Rightarrow hr \Rightarrow st$. With the notation established above, we may now restate results from Kochar et al.[7] relating properties of signatures to properties of network lifetimes.

Theorem 1: Let s_1 and s_2 be the signatures of two networks with n i.i.d. edges, and let T_1 and T_2 be their respective lifetimes. If $s_1 \leq_{st} s_2$ or $s_1 \leq_{hr} s_2$ or $s_1 \leq_{lr} s_2$, then $T_1 \leq_{st} T_2$ or $T_1 \leq_{hr} T_2$ or $T_1 \leq_{lr} T_2$, respectively.

The results above have been applied with profit to stochastic comparisons of k-out-of-n structures with system-wise or component-wise redundancy[7], to indirect majority systems of varying design[4] and to consecutive k-out-of-n systems with varying n[5]. We will apply these results again in an example in the concluding section.

4. The Linkage Between Dominations and Signatures

As noted above, the reliability polynomial of a network with signature s may be written as

$$h(p) = \sum_{j=1}^{n} \left(\sum_{i=n-j+1}^{n} s_i \right) \binom{n}{j} p^j q^{n-j} \tag{11}$$

or equivalently as

$$h(p) = \sum_{j=1}^{n} a_j \binom{n}{j} p^j q^{n-j}, \tag{12}$$

where a is as in (7). In linking the vector of dominations d to the notion of signature, we will first utilize the vector a of tail probabilities of the signature s. Let us consider the reliability polynomials in (1) and (12). Writing q^{n-j} in the latter expression as $(1-p)^{n-j}$ and expanding the term as a binomial, one may rewrite the polynomial in (12) as

$$h(p) = \sum_{j=1}^{n} \binom{n}{j} p^j \left(\sum_{i=0}^{n-j} \binom{n-j}{i} (-1)^i p^i \right)$$

$$= \sum_{r=1}^{n} \left(\sum_{j=1}^{r} a_j \binom{n}{j} \binom{n-j}{r-j} (-1)^{r-j} \right) p^r. \qquad (13)$$

Examining the expressions in (1) and (13), we see that the vectors d and a are related via the equations

$$d_r = \sum_{j=1}^{r} a_j \binom{n}{j} \binom{n-j}{r-j} (-1)^{r-j}, \qquad r = 1, \cdots n. \qquad (14)$$

Alternatively, the components of the domination and signature vectors satisfy the relationships

$$d_r = \sum_{j=1}^{r} \left(\sum_{i=n-j+1}^{n} s_i \right) \binom{n}{j} \binom{n-j}{r-j} (-1)^{r-j}, \qquad r = 1, \cdots, n. \qquad (15)$$

If we denote the linear relationship between d and a in (14) as $d = Ma$ and if we denote the linear relationship between a and s in (7) as $a = Ps$, where M is given by

$$M = \begin{pmatrix} \binom{n}{1}\binom{n-1}{0} & 0 & 0 & \cdots & 0 \\ -\binom{n}{1}\binom{n-1}{1} & \binom{n}{2}\binom{n-2}{0} & 0 & \cdots & 0 \\ \binom{n}{1}\binom{n-1}{2} & -\binom{n}{2}\binom{n-2}{1} & \binom{n}{3}\binom{n-3}{0} & \cdots & 0 \\ \vdots & \vdots & \vdots & \vdots & \vdots \\ \pm\binom{n}{1}\binom{n-1}{n-1} & \mp\binom{n}{2}\binom{n-2}{n-2} & \pm\binom{n}{3}\binom{n-3}{n-3} & \cdots & \binom{n}{n}\binom{n-n}{n-n} \end{pmatrix} \qquad (16)$$

and P is as given in (9), then we may write the relationship of interest as $s = P^{-1}M^{-1}d$. We now turn our attention to making this latter expression explicit.

Given that the inverse of P has already been identified, the work consists in finding an expression for M^{-1}. To facilitate our derivation, we will use the notation $(k)_j$ for nonnegative integers $j \le k$ to denote the permutation coefficient of k with respect to j. In other words, $(k)_j$ denotes the number of ways of selecting without replacement j items from k items, where order is important. With this notation, we claim that the inverse of M is the matrix M^* having ith row given by

$$(m_{i1}^*, \cdots, m_{ii}^*, 0, \cdots, 0) = \left(\underbrace{\frac{(i)_1}{(n)_1}, \frac{(i)_2}{(n)_2}, \cdots, \frac{(i)_i}{(n)_i}}_{i \text{ slots}}, \underbrace{0, \cdots, 0}_{n-i \text{ slots}} \right). \qquad (17)$$

In proving this claim, we will show that the inner product of the ith row of M^* with the jth column of M is 1 if $i = j$ and is 0 otherwise, that is, we will show that

$$\left(\frac{(i)_1}{(n)_1}, \cdots, \frac{(i)_i}{(n)_i}, 0 \cdots, 0 \right) \begin{pmatrix} 0 \\ \vdots \\ 0 \\ (-1)^0 \binom{n}{j} \binom{n-j}{0} \\ (-1)^1 \binom{n}{j} \binom{n-j}{1} \\ \vdots \\ (-1)^{n-j} \binom{n}{j} \binom{n-j}{n-j} \end{pmatrix} = \begin{cases} 1 & i = j \\ 0 & i \neq j \end{cases}, \qquad (18)$$

where the first $j - 1$ entries of the jth column of M are zero. There are three cases to consider: (a) $i \leq j - 1$, (b) $i \geq j + 1$, and (c) $i = j$.

It is clear that (18) holds in case (a). For case (b), the left side of (18) is

$$\binom{n}{i-m} \sum_{k=0}^{m} (-1)^k \frac{(i)_{i-m+k}}{(n)_{i-m+k}} \binom{n-i+m}{k}$$

$$= \binom{n}{i-m} \sum_{k=0}^{m} (-1)^k \frac{i \times (i-1) \times \cdots \times (m-k+1)}{n \times (n-1) \times \cdots \times (n-i+m-k+1)} \frac{(n-i+m)!}{k!(n-i+m-k)!}$$

$$= \binom{n}{i-m} \frac{(n-i+m)!i!}{n!m!} \sum_{k=0}^{m} (-1)^k \frac{m!}{k!(m-k)!}$$

$$= \binom{i}{m} \sum_{k=0}^{m} (-1)^k \binom{m}{k}$$

$$= 0,$$

where the last equality follows by the binomial theorem. We have so far shown that M^*M has all off-diagonal elements equal to zero. To see that the diagonal elements of M^*M are identically equal to one, we observe that in the $i = j$ case the left side of (18) becomes

$$\frac{(i)_i}{(n)_i} \times (-1)^0 \binom{n}{i} \binom{n-i}{0} = \frac{i!}{n \times \cdots \times (n-i+1)} \frac{n!}{i!(n-i)!} = 1;$$

this completes the demonstration of the validity of our claim.

With the forms of both M^{-1} and P^{-1} given in (17) and (10), it is easy to verify that the ith row of $(MP)^{-1}$ is given by

$$\left(\underbrace{-m^*_{n-i,1} + m^*_{n-i+1,1}, \cdots, -m^*_{n-i,n-i} + m^*_{n-i+1,n-i}}_{n-i \text{ slots}}, m^*_{n-i+1,n-i+1}, \underbrace{0\cdots,0}_{i-1} \right). \tag{19}$$

for $i = 1, \cdots, n$, where $m^*_{ij} = (i)_j/(n)_j$. Recalling that $s = (MP)^{-1}d$, we may summarize the above developments as the following theorem.

Theorem 2: Let d and s denote the domination and signature vectors for a given network of order n. Then for $i = 1, \cdots, n$ we have

$$s_i = \sum_{j=1}^{n-i} (-m^*_{n-i,j} + m^*_{n-i+1,j})d_j + m^*_{n-i+1,n-i+1}d_{n-i+1}. \tag{20}$$

That is, for $i = 1, \cdots, n$ we have

$$s_i = \sum_{j=1}^{n-i} \frac{-(n-i)_j + (n-i+1)_j}{(n)_j} d_j + \frac{(n-i+1)_{n-i+1}}{(n)_{n-i+1}} d_{n-i+1}. \tag{21}$$

We suggested, in Section 1, that the comparison of networks via their vector of dominations was unintuitive and, for complex networks, quite difficult. The reason for this is that the difference of two polynomials in standard form (that is, in the form displayed by (1)) is another polynomial in standard form. For two complex networks, the difference polynomial $\sum(d_{2r} - d_{1r})p^r$ will typically be of quite high degree. Because of the requirement $\sum d_r = 1$ on the domination vector of an arbitrary network, it can never be the case that all coefficients of the difference polynomial will have the same sign. Thus, determining whether one reliability polynomial is uniformly larger than another for all $0 < p < 1$ is a task equivalent to finding the roots of a high degree polynomial. But that algebraic problem is a quite famous one, a problem that was dramatically resolved by Evariste Galois. Finding roots of polynomials of degree greater than 4 is a problem that is not "solvable by radicals"; thus, closed-form expressions for the solutions of such problems are not possible in general.

Transforming this problem into the world of signatures changes things substantially. To see this more graphically, let us consider the comparison between the two G(9,27) networks pictured below.

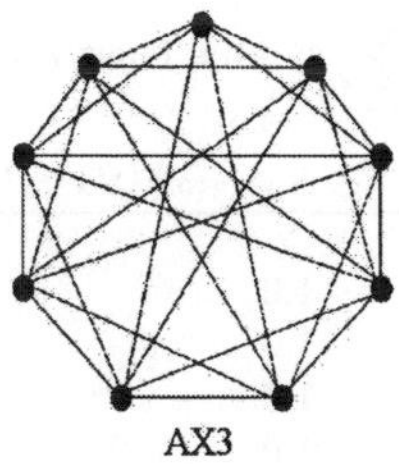 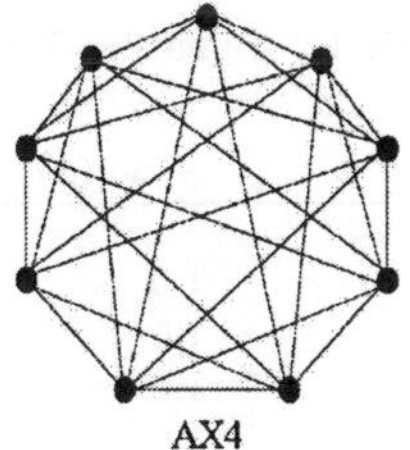

In standard form, the reliability polynomials for these two networks are found to be

$$h_{AX3}(p) = 419904p^{27} - 6021144p^{26} + 41705280p^{25} - 18489826p^{24}$$
$$+586821717p^{23} - 1413876060p^{22} + 2677774329p^{21}$$
$$-4074363810p^{20} + 5048856414p^{19} - 5135792742p^{18}$$
$$+4303029693p^{17} - 2967712776p^{16} + 1676975886p^{15}$$
$$-769265910p^{14} - 282176568p^{13} + 80853282p^{12}$$
$$+17445456p^{11} - 2667060p^{10} + 257634p^{9} - 11828p^{8}$$

$$h_{AX4}(p) = 414720p^{27} - 5934288p^{26} + 41015964p^{25} - 181453380p^{24}$$
$$+574666025p^{23} - 1381692972p^{22} + 2611463517p^{21}$$
$$-3965536554p^{20} - 4904464002p^{19} + 4979513718p^{18}$$
$$+4164454729p^{17} - 2867022480p^{16} + 1617256842p^{15}$$
$$-740601350p^{14} - 271201476p^{13} + 75576922p^{12}$$
$$+16709916p^{11} - 2550156p^{10} + 245898p^{9} - 11268p^{8}$$

While the uniform superiority of one of these networks over the other is certainly not obvious by inspection, one could by numerical means show that $h_{AX3}(p) \geq h_{AX4}(p)$ for all $p \in [0, 1]$. However, the comparison of the two signatures immediately yields this same conclusion, and in addition, a stronger one. From table 1 below, it is apparent that the signature of network AX3 is stochastically larger than that of AX4, yielding the uniform domination of AX3 over AX4 alluded to above. However, the comparison of the two signatures vectors yields an additional insight. The ratios of the two survival functions displayed in the last column of table 1 shows that AX3 dominates AX4 in the hazard rate ordering as well. We can thus rightly say that AX3 is not only better than AX4, it is actually quite a bit better!

The main thesis of this note can be summarized quite succinctly: Domination theory is a useful tool in making network reliability calculations. However, for the purpose of comparing the performance characteristics of two competing networks, reliabilities expressed in terms of signed dominations will tend to have little intuitive content and may be of limited use (except for the possibility of brute force computation). The utility of signatures in the comparison of networks of the same size immediately raises

     P. J. Boland, F. J. Samaniego and E. M. Vestrup

Table 1.   Signature Tail Probabilities $S(x) = \sum_{i=x}^{27} s_i$ And Their Ratios.

x	$S_{AX3}(x)$	$S_{AX4}(x)$	$S_{AX3}(x)/S_{AX4}(x)$
1	1.0	1.0	1.0
2	1.0	1.0	1.0
3	1.0	1.0	1.0
4	1.0	1.0	1.0
5	1.0	1.0	1.0
6	1.0	1.0	1.0
6	1.0	1.0	1.0
7	0.999970	0.999970	1.0
8	0.999787	0.999787	1.0
9	0.999149	0.999149	1.0
10	0.997367	0.997367	1.0
11	0.993612	0.993612	1.0
12	0.985922	0.985922	1.0
13	0.971744	0.971743	1.0000005
14	0.947220	0.947214	1.0000063
15	0.906907	0.906867	1.0000442
16	0.843421	0.843240	1.0002148
17	0.747317	0.746717	1.0008024
18	0.607883	0.606416	1.0024183
19	0.417560	0.415077	1.0059834
20	0.189140	0.186804	1.0125000
21	0.0	0.0	–
22	0.0	0.0	–
23	0.0	0.0	–
24	0.0	0.0	–
25	0.0	0.0	–
26	0.0	0.0	–
27	0.0	0.0	–

questions about the exact relationship between the domination and signature vectors. The functional relationship linking the signature vector with the vector of dominations is displayed above. This linkage allows one to combine the computational advantages of domination theory with the intuitive and interpretive qualities of signatures for the purpose of making comparisons among networks. The growing but still inconclusive literature on the existence, uniqueness and identification of uniformly optimal networks of a given size[2,3,8,12] should benefit from the application of these linked tools.

Acknowledgments

Work supported in part by grant DAAD 19-99-1-1082 from the U.S. Army Research Office.

References

1. A. Agrawal and R. E. Barlow, A survey of network reliability and domination theory, *Operations Research*, **32**, 478-492 (1984).
2. Y. Ath and M. Sobel, Some conjectured uniformly optimal reliable networks, *Probability in the Engineering and Informational Sciences*, **14**, 375-83 (2000).
3. F. T. Boesch, X. Li, and C. Suffel, On the existence of uniformly optimally reliable networks, *Networks*, **21**, 181-191 (1991).
4. P. J. Boland, Signatures of direct and indirect majority systems, *Journal of Applied Probability*, **38**, 597-603 (2001).
5. P.J. Boland and F.J. Samaniego, The Signature of a Coherent System and its Applications in Reliability, Technical Report #373, Department of Statistics, University of California, Davis (2001).
6. W. Feller, *Introduction to Probability Theory and its Applications, 3rd edition* (Wiley, New York, 1968).
7. S. Kochar, H. Mukerjee and F.J. Samaniego, The signature of a coherent system and its application to comparisons among systems, *Naval Research Logistics*, **46**, 507-23 (1999).
8. W. Myrvold, K.H. Cheung, L.B. Page, and J.E Perry, Uniformly most reliable networks do not always exist. *Networks*, **21**, 417-19 (1991).
9. F.J. Samaniego, On Closure of the IFR Class Under Formation of Coherent Systems, *IEEE Transactions on Reliability*, **R-34**, 69-72 (1985).
10. A. Satyanarayana and A. Prabhakar, A New Topological Formula and Rapid Algorithm for Reliability Analysis of Complex Networks, *IEEE Transactions on Reliability*, **R-27**, 82-100 (1978).
11. M. Shaked and J.G. Shanthikumar, *Stochastic Orders and Their Applications* (Academic Press, San Diego, 1994).
12. G. Wang, A proof of Boesch's conjecture, *Networks*, **24**, 277-284 (1994).

8

IMPROVED SIMULATION METHODS FOR SYSTEM RELIABILITY EVALUATION

Arne Bang Huseby and Morten Naustdal

Department of Mathematics, University of Oslo,
P.O. Box 1053 Blindern, N-0316 Oslo, Norway
E-mail: arne@math.uio.no, mortenna@math.uio.no

In this chapter we show how Monte Carlo methods can be improved significantly by conditioning on a suitable variable. In particular this principle is applied to system reliability evaluation. In relation to this an efficient algorithm for simulating a vector of independent Bernoulli variables given their sum is presented. By using this algorithm one can generate such a vector in $O(n)$ time, where n is the number of variables. Thus, simulating from the conditional distribution can be done just as efficient as simulating from the unconditional distribution. The special case where the Bernoulli variables are i.i.d. is also considered. For this case the reliability evaluation can be improved even further. In particular, we present a simulation algorithm which enables us to estimate the entire system reliability polynomial expressed as a function of the common component reliability. If the component reliabilities are not too different from each other, a generalized version of the improved conditional method can be used in combination with importance sampling.

1. Introduction

As a result of the availability of high-speed computers, the use of Monte Carlo methods have accelerated. In many cases, however, it is necessary to use various clever tricks in order to make the estimates converge faster. In particular this is true in situations where one needs to estimate something that involves events with very low probabilities. One such area is system reliability estimation. Since failure events often have very low probability, a large number of simulations is needed in order to obtain stable results. Sometimes, however, conditioning can be used to improve the convergence.

In the general case the principle of conditioning can be stated as follows:

105

Let $\boldsymbol{X} = (X_n, \ldots, X_n)$ be a random vector with a known distribution, and assume that we want to calculate the expected value of $\phi = \phi(\boldsymbol{X})$. We denote this expectation by h. Assume furthermore that the distribution of ϕ cannot be derived analytically in polynomial time with respect to n. Using Monte Carlo simulations, however, we can proceed by generating a sample of size N, of independent vectors $\boldsymbol{X}_1, \ldots, \boldsymbol{X}_N$, all having the same distribution as $\boldsymbol{X}$, and then estimate h with the simple *Monte Carlo* estimate:

$$\hat{h}_{MC} = \frac{1}{N} \sum_{r=1}^{N} \phi(\boldsymbol{X}_r), \qquad \text{where} \qquad \text{Var}\left(\hat{h}_{MC}\right) = \text{Var}(\phi)/N \qquad (1)$$

Now, assume that $S = S(\boldsymbol{X})$ is some other function whose distribution can be calculated analytically in polynomial time w.r.t. n. More specifically, we assume that S is a *discrete* variable with values in the set $\{s_1, \ldots, s_k\}$. We also introduce:

$$\theta_j = E[\phi \mid S = s_j], \qquad j = 1, \ldots, k. \qquad (2)$$

We then have:

$$h = \sum_{j=1}^{k} \theta_j \Pr(S = s_j) \qquad (3)$$

Instead of using N samples of $\boldsymbol{X}$ on the estimation of h as in (1), we divide the set into k groups, one for each θ_j, where the j-th group has size N_j, $j = 1, \ldots, k$, and $N_1 + \cdots + N_k = N$. The data in the j-th group are sampled from the conditional distribution of $\boldsymbol{X}$ given that $S = s_j$, and are used to estimate the conditional expectation θ_j, $j = 1, \ldots, k$. Denoting the data in the j-th group by $\{\boldsymbol{X}_{r,j} : r = 1, \ldots, N_j\}$, θ_j is estimated by:

$$\hat{\theta}_j = \frac{1}{N} \sum_{r=1}^{N_j} \phi(\boldsymbol{X}_{r,j}), \qquad j = 1, \ldots, k. \qquad (4)$$

By combining these estimates, we get the *conditional Monte Carlo estimate*:

$$\hat{h}_{CMC} = \sum_{j=1}^{k} \hat{\theta}_j \Pr(S = s_j), \qquad (5)$$

where

$$\operatorname{Var}(\hat{h}_{CMC}) = \sum_{j=1}^{k} \operatorname{Var}(\phi \mid S = s_j)[\operatorname{Pr}(S = s_j)]^2/N_j$$

We observe that the variance of the conditional estimate depends on the choices of the N_j's. Ideally we would like N_j to be large if the product of the conditional variance and the squared probability is large, and small otherwise, as this would yield the most efficient partition of the total sample with respect to minimizing the total variance. In practice, however, this is difficult, since the conditional variance is typically not known. In order to compare the result with the variance of the original Monte Carlo estimate, we let $Nj \approx N \cdot \operatorname{Pr}(S = s_j)$, $j = 1, \ldots, k$. With this choice we get:

$$\operatorname{Var}(\hat{h}_{CMC}) = \sum_{j=1}^{k} \operatorname{Var}(\phi \mid S = s_j) \operatorname{Pr}(S = s_n)/N = E[\operatorname{Var}(\phi \mid S)]/N$$

$$= \{\operatorname{Var}(\phi) - \operatorname{Var}[E[\phi \mid S)]\}/N \leq \operatorname{Var}(\phi)/N = \operatorname{Var}(\hat{h}_{MC}) \qquad (6)$$

From the above equation we see that the conditional estimate has smaller variance than the original Monte Carlo estimate provided that $\operatorname{Var}[E(\phi \mid S)]$ is positive. This quantity can be interpreted as a measure of how much information S contains relative to ϕ. Thus, when looking for good choices for S we should look for variables containing as much information about ϕ as possible. However, there are some other important points that needs to be considered. First of all S must have a distribution that can be derived analytically in polynomial time. Next, the number of possible values of S, i.e., k, must be polynomially limited by n. Finally, it must be possible to sample efficiently from the distribution of X given S.

The problem of sampling from conditional distributions has been studied recently by several authors. Lindqvist and Taraldsen[6] proposes a general method for sampling from conditional distributions in cases where S is a sufficient statistic. Of special relevance to the present chapter, is Broström and Nilsson[2] who consider the problem of sampling independent Bernoulli variables given their sum. See also Nilsson[8]. In the remaining part of this chapter we shall focus on applications in reliability theory, where X typically is a vector of independent Bernoulli variables. In relation to this we shall derive our own conditional sampling methods.

2. Applications of Conditional Monte Carlo Methods in Reliability Estimation

We now apply the above ideas in the context of system reliability. That is, we assume that X is a vector of independent Bernoulli variables, interpreted as the component state vector relative to a system of n components. A component is *failed* if its state variable is 0, and *functioning* if the state variable is 1. The function ϕ is also assumed to take values 0, 1, and interpreted as the *system state*. If ϕ is 0, the system is *failed*, anf if ϕ is 1, the system is *functioning*. The function ϕ is referred to as the structure function of the system, and is assumed to be a nondecreasing function of the component state vector, X. In general the calculation of ϕ for a given component state vector depends very much on the representation of this function. If e.g., the system is represented as some sort of communication network its state can usually be evaluated in $O(n)$ time or better. If on the other hand the system is specified in terms of a list of minimal path (or cut) sets, the evaluation can be very slow since the number of such sets in the worst cases grows exponentially with the number of components. For this study, however, we assume that ϕ can be calculated in $O(n)$ time for a given component state vector.

The expectation h is called the *system reliability*, and is equal to $\Pr(\phi = 1)$. For an introduction to reliability theory, we refer to Barlow and Proschan[1].

There are of course many different choices for the variable S. One possibility is to let $S = (\phi_L, \phi_U)$, where ϕ_L and ϕ_U are two simpler functions such that $\phi_L \leq \phi \leq \phi_U$. If these two functions are close approximations to ϕ, a lot can be gained. Several different methods for constructing upper and lower bound functions are studied in Vårli[9]. A general algorithm for sampling from the conditional distribution for X given ϕ_L and ϕ_U is also provided. For similar approaches, see Fishman[3] and Fishman[4].

In this chapter, however, S will be the sum of all the component state variables. For this alternative the random variable S takes values in the set $\{0, 1, \ldots, n\}$. Thus, the uncertain quantities we need to estimate, are:

$$\theta_s = E[\phi \mid S = s], \qquad s = 0, 1, \ldots, n . \tag{7}$$

In order to do this we need to have an efficient way of sampling from the conditional distribution of X given $S = s$, $s = 0, 1, \ldots, n$. Before we present this, we introduce the component reliabilities $p_i = \Pr(X_i = 1)$, $i = 1, \ldots, n$. We also introduce the following notation:

$$S_m = \sum_{i=m}^{n} X_i\,, \qquad m = 1,\ldots,n\,. \tag{8}$$

Observe that $S = S_1$ and that $S_n = X_n$. The distributions of $S_1,\ldots,S_m$ can be calculated recursively using the following formula (modified in the obvious way in the limiting cases where $s = 0$ or $s = m$):

$$\Pr(S_m = s) = p_m \Pr(S_{m+1} = s - 1) + (1 - p_m)\Pr(S_{m+1} = s). \tag{9}$$

Thus, we start out by determining the distribution of S_n, and then calculate the distribution of S_{n-1}, etc. A table containing all these distributions (including the distribution of $S = S_1$) can thus be derived in $O(n^2)$-time.

Having calculated all these distributions, we now turn to the algorithm for sampling from the distribution of $\boldsymbol{X}$ given $S = s$. The idea is simply to start out by sampling X_1 from the conditional distribution of $X_1 \mid S = s$. We then continue by sampling X_2 from $X_2 \mid S = s$, $X_1 = x_1$, where x_1 denotes the sampled outcome of X_1, and so on. This turns out to be easy noting that:

$$
\begin{aligned}
&\Pr(X_m = x_m \mid X_1 = x_1, K, X_{m-1} = x_{m-1}, S = s) \\
&= \frac{\Pr(X_m = x_m, S = s \mid X_1 = x_1, K, X_{m-1} = x_{m-1}}{\Pr(S = s \mid X_1 = x_1, K, X_{m-1} = x_{m-1})} \\
&= \frac{\Pr(X_m = x_m, S_{m+1} = s - \sum_{j=1}^{m} x_j)}{\Pr(S_m = ss - \sum_{j=1}^{m-1} x_j)} \\
&= \frac{p_m^{x_m}(1 - p_m)^{1 - x_m} \Pr(S_{m+1} = s - \sum_{j=1}^{m} x_j)}{\Pr(S_m = s - \sum_{j=1}^{m-1} x_j)}
\end{aligned}
\tag{10}
$$

Assuming that the distributions of $S_1,\ldots,S_m$ are all calculated before running the simulations, by using (9), we see that all the necessary conditional probabilities can be calculated when needed during the simulations without imposing additional computational complexity. Note that we only calculate those conditional probabilities we really need along the way during the simulations, not the entire set of all possible conditional probabilities corresponding to all possible combinations of values of the X_j's. Thus, in each simulation run we calculate n probabilities, one for each X_j. Moreover, each probability can be calculated using a fixed number of operations (independent of n). Hence, it follows that sampling from the conditional distribution of $\boldsymbol{X}$ given $S = s$, can be done in $O(n)$ time.

The CMC-algorithm can now be summarized as follows: Divide the N simulations into $(n+1)$ groups, one for each θ_s, where the s-th group has size N_s, $s = 0, 1, \ldots, n$. The data in the s-th group is sampled from the conditional distribution of X given that $S = s$, and is used to estimate the conditional expectation θ_s, $s = 0, 1, \ldots, n$. Denoting the data in the s-th group by $\{X_r, s : r = 1, \ldots, N_s\}$, θ_s is estimated essentially by using (4), and combined into the CMC-estimate by using (5).

A remaining problem is of course how to choose the sample sizes for each of the $(n+1)$ groups. One possibility is to let $N_s \approx N \cdot \Pr(S = s)$, $s = 0, 1, \ldots, n$. In this case, however, it is possible to improve the results slightly. Denote the size of the smallest path set by d, and the size of the smallest cut set by c. By examining the system, it is often easy to determine d and c. Given these two numbers it is easy to see that $\theta_s = 0$ for $s < d$, and $\theta_s = 1$ for $s > n - c$. Moreover, $\mathrm{Var}(\phi \mid S = s) = 0$ for $s < d$, or $s > n - c$. Hence, there is no point in spending simulations on estimating θ_s for $s < d$, or $s > n - c$, so we let $N_s = 0$ for $s < d$, or $s > n - c$. As a result, we have more simulations to spend on the remaining quantities.

An extreme situation occurs when the system is a k-out-of-n-system, i.e., when $\phi(X) = I(S \geq k)$. For such systems $d = k$, and $c = (n - k + 1)$. In this situation *all* the θ_s's are known. Specifically, $\theta_s = 0$ for $s < k$ and $\theta_s = 1$ for $s \geq k$. Thus, the CMC-estimate is equal to the true value of the reliability, and can be calculated without doing any simulations at all. The reason for this is of course that in this case ϕ depends on X only through S.

For all other nontrivial systems, however, it is easy to see that we always have that $d \leq n - c$. In order to ensure that we get improved results, we assume that $\Pr(d \leq S \leq n - c) > 0$, and let:

$$N_s \approx N \cdot \Pr(S = s) / \Pr(d \leq S \leq n - c), \qquad s = d, \ldots, n - c. \qquad (11)$$

Using a similar argument as we did in (6) we now get that:

$$\mathrm{Var}(\hat{h}_{CMC}) \leq \Pr(d \leq S \leq n - c) \, \mathrm{Var}(\hat{h}_{MC}) \qquad (12)$$

Hence, we see that if d and $(n - c)$ are close, i.e., there are few unknown θ_s's, the variance is reduced considerably.

3. System Reliability when All the Component State Variables Have Identical Reliabilities

If all the components in the system have the same reliability, i.e., $p_1 = \cdots = p_n = p$, it is possible to improve things even further. In this case we note that S has a binomial distribution. Moreover, the conditional distribution of X given S is given by:

$$\Pr(X = x \mid S = s) = \frac{p^{\sum_{i=1}^{n} x_i}(1-p)^{n-\sum_{i=1}^{n} x_i}}{\binom{n}{s} p^s (1-p)^{n-s}} = \frac{1}{\binom{n}{s}} \, , \qquad (13)$$

for all x such that $\sum_{i=1}^{n} x_i = s$.
From this it follows that:

$$\theta_s = E[\phi \mid S = s] = \sum_{\{x \mid \sum_{i=1}^{n} x_i = s\}} \phi(x) \Pr(X = x \mid S = s) = \frac{b_s}{\binom{n}{s}} \, , \qquad (14)$$

where $b_s = $ the number of path sets with s components, $s = 0, \ldots, n$.
Finally, the system reliability, h, expressed as a function of p, is:

$$h(p) = \sum_{s=0}^{n} b_s p^s (1-p)^{n-s}. \qquad (15)$$

Note that the desired quantities, $\theta_0, \ldots, \theta_n$, do not depend on p. Thus, by estimating these quantities, we get an estimate of the entire $h(p)$-function. Moreover, we see that θ_s can be interpreted as the fraction of path sets of size s among all sets of size s, for $s = 0, 1, \ldots, n$. Thus, θ_s can be estimated by sampling random sets of size s and calculating the frequency of path sets among the sampled sets. It turns out that this can be done very efficiently as follows:

The idea is to sample sequences of sets of increasing size by sampling from the component set, $C = \{1, \ldots, n\}$, without replacements. The only set of size 0 is of course $\emptyset$. If $\emptyset$ is a path set, we have a trivial system which is always functioning. Obviously, $\theta_0 = 1$ in this case. Moreover, the reliability of such a system is 1. If $\emptyset$ is not a path set, it follows that $\theta_0 = 0$. In both cases we do not need to sample anything to determine the value of θ_0. Thus, we can focus on estimating $\theta_1, \ldots, \theta_n$ by sampling sets of size s, for $s = 1, \ldots, n$.

Let $T = (T_1, \ldots, T_n)$ be vector for storing results from the simulations. Before we start the simulations, this vector is initialized as $(0, \ldots, 0)$. The simulation algorithm now runs as follows:

For $i = 1, \ldots, N$ **do:**

Step 1. Sample a component from the set C, say component i_1, and define $A_1 = \{i_1\}$. If A_1 is a path set, T_1 is incremented with 1.

Step 2. Sample a component from the set $C \setminus A_1$, say component i_2, and define $A_2 = \{i_1, i_2\}$. If A_2 is a path set, T_2 is incremented with 1.

. . .

Step n**.** Sample the last remaining component, say component i_n, and define $A_n = C$. If A_n is a path set, T_n is incremented with 1.

When all the simulations are carried out, the vector T contains the number of observed path sets of sizes $1, \ldots, n$. From this we get the resulting estimates of $\theta_1, \ldots, \theta_n$ simply as:

$$\theta_s = T_s/N, s = 1, \ldots, n. \tag{16}$$

When running this algorithm, much of the time is spent on the steps where the system state is evaluated. Thus, in order to minimize the running time of the algorithm, one would like to reduce the number of such evaluations. Since $A_1 \subset A_2 \subset \cdots \subset A_n$, it follows by the monotonicity of the structure function, ϕ, that if A_k is a path set, then so is A_{k+1}. Thus, we can stop evaluating the state of the system as soon as we have generated a path set, since we then know that all the remaining sets will be path sets as well.

It is easy to see that sampling components randomly without replacements is equivalent to sampling a random permutation, $(i_1, \ldots, i_n)$, of the component set $\{1, \ldots, n\}$. Such a permutation can easily be generated using an algorithm described in Knuth[5].

We observe that since all sets in a sequence are derived from the same permutation, the θ_s-estimates become correlated. However, the effect of this is more than compensated for since each θ_s-estimate now can be based on all N simulations, compared to just a subset of size N_s as was the case in the previous section.

We close this section by illustrating the effect of the improved method on two different examples. In the first example we consider a simple bridge structure with 5 components shown in Figure 1.

The components of the system are the edges of the network, denoted by $1, 2, \ldots, 5$. The system is functioning if the terminal nodes, S and T, can communicate through the network.

Our objective is to estimate the reliability h as a function of the common component reliability p. To illustrate the strength of the CMC-estimate, we

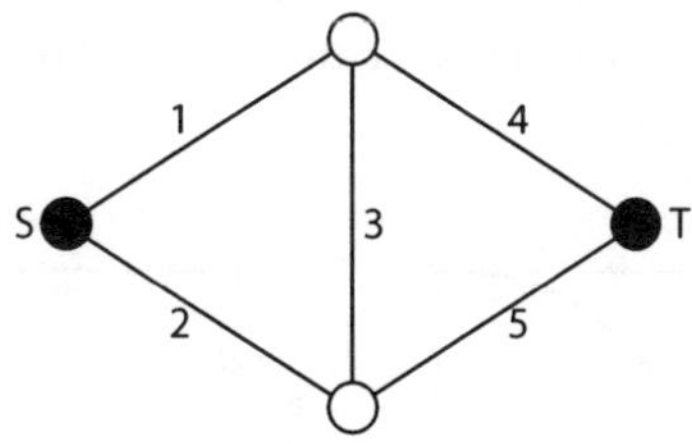

Fig. 1.   A bridge structure.

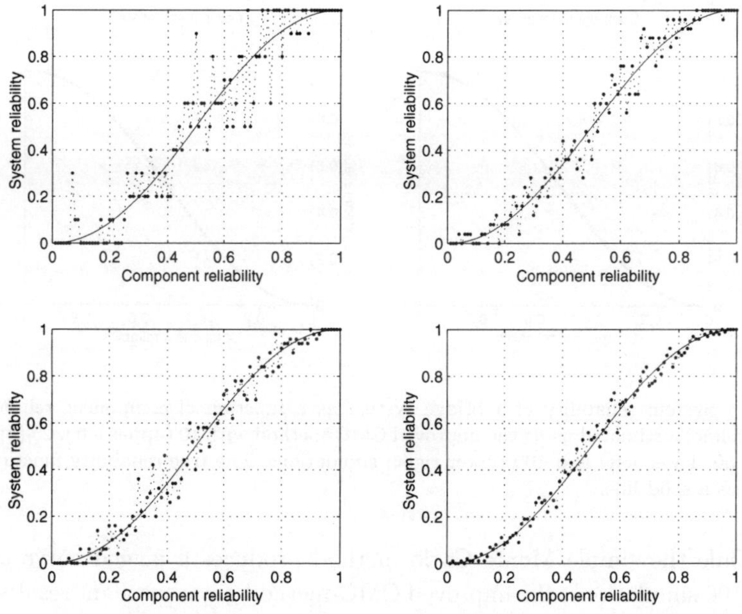

Fig. 2.   System reliability of a bridge system as a function of component reliability (dotted line), estimated using a simple Monte Carlo method with 10 (upper left), 25 (upper right), 50 (lower left) and 100 (lower right) simulations. The true reliability function is shown as a solid line.

will compare it to a simple Monte Carlo estimate. When using the latter method, one has to estimate h separately for each value of p.

In Figure 2 we have plotted the estimated values of h for $p = 0.01, \ldots, 0.99$ obtained after 10, 25, 50 and 100 simulations for each p-value.

By using the improved method, we only need to estimate the θ_s's. Based on these estimates the entire curve can be calculated. The corresponding results are shown in Figure 3.

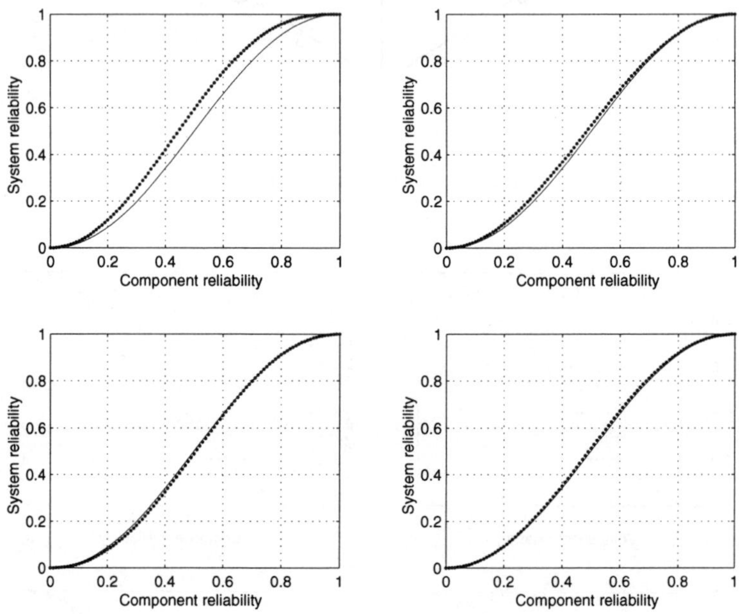

Fig. 3. System reliability of a bridge system as a function of component reliability (dotted line), estimated using the improved CMC-method with 10 (upper left), 25 (upper right), 50 (lower left) and 100 (lower right) simulations. The true reliability function is shown as a solid line.

While the simple Monte Carlo method produces a jagged curve even after 100 simulations, the improved CMC-method gives excellant results already after 50 runs. In fact the improved method always produces a smooth S-shaped curve which rapidly converges to the true curve.

We recall that the improved method actually produces perfect results without any simulations for k-out-of-n-systems. We now consider a more general class of systems called *threshold systems*, which have the property that the structure function can be written in the following form:

$$\phi(\boldsymbol{X}) = \Big(\sum_{i=1}^{n} a_i X_i \geq b\Big) \tag{17}$$

where $a_1, \ldots, a_n$ and b are positive numbers. Notice that if $a_1 = \cdots = a_n =$

1 and $b = k$, we get a k-out-of-n-system. The system considered in the second example, is a threshold with 20 components, and with the following structure function:

$$\phi(X) = I(2X_1 + 2X_2 + 3X_3 + 3X_4 + 3X_5 + 3X_6 + 4X_7 + 4X_8$$
$$+4X_9 + 4X_{10} + 4X_{11} + 4X_{12} + 4X_{13} + 5X_{14} + 5X_{15}$$
$$+5X_{16} + 5X_{17} + 5X_{18} + 5X_{19} + 6X_{20} \geq 49) \tag{18}$$

The resulting estimates using the simple Monte Carlo method are shown in Figure 4, while the corresponding results for the improved CMC-method are shown in Figure 5.

For this type of system we see that the improved method actually produces nearly perfect results already after 10 simulations, while the corresponding results for the simple Monte Carlo method, is not satisfactory even after 100 simulations. In addition one must remember that the simple Monte Carlo method requires estimates for each of the 99 values of p. Thus, this method actually uses 9900 simulations with significantly poorer results than we get with only 10 simulations using the improved CMC-method.

4. General Sequential Sampling Methods

The sampling method presented in the previous section was an example of what we shall refer to as a *sequential sampling method*. The strength of this method makes it tempting to investigate whether or not this idea can be generalized to a situation where the components have different reliabilties. A generalized sequential sampling method can be described as follows:

Let $C = \{1, \ldots, n\}$ be a set of components. We then sample an ordered sequence, $(i_1, \ldots, i_n)$ of components without replacements as follows: Assume that we have sampled the s first components in the sequence, i.e., $i_1, \ldots, i_s$, and denote the corresponding unordered set $\{i_1, \ldots, i_s\}$ by A_s. Then the next component is then sampled from the set of remaining components, i.e., $C \setminus A_s$, with probability:

$$\alpha_{A_s, i} = \Pr(\text{The } (s+1)\text{-th sampled component is } i \mid A_s),$$
$$\text{for all } i \in C \setminus A_s. \tag{19}$$

The probability of a sampling a given ordered sequence, $(i_1, \ldots, i_n)$ is then:

$$\alpha_{A_0, i_1} \cdot \alpha_{A_1, i_2} \cdots \alpha_{A_{n-1}, i_n} \tag{20}$$

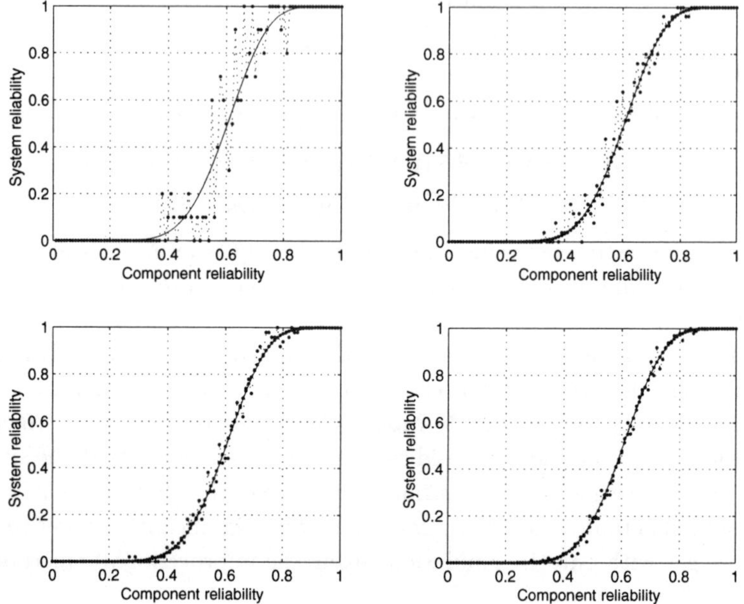

Fig. 4.　System reliability of a threshold system as a function of component reliability (dotted line), estimated using a simple Monte Carlo method with 10 (upper left), 25 (upper right), 50 (lower left) and 100 (lower right) simulations. The true reliability function is shown as a solid line.

where $A_0 = \emptyset$.

The sequential sampling method is characterized by the sampling probability distributions used in each sampling step, i.e., by the $\alpha_{A_s,i}$'s.

Using such a sampling method, the probability that after having sampled s components, we have sampled the set A, where $|A| = s$, is given by:

$$\Pr(A_s = A) = \sum_{(i_1,\ldots,i_s) \in \pi(A)} \alpha_{\emptyset,i_1} \cdots \alpha_{\{i_1\},i_2} \cdots \alpha_{\{i_1,K,i_{s-1}\},i_s}, \qquad (21)$$

where $\pi(A)$ denotes the set of all permutations of A.

Now, ideally we want the sequential sampling method to produce the sets $A_1,\ldots,A_n$ with the "correct" probabilities. That is, we want the $\alpha_{A_s,i}$'s to be chosen such that:

$$\Pr(A_s = A) = \Pr(X = x(A) \mid S = s), \qquad s = 1,\ldots,n \qquad (22)$$

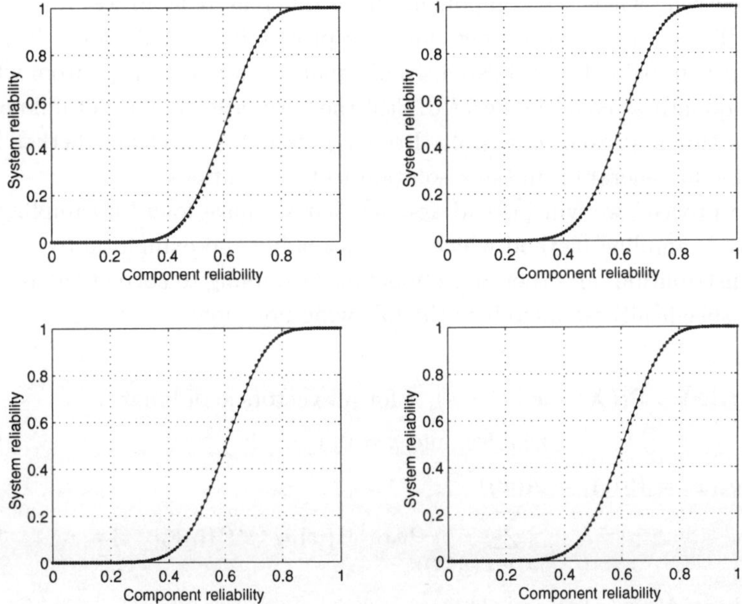

Fig. 5.   System reliability of a threshold system as a function of component reliability (dotted line), estimated using the improved CMC-method with 10 (upper left), 25 (upper right), 50 (lower left) and 100 (lower right) simulations. The true reliability function is shown as a solid line.

where $\boldsymbol{x}(A)$ denotes the vector $\boldsymbol{x} = (x_1, \ldots, x_n)$ such that $x_i = 1$ if $i \in A$, and 0 otherwise. If such a sampling method can be found, we could use the same method as in the previous section in order to obtain estimates for $\theta_0, \ldots, \theta_n$.

In the special case considered in the previous section where all the components have equal reliability, this is easily accomplished by using a simple *uniform sampling method*, i.e., by letting:

$$\alpha_{A_s,i} = \frac{1}{n - |A_s|} = \frac{1}{n - s}\,, \qquad \text{for all } i \in C \setminus A_s \tag{23}$$

With this choice we get by applying (13) that:

$$\Pr(A_s = A) = \frac{s!}{n(n-1)\cdots(n-s+1)} = \frac{1}{\binom{n}{s}} \tag{24}$$

$$= \Pr(\boldsymbol{X} = \boldsymbol{x}(A) \mid S = s), \qquad s = 1, \ldots, n\,.$$

However, when the components have different reliabilities, it is much more difficult to find the right sampling probability distributions. In principle it is possible to establish a set of equations for the $\alpha_{A_s,i}$'s from which these quantities could be derived. Still the very large number of unknowns makes this approach unfeasible. In fact it is not even clear whether it is possible in general to find any solution to the equations.

To proceed we will instead assume that we have found a sampling sequential sampling method which is a reasonable approximation to the correct distribution, and then use importance sampling to correct the results. More specifically we introduce the following notation:

$$f_s(\boldsymbol{x}) = \Pr(\boldsymbol{X} = \boldsymbol{x} \mid S = s), \quad \text{for all vectors } \boldsymbol{x} \text{ of binary}$$
$$\text{variables, and } s = 0, 1, \ldots, n. \tag{25}$$

$$g_s(\boldsymbol{x}) = \Pr(A_s = A(\boldsymbol{x}))$$
$$= \sum_{(i_1,\ldots,i_s)\in\pi(A(\boldsymbol{x}))} \alpha_{\emptyset,i_1} \cdot \alpha_{\{i_1\},i_2} \cdots \alpha_{\{i_1,K,i_{s-q}\},i_s}, \tag{26}$$

where $A(\boldsymbol{x}) = \{i : x_i = 1\}$.

We assume that $g_s(x)$ is a "close" approximation to $f_s(\boldsymbol{x})$, and that $g_s(x) \neq 0$ for all $\boldsymbol{x}$. Under this assumption an unbiased estimator for the unknown quantity θ_s is given by the following importance sampling estimator:

$$\hat{\theta}_{s,I} = \frac{1}{N} \sum_{i=1}^{N} \phi(\boldsymbol{x}_{i,s}) \frac{f_s(\boldsymbol{x}_{i,s})}{g_s(\boldsymbol{x}_{i,s})}, \qquad s = 1, \ldots, n \tag{27}$$

where $\boldsymbol{x}_{1,s}, \ldots, \boldsymbol{x}_{N,s}$ are generated from the distribution g_s.

A serious difficulty with this method is that calculating $g_s(\boldsymbol{x})$ implies computing a sum of $s!$ terms. If s is large, this takes a very long time, and as a result the efficiency of the algorithm is destroyed. In order to avoid this problem, we will instead extend our stochastic experiments by considering ordered samples.

An ordered set, $A^o = (i_1, \ldots, i_s)$, sampled according to the sequential method, has the following probability of occurring:

$$\Pr(A^o = (i_1, \ldots, i_s)) = \alpha_{A_0,i_1} \cdot \alpha_{A_1,i_2} \cdots \alpha_{A_{s-1},i_s}, \tag{28}$$

where $A_0 = \emptyset$, $A_1 = \{i_1\}$, $\ldots$, $A_{s-1} = \{i_1, \ldots, i_{s-1}\}$.
We denote this probability by $g_s(A^o)$.

It is easy to see that the matching "correct" probability, which we denote by $f_s(A^o)$, is given by:

$$f_s(A^o) = (s!)^{-1} \cdot f_s(\boldsymbol{x}(A^o)) \tag{29}$$

where $\boldsymbol{x}(A^o)$ is defined in the same way as for unordered sets.

An unbiased importance sampling estimator for θ_s is then given by:

$$\hat{\theta}_{s,I} = \frac{1}{N} \sum_{i=1}^{N} \phi(\boldsymbol{x}(A_{i,s}^o)) \frac{f_s(A_{i,s}^o)}{g_s(A_{i,s}^o)} \, , \qquad s = 1, \ldots, n \tag{30}$$

where $A_{1,s}^o, \ldots, A_{N,s}^o$ are generated from the distribution $g_s(A^o)$.
The variance of this estimator is:

$$\mathrm{Var}(\hat{\theta}_{s,I}) = \frac{1}{N} \, \mathrm{Var}\left(\phi(A_s^o) \frac{f_s(A_s^o)}{g_s(A_s^o)} \right) \tag{31}$$

where A_s^o is distributed according to the sequential sampling distribution, g_s. From this it follows that this variance is bounded, and that it converges towards 0 as N goes to infinity. Using standard methods it can easily be shown that the estimator will converge almost surely to the correct value.

By choosing the sequential sampling distribution, g_s, in a clever way one may reduce the variance of the importance sampling estimator. Still finding the optimal distribution is typically a difficult task. This is especially true in this case, since we need a sampling distribution which produces stable results for all the unknown quantities simultaneously. A reasonable strategy, however, can be to look for distributions which are good approximations to f_s. In particular one would like the ratio between f_s and g_s to be as stable as possible. This ratio is referred to as the *importance function*, and can be interpreted as a correction factor compensating for the fact that we sample from an incorrect distribution.

We close this section by presenting an example of a sequential sampling method. The idea behind this approach is to use a sampling distribution where components with high reliabilities have a greater chance of being selected than those with low reliabilities. Specifically, we use the following distributions:

$$\alpha_{A_s,i} = \frac{p_i}{\sum_{j \notin A_s} p_j} \, , \qquad \text{for all } i \in C \setminus A_s \tag{32}$$

We observe that if all the component reliabilities are equal, this method reduces to the uniform sampling method.

Now, assume that $A_s^o = (i_1, \ldots, i_s)$, and let $A_0 = \emptyset$, $A_1 = \{i_1\}$, $A_1 = \{i_1, i_2\}, \ldots, A_{s-1} = \{i_1, \ldots, i_{s-1}\}$. Then g_s can be written as:

$$g_s(A_s^o) = \frac{\prod_{i \in A_s^o} p_i}{\prod_{k=0}^{s-1} \sum_{j \notin A_k} p_j} \tag{33}$$

After a few intermediate steps, we arrive at the following importance function:

$$\frac{f_s(A_s^o)}{g_s(A_s^o)} = \frac{\left(\prod_{k=0}^{s-1} \frac{1}{n-k} \sum_{j \notin A_k} p_j\right) \left(\prod_{i \notin \{A_s^o\}} (1 - p_i)\right)}{\binom{n}{s}^{-1} \Pr(S = s)} \tag{34}$$

We observe that the first factor of the numerator is a product of s average component reliabilities, where the averages are taken over the sets $(C \setminus A_0), (C \setminus A_1), \ldots, (C \setminus A_s)$. This averaging tends to make the importance function more stable. The second factor, however, may vary a lot. Assume e.g., that the system under consideration contains ten components of which five of them have reliabilities equal to 0.9, while the remaining components have reliabilities 0.99. If $s = 5$, we see that the second factor of the numerator varies between 10^{-5} and 10^{-10}. As a result the variance of the importance function becomes very large.

The above sampling method along with some others are studied in more detail in Naustdal[7]. The main results from this study is that the proposed methods works very well as long as the component reliabilities are not too different. However, when there are considerable differences between these numbers, the importance function becomes unstable, and the resulting estimates converges much slower.

5. Conclusions

In this chapter we suggest several improved simulation methods with application to system reliability estimation. For the case where all the components have the same reliability, the best approach is to apply the sequential sampling method described in Section 3. If the component reliabilities are not too different, the proportional sequential sampling method described in Section 4 combined with importance sampling produces good results. When there are significant differences between these reliabilities, however, the unsequential conditional Monte Carlo method suggested in Section 2 appears to be better. Still we believe that the importance sampling methods can be improved further by finding sampling methods which produces closer approximations to the "correct" distributions.

Acknowledgments

The authors are grateful to Professor Bo Lindqvist and Professor Bent Natvig for helpful discussions.

References

1. R. E. Barlow and F. Proschan, *Statistical theory of reliability and life testing. Probability models* (Holt, Rinehart and Winston, New York, 1975).
2. G. Broström, and L. Nilsson, Acceptance/rejection sampling from the conditional distribution of independent discrete random variables, given their sum, *Statistics* **34**, 247-258 (2000).
3. G. S. Fishman, A Monte Carlo sampling plan for estimating network reliability, *Operations Research* **34**, 581-594 (1986).
4. G. S. Fishman, A comparison of four Monte Carlo methods for estimating the probability of $s - t$ connectedness, *IEEE Transactions on Reliability*, **R-35**, 145-154 (1986).
5. D. E. Knuth, *The Art of Computer Programming, Vol. 2, Second Edition, Seminumerical Algorithms*, (Addison-Wesley, Reading, Mass., 1981).
6. B. H. Lindqvist and G. Taraldsen, Monte Carlo conditioning on a sufficient statistic, Preprint Statistics No. 9/2001, Norwegian University of Science and Technology.
7. M. Naustdal, Improved conditional Monte Carlo methods in reliability evaluations, (in Norwegian), Master thesis, University of Oslo (2001).
8. L. Nilsson, On the simulation of conditional distributions in the Bernoulli case, Research report 1997-14, Umeå University, Sweden.
9. I. D. Vårli, Improved Monte Carlo methods in reliability evaluations, (in Norwegian), Master thesis, University of Oslo (1996).

9

RELIABILITY OF MULTI-STATE SYSTEMS: A HISTORICAL OVERVIEW

Gregory Levitin and Anatoly Lisnianski

The Israel Electric Corporation, Reliability Department
P.O.Box 10, Haifa 31000, Israel
E-mail: levitin@iec.co.il, anatoly-l@iec.co.il

Igor Ushakov

San Diego, USA
E-mail: iushakov@mail.com

1. Main Directions of Research in Multi-State System Reliability

Multi-state system (MSS) reliability analysis relates to systems for which one cannot formulate an "all or nothing" type of failure criterion. Such systems are able to perform their task with partial performance (intensity of the task accomplishment). Failures of some system components lead only to the degradation of the system performance.

The methods of MSS reliability assessment are based on four different approaches: an extension of the Boolean models to the multi-valued case, the stochastic process (mainly Markov and semi-Markov) approach, the universal generating function approach, and the Monte-Carlo simulation technique.

The approach based on the extension of Boolean models is historically the first method that was developed and applied for the MSS reliability evaluation. It is based on the natural expansion of the Boolean methods (that were well established for the binary-state system reliability analysis) to the multi-state systems. The main difficulties in the MSS reliability analysis is the "dimension curse" since each system component can have many different states (not only two states as existed in the binary-state system).

This makes the structure function approach overworked and time consuming. Some achievements in the approach have dramatically improved it, simplified the model, and decreased the computation burden. As a result, the approach has become applicable for the reliability assessment of many real world MSS. However when using the structure function approach one cannot find MSS reliability measures that are based on a distribution of the system's sojourn time in the set of unacceptable states.

The stochastic process methods that are widely used for the MSS reliability analysis are more universal. In fact, this approach was successfully used for the reliability assessment of multi-state power systems and some types of communication systems even before MSS was theoretically defined. The stochastic process method can be applied only to relatively small MSS because the number of system states increases dramatically with the increase in the number of system components. From the computational point of view, this restriction is more severe for the semi-Markov process technique that is based on solving a system of integral equations than for the Markov process technique that is based on solving a system of differential equations.

The computational burden is the crucial factor when one solves optimization problems where the reliability measures have to be evaluated for a great number of possible solutions along the search process. This makes using the two above mentioned methods in reliability optimization problematic. On the contrary, the universal generating function (UGF) technique is fast enough. This technique allows one to find the entire MSS performance distribution based on the performance distributions of its components by using a rapid algebraic procedure. An analyst can use the same recursive procedures for MSS with a different physical nature of performance and different types of component interaction.

Even though almost every real world MSS can be represented by the Monte-Carlo simulation for the reliability assessment, the main disadvantages of this approach are the time and expenses involved in the development and execution of the model. Since there is not any difference in using this approach for the binary-state system and the MSS, we do not consider the Monte-Carlo simulation in this paper.

2. History of Ideas in MSS Reliability

The MSS was introduced in the middle of the 1970's in [Murchland (1975)], [El-Neweihi et al., 1978], [Barlow and Wu (1978)], [Ross (1979)]. In these works, the basic concepts of MSS reliability were primarily formulated, the system structure function was defined, and its properties were initially studied. Notions of minimal cut set and minimal path set were suggested in the MSS context, as well as notions of coherence and component relevancy. Natvig (1982) subsequently generalized the results obtained in these works. Application of the basic notions to MSS reliability evaluation was demonstrated in [Hudson and Kapur (1982)] on a simple example of a domestic hot water system. In [Griffith (1980)], the coherence definition was generalized and different types of coherence were studied. The reliability importance was extended to MSS in [Butler (1979)], [Griffith (1980)], [Finkelstein (1994)]. Concepts of MSS importance were also discussed in [Block and Savits (1982)], where a decomposition theorem for MSS structure functions was proved.

Natvig (1985) and El-Neweihi and Prochan (1984), Reinshke and Ushakov (1988) summarized the achievements that were attained up to the mid 1980's. In their papers one can find state of the art in the field of MSS reliability for this stage.

In the initial studies, it was assumed that the values indicating the performance levels of the system components and of the entire MSS are totally ordered. However, this assumption is not satisfactory for many practical applications because some system (component) states are not comparable. In [Yu, Koren and Guo (1994)] the order assumption was relaxed, and a partially ordered state set was introduced in order to define a generalized multi-state monotone coherent system. Using this approach allows one to assess the reliability of multi-objective MSS that has several different types of performance.

An alternative approach that does not require ordering a set of states was proposed by Caldarola (1980). He assigns a Boolean variable to each component state. This variable indicates whether the component is at this particular state or not. Since a number of Boolean variables may belong to the same component, the variables are no longer independent. A special technique called Boolean algebra with restrictions on the variables is used to deal with this dependence. Using this method one reduces the MSS to a binary system for which analysis tools are well established. In [Wood (1985)] and [Veeraraghavan and Trivedi (1994)] this method was further developed. In [Wood (1985)] a factoring solution was used to deal with the dependence among Boolean variables in order to obtain the probability of each MSS state. A modeling technique was suggested which applies the existing binary algorithms (block diagrams and fault trees) to the MSS.

The Boolean methods are characterized by the great number of variables one has to deal with and by the existence of dependencies among the variables characterizing the same component. Several methods were developed in order to evaluate MSS reliability under these conditions. A form of the inclusion-exclusion formula was used in [Veeraraghavan and Trivedi, (1994)] and [Veeraraghavan and Trivedi (1993)] to obtain the expressions for state probabilities. Zang, Sun and Trivedi (1999) suggested a Binary Decision Diagram (BDD) approach in which each state of the multi-state components is represented by a Boolean variable, and the entire MSS behavior is represented by a series of multi-state fault trees. Boolean algebra with restrictions on variables is used to study the dependence among these Boolean variables, and a new BDD operation is proposed to realize this Boolean algebra. The nature of the BDD technique allows the sum of the disjoint products to be implicitly represented, which avoids using a great deal of storage space and reduces the computation burden.

In order to provide a comprehensive description of states and state transitions in the MSS and its components, Garriba et al. (1980) introduced the concept of equivalent behavior. The analysis of the multiple-valued logic tree is aimed at eliciting prime implicants. These prime implicants are the multiple-valued logic analogue of minimal cut sets encountered in binary fault trees. The prime implicants were also successfully used in the dependability analysis of software controlled systems [Yau, Apostolakis and Guarro (1998)].

The problem of a huge number of MSS states stimulated efforts to approximate the estimates and to obtain reliability bounds. The fundamental research in the field of MSS availability bounding was presented in [Funnemark and Natvig (1985)], [Natvig (1986)] and [Natvig (1993)]. The theory was applied to an electrical power generation subsystem for two nearby oil-rigs in [Natvig et Al. (1986)]. Butler (1982) proposed some effective methods for obtaining the stochastic bounds on MSS reliability by using the modular decomposition in which a large and complex system is decomposed into simple subsystems. Shinmori, Ohi, Hagihara and Nishida (1989) proposed two extended classes of MSS and clarified mathematical aspects of modular decomposition.

For the binary-state systems, the limit reliability function can be found by assuming that the number of the system components tends to infinity. This is closely related to the limit theorems in the extreme value theory discussed in many publications. The solution for binary systems is well known and given in [Barlow and Proschan (1975)]. Based on this basic approach, the results on limit reliability functions were extended to MSS by Kolowrocki (1999). Theorems on asymptotic reliability functions of multi-state series and parallel systems were applied by Kolowrocki (2000) for reliability evaluation for some types of telecommunication networks and rope systems .

Pouret, Collet and Bon (1999) proposed a method for a two-sided estimation of the MSS unavailability based on the binary model, which can be assessed by using the usual reliability software tools.

The determination of the MSS structure function and reliability measures for applying the Boolean methods is not a trivial problem in many practical cases. For such cases Boedidheimer and Kapur (1994) presented a methodology of customer structure function development. According to this methodology, the customer defines the appropriate number of states for components and the entire system. He also defines the performance measures for the given MSS and the function that relates the component states to the system state by specifying when a change in the state of any one of the component forces a change in the state of the system.

As was mentioned above, the stochastic processes method was applied in practice for the MSS reliability analysis independent of developing the MSS reliability theory. Ringlee and Wood (1969) presented the frequency and duration method that is very effective for evaluating the probability of capacity deficiency in power systems. Endrenyi (1979) and Billinton and Allan (1996) presented successful applications of stochastic methods to the electric power system reliability evaluation. Natvig and Streller (1984) first applied the stochastic process approach in the MSS reliability context. In this work the steady-state behavior of monotone MSS was analyzed by applying the theory for stationary and synchronous processes with the embedded point process. The modern theory of stochastic processes provides a more advanced probabilistic framework [Aven and Jensen (1999)] that allows one to formulate general failure models, to obtain formulas for computing various performance measures, and to determine the optimal replacement policies in complex situations. The MSS reliability measures were systematically studied by using the stochastic processes by Aven (1993), Korczak (1997) and Brunelle and Kapur (1999). Korczak (1988) applied the semi-Makov process technique in order to analyze the MSS behavior under a randomly fluctuating environment. Based on the stochastic processes theory Vaurio, (1984) developed models for the reliability and availability evaluation for multi-state components with general failure and repair time distributions.

The random process method was used for finding the MSS reliability bounds in [Lindqvist (1987)]. Utkin (1993) then used this method for MSS uncertainty importance evaluation. Lubkov and Stepanyanc (1978) and Volik et al. (1988) applied Markov reward models for the development of effective computation algorithms for finding the frequency-type MSS reliability indices.

Since the number of MSS states increases very rapidly with the increase in the number of its components, the stochastic process methods can be applied only to relatively small systems. The system structure function approach is more suitable for analysis of large MSS. The idea to combine the Markov processes and coherent structure function approaches was proposed by Xue and Kai (1995). In their paper, the Markov process technique was used to describe the dynamic characteristics of the component state transitions, and the system structure function was used for analysis of the dynamic characteristics of the entire MSS. Amari and Misra (1997) essentially corrected some results of this paper. A Markov model of a single multi-state component was developed in [Pham, Amari and Misra (1997)]. One of the additional possibilities to evaluate asymptotic unavailability for large Markov models was suggested by Bouissou and Lefebvre (2002). The main idea is to use only a local description of the MSS. The Markov chain is not actually constructed, but the knowledge of the rules, which govern its evolution makes it possible to develop a state-space graph step by step.

The universal generating function (UGF) was introduced by Ushakov (1986). More detailed mathematical foundations of the method were presented in [Gnedenko and Ushakov (1995)], [Reinshke and Ushakov (1998)], and in [Ushakov (2000)]. In [Lisnianski et al. (1996)] the method was applied to the power system reliability analysis. In the series of works [Levitin et al. (1998)], [Levitin and Lisnianski (1998)], [Lisnianski et al. (2000)] different operators providing the determination of the entire MSS performance distribution based on performance distributions of system components were defined for MSS with series, parallel, series-parallel, and bridge structure. The computational efficiency of corresponding procedures was demonstrated.

An important direction of MSS research is the development of MSS optimization algorithms aimed at solving different application problems. One of the first MSS reliability optimization problems was formulated and solved by El-Neweihi, Proschan and Setharaman (1988). It was the allocation of multi-state components into k series systems that maximizes the expected number of systems functioning at level α or higher or the probability that at least one of the components functions at level α or higher. The majorization and Schur functions were used as mathematical tools. Meng (1996) extended the problem of optimal components interchange from binary systems to MSS. Chen, Meng and Zuo (1999) solved the problem of total MSS cost minimization subject to system state probability requirements by using the integer non-linear programming formulation and the shortest path technique.

The development of the UGF method that allows for system performance distribution and, thereby its reliability index to be evaluated based on a fast procedure, opened new possibilities for solving MSS reliability optimization problems. Based on the UGF technique, the MSS reliability can be obtained as a function of the system structure (topology and number of components) and the performance and reliability characteristics of its components. Therefore, numerous optimization problems can be formulated in which the optimal composition of all or part of the factors influencing the entire MSS reliability has to be found subject to different constraints (e.g. system cost).

In order to solve practical problems in which a variety of products exist in the market and analytical dependencies are unavailable for the cost of the system components, the reliability engineer should have an optimization methodology in which each product (a version of the system component) is characterized by its productivity, reliability, and price. To find the optimal system structure, one should choose the appropriate versions from a list of available products for each type of equipment, as well as the number of parallel components of these versions. The objective is to minimize the total cost of the system subject to the requirement of meeting the demand with the desired level of reliability. In this case, an optimal solution should comprise both reliability and cost estimations [Levitin and Lisnianski (2001a)].

In practice, the designer often has to include additional components in the existing system. It may be necessary, for example, to modernize a system according to new demand levels or according to new reliability requirements. This is a problem of optimal single stage MSS expansion [Levitin et al. (1997)]. In this case, one has to decide which components should be added to the existing system and to which component they should be added. During the MSS lifetime, the demand and reliability requirements can change. To provide a desired level of MSS performance, management should develop the multistage expansion plan. For the problem of optimal multistage MSS expansion [Levitin (2000)], it is important to answer not only the question of what must be included into the system, but also the question of when.

By optimizing the maintenance policy, one tries to achieve the desired level of system availability with minimal cost. An optimal policy of the MSS maintenance can be developed which would answer the following questions: 'which components should be the focus of the maintenance activity?' and 'what should the intensity of this activity be?' [Levitin and Lisnianski (2000a)].

Since both incorporating redundant components and providing maintenance actions are aimed at improving the MSS availability, the question arises as to which one is more effective. In other words, should the designer prefer a structure with more redundant components and less investment in maintenance or vise versa? The optimal compromise should minimize the MSS cost while providing its desired reliability. The joint maintenance and redundancy optimization problem [Levitin and Lisnianski (1998)] is to find this optimal compromise by taking into account the differences in reliability and performance rates of the components composing the MSS.

Finally, the most general optimization problem is optimal multistage MSS modernization subject to reliability and performance requirements [Levitin and Lisnianski (1999)]. In order to solve this problem, one should develop a minimal-cost modernization plan that includes maintenance, component modernization, and system expansion actions. The objective is to provide the desired reliability level while meeting the increasing demand during the whole lifetime of the MSS.

Algorithms for reliability assessment for MSS composed of devices with two failure modes were considered in [Levitin and Lisnianski (2001b)]. The special case of such systems, named weighted voting systems and classifiers were studied in [Nordmann and Pham (1999)], [Levitin and Lisnianski (2001c)].

An important type of technical system, namely the consecutively-connected system, is also successfully studied within the MSS framework. The linear consecutive k-out-of-r-from-n:F system has n ordered components and fails if at least k out of r consecutive components fail. This system was formally introduced by Griffith (1986), but had been mentioned previously by Tong (1985), Saperstein (1972), Saperstein (1973), Naus (1974) and Nelson (1978) in connection with tests for non-random clustering, quality control and inspection procedures, service systems, and radar problems. Different algorithms for evaluating the system reliability were suggested in [Papastavridis and Sfakianakis (1991)], [Sfakianakis et al. (1992)], [Cai (1994)], [Psillakis (1995)].

When $r = n$, one has the well studied simple k-out-n:F system. When

$k = r$, one has the consecutive k-out-of-n:F system, which was introduced by Chiang and Niu (1981) and by Bollinger (1982) and Bollinger and Salvia (1987). The simple k-out-n:F system was generalized to the multi-state case by Wu and Chen (1994), where system components have two states but can have different integer values of a nominal performance rate. The general models in which the components can have an arbitrary number of states is developed in [Huang, Zuo and Wu (2000)]. The multi-state generalization of the consecutive k-out-of-n:F system was first suggested by Hwang and Yao (1989) as a generalization of the linear consecutive-k-out-of-n:F system and linear consecutively-connected system with 2-state components, studied by Shanthikumar (1982) and Shanthikumar (1987). Algorithms for linear multistate consecutive k-out-of-n:F system reliability evaluation were developed by Hwang and Yao (1989), Kossow and Preuss (1995), Zuo and Liang (1994). The generalization of the consecutive k-out-of-r-from-n:F system to the multi-state case (named the linear multi-state sliding window system) was suggested by Levitin (2002).

The study of the multi-state network reliability began by evaluating the probability that the maximal flow between a specified pair of network nodes satisfies a demand when the edge capacities are discrete random variables. Doulliez and Jamoulle (1972) suggested the state space decomposition approach for solving the problem of finding the probability distribution of the maximal flow. This approach was improved by Alexopoulos (1995). Evans (1976) investigated some theoretical properties of the problem and Somers (1982) studied its approximate solutions and bounds. Different algorithms were suggested for evaluating the upper boundary points for flows corresponding to different network cutsets [Jane, Lin and Yuan (1993), Lin, Jane and Yuan (1995), Yeh (1998), Yeh (2001)].

Varshney, Joshi and Chang [1994] suggested a multi-state network reliability index that comprised both a network connectivity measure and estimates of the source-sink connection capacity. The reliability of multi-state communication networks characterized by signal delivery delays was considered by Chiou and Li (1986).

The full list of the literature in the field of MSS reliability is available in the Internet at http://ie.technion.ac.il/~levitin/MSS.html .

References

1. Alexopoulos, C. (1995), "A note on state-space decomposition methods for analyzing stochastic flow networks", *IEEE Transactions on Reliability*, **44**, 354-357.

2. Amari, S. and Misra, B. (1997), "Comment on: Dynamic reliability of coherent multistate systems", *IEEE Transactions on Reliability*, **46**, 460-461.

3. Aven, T. (1990), "Availability evaluation of flow networks with varying throughput-demand and deferred repairs", *IEEE Transactions on Reliability*, **38**, 499-505.

4. Aven, T. (1993), "On performance measures for multistate monotone systems", *Reliability Engineering and System Safety*, **41**, 259-266.

5. Aven, T. and Jensen, U. (1999), *Stochastic Models in Reliability*, Springer, NY, Berlin.

6. Barlow,R. and Wu, A. (1978), "Coherent systems with multistate elements", *Math. Operation Research* **3**, 275-281.

7. Billinton, R. and Allan, R. (1996), *Reliability Evaluation of Power Systems*, Plenum Press, NY.

8. Block, H. and Savits, T. (1982), "A Decomposition of Multistate monotone system", *J. Applied Probability* **19**, 391-402.

9. Boedidheimer, R. and Kapur, K. (1994), "Customer-Driven Reliability Models for Multi-state Coherent Systems", *IEEE Transactions on Reliability*, **43**(1), 46-50.

10. Bollinger, R. (1982), "Direct computations for consecutive-k-out-of-n:F systems", *IEEE Transactions on Reliability* **31**, 444-446.

11. Bollinger, R. and Salvia, A. (1987), "Consecutive-k-out-of-n:F networks", *IEEE Transactions on Reliability* **36**, 53-56.

12. Bouissou, M. and Lefebvre, Y. (2002), "A Path-based Algorithm to Evaluate Asymptotic Unavailability for Large Markov Models", *Proceedings Annual Reliability and Maintainability Symposium*, Seattle, Washington, USA, 32-39.

13. Brunelle, R. and Kapur, K. (1999), "Review and classification of reliability measures for multistate and continuum models", *IEE Transactions* **31**, 1171-1180.

14. Butler, D. (1979), "A complete importance ranking for components of binary coherent systems with extensions to multistate systems", *Naval Research Logistics Quarterly* **26**, 565-578.

15. Butler, D. (1982), "Bounding the reliability of multi-state systems", *Operations Research*, **30**, 3, 530-544.

16. Cai, J. (1994), "Reliability of a large consecutive-k-out-of-r-from-n:F system with unequal element-reliability ", *IEEE Transactions on Reliability*, **43**, 107-111.

17. Caldarola, L. (1980), "Coherent systems with multi-state elements", *Nuclear Engineering and Design*, **58**, 127-139.

18. Chen, C., Meng, M. and Zuo, M. (1999), "Selective Maintenance Optimization for Multi-state Systems", *Proceedings of the 1999 IEEE Canadian Conference on Electrical and Computer Engineering*, Edmonton, Alberta, Canada May 9-12 1999.

19. Chiang, D. and Niu, S. (1981), "Reliability of consecutive-k-out-of-n:F systems", *IEEE Transactions on Reliability* **30**, 87-89

20. Chiou, S. and Li, V. (1986), "Reliability analysis of a communication net-

work with multimode elements", *IEEE Journal of Selected Areas in Communications* , **4**, 1156-1161.

21. Doulliez, P. and Jamoulle, E. (1972), "Transportation networks with random arc capacities", *RAIRO*, **3**, 45-60.

22. El-Neweihi, E., Proschan, F. and Setharaman, J. (1978), "Multistate coherent systems", *J. Applied Probability* **15**, 675-688.

23. El-Neweihi, E. and Prochan, F. (1984), "Degradable systems: A survey of multi-state system theory", *Comm. Statist.* **13**(4), 405-432.

24. El-Neweihi, E., Proschan, F. and Setharaman J., (1988), "Optimal Allocation of Multistate Components" in *Handbook of statistics, vol. 7*, Krishnaiah P. (eds), Elsevier Science Publishers, 427-432.

25. Endrenyi, J. (1979), *Reliability Modeling in Electric Power Systems*, John Wiley & Sons, NY.

26. Evans, J. (1976), "Maximum flow in probabilistic graphs: The discrete case", *Networks*, **6**, 161-183.

27. Finkelstein, M. (1994), "Once more on measures of importance of system components", Microelectron. Reliab., **34**, 1431-1439.

28. Funnemark, E. and Natvig, B. (1985), "Bounds for the availabilities in a fixed time interval for multistate monotone systems". *Adv. Appl. Prob.* **17**, 638-655.

29. Garriba, S, Mussio, P. and Naldi, F. (1980). Multivalued logic in the representation of engineering systems, *in Synthesis and analysis methods for safety and reliability studies*, Apostolakis G, Garriba S, Volta G. eds., Plenum Press, 183-197.

30. Gnedenko, B. and Ushakov, I. (1995), *Probabilistic Reliability Engineering*, John Wiley & Sons, NY/Chichester/Brisbane.

31. Griffith, W. (1980), "Multistate reliability models", *J. Applied Probability* **17**, 735-744.

32. Griffith, W. (1986), "On consecutive k-out-of-n failure systems and their generalizations", A. P. Basu (ed), *Reliability and Quality Control*, Elsevier (North-Holland), 157-165.

33. Huang, J. and Zuo, M. (2000), "Generalized Multi-State k-out-of-n:G Systems", *IEEE Transactions on Reliability* 49(1), 105-111.

34. Hudson, J. and Kapur, K. (1982), "Reliability theory for multistate systems with multistate components", *Microelectron. Reliability* **22**, 1-7.

35. Hudson, J. and Kapur, K. (1985), "Reliability bounds for multistate systems with multistate components", *Operations Research* **33**, 735-744.

36. Hwang, F. and Yao, Y. (1989), "Multistate consecutively-connected systems", *IEEE Transactions on Reliability* **38**, 472-474.

37. Jane, C., Lin, J. and Yuan, J. (1993), "Reliability evaluation of a limited-flow network in terms of minimal cutsets", *IEEE Transactions on Reliability*, **42**, 354-368.

38. Kolowrocki, K. (1999), "On limit reliability functions of large systems", *Statistical and Probabilistic Models in Reliability*, eds. Ionescu, D. and Limnios, N., Birkhauser, Boston, 153-183.

39. Kolowrocki, K. (2000a), "An asymtotic approach to Multi-state systems

reliability evaluation", *Recent Advances in Reliability Theory, Methodology, Practice and Inference*, eds. Limnios N., Nikulin M., Birkhauser, Boston, Berlin, 163-180.

40. Kolowrocki, K. (2000b), "Asymptotic approach to reliability evaluation of rope transportation system", *Reliability Engineering and System Safety* **71**, 57-64.

41. Korczak, E. (1988), "Reliability model of multistate system with randomly varying structure", *Proceedings of 7th symposium on reliability in electronics, vol. 1, August, 29-September, 2*, Budapest, Hungary, 396-402.

42. Korczak, E. (1997), "Reliability analysis of non-repaired multistate systems", *Advances in Safety and Reliability. Proceedings of the ESREL'97 International Conference on Safety and Reliability, 17-20 June, Lisbon, Portugal, vol. 3*, eds. Guedes, C. and Soares, Pergamon, London, 2213-2220.

43. Kossow, A., and Preuss, W. (1995), "Reliability of linear consecutively-connected systems with multistate elements", *IEEE Transactions on Reliability* **44**, 518-522.

44. Levitin, G., Lisnianski, A. and Elmakis, D. (1997), "Structure optimization of power system with different redundant elements", *Electric Power Systems Research* **43**, 19-27.

45. Levitin, G, Lisnianski, A., Ben-Haim, H. and Elmakis, D. (1998), "Redundancy optimization for series-parallel multi-state systems", *IEEE Transactions on Reliability* **47**(2), 165-172.

46. Levitin, G. and Lisnianski, A. (1998), "Joint redundancy and maintenance optimization for multistate series-parallel systems", *Reliability Engineering & System Safety* **64**, 33-42.

47. Levitin, G. and Lisnianski, A. (1999), "Optimal multistage modernization of power system subject to reliability and capacity requirements ", *Electric Power Systems Research* **50**, 183-190.

48. Levitin, G. (2000), "Multi-state Series-Parallel System Expansion Scheduling Subject to Availability Constraints", *IEEE Transactions on Reliability* **49**(1), 71-79.

49. Levitin, G. and Lisnianski, A. (2000a), "Optimization of imperfect preventive maintenance for multi-state systems", *Reliability Engineering & System Safety* **67**, 193-203.

50. Levitin, G. and Lisnianski, A. (2000b), "Optimal replacement scheduling in multi-state series-parallel systems (short communication)", *Quality and Reliability Engineering*,**16**, 157-162.

51. Levitin, G. and Lisnianski, A. (2001a), "A new approach to solving problems of multi-state system reliability optimization", *Quality and Reliability Engineering International* **47**, 93-104.

52. Levitin G. and Lisnianski A. (2001b), "Structure Optimization of Multi-state System with Two Failure Modes", *Reliability Engineering & System Safety* **72**, 75-89.

53. Levitin G., Lisnianski A. (2001c), "Reliability optimization for weighted voting system", *Reliability Engineering & System Safety* **71**, 131-138.

54. Levitin, G. (2002), "Optimal allocation of elements in linear multistate slid-

ing window system", *Reliability Engineering & System Safety*, **76**, 247-255.

55. Lin J., Jane C. and Yuan, J. (1995), "On reliability evaluation of a capacitated-flow network in terms of minimal pathsets", *Networks*, 25, 131-138.

56. Lindqvist, B. (1987), "Monotone and associated Markov chains, with applications to reliability theory", *J. Appl. Prob.* **24**, 679-695.

57. Lindqvist, B. and Langseth, H. (1998), "Uncertainty bounds for a monotone multistate system", *Probability in the Engineering and Informational Sciences* **12**, 239-260.

58. Lisnianski, A. (2002), "Estimation of boundary points for continuum-state system reliability measures", *Reliability Engineering and System Safety* **74**, 81-88

59. Lubkov, N. and Stepanyanc, A. (1978), "Reliability analysis of technical systems based on multi-level models", *Theoretical issues of control systems development for ships*, Eds. Trapeznikov, V., Nauka, Moscow, 193-206, (in Russian).

60. Meng, F. (1996), "More on optimal allocation of components in coherent systems", *J. Appl. Prob.* **33**, 548-556.

61. Murchland, J. (1975), "Fundamental concepts and relations for reliability analysis of Multistate systems. Reliability and Fault Tree Analysis", *Theoretical and Applied Aspects of System Reliability*, SIAM, 581-618.

62. Natvig, B. (1982), "Two Suggestions of How to define a Multistate Coherent System", *Adv. Appl. Probab.*, **14**, 434-455.

63. Natvig, B. and Streller, A. (1984), "The steady-state behavior of multistate monotone systems", *J. Applied Probability* **21**, 826-835.

64. Natvig, B. (1985), "Multistate coherent systems", *Encyclopedia of Statistical Sciences*, vol. 5, eds. Johnson, N. and Kotz, S., Wiley&Sons, NY.

65. Natvig, B., (1986), Improved upper bounds for the availabilities in a fixed time interval for multistate monotone systems. *Adv. Appl. Prob.* **18**, 577-579.

66. Natvig, B., Sørmo, S., Holen, A. and Høgåsen, G. (1986), "Multistate reliability theory - a case study". *Adv. Appl. Prob.* **18**, 921-932.

67. Natvig, B. (1993), "Strict and exact bounds for the availabilities in a fixed time interval for multistate monotone systems", *Scand. J. Statist* **20**, 171-175.

68. Naus, J. (1974), "Probabilities for a generalized birthday problem", *J. Amer. Statistical Assoc* **69**, 810-815.

69. Nelson, J. (1978), "Minimal-order models for false-alarm calculations on sliding windows", *IEEE Trans. Aerospace Electronic Systems* **14**, 351-363.

70. Nordmann L. and Pham H. (1999), "Weighted voting systems", *IEEE Transactions on Reliability*, **48**, 42-49.

71. Papastavridis, S. and Sfakianakis, M. (1991), "Optimal-arrangement & Importance of the elements in a consecutive-k-out-of-r-from-n:F system", *IEEE Transactions on Reliability* **40**, 277-279.

72. Pham, H., Amari, S. and Misra, R. (1997), "Availability and Mean Life Time prediction of ultistage degraded system with partial repairs", *Reliability*

Engineering and System Safety, **56**, pp 169-173.

73. Pouret, O., Collet, J. and Bon, J-L. (1999), "Evaluation of the unavailability of a multistate-component system using a binary Model", *Reliability Engineering & System Safety* **66**, 13-17.

74. Psillakis, Z. (1995), "A simulation algorithm for computing failure probability of a consecutive-k-out-of-r-from-n:F system", *IEEE Transactions on Reliability* **44**, 523-531.

75. Ringlee, R. and Wood, A. (1969), "Frequency and duration methods for power system reliability calculations", *IEEE Transactions on Power Apparatus and Systems* **88**, 1216-1223.

76. Reinshke K. and Ushakov I. (1988), *Application of Graph Theory for Reliability Analysis*, (in Russian), Radio i Sviaz, Moscow.

77. Ross, S. (1979), "Multivalued state component systems", *Annals of Probability* **7**, 379-383.

78. Saperstein, B. (1972), "The generalized birthday problem", *J. Amer. Statistical Assoc.* **67**, 425-428.

79. Saperstein, B. (1973), "On the occurrence of n successes within N Bernoulli trails", *Technometrics* **15**, 809-818.

80. Sfakianakis, M., Kounias, S. and Hillaris, A. (1992), "Reliability of consecutive-k-out-of-r-from-n:F systems", *IEEE Transactions on Reliability* **41**, 442-447.

81. Shanthikumar, J. (1982), "A recursive algorithm to evaluate the reliability of a consecutive-k-out-of-n:F system", *IEEE Transactions on Reliability* **31**, 442-443.

82. Shanthikumar, J. (1987), "Reliability of systems with consecutive minimal cutsets", *IEEE Transactions on Reliability*, **R-36**, 546-550.

83. Shinmori, S., Ohi, F., Hagihara, H., and Nishida, T. (1989), "Modules for Two Classes of Multi-State Systems", *The Transactions of the IEICE*, **E-72**, No. 5, May, 600-608.

84. Somers, J. (1982), "Maximum flow in networks with small number of random arc capacities", *Networks*, **12**, 241-253.

85. Tong, Y. (1985), "A rearrangement inequality for the longest run with an application in network reliability", *J. Applied Probability* **22**, 386-393.

86. Ushakov, I. (1986), " A universal generating function", *Soviet Journal Computer Systems Science* **24**, 37-49.

87. Ushakov, I. (2000), "The method of generalized generating sequences", *European Journal of Operating Research* **125**(2), 316-323.

88. Utkin, L. (1993), "Uncertainty importance of multistate system components", *Microelectron. Reliability* **33** (13), 2021-2029.

89. Varshney, P., Joshi, A. and Chang, P. (1994), "Reliability modeling and performance evaluation of variable link-capacity networks", *IEEE Transactions on Reliability*, **43**, 378-382.

90. Vaurio, J. (1984), "Reliability and Availability Equations for Multi-state components", *Reliability Engineering* **7**, 1-19.

91. Veeraraghavan, M. and Trivedi, K. (1994), "A Combinatorial Algorithm for Performance and Reliability Analysis Using Multistate Models", *IEEE*

Transactions on Computers **43**, 229-234.

92. Veeraraghavan, M. and Trivedi, K. (1993), "An Approach for Combinatorial Performance and Availability Analysis", *Reliable Distributed Systems, Proceedings 12-th Symposium,* 24-33.

93. Volik, B. et al. (1988). *Methods of analysis and synthesis of control systems structures,* Moscow, Energoatomizdat, (Russian).

94. Wood, A. (1985), "Multistate Block Diagrams and Fault Trees", *IEEE Transactions on Reliability* **34**, 236-240.

95. Wu, J. and Chen, R. (1994), "An algorithm for computing the reliability of weighted-k-out-of-n systems", *IEEE Transactions on Reliability* **43**, 327-328.

96. Xue, J. and Yang K. (1995), "Dynamic Reliability Analysis of Coherent Multi-state Systems", *IEEE Transactions on Reliability* **44**(4), 683-688.

97. Yau, M., Apostolakis, G. and Guarro, S. (1998), "The use of prime imlicants in dependability analysis of software controlled systems", *Reliability Engineering and System Safety* **62**, 23-32.

98. Yeh, W. (1998), "A Revised Layered-Network Algorithm to Search for all d-Minpaths of a Limited-Flow Acyclic Network", *IEEE Transactions on Reliability,* **47**, 436-442.

99. Yeh, W. (2001), "A simple approach to search for all d-MCs of limited-flow network", *Reliability Engineering & System Safety,* **71,** 15-19.

100. Yu, K., Koren, I. and Guo, Y. (1994), "Generalized multistate monotone coherent systems", *IEEE Transactions on Reliability* **43**(2), 242-250.

101. Zang X., Sun H. and Trivedi K. (1999), "A BDD-Based Algorithm for Reliability Analysis of Phased-Mission Systems", *IEEE Transactions on Reliability* **48**(1), 50-60.

102. Zuo, M., and Liang, M. (1994), "Reliability of multistate consecutively-connected systems", *Reliability Engineering & System Safety* **44**, 173-176.

Part IV

STOCHASTIC MODELLING AND OPTIMIZATION IN RELIABILITY

10

WAITING TIMES ASSOCIATED WITH THE SUM OF SUCCESS RUN LENGTHS

D. L. Antzoulakos, S. Bersimis and M. V. Koutras

Department of Statistics and Insurance Science, University of Piraeus
Piraeus 18534, Greece

In the present chapter, we study the distribution of a waiting time statistic associated with the sum of success run lengths in a sequence of binary trials. More specifically, we start accumulating the exact run lengths of subsequences (strings) consisting of k or more consecutive successes (k is a given positive integer) and then we focus on the waiting time until the sum exceeds a predetermined level. The investigation of the waiting time distribution of the statistic of interest is achieved by the aid of appropriate Markov chain embedding techniques.

1. Introduction

The idea of using the concept of run to establish reasonable stopping criteria in acceptance sampling is quite old and may be attributed to Mosteller[14] and Wolfowitz[20]. From that time on, numerous run-based sampling plans have appeared in quality control literature. For a complete history on this subject the reader might wish to consult Hald[8] or the recent article by Schmueli and Cohen[18] and the references quoted therein.

Prairie, Zimmer and Brookhouse[17] discussed relevant plans under sampling with replacement wherein the outcomes from test to test consist a sequence of Bernoulli trials. Their results were further extended by Balakrishnan, Balasubramanian and Viveros[3] who derived the acceptance and rejection probabilities, operating characteristic curve and average sample number of the plans.

In the aforementioned approaches, the stopping criteria engaged in the plans take into account either the *number* of runs observed in the experimental sequence or the *number* of runs whose length exceeds a predetermined level or the length of the *longest run* observed so far. None of them makes use of the *exact lengths* of runs observed in the outcome sequence. It seems quite plausible to consider alternative run-related statistics which keep score of the total number of sampled units that appear in unusually long blocks of "non-conforming" items (defective units or units beyond the control limits of a specific control chart). This can be formally accomplished as follows. Let k be any positive integer and flag as "non-conforming block" a succession of at least k non-conforming items. The number of items involved in non-conforming blocks are accumulated as we proceed with the inspection and if this number exceed a prespecified level, say r, we arrive at a conclusion whether a corrective action should be taken or not. More specifically, if we denote by T_r the waiting time until the total number of defects involved in non-conforming blocks exceeds r, then a reasonable decision scheme is the following: "If $T_r \leq c$ then declare the process out of control". Alert readers may easily understand that, in practical applications, c and r represent "control parameters" chosen as to achieve a compromise between sampling costs and desirable statistical properties of the associated inferential procedures. They may further observe that c will be determined according to the allowable level (probability) of reaching a wrong decision through the aforementioned rule. A big step in achieving that would be to find the exact distribution of the statistic T_r and investigate its characteristics. This is what the present chapter focuses on.

In Section 2 we introduce the basic definitions and notations that will be used in our exposition. Section 3 contains some general results pertaining to waiting time distributions associated with discrete enumerating random variables. More specifically, we provide first a brief discussion on the class of Markov chain embeddable variables of polynomial type (MVP) which were recently introduced by Antzoulakos, Bersimis and Koutras[2] and can be efficiently used for the evaluation of the exact distribution of enumerating random variables. Then we consider the respective waiting time random variable and investigate its distribution. In Section 4 we consider sequences of Bernoulli trials and focus on the distribution of the waiting time T_r until the sum of run lengths of subsequences consisting of k or more consecutive successes, exceeds r. We verify first that T_r is a waiting random variable associated with an MVP enumerating statistic and then we use the general results given in Section 3 to proceed to the investigation of its distribution.

2. Definitions and Notation

Let V_n (n a non-negative integer) be a non-negative finite integer-valued random variable, and denote by

$$f_n(\nu) = \Pr(V_n = \nu), \quad n \geq 0, \quad \nu = 0, 1, \ldots$$

its probability mass function and by

$$\varphi_n(z) = E[z^{V_n}] = \sum_{\nu=0}^{\infty} \Pr(V_n = \nu)z^{\nu}, \quad n \geq 0,$$

$$\Phi(z, w) = \sum_{n=0}^{\infty} \varphi_n(z)w^n = \sum_{n=0}^{\infty} \sum_{\nu=0}^{\infty} \Pr(V_n = \nu)z^{\nu}w^n$$

its single and double probability generating function, respectively. We conventionally set $\varphi_0(z) = 1$, i.e. we assume that

$$\Pr(V_0 = \nu) = \begin{cases} 1, & \text{if } \nu = 0 \\ 0, & \text{if } \nu = 1, 2, \ldots \end{cases}.$$

Next, let us denote by T_r (r a non-negative integer) the waiting time random variable defined by

$$T_r = \min\{n \geq 0 : V_n \geq r\}$$

and let

$$h_r(n) = \Pr(T_r = n), \quad n = 0, 1, \ldots$$

be its probability mass function. The single and double probability generating function of T_r will be denoted by $H_r(w)$ and $H(z, w)$, respectively; i.e.

$$H_r(w) = E[w^{T_r}] = \sum_{n=0}^{\infty} h_r(n)w^n, \quad r \geq 0$$

$$H(z, w) = \sum_{r=0}^{\infty} H_r(w)z^r = \sum_{r=0}^{\infty} \sum_{n=0}^{\infty} h_r(n)w^n z^r.$$

Note that, due to the previous definitions and conventions, we may write

$$\Pr(T_0 = n) = \begin{cases} 1, & \text{if } n = 0 \\ 0, & \text{if } n = 1, 2, \ldots \end{cases}$$

which implies that $H_0(w) = 1$.

3. General Results

The Markov chain approach for the investigation of problems associated with runs was first introduced by Fu and Koutras[6]. They developed a unified method for capturing the exact distribution of the number of runs of specified length by employing a Markov chain embedding technique. As a matter of fact, the approach taken there was much more general and can practically cover most of the cases where the interest focuses on the distribution of an enumeration variable in a sequence of binary (or multistate) trials. Koutras and Alexandrou[11] introduced the *Markov chain embeddable variables of binomial type* and demonstrated how the run related problems dealt in Fu and Koutras[6] can be tackled in a much more efficient fashion. This class is not wide enough to accommodate the distribution of the enumerating statistic associated with the waiting time distribution we are focusing on in the present chapter. Fortunately, the more general class of *Markov chain embeddable variables of polynomial type*, which were recently introduced and studied by Antzoulakos, Bersimis and Koutras[2] offers sufficient tools for studying the waiting time distribution of interest here.

Before advancing to the main part of this section, we deem it necessary to provide a brief outline of the aforementioned Markov chain imbedding techniques as some of its machinery is essential for the derivation of our results. For more details we refer to Fu and Koutras[6], Koutras and Alexandrou[11], Koutras[12], Antzoulakos, Bersimis and Koutras[2] and Balakrishnan and Koutras[4].

Definition 1: A non-negative finite integer-valued random variable V_n (n a non-negative integer) will be called *Markov chain embeddable variable of polynomial type (MVP)* if there exists a Markov chain $\{Y_t, t \geq 0\}$ defined on a discrete state space Ω such that

(a) Ω can be partitioned as

$$\Omega = \bigcup_{\nu \geq 0} C_\nu, \quad C_\nu = \{c_{\nu,0}, c_{\nu,1}, ..., c_{\nu,s-1}\}$$

(b)   there exists a positive integer m such that, for $t \geq 1$

$$\Pr(Y_t \in C_u | Y_{t-1} \in C_\nu) = 0 \quad \text{for all} \quad u \neq \nu, \nu + 1, ..., \nu + m$$

(c)   the probability mass function of V_n can be captured by considering the projection of the probability space of Y_n onto C_ν, i.e.

$$\Pr(V_n = \nu) = \Pr(Y_n \in C_\nu), \quad n \geq 0, \quad \nu \geq 0.$$

It is worth mentioning that conditions (a), (c) of Definition 1 guarantee that the random variable V_n is a Markov chain embeddable variable in the sense introduced in Fu and Koutras[6]. In addition, the special case $m = 1$ of condition (b) reduces the definition to the class of Markov chain embeddable variables of binomial type considered by Koutras and Alexandrou [11]. Hence, the *MVP* class is a subclass of the first family and a superclass of the latter.

The distribution of a *MVP* is completely determined by the following quantities:

- the initial probabilities

$$\pi_\nu = (\Pr(Y_0 = c_{\nu,0}), \Pr(Y_0 = c_{\nu,1}), \ldots, \Pr(Y_0 = c_{\nu,s-1})), \nu \geq 0$$

- the *within* subclass C_ν one step transition matrices

$$A_{t,0}(\nu) = (\Pr(Y_t = c_{\nu,j'} \mid Y_{t-1} = c_{\nu,j}))_{s \times s}, t \geq 1, \nu \geq 0$$

- the *between* subclasses C_ν and $C_{\nu+i}$ one step transition matrices

$$A_{t,i}(\nu) = (\Pr(Y_t = c_{\nu+i,j'} \mid Y_{t-1} = c_{\nu,j}))_{s \times s}, 1 \leq i \leq m, t \geq 1, \nu \geq 0.$$

More specifically as Antzoulakos, Bersimis and Koutras[2] indicated, if we introduce the probability vectors

$$\mathbf{f}_t(\nu) = (\Pr(Y_t = c_{\nu,0}), \Pr(Y_t = c_{\nu,1}), \ldots, \Pr(Y_t = c_{\nu,s-1})), \quad t \geq 0, \quad \nu \geq 0$$

then the next recurrence holds true

$$\mathbf{f}_t(\nu) = \sum_{i=0}^{m} \mathbf{f}_{t-1}(\nu - i) A_{t,i}(\nu - i), \quad t \geq 1, \quad \nu \geq 0 \tag{3.1}$$

($\mathbf{f}_t(\nu) = \mathbf{0}$ and $A_{t,i}(\nu) = \mathbf{0}$ if $\nu < 0$).

Relation (3.1) in conjuction with the initial probabilities $\pi_\nu = \mathbf{f}_0(\nu)$ offer a simple computational scheme for the evaluation of the probability mass function of V_n. More specifically, if we launch the recurrence scheme (3.1) and evaluate all the probability vectors $\mathbf{f}_n(\nu)$, $\nu \geq 0$, the probability mass function of V_n can be easily evaluated by the aid of the formula

$$\Pr(V_n = \nu) = \mathbf{f}_n(\nu)\mathbf{1}', \quad n \geq 0, \quad \nu \geq 0 \tag{3.2}$$

where $\mathbf{1} = (1, 1, \ldots, 1)$ is the (row) vector of $\mathbb{R}^s$ containing 1 in all its entries. It is sufficient for our purposes and also of a great simplicity to assume that $\pi_\nu = \mathbf{0}$, $\nu \geq 1$ and $\pi_0 \mathbf{1}' = \mathbf{f}_0(0)\mathbf{1}' = 1$; this convention is in fact equivalent to the condition $\Pr(V_0 = 0) = 1$.

Let now T_r, $r \geq 1$, be the waiting time for the first occurrence of the event $\{V_n \geq r\}$. The next theorem provides a formula for the evaluation of the probability mass function of T_r.

Theorem 2: The probability mass function $h_r(n)$ of T_r, $r \geq 1$, is given by

$$h_r(n) = \sum_{i=1}^{m} \sum_{j=1}^{i} \mathbf{f}_{n-1}(r-j) A_{n,i}(r-j)\mathbf{1}', \quad n \geq 1.$$

Proof. The probability mass function $h_r(n)$ of T_r can be expressed as

$$\Pr(T_r = n) = \sum_{i=1}^{m} \sum_{i'=0}^{m-i} \Pr(Y_n \in C_{r+i'}, \ Y_{n-1} \in C_{r-i})$$

$$= \sum_{i=1}^{m} \sum_{j=i}^{m} \Pr(Y_n \in C_{r-i+j}, \ Y_{n-1} \in C_{r-i})$$

and replacing $\Pr(Y_n \in C_{r-i+j}, \ Y_{n-1} \in C_{r-i})$ by the sum $\sum_{k=0}^{s-1} \Pr(Y_n \in C_{r-i+j}|Y_{n-1} = c_{r-i,k}) \Pr(Y_{n-1} = c_{r-i,k})$ we obtain

$$\Pr(T_r = n) = \sum_{i=1}^{m} \sum_{j=i}^{m} \sum_{k=0}^{s-1} \Pr(Y_n \in C_{r-i+j}|Y_{n-1} = c_{r-i,k}) \Pr(Y_{n-1} = c_{r-i,k})$$

Observe next that

$$\Pr(Y_{n-1} = c_{r-i,k}) = \mathbf{f}_{n-1}(r-i)\mathbf{e}'_{k+1}$$

where $\mathbf{e}_k = (0, ..., 1, ..., 0)$ denotes the k-th unit row vector of $\mathbb{R}^s$, and

$$\Pr(Y_n \in C_{r-i+j} \mid Y_{n-1} = c_{r-i,k}) = \sum_{\ell=1}^{s} \mathbf{e}_{k+1} A_{n,j}(r-i)\mathbf{e}'_\ell = \mathbf{e}_{k+1} A_{n,j}(r-i)\mathbf{1}'.$$

Therefore we may write

$$\Pr(T_r = n) = \sum_{i=1}^{m} \sum_{j=i}^{m} \sum_{k=0}^{s-1} \mathbf{e}_{k+1} A_{n,j}(r-i)\mathbf{1}'\mathbf{f}_{n-1}(r-i)\mathbf{e}'_{k+1}$$

and taking into account that

$$\sum_{k=0}^{s-1} \mathbf{e}_{k+1} A_{n,j}(r-i)\mathbf{1}'\mathbf{f}_{n-1}(r-i)\mathbf{e}'_{k+1} = \sum_{k=0}^{s-1} \mathbf{f}_{n-1}(r-i)\mathbf{e}'_{k+1}\mathbf{e}_{k+1} A_{n,j}(r-i)\mathbf{1}'$$

$$= \mathbf{f}_{n-1}(r-i)\sum_{k=0}^{s-1} \mathbf{e}'_{k+1}\mathbf{e}_{k+1} A_{n,j}(r-i)\mathbf{1}'$$

$$= \mathbf{f}_{n-1}(r-i)I A_{n,j}(r-i)\mathbf{1}'$$

$$= \mathbf{f}_{n-1}(r-i)A_{n,j}(r-i)\mathbf{1}'$$

(where I is the identity $s \times s$ matrix) we deduce

$$\Pr\left(T_r = n\right) = \sum_{i=1}^{m}\sum_{j=i}^{m} \mathbf{f}_{n-1}(r-i)A_{n,j}(r-i)\mathbf{1}'.$$

from which Theorem 2 follows immediately. $\qquad\square$

If the within and between states transition probability matrices are independent of t, ν that is $A_{t,i}(\nu) = A_i$ for all $t \geq 1$ and $\nu \geq 0$ then the double probability generating probability function of V_n is given by (see Antzoulakos, Bersimis and Koutras[2]).

$$\boldsymbol{\Phi}(z,w) = \sum_{n=0}^{\infty} \varphi_n(z)w^n = \pi_0 \left(I - w\sum_{i=0}^{m} A_i z^i\right)^{-1} \mathbf{1}' \qquad (3.3)$$

The next theorem describes an analogous result for the double generating function of the waiting time random variable T_r.

Theorem 3: If $A_{t,i}(\nu) = A_i$ for all $t \geq 1$ and $\nu \geq 0$ then the double probability generating function of T_r is given by

$$H(z,w) = 1 + w\pi_0 \sum_{i=1}^{m}\sum_{j=1}^{i} z^j \left(I - w\sum_{\ell=0}^{m} A_\ell z^\ell\right)^{-1} A_i \mathbf{1}'.$$

Proof. The double probability generating function $H(z,w)$ of T_r can be expressed as

$$H(z,w) = 1 + \sum_{r=1}^{\infty}\sum_{n=1}^{\infty} h_r(n)w^n z^r$$

and replacing $h_r(n)$ by the formula deduced in Theorem 2 we may write

$$H(z,w) = 1 + \sum_{r=1}^{\infty}\sum_{n=1}^{\infty}\sum_{i=1}^{m}\sum_{j=1}^{i} \mathbf{f}_{n-1}(r-j)A_i\mathbf{1}'w^n z^r$$

$$= 1 + \sum_{i=1}^{m}\sum_{j=1}^{i}\sum_{n=0}^{\infty}\sum_{r=j}^{\infty} \mathbf{f}_n(r-j)A_i\mathbf{1}'w^{n+1}z^r$$

$$= 1 + \sum_{i=1}^{m}\sum_{j=1}^{i}\sum_{n=0}^{\infty}\sum_{r=0}^{\infty} \mathbf{f}_n(r)A_i\mathbf{1}'w^{n+1}z^{r+j}$$

$$= 1 + \sum_{i=1}^{m}\sum_{j=1}^{i} wz^j \left(\sum_{n=0}^{\infty}\sum_{r=0}^{\infty} \mathbf{f}_n(r)z^r w^n\right) A_i\mathbf{1}'.$$

The final formula for $H(z,w)$ is readily ascertainable from the identity (c.f.(3.3))

$$\left(\sum_{n=0}^{\infty}\sum_{r=0}^{\infty} \mathbf{f}_n(r)w^n z^r\right)\mathbf{1}' = \sum_{n=0}^{\infty}\varphi_n(z)w^n = \pi_0\left(I - w\sum_{\ell=0}^{m}A_\ell z^\ell\right)^{-1}\mathbf{1}' \quad \square$$

It is worth mentioning that for $m = 1$ Theorems 2 and 3 reduce to the results established by Koutras[12] for Markov chain embeddable variables of binomial type.

4. Waiting Time Associated with Success Run Lengths

In this section, we conduct a detailed study of a waiting time distribution, associated with the sum of exact lengths of subsequences (strings) consisting of k or more consecutive successes (k is a given positive integer). This random variable will be first coupled with an enumerating random variable that falls within the class of a Markov chain embeddable variable of polynomial type (MVP). Then, the tools established in the previous section will be exploited for the investigation of the distribution of interest. Consider a sequence of Bernoulli trials $Z_1, Z_2, \ldots$ with success probabilities $p_t = \Pr(Z_t = 1)$, and failure probabilities $q_t = \Pr(Z_t = 0) = 1 - p_t,\quad t \geq 1$, and let n, k be two positive integers. For $k \leq t \leq n$ we define

$$U_t = \begin{cases} k + \ell, & \text{if } Z_{t-k-\ell+1} = Z_{t-k-\ell+2} = \ldots = Z_t = 1,\ Z_{t-k-\ell} = Z_{t+1} = 0 \\ 0, & \text{otherwise} \end{cases}$$

where ℓ is a non-negative integer (convention: $Z_0 = Z_{n+1} = 0$). The sum of the exact lengths of substrings of the sequence $Z_1, Z_2, ..., Z_n$ containing k or more consecutive successes, can be expressed in the form of a sum of random variables, as follows

$$V_n = \sum_{t=k}^{n} U_t, \quad n \geq 1.$$

It is clear that the support of V_n is $\{0\}$ if $n \leq k$ and $\{0, k, k+1, ..., n\}$ if $n > k$. Antzoulakos, Bersimis and Koutras[2] proved that V_n can be viewed as an *MVP* with $\pi_0 = (1, 0, ..., 0)$, and non vanishing transition probability matrices

$$A_{t,0} = \begin{bmatrix} q_t & p_t & 0 & \cdot & 0 & 0 & 0 \\ q_t & 0 & p_t & \cdot & 0 & 0 & 0 \\ \cdot & \cdot & \cdot & \cdot & \cdot & \cdot & \cdot \\ q_t & 0 & 0 & \cdot & 0 & p_t & 0 \\ q_t & 0 & 0 & \cdot & 0 & 0 & 0 \\ q_t & 0 & 0 & \cdot & 0 & 0 & 0 \end{bmatrix}_{(k+1)\times(k+1)} \tag{4.1}$$

$$A_{t,1} = \begin{bmatrix} 0 & 0 & \cdot & 0 & 0 \\ 0 & 0 & \cdot & 0 & 0 \\ \cdot & \cdot & \cdot & \cdot & \cdot \\ 0 & 0 & \cdot & 0 & 0 \\ 0 & 0 & \cdot & 0 & p_t \end{bmatrix}_{(k+1)\times(k+1)} \qquad A_{t,k} = \begin{bmatrix} 0 & 0 & \cdot & 0 & 0 \\ 0 & 0 & \cdot & 0 & 0 \\ \cdot & \cdot & \cdot & \cdot & \cdot \\ 0 & 0 & \cdot & 0 & p_t \\ 0 & 0 & \cdot & 0 & 0 \end{bmatrix}_{(k+1)\times(k+1)}$$

(the matrices $A_{t,2}, A_{t,3}, ..., A_{t,k-1}$ have all their entries equal to zero). Accordingly, the probability mass function of V_n can be easily obtained through the formulae (see (3.1),(3.2))

$$\Pr(V_n = \nu) = \mathbf{f}_n(\nu)\mathbf{1}', \quad n \geq 0, \quad \nu \geq 0$$

and

$$\mathbf{f}_n(\nu) = \mathbf{f}_{n-1}(\nu)A_{n,0} + \mathbf{f}_{n-1}(\nu-1)A_{n,1} + \mathbf{f}_{n-1}(\nu-k)A_{n,k}, \quad n \geq 1, \quad \nu \geq 0.$$

Consider next the random variable T_r, $r \geq 0$, defined as

$$T_r = \min\{n \geq 0 : V_n \geq r\}.$$

The random variable T_r denotes the waiting time for the first instance where the sum of the exact lengths of substrings of the sequence $Z_1, Z_2, ...$ containing k or more consecutive successes is equal to or exceeds r, that is,

T_r describes the event *"the waiting time until the random variable V_n equals or exceeds the value r for first time"*.

The above definitions are best illustrated by the following example. Let us consider the sequence of 30 Bernoulli outcomes shown below (success have been pictured by the symbol ($\square$) while failures by ($\blacksquare$)). For $k = 3$ we have that $V_{30} = 12$, which means that the sum of exact lengths of subsequences (strings) consisting of $k = 3$ or more consecutive successes equals 12.

As far as the waiting time random variables T_r are concerned, it can be easily checked that $T_1 = T_2 = T_3 = 9$, $T_4 = T_5 = T_6 = 13$, $T_7 = 14$, $T_8 = T_9 = T_{10} = 21$, $T_{11} = 22$ and $T_{12} = 23$.

The general approach presented in Section 3 can be easily adopted to the specific problem introduced in the present section, thereby obtaining a functional framework for the study of T_r's distribution. To start with, the next proposition establishes a formula for the evaluation of the probability mass function of T_r.

Proposition 4: *The probability mass function $h_r(n)$ of T_r, $r \geq 1$, is given by*

$$h_r(n) = \Pr(T_r = n) = \mathbf{f}_{n-1}(r-1)p_n\mathbf{e}'_k + \sum_{i=1}^{k} \mathbf{f}_{n-1}(r-i)p_n\mathbf{e}'_{k-1}$$

Proof. A direct application of Theorem 2 reveals that

$$h_r(n) = \mathbf{f}_{n-1}(r-1)A_{n,1}\mathbf{1}' + \sum_{i=1}^{k} \mathbf{f}_{n-1}(r-i)A_{n,k}\mathbf{1}', \quad n \geq 1$$

and the proof of the proposition is immediate. $\square$

From now on we assume that the sequence $Z_1, Z_2, \ldots$ consists of i.i.d. Bernoulli trials, i.e. $p_t = p$, $q_t = q$, for all $t = 1, 2, \ldots$. In this case the matrices $A_{t,i}$, $i = 0, 1, 2, \ldots, k$, do not depend on t that is $A_{t,i} = A_i$, and their form can be easily identified from (4.1) by replacing all p_t's by p and all q_t's by q.

The following proposition provides a formula for the evaluation of the double probability generating function of T_r.

Proposition 5: *The double probability generating function $H(z,w)$ of T_r is given by*

$$H(z,w) = 1 + wz\pi_0 \left[I - w(A_0 + zA_1 + z^k A_k) \right]^{-1} p\mathbf{e}'_k +$$

$$wz \left(\frac{z^k - 1}{z - 1} \right) \pi_0 \left[I - w(A_0 + zA_1 + z^k A_k) \right]^{-1} p\mathbf{e}'_{k-1}.$$

Proof. The proof follows immediately from Theorem 3 if we take into account that $A_i = \mathbf{0}$, for $i = 2, 3, ..., k - 1$. $\square$

The outcome of Proposition 5 can be used to express $H(z,w)$ as a rational function. To achieve this goal we shall use the probability generating function of a run-related statistic, more specifically the waiting time X_k until k consecutive successes (success run of length k) are observed for the first time in the sequence of Bernoulli trials $Z_1, Z_2, ...$. The random variable X_k can be formally defined as

$$X_k = \min\{n : Z_{n-k+1} = ... = Z_n = 1\}$$

and its probability generating function is given by

$$E[w^{X_k}] = G(w) = \sum_{n=0}^{\infty} g(n)w^n = \frac{(wp)^k(1 - wp)}{1 - w + (wq)(wp)^k}.$$

The distribution of X_k is called *the geometric distribution of order k* and has been extensively studied in the last two decades, see e.g. Philippou and Muwafi[15], Philippou, Georgiou and Philippou[16], Hahn and Gage[7], Aki, Kuboki and Hirano[1], Uppuluri and Patril[19]. For more details on this distribution and its applications in several sciences the interest reader may refer to Johnson, Kotz and Kemp[10] or Balakrishnan and Koutras[4].

The next corollary indicates that the double generating function $H(z,w)$ can be expressed in terms of the single generating function $G(w)$ of X_k.

Corollary 6: *The double probability generating function $H(z,w)$ of T_r may be expressed as*

$$H(z,w) = \frac{1 + z(G(w) - wp) + (1 - wp)G(w) \sum_{i=2}^{k-1} z^i + (1 - w)G(w)z^k}{1 - z(wp) - z^k(wq)G(w)}.$$

Proof. The proof follows from Proposition 5 by carrying out the inversion of the matrix $\left[I - w(A_0 + zA_1 + z^k A_k) \right]$. It is worth stressing that only specific entries of the inverse matrix are needed and therefore one could avoid calculating the whole matrix. $\square$

The outcome of Corollary 6 can be exploited for the derivation of a recursive scheme for the evaluation of the single generating function $H_r(w)$ of T_r. More specifically we have the following.

Corollary 7: *The probability generating function $H_r(w)$ of T_r satisfies the recursive scheme*

$$H_r(w) = (wp)H_{r-1}(w) + (wq)G(w)H_{r-k}(w), \quad r \geq k+1$$

with initial conditions $H_r(w) = G(w)$, for $1 \leq r \leq k$.

Proof. Using Corollary 6 we may write

$$\left(1 - z(wp) - z^k(wq)G(w)\right) \sum_{r=0}^{\infty} H_r(w)z^r$$

$$= 1 + z(G(w) - wp) + (1 - wp)G(w) \sum_{i=2}^{k-1} z^i + (1 - w)G(w)z^k.$$

The desired results follows easily by equating coefficients of z^r in both sides of the above identity. $\square$

It is worth mentioning that, since $H_r(w) = G(w)$, for $1 \leq r \leq k$, the distribution of T_r, $r = 1, 2, ..., k$, coincides with that of X_k, while for $k + 1 \leq r \leq 2k$ a repeated application of the recursive scheme of Corollary 7 yields

$$H_r(w) = G(w) \left((wq)G(w) \sum_{i=0}^{r-k-1} (wp)^i + (wp)^{r-k} \right). \tag{4.2}$$

This identity reveals that $H_{k+1}(w)$ coincides with the probability generating function of the waiting time for the second overlapping success run of length k (see, e.g. Hirano, Aki, Kashiwagi and Kuboki[9], or Balakrishnan and Koutras[4]).

Formula (4.2) can also be established directly if we observe that the event $\{T_r = n\}$, for $k + 1 \leq r \leq 2k$ corresponds to outcomes of the sampling space of the form

$$\underbrace{\overbrace{...11...1}^{k}}_{X_k} \overbrace{11...1}^{r-k} \quad \text{or} \quad \underbrace{\overbrace{...11...1}^{k}}_{X_k} \overbrace{11...1}^{i} 0 \underbrace{\overbrace{...11\cdots1}^{k}}_{X_k}, \quad 0 \leq i \leq r-k-1.$$

Another interesting by-product of Corollary 7 is that for $r \geq k + 1$, the waiting times T_r obey the following recursive scheme

$$T_r = \begin{cases} T_{r-1} + 1, & \text{with probability } p \\ \\ T_{r-k} + 1 + X_k, & \text{with probability } q \end{cases} \tag{4.3}$$

where T_{r-k} and X_k are independent random variables.

We shall now exploit the outcome of Corollary 7 to derive a recursive relation for the probability mass function of T_r.

Corollary 8: *The probability mass function $h_r(n)$ of T_r satisfies the recursive scheme*

$$h_r(n) = h_r(n-1) + p[h_{r-1}(n-1) - h_{r-1}(n-2)]$$
$$-qp^k[h_r(n-k-1) - h_{r-k}(n-k-1)]$$
$$+qp^{k+1}[h_{r-1}(n-k-2) - h_{r-k}(n-k-2)]$$

with initial conditions $h_r(n) = g(n) = \Pr(X_k = n)$, for $1 \leq r \leq k$.

Proof. The formula established in Corollary 7 can be equivalently written as

$$\left(1 - w + w^{k+1}qp^k\right) \sum_{n=0}^{\infty} h_r(n)w^n = (wp)(1 - w + w^{k+1}qp^{k+1}) \sum_{n=0}^{\infty} h_{r-1}(n)w^n$$

$$+(wq)(wp)^k(1 - wp) \sum_{n=0}^{\infty} h_{r-k}(n)w^n.$$

Equating the coefficients of w^n in both sides of the above relation we may easily verify the stated recursive scheme. The initial conditions follow immediately from the fact that $H_r(w) = G(w)$, for $1 \leq r \leq k$. $\square$

Figure 1 presents the probability mass function $h_r(n)$ of the random variable T_r for various combinations of the parameters r, k, p.

Having at hand a recursive scheme for the probability mass function of T_r it is not difficult to establish a recursive scheme for the cumulative distribution function or the tail probabilities of the same random variable. Such a scheme is described in the next corollary.

Corollary 9: *The tail probability function $\bar{h}_r(n) = \Pr(T_r > n)$ of T_r sat-*

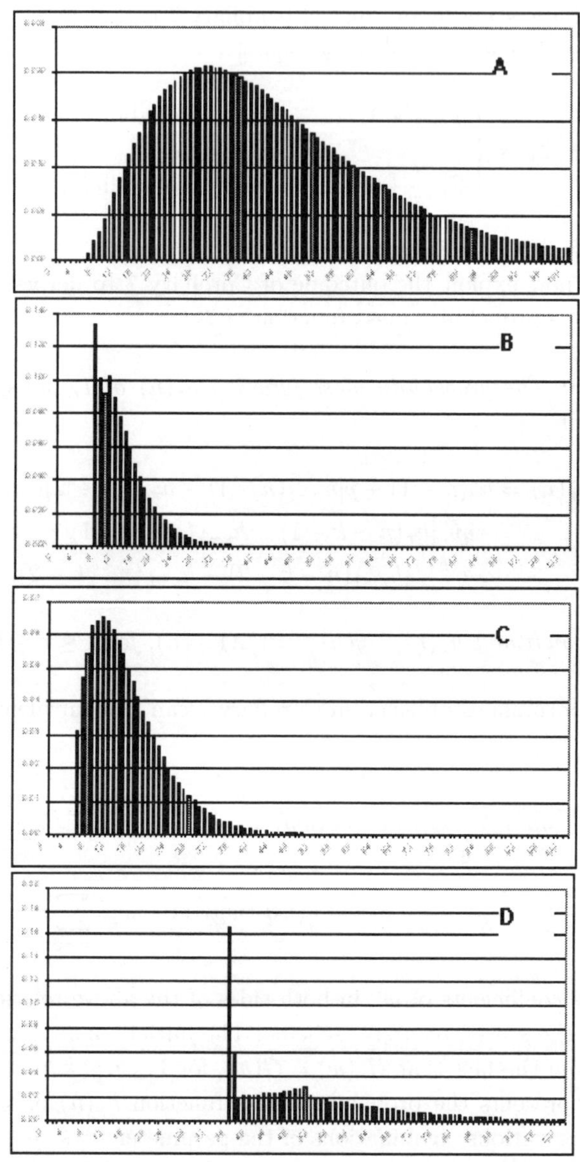

Fig. 1.　The Probability Mass Function $h_r(n)$ of the random variable T_r for various combinations of the parameters r, k, p. A: $p = 0.30, k = 2, r = 6$, B: $p = 0.75, k = 3, r = 7$ C: $p = 0.50, k = 2, r = 5$, D: $p = 0.95, k = 15, r = 35$

isfies the recurrence relation

$$\overline{h}_r(n) = 2\overline{h}_r(n-1) - \overline{h}_r(n-2)$$
$$-p[2\overline{h}_{r-1}(n-2) - \overline{h}_{r-1}(n-1) - \overline{h}_{r-1}(n-3)]$$
$$+qp^k[\overline{h}_r(n-k-2) - \overline{h}_r(n-k-1)]$$
$$+qp^k[\overline{h}_{r-k}(n-k-1) - \overline{h}_{r-k}(n-k-2)]$$
$$-qp^{k+1}[\overline{h}_{r-1}(n-k-3) - \overline{h}_{r-1}(n-k-2)]$$
$$-qp^{k+1}[\overline{h}_{r-k}(n-k-2) - \overline{h}_{r-k}(n-k-3)]$$

with initial conditions $\overline{h}_r(n) = \overline{g}(n) = \Pr(X_k > n)$, *for* $1 \leq r \leq k$.

Proof. This can be easily carried out either from the outcome of Corollary 8 upon replacing $h_r(n)$ by $\overline{h}_r(n-1) - \overline{h}_r(n)$ or from the outcome of Corollary 7 on observing that the generating function of $\overline{h}_r(n)$ may be expressed in terms of $H_r(w)$ by

$$\sum_{n=0}^{\infty} \overline{h}_r(n)w^n = \frac{1 - H_r(w)}{1 - w}. \qquad \square$$

Apparently, the last corollary is quite useful for the investigation of the operating characteristic curve of the sampling plan described in Section 1.

Although it does not seem feasible to derive a neat explicit expression for the mean of T_r, the numerical evaluation of $\mu_r = E[T_r]$ can be easily carried out by a recursive scheme, as the next corollary indicates.

Corollary 10: *The means* $\mu_r = E[T_r]$, $r = 1, 2, \ldots$ *of the random variables* T_r *satisfy the recursive scheme*

$$\mu_r = \frac{1}{p^k} + p\mu_{r-1} + q\mu_{r-k}, \quad r \geq k+1$$

with initial conditions $\mu_r = (1 - p^k)/p^k$, *for* $1 \leq r \leq k$.

Proof. Applying the mean value operator in (4.3) we may easily gain the equality

$$\mu_r = 1 + p\mu_{r-1} + q(\mu_{r-k} + E[X_k]), \quad r \geq k+1$$

and the proof is completed by observing that $\mu_r = E[X_k]$, $1 \leq r \leq k$ (c.f. discussion following Corollary 7) and recalling the well known result (see, e.g. Balakrishnan and Koutras[4])

$$E[X_k] = \frac{1 - p^k}{p^k}. \qquad \square$$

It is noteworthy, that $E[T_r]$ is in fact the average sample number (ASN) of the sampling inspection plan described in Section 1.

In closing this section we mention that the expression established in Corollary 6 for $H(z,w)$ may also be extracted directly, using the formula

$$H(z,w) = \frac{z(1-w)\Phi(z,w) - 1}{z-1} \tag{4.4}$$

which describes the connection between the double generating functions of the waiting time variable T_r and the associated enumerating statistic V_n. This equality can be formally verified by the aid of the obvious relation

$$\Pr(T_r > n) = \Pr(V_n < r).$$

An analogous identity has been mentioned in Feller[5] in connection with the occurrence of recurrent events; see also Koutras[13]. Antzoulakos, Bersimis and Koutras[2] proved that $\Phi(z,w)$ takes on the form

$$\Phi(z,w) = \frac{P_1(z,w)}{P_2(z,w)}$$

where

$$P_1(z,w) = 1 - wpz - (wp)^k\left(1 - z^k\right) - (wp)^{k+1}\left(z^k - z\right)$$

$$P_2(z,w) = 1 - w(1+pz) + w^2pz + w^{k+1}qp^k(1 - z^k) + w^{k+2}qp^{k+1}(z^k - z).$$

Substituting these last expressions in (4.4) we may easily reestablish, after some lengthy but quite straightforward manipulations, the result mentioned in Corollary 6.

Acknowledgments

S.Bersimis and M.V.Koutras were supported by General Secretary of Research and Technology of Greece under PENED 2001.

References

1. Aki, S., Kuboki, H. and Hirano, K., On discrete distributions of order k, *Annals of the Institute of Statistical Mathematics*, **36**, 431-440, (1984).
2. Antzoulakos, D., Bersimis, S. and Koutras, M. V., On the distribution of the total number of run lengths. Submitted for publication, (2002).
3. Balakrishnan, N., Balasubramanian, K. and Viveros, R., On sampling inspection plans based on the theory of runs, *The Mathematical Scientist*, **18**, 113-126, (1993).

4. Balakrishnan, N. and Koutras, M. V., *Runs and Scans with Applications*, John Wiley & Sons, New York, (2002).

5. Feller, W. *An Introduction to Probability Theory and its Applications, Vol. I*, Third Edition, John Wiley and Sons, New York, (1968).

6. Fu, J. C. and Koutras, M. V., Distribution theory of runs: a Markov chain approach, *Journal of the American Statistical Association*, **89**, 1050-1058, (1994).

7. Hahn, G. J. and Gage, J. B., Evaluation of a start-up demonstration test, *Journal of Quality Technology*, **15**, 103-105, (1983).

8. Hald, A., *Statistical Theory of Sampling Inspection by Attributes*, Academic Press, London, (1981).

9. Hirano, K., Aki, S., Kashiwagi, N. and Kuboki, H., On Ling's binomial and negative binomial distributions of order k, *Statistics and Probability Letters*, **11**, 503-509, (1991).

10. Johnson, N. L., Kotz, S. and Kemp, A. W., *Univariate Discrete Distributions*, Second Edition, John Wiley & Sons, New York, (1992).

11. Koutras, M. V. and Alexandrou, V. A., Runs, scans and urn model distributions: A unified Markov chain approach, *Annals of the Institute of Statistical Mathematics*, **47**, 743-766, (1995).

12. Koutras, M. V., Waiting time distributions associated with runs of fixed length in two-state Markov chains, *Annals of the Institute of Statistical Mathematics*, **49**, 123-139, (1997).

13. Koutras, M. V., Waiting times and number of appearances of events in a sequence of discrete random variables, in *Advances of Combinatorial Methods and Applications to Probability and Statistics* (ed. N. Balakrishnan), 363-384, Birkhauser, Boston, (1997).

14. Mosteller, F., Note on an application of runs to quality control charts, *Annals of Mathematical Statistics*, **12**, 228-232, (1941).

15. Philippou, A. N., and Muwafi, A. A., Waiting for the k-th consecutive success and the fibonacci sequence of order k, *The Fibonacci quarterly*, **20**, 28-32, (1982).

16. Philippou, A. N., Georgiou, C. and Philippou, G. N., A generilized geometric distribution and some of its properties, *Statistics and Probability Letters*, **1**, 171-175, (1983).

17. Prairie, R. R., Zimmer, W. J., and Brookhouse, J. K., Some acceptance sampling plans based on the theory of runs, *Technometrics*, **4**, 177-185, (1962).

18. Schmueli, G. and Cohen, A., Run-related probability functions applied to sampling inspection, *Technometrics*, **42**, 188-202, (2000).

19. Uppuluri, V. R. R., and Patil, S. A., Waiting times and generilized Fibonacci sequences, *The Fibonacci quarterly*, **21**, 242-249, (1983).

20. Wolfowitz, J., On the theory of runs with some applications to quality control, *Journal of the American Statistical Association*, **14**, 280-288, (1943).

MAXIMUM LIKELIHOOD PREDICTION OF FUTURE RECORD STATISTIC

Prasanta Basak[a]

Penn State Altoona
Altoona, PA 16601-3760
E-mail: fkv@psu.edu

N. Balakrishnan

McMaster University
Hamilton, Canada
E-mail: bala@mcmail.cis.macmaster.ca

Let $X_1, X_2, \cdots$ be a sequence of independent and identically distributed random variables with absolutely continuous distribution function. For $n \geq 1$, we denote the order statistics of X_1, X_2, $\cdots$, X_n by $X_{1,n} \leq X_{2,n} \leq \cdots \leq X_{n,n}$. Define $L(1) = 1, L(n+1) = \min\{j : j > L(n), X_j > X_{j-1,j-1}\}$, and $X(n) = X_{L(n),L(n)}, n \geq 1$. The sequence $\{X(n)\}$ ($\{L(n)\}$) is called upper record statistics (times). In this chapter, we deal with maximum likelihood prediction of record statistics. In particular, having observed a sequence of record statistics from an absolutely continuous population, we consider the maximum likelihood prediction of a future record statistic. The conditional median predictor (CMP) is also considered. Some desirable properties of maximum likelihood predictor (MLP) such as unbiasedness, consistency, and efficiency are discussed. The properties of MLP are examined in detail for the exponential and uniform populations. The conditional distribution is used to derive the CMP for the exponential and uniform populations. Finally, we compare the CMP with median unbiased predictor (MUP), MLP, and best linear unbiased predictor (BLUP). The comparisons indicate that the CMP performs well in terms of mean squared prediction error.

---

[a]The research was done when the author was on sabbatical leave at McMaster University, Hamilton, Ontario, Canada.

1. Introduction

Let $X_1, X_2, \cdots$ be independent and identically distributed (iid) random variables (rvs) with absolutely continuous distribution function (df) $F(x; \theta)$ and corresponding probability density function (pdf) $f(x; \theta)$, where θ is possibly a vector-valued parameter. For $n \geq 1$, we denote the order statistics of $X_1, X_2, \cdots, X_n$ by $X_{1,n} \leq X_{2,n} \leq \cdots \leq X_{n,n}$. Define

$$L(1) = 1, L(n+1) = \min\{j : j > L(n), X_j > X_{j-1,j-1}\}, \text{ and}$$
$$X(n) = X_{L(n),L(n)}, n \geq 1.$$

The sequence $\{X(n)\}$ $(\{L(n)\})$ is called upper record statistics (times). Some key references for record statistics are Ahsanullah[2], and Arnold, Balakrishnan, and Nagaraja[3], Nevzorov and Balakrishnan[9], and Nevzorov[10]. Ahsanullah[1], and Dunsmore[4] have dealt with prediction of future records from exponential distibution. Dunsmore[4] also considered Bayesian prediction intervals. Recently, Raqab and Nagaraja[11] used order statistics $X_{1,n}, X_{2,n}, \cdots, X_{r,n}$ to predict the future order statistic $X_{s,n}$ for $1 \leq r < s \leq n$. In this chapter, we are concerned with the prediction of the record statistic $X(m)$, $m > n$, based on the first n record statistics, $\mathbf{X} = (\mathbf{X}(1), \mathbf{X}(2), \cdots, \mathbf{X}(n))$, using maximum likelihood techniques.

Since F is continuous, the conditional distribution of $X(m)$ given $\mathbf{X}$ is, by Markovian property of record statistics, just the distribution of the $X(m)$ given $X(n)$, i.e.,

$$f_{X(m)|\mathbf{X}}(y|\mathbf{x}) = \mathbf{f}_{\mathbf{X(m)}|\mathbf{X(n)}}(\mathbf{y}|\mathbf{x}). \tag{1}$$

In view of (1), the best unbiased predictor (BUP) of $X(m)$, $E[X(m)|\mathbf{X}]$, is nothing but $E[X(m)|X(n)]$. If the parameter θ is unknown, the parameter in this conditional expectation will have to be estimated. In Section 2, we derive MLP when observations are coming from continuous location-scale family of distributions. One can use the idea of median unbiasedness (see Takada [14]) to suggest a predictor of $X(m)$. A statistic $t(\mathbf{X})$ is a MUP of $X(m)$ if for all θ

$$P_\theta\left[t(\mathbf{X}) \leq X(m)\right] = P_\theta\left[t(\mathbf{X}) \geq X(m)\right].$$

Takada[14] showed that, for the location-scale family, a specially chosen MUP is better than the BUP of $X(m)$ under Pitman's measure of closeness (PMC). For literature on the PMC criterion see, for example, the papers in Keating et al. [7]. For predicting $X(m)$, $t_1(\mathbf{X})$ is said to be better than $t_2(\mathbf{X})$ under PMC, if for all θ,

$$P_\theta\left[|t_1(\mathbf{X}) - X(m)| \leq |t_2(\mathbf{X}) - X(m)|\right] \geq \frac{1}{2}. \tag{2}$$

A related predictor is the conditional median predictor (CMP). A statistic $X^C(m)$ will be called a CMP of $X(m)$ if it is the median of the conditional distribution of $X(m)$ given $X(n)$. Observe that a CMP is also an MUP.

In Section 2, we discuss some properties of the MLP such as unbiasedness, consistency, and efficiency. We examine there in detail, the properties of the MLP for the exponential and uniform populations. In Section 3, we give MUP of $X(m)$ for the exponential and uniform populations and introduce the CMP of $X(m)$. In Section 4, a modified CMP is compared with the MUP, MLP, and BLUP in terms of mean squared error (MSE), for the exponential and uniform populations. Numerical evidence suggests that in addition to being easy to compute, the modified CMP also performs extremely well.

Throughout the chapter, we will use the following notations:

$F^{-1}(p)$: $\sup\{x : F(x) \le p\}$, for $0 < p < 1$, quantile function of F

$\mathrm{Mod}(X)$: Mode of F, when it is unimodal

$X \stackrel{d}{=} Y$: X and Y are identically distributed

$U(a,b)$: Uniform rv over the interval (a,b)

$\chi^2(r)$: Chi-square distribution with r degrees of freedom

$F(r,s)$: F-distribution with r and s degrees of freedom

$X^*(m)$: BUP of $X(m)$

$X^B(m)$: BLUP of $X(m)$

$X^C(m)$: CMP of $X(m)$

$X^M(m)$: MUP of $X(m)$

$X^L(m)$: MLP of $X(m)$

$\hat{\theta}$: Uniformly minimum variance unbiased estimator (UMVUE) of θ

$R(x)$: $-\log(1 - F(x))$, the cumulative hazard rate of X

$r(x)$: $R'(x) = f(x)/(1 - F(x))$, the hazard rate of X

2. Maximum Likelihood Predictor

Let $\mathbf{X} = (X(1), X(2), \cdots, X(n))$ denote the first n record statistics from a population with pdf $f(x) = f(x; \theta)$, $x \in [l, u]$, $\theta \in \mathbf{D}$, a k-dimensional parameter vector. We consider the prediction of $X(m)$, $m > n$, having observed $\mathbf{X}$. Let $\mathbf{x} = (x_1, x_2, \cdots, x_n)$ with $x_1 < x_2 \cdots < x_n$. The predictive likelihood function (PLF) of $X(m)$ and θ is given by

$$L = L(x_m, \theta; \mathbf{x}) = \prod_{j=1}^{n} r(x_j) \frac{[R(x_m) - R(x_n)]^{m-n-1}}{\Gamma(m-n)} f(x_m), \quad x_n < x_m.$$

$$(3)$$

If $X^L(m) = t(\mathbf{X})$ and $\theta^* = \mathbf{u}(\mathbf{X})$ are statistics for which

$$L(t(\mathbf{x}), \mathbf{u}(\mathbf{x}); \mathbf{x}) = \sup_{\mathbf{x(m)}, \theta} \mathbf{L}(\mathbf{x(m)}, \theta; \mathbf{x}),$$

then we will call $t(\mathbf{X})$ the MLP of $X(m)$, and $u(\mathbf{X})$ the predictive maximum likelihood estimator (PMLE) of θ.

2.1. *Parameter Known*

By continuity of f and F, L converges to zero both as $x_m \downarrow x_n$ and as $x_m \uparrow u$. Also, $L > 0$ on $\mathbf{D}$. This means that if there exists a unique solution, $X^L(m)$, of the likelihood equation $\partial \log L / \partial x_m = 0$, then $X^L(m)$ must be the unique MLP of $X(m)$. The log-likelihood is given by

$$\log L = \sum_{j=1}^{n} \log r(x_j) + (m-n-1) \log(R(x_m) - R(x_n)) - \log \Gamma(m-n) + \log f(x_m).$$

The likelihood equation is given by

$$\frac{\partial \log L}{\partial x_m} = (m - n - 1)\frac{r(x_m)}{R(x_m) - R(x_n)} + \frac{f'(x_m)}{f(x_m)}$$

$$= r(x_m)\left[\frac{m - n - 1}{R(x_m) - R(x_n)} + \frac{f'(x_m)}{f^2(x_m)}(1 - F(x_m))\right] = 0. \quad (4)$$

From equation (4) or from the Markovian property of record statistics, it readily follows that when θ is known, the MLP of $X(m)$, if it exists, is a function of $X(n)$ and the known value of θ.

Remark 1: The PLF in (3) can be rewritten as

$$L(x_m, \theta; \mathbf{x}) = \mathbf{f}_{\mathbf{X(m)|X(n)}}(\mathbf{x_m}|\mathbf{x_n}, \theta) \cdot \mathbf{f}(\mathbf{x}, \theta).$$

So, when θ is known, MLP $X^L(m)$ for $X(m)$ is mode of the conditional distribution of $X(m)$, given $X(n) = x_n$.

Result 1. When the parameter θ is known, the MLP $X^L(m)$ of $X(m)$ is unbiased if the mode and the mean of the conditional distribution of $X(m)$, given $X(n)$, coincide.

Proof. Note that

$$E[X^L(m) - X(m)] = 0 \Leftrightarrow E[X^L(m)] - E[E(X(m)|X(n))] = 0$$

$$\Leftrightarrow E[X^L(m) - E(X(m)|X(n))] = 0.$$

We will consider two cases separately: (i) $m > n + 1$, and (ii) $m = n + 1$.
Case (i) $m > n + 1$: In this case, the function $\frac{m-n-1}{R(x_m)-R(x_n)}$ as a function of x_m on $[x_n, u)$ is continuous, decreasing, converges to $+\infty$ as $x_m \uparrow u$ and converges to 0 as $x_m \downarrow x_n$. Thus, from equation (4), we see that a unique MLP for $X(m)$ exists if

$$\frac{f'(x)}{f^2(x)} = -\frac{\partial\left(\frac{1}{f(x)}\right)}{\partial x} \text{ is non-increasing on } [x_n, u).$$

Case (ii) $m = n + 1$: Again by reasoning similar to that in Case (i), we find that a unique MLP for $X(m)$ exists if $f'(x)/f^2(x)$ is decreasing on $[x(n), u)$.

Remark 2: The exponential; gamma with shape parameter ≥ 1; Weibull with shape parameter ≥ 1; logistic; normal; half-normal; Student's t; Cauchy; Pareto; and power function all satisfy above condition. For PF_2 densities for which f'' exists, the above condition is satisfied. Hence, for such densities, a unique MLP for $X(m)$ exists. The converse is false since Cauchy is not PF_2 density even though it produces a unique MLP for $X(m)$.

A condition for $X^L(m)$ to be a consistent predictor of $X(m)$ can be given through its MSE.

Result 2. Suppose that $m = [n\lambda_m] + 1; 0 < \lambda_m < 1, f(F^{-1}(\lambda_m)) > 0$, and $E|X|^\epsilon < \infty$ for some $\epsilon > 0$. Then, the MLP $X^L(m)$ of $X(m)$ is consistent if $E[X^L(m) - E(X(m)|X(n))]^2 \to 0$ as $n \to \infty$.

Proof. Consider

$$E[X^L(m) - X(m)]^2$$

$$= E\left\{[X^L(m) - E(X(m)|X(n))] + [E(X(m)|X(n)) - X(m)]\right\}^2$$

$$= E[X^L(m) - E(X(m)|X(n))]^2 + E[X(m) - E(X(m)|X(n))]^2$$
$$-2E\left\{[X^L(m) - E(X(m)|X(n))]\,[X(m) - E(X(m)|X(n))]\right\}$$

$$- E[X^L(m) - E(X(m)|X(n))]^2$$
$$+E\left\{E\{[X(m) - E(X(m)|X(n))]^2|X(n)\}\right\}$$

$$= E[X^L(m) - E(X(m)|X(n))]^2 + E[Var(X(m)|X(n))] \qquad (5)$$

(the term $E\left\{[X^L(m) - E(X(m)|X(n))]\,[X(m) - E(X(m)|X(n))]\right\}$ is zero can be seen by conditioning on $\mathbf{X}$, using (1), and the fact that $X^L(m)$ is a function of $\mathbf{X}$.)

From the assumptions, it follows (from Shorack and Wellner[12], page 475) that

$$E[n^{k/2}|X(m) - F^{-1}(\lambda_m)|^k] \to E|Y|^k \text{ as } n \to \infty \qquad (6)$$

iff $E|X|^\epsilon < \infty$ for some $\epsilon > 0$, where $Y \sim N\left(0, \frac{\lambda_m(1-\lambda_m)}{f^2(F^{-1}(\lambda_m))}\right)$.

With $k = 2$, (6) implies $Var[X(m)] \to 0$ as $n \to \infty$. Since $Var[X(m)] = E[Var(X(m)|X(n))] + Var[E(X(m)|X(n))]$, $Var[X(m)] \geq E[Var(X(m)|X(n))]$.

Hence if assumptions are valid, then $E[Var(X(m)|X(n))] \to 0$ as $n \to \infty$. Therefore, from (5), the result follows.

The following inequality follows from (5):

$$E[X^L(m) - X(m)]^2 \geq E[X(m) - E(X(m)|X(n))]^2. \qquad (7)$$

In view of (5), the equality holds in (7) iff $E[X^L(m) - E(X(m)|X(n))]^2 = 0$. Therefore $X^L(m)$ is efficient, i.e., has the smallest MSE if $X^L(m) = E[X(m)|X(n)]$. So, Result 1 also gives us a sufficient condition for the MLP to be the BUP.

2.2. *Parameter Unknown*

In practice, we usually have the parameter to be unknown. In that case, we may maximize the PLF using standard methods, if a maximum does exist and the MLP of $X(m)$ may be determined along with PMLE of θ.

Remark 3: When F belongs to a location-scale family with location parameter μ (real) and the scale parameter $\sigma(> 0)$, the best linear unbiased predictor (BLUP) $X^L(m)$ of $X(m)$ is given by (see Arnold, Balakrishnan, and Nagaraja[3], (page 150))

$$X^L(m) = \mu^* + \alpha_m \sigma^* + \omega' \Sigma^{-1}\left(\mathbf{X} - \mu^* \mathbf{1} - \sigma^* \alpha\right),$$

where μ^*, σ^* are the BLUEs of μ and σ, $\mathbf{1}$ is vector of 1's, α is the vector of means of record statistics from the standard distribution, Σ is the dispersion matrix of standard record statistics and $\omega' = (\sigma_{1m}, \sigma_{2m}, \cdots, \sigma_{nm})$, $\sigma_{im} = \text{Cov}[X(i), X(m)]$. The MSE of $X^L(m)$ can be found once we get the variances and covariances among $\mathbf{X}$, μ^*, and σ^*.

We can apply the result similar to the one obtained by Ishii [5] and mentioned in Takada[13]. It states that an unbiased predictor $X^*(m)$ of $X(m)$ is its BUP if and only if

$$E_\theta[(X(m) - X^*(m)) \cdot \gamma(\mathbf{X})] = \mathbf{0} \tag{8}$$

for all θ and any $\gamma(\mathbf{X})$ that is an unbiased estimator of zero.

Example 4: The Exponential distribution

We have presented above sufficient conditions under which a unique MLP for $X(m)$ exists when θ is known. The exponential distribution is one of the distributions for which the conditions are satisfied. Let n records $x_1, x_2, \cdots, x_n$ be observed from the pdf $f(x; \mu, \sigma) = \frac{1}{\sigma}e^{-(x-\mu)/\sigma}$, $x \geq \mu$, $\sigma > 0$. For the sake of brevity, we will cover only the case of unknown scale parameter σ, and without loss of generality, we take $\mu = 0$. The case when both parameters are unknown can be handled with little additional difficulty. We examine the properties of $X^L(m)$ both when the parameter is known and unknown. The log PLF of $X(m)$ in this case is given by

$$-m \cdot \log \sigma - \log \Gamma(m - n) + (m - n - 1) \cdot \log(x_m - x_n) - \frac{x_m}{\sigma}. \tag{9}$$

Parameter σ known: The log PLF has unique maximum relative to x_m and the MLP $X^L(m)$ of $X(m)$ is given by

$$X^L(m) = X(n) + (m - n - 1)\sigma. \tag{10}$$

Now, $X^L(m)$ is a biased predictor of $X(m)$, with the bias being

$$E[X(m) - X^L(m)] = \sigma.$$

The MSE of MLP $X^L(m)$ is given by

$$\text{MSE}(X^L(m)) = E[X(m) - X^L(m)]^2 = (m - n + 1)\sigma^2,$$

which shows that $X^*(m)$ is not consistent.

Parameter σ unknown: When the parameter is unknown, the log PLF of $X(m)$ and σ is again given by (9). It is easy to see that this function has unique maximum relative to x_m and σ, for any $m \geq n+1$. When $m = n+1$, the maximum occurs on the boundary of the region, $x_m = x_n$, where L is positive. The MLP and PMLE are easily found to be

$$X^L(m) = \frac{m}{n+1}x_n, \qquad \sigma^* = \frac{1}{n+1}x_n.$$

Here also, $X^L(m)$ and σ^* are biased predictors of $X(m)$ and σ, respectively. The biases are $E[X(m) - X^L(m)] = \frac{m}{n+1}\sigma$, and $E[\sigma - \sigma^*] = \frac{1}{n+1}\sigma$, respectively.

The MSE of MLP $X^L(m)$ is given by

$$\mathrm{MSE}(X^L(m)) = E[X(m) - X^L(m)]^2 = \frac{m(m-n+1)}{n+1}\sigma^2. \qquad (11)$$

The BLUP $X^B(m)$ of $X(m)$ can easily be calculated with the help of Kaminsky and Nelson[6] to be

$$X^B(m) = X(n) + (m-n)\sigma^{**} = \frac{m}{n}X(n),$$

where $\sigma^{**} = \frac{1}{n}X(n)$ is the BLUE of σ.

This gives the MSE of the BLUP $X^B(m)$ to be

$$\mathrm{MSE}(X^B(m)) = E[X(m) - X^B(m)]^2 = \frac{m(m-n)}{n}\sigma^2. \qquad (12)$$

So, the efficiency of MLP $X^L(m)$ relative to BLUP $X^B(m)$, denoted by $\mathrm{Eff}[X^L(m), X^B(m)]$, is then given by

$$\mathrm{Eff}[X^L(m), X^B(m)] = \frac{\mathrm{MSE}(X^B(m))}{\mathrm{MSE}(X^L(m))} = \frac{(n+1)(m-n)}{n(m-n+1)}.$$

It is then clear that

$$\mathrm{Eff}[X^L(m), X^B(m)] \begin{cases} > 1 \text{ if } m > 2n, \\[2mm] = 1 \text{ if } m = 2n, \\[2mm] < 1 \text{ if } m < 2n. \end{cases}$$

Thus, the MLP of $X(m)$ is better than, as good as, or not as good as the BLUP of $X(m)$.

Example 5: The Uniform distribution

Let n records $x_1, x_2, \cdots, x_n$ be observed from $U(0, \sigma)$ with pdf $f(x; \sigma) = \frac{1}{\sigma}$, $0 \le x \le \sigma$.

Parameter σ known: For known σ, the MLP of $X(m)$ is unique and we examine its unbiasedness and consistency properties.

We know that, given $X(n) = x_n$, $X(m)$ behaves like $(m-n)$th record statistic from $U(x_n, \sigma)$. Hence, the variable $\{[X(m)|X(n) = x_n] - x_n\}/[\sigma - x_n]$ behaves like $(m-n)$th record statistic from $U(0, 1)$. ¿¿From this, it follows that,

$$E[X(m)|X(n)] = X(n) + (\sigma - X(n))\frac{2^{m-n} - 1}{2^{m-n}}$$

$$= \frac{X(n) + (2^{m-n} - 1)\sigma}{2^{m-n}}, \qquad (13)$$

and

$$X^L(m) = \text{Mod}\,[X(m)|X(n)] = \sigma. \tag{14}$$

If we take expectation in (13), we see that the RHS of (13) is indeed unbiased for $X(m)$. Similarly, (14) shows that $X^L(m)$, the MLP of $X(m)$, is always positively biased. Also, it should be observed that

$$E\left\{X^L(m) - E[X(m)|X(n)]\right\}^2 = 2^{2(n-m)}E\left[\sigma - X(n)\right]^2. \tag{15}$$

So, as $m - n \to \infty$, the RHS of (15) tends to 0 because $E\left[\sigma - X(n)\right]^2$ is bounded. This makes $X^L(m)$ a consistent predictor of $X(m)$.

Parameter σ unknown: For unknown σ, the PLF of $X(m)$ and σ is given by

$$\log L = \sum_{j=1}^{n} -\log(\sigma - x_j) + (m - n - 1)\cdot\log\left\{\log(\sigma - x_n) - \log(\sigma - x_m)\right\}$$
$$- \log\sigma - \log\Gamma(m - n),$$
$$x_1 < x_2 < \cdots < x_n < x_m < \sigma.$$

Observe that $\log L$ is monotonically increasing in x_m and monotonically decreasing in σ, and the MLP of $X(m)$ and PMLE of σ are given by

$$X^L(m) = X(n) = \hat{\sigma}.$$

So, we have, using (13),

$$E\left[X^L(m) - E[X(m)|X(n)]\right] = E\left[\frac{2^{m-n} - 1}{2^{m-n}}(X(n) - \sigma)\right]$$
$$= \frac{2^{m-n} - 1}{2^{m-n}}\left(1 - 2^{-n} - 1\right)\sigma$$
$$= -\frac{2^{m-n} - 1}{2^m}\sigma,$$

which shows that $X^L(m)$ is always negatively biased.

3. Conditional Median Predictor

Takada[14] introduced a MUP in an invariant prediction problem for the location-scale family. He has shown that the MUP of the form,

$$X^M(m) = X^*(m) + \hat{\sigma}\cdot\text{Med}\left[\frac{X(m) - X^B(m)}{\hat{\sigma}}\right], \tag{16}$$

where $X^*(m) = $ BUP of $X(m)$, and $\hat{\sigma} = $ MVUE of σ, is better than $X^B(m)$, and the best invariant predictor in terms of the PMC as eluded in

(2). We now propose the median of the conditional distribution of $X(m)$ given $X(n)$ as a predictor of $X(m)$. For the exponential and uniform populations, we derive the CMP $X^C(m)$ and Takada-type MUP $X^M(m)$ of $X(m)$.

Example 6: The Exponential distribution (contd.)

   Exp(σ) parent with known σ

   From (10), we have $X^B(m) = X(n) + (m - n - 1)\sigma$. Then, from (16) a MUP $X^M(m)$ of $X(m)$ is given by

$$X^M(m) = X^B(m) + \sigma \cdot \text{Med}\left[\frac{X(m) - X^B(m)}{\sigma}\right]$$

$$= X(n) + \sigma \cdot \text{Med}[Z(m - n)], \tag{17}$$

where $Z(m - n)$ is the $(m - n)$th record from Exp(1). Now we look at the CMP $X^C(m)$ of $X(m)$. Since $[X(m) - x | X(n) = x] \overset{d}{=} \sigma \cdot Z(m - n)$, the CMP $X^C(m)$ of $X(m)$ is given by

$$X^C(m) = X(n) + \sigma \cdot \text{Med}\left[Z(m - n)\right]. \tag{18}$$

We conclude from Eqs. (17) and (18) that the MUP $X^M(m)$ and the CMP $X^C(m)$ are identical.

   Exp(σ) parent with unknown σ

   When σ is unknown, the UMVUE of σ is given by $X(n)/n$. Let $X^B(m) = X(n) + (m - n - 1)\hat{\sigma}$, and $\gamma(\mathbf{X})$ be any unbiased estimator of zero based on $\mathbf{X}$. Then,

$$E\left[(X(m) - X^B(m)) \cdot \gamma(\mathbf{X})\right] = E\left[(X(m) - X(n) - (m - n - 1)\hat{\sigma}) \cdot \gamma(\mathbf{X})\right]$$

$$= E\left[(X(m) - X(n)) \cdot \gamma(\mathbf{X})\right]$$

$$-(m - n - 1) \cdot E\left[(\hat{\sigma} - \sigma) \cdot \gamma(\mathbf{X})\right].$$

By conditioning on $\mathbf{X}$ and using (1), and the fact that $X(m) - X(n)$ and $X(n)$ are independent, it follows that the first term is zero. From Theorem 2.1.1 of Lehmann [8], $\hat{\sigma}$ and $\gamma(\mathbf{X})$ are uncorrelated. So, we can conclude that

$$E\left[(X(m) - X^B(m)) \cdot \gamma(\mathbf{X})\right] = 0.$$

   Hence, from (8) it follows that $X^*(m)$ is the BUP of $X(m)$. From (16), we obtain on simplification,

$$X^M(m) = X(n) + \hat{\sigma} \cdot \text{Med}\left[\frac{X(m) - X(n)}{\hat{\sigma}}\right]. \tag{19}$$

For the distribution of $\frac{X(m)-X(n)}{\hat{\sigma}} = \frac{X(m)-X(n)}{X(n)/n}$, observe that $X(m) - X(n)$ and $X(n)$ are independently distributed with $2(X(m) - X(n))/\sigma \sim \chi^2(2(m-n))$, and $2X(n)/\sigma \sim \chi^2(2n)$. Therefore, $\frac{n(X(m)-X(n))}{(m-n)X(n)} \sim F(2(m-n), 2n)$. As a result, the term $\mathrm{Med}[(X(m) - X(n))/\hat{\sigma}]$ appearing in (19) can be expressed readily in terms of percentiles of F-distribution.

With the conditional median approach, we can obtain modified CMP $X^{MC}(m)$ of $X(m)$ by

$$X^{MC}(m) = X(n) + \hat{\sigma} \cdot \mathrm{Med}[Z(m - n)]. \tag{20}$$

Example 7: The Uniform distribution (contd.)

$U(0, \sigma)$ **parent with known** σ

Eq. (13) gives $X^*(m)$, the BUP of $X(m)$. Since

$$\mathrm{Med}\left[\frac{X(m) - X^*(m)}{\sigma}\right] = \frac{1}{\sigma}\mathrm{Med}\left[X(m) - \frac{X(n)}{2^{m-n}}\right] - \frac{2^{m-n} - 1}{2^{m-n}},$$

from Eq. (17), the corresponding MUP $X^M(m)$ of $X(m)$ is given by

$$X^M(m) = X^B(m) + \sigma \cdot \mathrm{Med}\left[\frac{X(m) - X^B(m)}{\sigma}\right]$$

$$= \frac{X(n) + (2^{m-n} - 1)\sigma}{2^{m-n}} + \mathrm{Med}\left[X(m) - \frac{X(n)}{2^{m-n}}\right] - \frac{2^{m-n} - 1}{2^{m-n}}\sigma$$

$$= \frac{X(n)}{2^{m-n}} + \mathrm{Med}\left[X(m) - \frac{X(n)}{2^{m-n}}\right]. \tag{21}$$

We need to determine the distribution of $X(m) - X(n)/2^{m-n}$ to evaluate $X^M(m)$. For this, we begin with the joint pdf of $X(n)$ and $X(m)$ and then make the transformation $V = X(m) - X(n)/2^{m-n}$, $W = X(n)$, and integrate out W. We obtain the joint pdf of V and W as follows:

$$f(v, w) = \frac{\left[\log\left(\frac{\sigma}{\sigma - w}\right)\right]^{n-1}}{\Gamma(n)(\sigma - w)} \cdot \frac{\left[\log\left(\frac{\sigma - w}{\sigma - 2^{n-m}w - v}\right)\right]^{m-n-1}}{\Gamma(m - n)\sigma}$$

$$= \sum_{j=0}^{m-n-1} \frac{\left(\log\frac{\sigma}{\sigma - w}\right)^{m-j-2}\left(\log\frac{\sigma}{\sigma - 2^{n-m}w - v}\right)^{j}}{\Gamma(n)\Gamma(j + 1)\Gamma(m - n - j)(\sigma - w)\sigma}. \tag{22}$$

We do not get a closed form expression for the cdf of V on integrating (22). The expression in (22) can be used to determine numerically the median of V. For $n = m - 1$, $\mathrm{Med}(V) = \frac{\sigma}{2}$.

$U(0, \sigma)$ **parent with unknown** σ

It can be shown that $X(n)$ is a complete sufficient statistic for σ, and hence $\hat{\sigma} = \frac{X(n)}{1-2^{-n}}$ is the UMVUE of σ. Let $X^B(m) = \frac{1-2^{-m}}{1-2^{-n}}X(n) = (1-2^{-m})\hat{\sigma}$, and $\gamma(\mathbf{X})$ be any unbiased estimator of zero based on $\mathbf{X} = (X(1), X(2), \cdots, X(n))$. Then

$$E\left[(X(m) - X^B(m)) \cdot \gamma(\mathbf{X})\right] = E\left[\left(X(m) - \frac{1-2^{-m}}{1-2^{-n}}X(n)\right) \cdot \gamma(\mathbf{X})\right]$$

$$= E\left\{E\left[\left(X(m) - \frac{1-2^{-m}}{1-2^{-n}}X(n)\right) \cdot \gamma(\mathbf{X})\right]\Big|\, \mathbf{X}\right\}$$

$$= E\left\{\gamma(\mathbf{X}) \cdot E\left[(X(m)|X(n)) - \frac{1-2^{-m}}{1-2^{-n}}X(n)\right]\right\}.$$

The last part is obtained by using the Markovian property of record statistics. Eq. (13) shows that $E[X(m)|X(n)]$ is linear in $X(n)$. Hence, the factor of $\gamma(\mathbf{X})$ is linear in $\hat{\sigma}$, say $c\hat{\sigma} + d$. Since $\hat{\sigma}$ and $\gamma(\mathbf{X})$ are uncorrelated and $E(\gamma(\mathbf{X})) = 0$, we conclude that $E\left[\left(X(m) - \frac{1-2^{-m}}{1-2^{-n}}X(n)\right) \cdot \gamma(\mathbf{X})\right] = 0$. Hence, from (8) it follows that $X^*(m) = \frac{1-2^{-m}}{1-2^{-n}}X(n)$ is the BUP of $X(m)$.

The MUP $X^M(m)$ of $X(m)$ can now be obtained from (16) as follows:

$$X^M(m) = \frac{1-2^{-m}}{1-2^{-n}}X(n) + \frac{X(n)}{1-2^{-n}} \cdot \mathrm{Med}\left[\frac{X(m) - \frac{1-2^{-m}}{1-2^{-n}}X(n)}{\frac{X(n)}{1-2^{-n}}}\right]$$

$$= X(n) \cdot \mathrm{Med}\left[\frac{X(m)}{X(n)}\right].$$

The CMP $X^C(m)$ of $X(m)$, when σ is known, is found to be

$$X^C(m) = X(n) + (\sigma - X(n)) \cdot \mathrm{Med}\left[\frac{X(m) - X(n)}{\sigma - X(n)}\Big|\, X(n)\right]$$

$$= X(n) + (\sigma - X(n)) \cdot \mathrm{Med}\left[U(m-n)\right], \tag{23}$$

because $\left[\frac{X(m)-X(n)}{\sigma-X(n)}\Big|\, X(n) = x\right] = U(m-n)$, where $U(m-n)$ is the $(m-n)$th record statistic from $U(0,1)$.

For $n = m-1$, (23) reduces to $X^C(m) = \frac{X(n)+\sigma}{2}$, which is identical to (21).

When σ is unknown, we replace σ in (23) by $\hat{\sigma} = \frac{X(n)}{1-2^{-n}}$. Then, the modified CMP $X^{MC}(m)$ of $X(m)$ can be written as

$$X^{MC}(m) = X(n) \cdot \left(1 + \frac{1}{2^n - 1} \cdot \mathrm{Med}\left[U(m-n)\right]\right).$$

4. Numerical Comparisons

In this section, we carry out a numerical study to compare the performance of the modified CMP with those of the BLUP, MLP and MUP of $X(m)$ for the exponential and uniform populations in terms of their MSEs.

Example 8: The Exponential distribution (contd.)

We consider $n = 2, 3, 4, 5$ record statistics from $\text{Exp}(\sigma)$ parent with unknown σ. Observe that $X(m) - X(n)$ is independent of $\hat{\sigma} = \text{UMVUE of } \sigma = X(n)/n$, $E(\hat{\sigma}^2) = (1 + 1/n)\sigma^2$, and $\hat{\sigma}$ is an unbiased estimator of σ.

For the modified CMP $X^{MC}(m)$ of $X(m)$ given in (20),

$$\text{MSE}\left(X^{MC}(m)\right) = E\left[X(m) - X^{MC}(m)\right]^2$$
$$= \left[(m - n) + (m - n)^2 + \tfrac{n+1}{n}q_1^2 - 2(m - n)q_1\right]\sigma^2,$$

where $q_1 = \text{Med}(Z(m - n)) = \frac{1}{2}\chi^2_{1/2}(2(m - n))$ and hence can be obtained from the χ^2-tables easily. It is actually half of the median of $\chi^2(2(m - n))$.

The MSE of the BLUP $X^B(m)$ is already obtained and is given by (see (12))

$$\text{MSE}(X^B(m)) = \frac{m(m - n)}{n}\sigma^2.$$

For the MLP $X^L(m)$ of $X(m)$ given in (10), MSE is given by (see (11))

$$\text{MSE}(X^L(m)) = \frac{m(m - n + 1)}{n + 1}\sigma^2.$$

For the MUP $X^M(m)$ of $X(m)$ given in (19),

$$\text{MSE}\left(X^M(m)\right) = E\left[X(m) - X^{MC}(m)\right]^2$$
$$= \left[(m - n) + (m - n)^2 + \tfrac{n+1}{n}q_2^2 - 2(m - n)q_2\right]\sigma^2,$$

where $q_2 = \text{Med}\left((X(m) - X(n))/\hat{\sigma}\right) = (m - n)F_{1/2}(2(m - n), 2n)$, and hence can be obtained easily from the F-tables. In other words, it is $(m - n)$ times the median of $F(2(m - n), 2n)$.

In Table 1, we present the MSE's, and the relative efficiencies of the modified CMP relative to other estimators – BLUP, MLP, MUP – we have considered above for the exponential parent.

From Table 1, we see that the efficiency of modified CMP $X^{MC}(m)$ relative to BLUP and MUP $X^M(m)$ is very high for all values of n and m considered here. The performance of modified CMP is better than the BLUP for all values of n and m considered (column of Eff1) and is better than the MUP in almost all the cases (column of Eff3). The modified CMP is as easy to compute as the MUP since the values of the medians of χ^2 and F-distributions are tabulated (or can be easily evaluated). The efficiency of modified CMP $X^{MC}(m)$, when compared to the MLP is found also to exceed 1 in many cases, and in a number of cases it is also found to be worse (column of Eff2).

Table 1.   MSE's and Relative Efficiencies of Predictors of $X(m)$ for the $Exp(\sigma)$ Parent.

		MSEs in terms of σ^2				Efficiency of Modified CMP		
n	m	$X^{MC}(m)$	$X^B(m)$	$X^L(m)$	$X^M(m)$	Eff1	Eff2	Eff3
2	3	1.334	1.500	2.000	1.373	1.124	1.499	1.029
2	4	3.512	4.000	4.000	4.000	1.139	1.139	1.139
2	5	6.681	7.500	6.667	8.106	1.122	0.998	1.213
2	6	10.850	12.000	10.000	13.702	1.106	0.922	1.263
2	7	16.017	17.500	14.000	20.790	1.093	0.874	1.298
2	8	22.184	24.000	18.667	29.370	1.082	0.841	1.323
2	9	29.351	31.500	24.000	39.442	1.073	0.818	1.343
3	4	1.254	1.333	2.000	1.251	1.063	1.595	0.998
3	5	3.042	3.333	3.750	3.196	1.096	1.233	1.051
3	6	5.490	6.000	6.000	6.000	1.093	1.093	1.093
3	7	8.602	9.333	8.750	9.670	1.085	1.017	1.124
3	8	12.381	13.333	12.000	14.207	1.077	0.969	1.147
3	9	16.826	18.000	15.750	19.612	1.070	0.936	1.166
4	5	1.214	1.250	2.000	1.202	1.029	1.647	0.990
4	6	2.808	3.000	3.600	2.866	1.069	1.282	1.021
4	7	4.894	5.250	5.600	5.129	1.073	1.444	1.048
4	8	7.479	8.000	8.000	8.000	1.070	1.070	1.070
4	9	10.563	11.250	10.800	11.479	1.065	1.022	1.087
5	6	1.190	1.200	2.00	1.176	1.008	1.680	0.988
5	7	2.667	2.800	3.500	2.687	1.050	1.312	1.008
5	8	4.536	4.800	5.333	4.658	1.058	1.176	1.027
5	9	6.804	7.200	7.500	7.095	1.058	1.102	1.043

$$\text{Eff1} = \frac{MSE(X^B(m))}{MSE(X^{MC}(m))}, \quad \text{Eff2} = \frac{MSE(X^L(m))}{MSE(X^{MC}(m))}, \quad \text{Eff3} = \frac{MSE(X^M(m))}{MSE(X^{MC}(m))}.$$

Example 9: The Uniform distribution (contd.)

Here too, like in the exponential case, we concentrate on the case when σ is unknown and compare the different predictors of $X(m)$. The MSE's of the different estimators are given by

$$\text{MSE} = \left[k^2 \left(1 - 2^{1-n} + 3^{-n} \right) + \left(1 - 2^{1-m} + 3^{-m} \right) - 2k \left\{ 2^{n-m} \left(3^{-n} - 4^{-n} \right) + \left(1 - 2^{-n} \right) \left(1 - 2^{-m} \right) \right\} \right] \sigma^2,$$

where

$$
k = \begin{cases}
1 + \frac{1}{2^n - 1} \cdot \text{Med}[U(m-n)] & \text{for modified CMP} \\[2ex]
\frac{1 - 2^{-m}}{1 - 2^{-n}} & \text{for BLUP} \\[2ex]
1 & \text{for MLP} \\[2ex]
\text{Med}\left[\frac{X(m)}{X(n)}\right] & \text{for MUP.}
\end{cases}
$$

In Table 2, we present the MSE's of the above predictors, and the relative efficiencies of the modified CMP relative to other estimators we have considered above for the uniform parent.

Table 2. MSE's and Relative Efficiencies of Predictors of $X(m)$ for the $U(0, \sigma)$ Parent.

		MSEs in terms of σ^2				Efficiency of Modified CMP		
n	m	$X^{MC}(m)$	$X^B(m)$	$X^L(m)$	$X^M(m)$	Eff1	Eff2	Eff3
2	3	0.031	0.031	0.037	0.030	1.000	1.200	0.957
2	4	0.056	0.054	0.068	0.051	0.959	1.205	0.891
2	5	0.071	0.069	0.087	0.063	0.967	1.234	0.893
2	6	0.078	0.077	0.099	0.071	0.980	1.256	0.903
2	7	0.082	0.081	0.105	0.075	0.988	1.270	0.910
2	8	0.084	0.084	0.108	0.077	0.994	1.278	0.915
2	9	0.085	0.085	0.109	0.078	0.997	1.281	0.919
3	4	0.010	0.010	0.012	0.011	1.000	1.225	1.042
3	5	0.018	0.018	0.023	0.018	0.978	1.262	0.999
3	6	0.023	0.022	0.029	0.023	0.983	1.290	0.998
3	7	0.025	0.025	0.033	0.025	0.989	1.306	1.001
3	8	0.027	0.026	0.035	0.027	0.994	1.315	1.001
3	9	0.027	0.027	0.036	0.027	0.997	1.320	0.997
4	5	0.0034	0.0034	0.0041	0.0037	1.000	1.200	1.082
4	6	0.0061	0.0060	0.0075	0.0065	0.988	1.242	1.065
4	7	0.0077	0.0076	0.0097	0.0082	0.991	1.264	1.066
4	8	0.0086	0.0085	0.0110	0.0092	0.995	1.275	1.065
4	9	0.0091	0.0091	0.0116	0.0096	0.997	1.281	1.058
5	6	0.0012	0.0012	0.0014	0.0013	1.000	1.163	1.093
5	7	0.0021	0.0021	0.0025	0.0023	0.993	1.200	1.091
5	8	0.0027	0.0027	0.0032	0.0029	0.995	1.217	1.093
5	9	0.0030	0.0030	0.0037	0.0032	0.997	1.224	1.087

Table 2 shows that the efficiency of the modified CMP is very high relative to the MLP. The efficiency compares very well relative to MUP and exceeds 1 as the values of n and m increase. It is slightly less efficient than but close to BLUP in most cases.

5. Conclusions

In this chapter, we have studied different predictors of a future record statistic based on observing a number of past records. From this study, it is evident that the modified CMP is quite easy to compute and it compares very well with all other predictors in terms of MSE. We also should note here that predictive likelihood equations usually cannot be solved to obtain closed-form expression for the MLP.

References

1. M. Ahsanullah, Linear prediction of record values for the two parameter exponential distribution, *Annals of the Institute of Statistical Mathematics*, **32**, 363-368 (1980).
2. M. Ahsanullah, *Record Statistics*. Nova Science Publishers, Commack, NY (1995).
3. B. C. Arnold, N. Balakrishnan and H. N. Nagaraja, *Records*, John Wiley & Sons, New York (1990).
4. I. R. Dunsmore, The future occurence of records, *Annals of the Institute of Statistical Mathematics*, **35**, 267-277 (1983).
5. G. Ishii, *Tokeiteki Yosoku* (Statistical Prediction), in Basic Sugaku n. **7**, Gendai-Sugakusha. Tokyo (1978).
6. K. S. Kaminsky and P. I. Nelson, Best linear unbiased prediction of order statistics in location and scale families, *Journal of the Amererican Statistical Association*, **70**, 145-150 (1975).
7. J. P. Keating, R. L. Mason, C. R. Rao P. K. Sen, Pitman's measure of closeness, *Communications in Statistics – Theory and Methods*, Special Issue, **20** (11) (1991).
8. E. L. Lehmann, *Theory of Point Estimation*, John Wiley & Sons, New York (1983).
9. V. B. Nevzorov and N. Balakrishnan, A Record of Records, In *Handbook of Statistics – Vol. 16: Order Statistics: Theory & Methods*, (eds., N. Balakrishnan and C. R. Rao), 515-570 (1999).
10. V. B. Nevzorov, *Records: Mathematical Theory*, Translations of Mathematical Monographs, American Mathematical Society, Providence Rhode Island (2001).
11. M. Z. Raqab and H. N. Nagaraja, On some predictors of future order statistics, *Metron*, **53**, 185-204 (1995).
12. G. R. Shorack and J. A. Wellner, *Empirical Processes with Applications to Statistics*, John Wiley & Sons, New York (1986).
13. Y. Takada, Relation of the best invariant prediction and the best unbiased

predictor in location and scale families, *Annals of Statistics*, **9**, 917-921 (1981)

14. Y. Takada, Median unbiasedness in an invariant prediction problem, *Statistics & Probability Letters*, **12**, 281-283 (1991).

12

DETECTING CHANGE POINTS ON DIFFERENT INFORMATION LEVELS

Tina Herberts and Uwe Jensen

Department of Stochastics, University of Ulm
89069 Ulm, Germany
E-mail: herberts@mathematik.uni-ulm.de, jensen@mathematik.uni-ulm.de

This chapter deals with the optimal detection of a change point in the intensity of a nonhomogeneous Poisson process N. We consider different information levels, namely sequential observation of N, retrospective analysis after having observed N up to some fixed time t^* and a combination of both observation schemes. The corresponding detection problems are viewed as optimal stopping problems which only differ in the filtration used, and they are solved simultaneously by means of a semimartingale approach.

1. Introduction

Problems involving the detection of a change point in the characteristics of a point process occur quite frequently in applications. For example, in quality control a production process might produce defective items more frequently from some change point on which is to be detected. In reliability theory the parameters describing the wear of a technical system may change due to different working conditions of the system. In risk theory, premium and claim processes may have changing characteristics due to different environmental conditions, and the insurance company is interested in intervening when the conditions deteriorate.

In this chapter we consider the following change point problem: Let $N = (N_t), t \in \mathbb{R}_+$, be a counting process with (natural) intensity

$$\lambda_t = \mu_0(t) I_{\{\rho > t\}} + \mu_1(t) I_{\{\rho \leq t\}},$$

where μ_0 and μ_1 are supposed to be continuous deterministic functions with $0 < \mu_0(t) \leq \mu_1(t)$ for all $t \in \mathbb{R}_+$ and where the change point ρ is assumed

177

to be exponentially distributed with parameter λ. This means that if ρ is given, N is a Poisson process with intensity $\mu_0(\cdot)$ up to ρ and it is a Poisson process with intensity $\mu_1(\cdot)$ after ρ. Here, μ_0, μ_1 and λ are assumed to be known. The case of unknown parameters is dealt with in Herberts and Jensen[5].

Now only the process N is observed, but not ρ, and the problem is to detect the unobservable change point ρ as well as possible by means of the observation of N. This means that a stopping time ζ has to be determined such that ζ maximizes the expectation of some gain process. Here we consider the following valuation structure: There is a gain of c_i per unit of time the process is in state $i \in \{0,1\}$ ("0" before ρ occurs and "1" afterwards), and the costs for stopping while the process is in state i are given by k_i, $i = 0,1$, where $c_0 \geq 0 > c_1$ and $k_0 \geq k_1 \geq 0$. If the process is stopped working at t we have the following gain:

$$Z_t = c_0(t \wedge \rho) + c_1(t - \rho)^+ - k_0(1 - Y_t) - k_1 Y_t,$$

where $Y_t = I_{\{\rho \leq t\}}$, $x \wedge y = \min\{x,y\}$ and $x^+ = \max\{x,0\}$. This gain process is a generalization of the one introduced by Shiryaev[11]. To balance the risks of a false alarm on the one hand and of detecting the change point too late on the other hand, he considered the risk

$$R_t = P(\rho > t) + cE[(t - \rho)^+]$$

for some constant $c > 0$ in a model different from ours. Choosing $c_0 = 0, c_1 = -c, k_0 = 1$ and $k_1 = 0$, we obtain $EZ_t = -R_t$. Of course, other stopping criteria are also possible and can be handled in a similar way. Now the task is to determine a stopping time ζ with respect to some given filtration $\mathbb{F}$ such that

$$EZ_\zeta = \sup\{EZ_\tau : \tau \in C^{\mathbb{F}}\},$$

i.e., ζ maximizes the expected reward in the class

$$C^{\mathbb{F}} = \{\tau : \tau \text{ is an } \mathbb{F}\text{-stopping time}, \tau < \infty, EZ_\tau > -\infty\}.$$

In the following we will call this problem the maximization problem (MAX).

Change point problems can be divided into sequential procedures and ex post analysis. When using sequential procedures, at time t one has to decide whether to stop some process or not, depending only on the information gathered up to t. In the case of ex post analysis, more information is available at time $t < t^*$, namely, one already has the full information up to some fixed time t^*. Usually, different methods are used for these two different observation schemes. Here we will introduce a general concept such that the sequential and the ex post analysis as well as a combination of both can be dealt with by the same approach. The idea is that all of these problems are optimal stopping problems, and one only has to change the filtration in order to get from one problem to the other. The filtrations we consider are the following:

(i) Sequential observation and decision: The information is described by the point process filtration $\mathbb{F}^N = (\mathcal{F}_t^N), t \in \mathbb{R}_+$, given by

$$\mathcal{F}_t^N = \sigma(N_s, 0 \le s \le t).$$

(ii) Ex post analysis: The corresponding filtration is given by $\mathbb{A} = (\mathcal{A}_t), t \in \mathbb{R}_+$, where

$$\mathcal{A}_t \equiv \mathcal{F}_{t^*}^N = \sigma(N_s, 0 \le s \le t^*) \ \forall t \in \mathbb{R}_+.$$

Actually this is a generalization of ex post analysis since here we also consider $t > t^*$, i.e., at time t^* one is not only interested in making inference about a change point in the past but also in the future.

(iii) A combination of the former two observation schemes: The point process is observed up to t^* and at t^* it is decided if a change has already taken place and when. If we conclude that no change has occurred so far, then the point process is observed sequentially from t^* on. The information is therefore described by $\mathbb{G} = (\mathcal{G}_t), t \in \mathbb{R}_+$, defined by

$$\mathcal{G}_t = \sigma(N_s, 0 \le s \le t \vee t^*),$$

where $x \vee y = \max\{x, y\}$.

The task is to solve the optimization problem (MAX) for each of these filtrations. This will be done by means of a smooth semimartingale representation of the gain process (Z_t). For a detailed exposition we refer to Herberts[4]. The case of constant intensities μ_i, $i = 0, 1$, is also dealt with in Herberts and Jensen[6].

2. Preliminaries

In order to describe the model, we introduce some notation.

Let $(\Omega, \mathcal{F}, P)$ be a complete probability space endowed with a filtration $\mathbb{F} = (\mathcal{F}_t), t \in \mathbb{R}_+$, which obeys the "usual conditions" concerning right-continuity and completeness. In the following, the term P-a.s is suppressed, i.e., relations such as $\subseteq, =$ or $\le, =$ between measurable sets and random variables, respectively, are always supposed to hold with probability one. Let $Z = (Z_t), t \in \mathbb{R}_+$, be a real right-continuous stochastic process, and let $\mathcal{M}_0^{\mathbb{F}}$ be the set of real $\mathbb{F}$-martingales M having paths which are right-continuous with left-hand limits and which start in $M_0 = 0$. Then Z is called a *smooth $\mathbb{F}$-semimartingale ($\mathbb{F}$-SSM)* if it has a decomposition

$$Z_t = Z_0 + \int_0^t f_s ds + M_t,$$

where $E|Z_0| < \infty$, $M = (M_t) \in \mathcal{M}_0^{\mathbb{F}}$ and $(f_t), t \in \mathbb{R}_+$, is a real, $\mathbb{F}$-progressively measurable process satisfying $E\left(\int_0^t |f_s|ds\right) < \infty$ for all $t \in \mathbb{R}_+$. We denote such a process by $Z = (f, M)$. This decomposition in the drift or regression part $\int_0^t f_s ds$ and an additive random fluctuation (martingale) part (satisfying $EM_t = 0$ for $t \geq 0$) can be used for solving optimal stopping problems like (MAX) where the task is to maximize the expectation EZ_τ of a smooth semimartingale Z in the class $C^{\mathbb{F}} = \{\tau : \tau \text{ is an } \mathbb{F}\text{-stopping time}, \tau < \infty, EZ_\tau > -\infty\}$. In order to determine an optimal stopping rule, one has to impose conditions on the structure of $Z = (f, M)$. If

$$\{f_t \leq 0\} \subseteq \{f_{t+h} \leq 0\} \; \forall t, h \in \mathbb{R}_+ \text{ and } \bigcup_{t \in \mathbb{R}_+} \{f_t \leq 0\} = \Omega, \qquad (1)$$

then the *monotone case* is said to hold true. The stopping time

$$\zeta = \inf\{t \in \mathbb{R}_+ : f_t \leq 0\}$$

is called *ILA-stopping rule* (infinitesimal-look-ahead).

Theorem 1: Let $Z = (f, M)$ be an $\mathbb{F}$-SSM and ζ the ILA-stopping rule. If the monotone case holds true and the martingale M is uniformly integrable, then ζ is an optimal stopping time, i.e., $EZ_\zeta = \sup\{EZ_\tau : \tau \in C^{\mathbb{F}}\}$.

For a proof, see Aven and Jensen[2].

For switching from some filtration to a subfiltration, the following projection theorem will be used (see van Schuppen[12], Brémaud[3], pp. 87, 108 and Kallianpur[8], p. 202):

Theorem 2: Let $Z = (f, M)$ be an $\mathbb{F}$-SSM and $\mathbb{H}$ a subfiltration of $\mathbb{F}$, and let the process $\hat{Z}$ be given by a cadlag version of $\hat{Z}_t = E[Z_t|\mathcal{H}_t]$. Then $\hat{Z}$ is an $\mathbb{H}$-SSM with representation

$$\hat{Z}_t = \hat{Z}_0 + \int_0^t \hat{f}_s ds + \hat{M}_t, \qquad (2)$$

where $\hat{f}$ is $\mathbb{H}$-progressively measurable with $\hat{f}_t = E[f_t|\mathcal{H}_t]$ for almost all $t \in \mathbb{R}_+$ and $\hat{M} \in \mathcal{M}_0^{\mathbb{H}}$. If in addition Z_0 and $\int_0^\infty |f_s|ds$ are square integrable and M is bounded in L^2 then the same properties are satisfied for the corresponding terms $\hat{Z}_0, \int_0^\infty |\hat{f}_s| ds$ and $\hat{M}$.

3. Optimal Detection

In this section we will use Theorem 1 in order to solve the detection problem (MAX) described in the introduction. This problem is solved for the different observation filtrations $\mathbb{F}^N$, $\mathbb{A}$ and $\mathbb{G}$ defined before, i.e., we deal with sequential observation, ex post analysis and a combination of these two observation schemes. First we will determine an SSM representation of the reward process Z. Since Z depends on ρ we have to consider its SSM representation with respect to some filtration $\mathbb{F}$ to which also ρ is adapted. In order to determine the optimal stopping time in the class $C^{\mathbb{H}}$, where $\mathbb{H} = (\mathcal{H}_t), t \in \mathbb{R}_+$, denotes one of the filtrations $\mathbb{F}^N$, $\mathbb{A}$ and $\mathbb{G}$, we will then turn to the conditional expectation $\hat{Z}_t = E[Z_t|\mathcal{H}_t]$ and derive the $\mathbb{H}$-SSM representation of $\hat{Z} = (\hat{Z}_t), t \in \mathbb{R}_+$, by means of the Projection Theorem (Theorem 2). Since each of the filtrations $\mathbb{F}^N$, $\mathbb{A}$ and $\mathbb{G}$ has to be a subfiltration of $\mathbb{F} = (\mathcal{F}_t), t \in \mathbb{R}_+$, we choose

$$\mathcal{F}_t = \sigma(Y_u, 0 \le u \le t, N_v, 0 \le v \le (t \vee t^*)),$$

where we denote $Y_t = I_{\{\rho \le t\}}$ again. For deriving an $\mathbb{F}$-SSM representation of Z we rewrite it as

$$Z_t = -k_0 + \int_0^t (c_0 + (c_1 - c_0)Y_s)ds + (k_0 - k_1)Y_t. \tag{3}$$

We will determine the $\mathbb{F}$-SSM representation of Z by finding such a representation of Y. This representation will be obtained in Theorem 4 by means of

$$D(s, h) = \frac{1}{h} E[Y_{s+h} - Y_s|\mathcal{F}_s].$$

It will turn out that for calculating $D(s, h)$ we will need the conditional expectations $\hat{Y}_t^{\mathbb{F}^N} = P(\rho \le t|\mathcal{F}_t^N)$ and $\hat{Y}_t^{\mathbb{A}} = P(\rho \le t|\mathcal{F}_{t^*}^N)$ of ρ which are an update of the prior exponential distribution given the available information. Formulas for $\hat{Y}^{\mathbb{F}^N}$ and $\hat{Y}^{\mathbb{A}}$ as well as the likelihood function of the counting process N, which is needed for these formulas, are derived in the next lemma. In the following we denote $\bar{\mu}_0(k) = \prod_{j=1}^k \mu_0(T_j)$ and $\bar{\mu}_1(k, l) = \prod_{j=k+1}^l \mu_1(T_j)$, where we set $\bar{\mu}_1(k, l) = 1$ for $k \ge l$.

Lemma 3:

(i) The likelihood function of $(N_s), 0 \leq s \leq t$, is given by

$$L_N(t) = \int_0^t \bar{\mu}_0(N_s) e^{-\int_0^s \mu_0(u)du} \bar{\mu}_1(N_s, N_t) e^{-\int_s^t \mu_1(u)du} \lambda e^{-\lambda s} ds$$

$$+ \bar{\mu}_0(N_t) \exp\left\{ -\int_0^t \mu_0(u)du - \lambda t \right\}.$$

(ii) $\hat{Y}^{\mathbb{A}}$ has the $\mathbb{A}$-SSM representation

$$\hat{Y}_t^{\mathbb{A}} = \int_0^t \hat{g}_s^{\mathbb{A}} ds,$$

where

$$\hat{g}_s^{\mathbb{A}} = \begin{cases} \frac{1}{L_N(t^*)} \bar{\mu}_0(N_s) \bar{\mu}_1(N_s, N_{t^*}) e^{-(\int_0^s \mu_0(u)du + \int_s^{t^*} \mu_1(u)du)} \lambda e^{-\lambda s}, & s \leq t^*, \\ \frac{1}{L_N(t^*)} \bar{\mu}_0(N_{t^*}) e^{-\int_0^{t^*} \mu_0(u)du} \lambda e^{-\lambda s}, & s > t^*. \end{cases}$$

(iii) For $\hat{Y}^{\mathbb{F}^N}$ the following formula holds:

$$\hat{Y}_t^{\mathbb{F}^N} = 1 - \frac{\bar{\mu}_0(N_t) \exp\{ -\int_0^t \mu_0(s)ds - \lambda t \}}{L_N(t)}, \quad t \in \mathbb{R}_+.$$

Proof: (i) Given $\rho = u$, (N_s) is a Poisson process with intensity $\mu_0(s)$ for $s < u$, and $(\bar{N}_s) = (N_{s+u} - N_u)$, $s \in \mathbb{R}_+$, is a Poisson process with intensity $\mu_1(s + u)$, $s \in \mathbb{R}_+$. Therefore the likelihood function of $(N_s), 0 \leq s \leq t$, can be determined by means of the likelihood functions of Poisson processes with rate $\mu_0(\cdot)$ and $\mu_1(\cdot)$, respectively. The likelihood function of a Poisson process $\tilde{N}$ with intensity $\alpha(\cdot)$ and $\tilde{N}_t = \sum_{j=1}^{\infty} I_{\{\tilde{T}_j \leq t\}}$, based on the observation of this process up to t, is proportional to

$$L_{\tilde{N}}(t) = \exp\left\{ \int_0^t \log \alpha(s) d\tilde{N}_s - \int_0^t \alpha(s) ds \right\} = \left(\prod_{j=1}^{\tilde{N}_t} \alpha(\tilde{T}_j) \right) e^{-\int_0^t \alpha(s)ds}$$

(see Liptser and Shiryaev[9], Theorem 19.7). Thus the likelihood function of $(N_s), 0 \leq s \leq t$, is given by

$$L_N(t) = \int_0^t L_N(t|\rho = s) \lambda e^{-\lambda s} ds + L_N(t|\rho > t) P(\rho > t)$$

$$= \int_0^t \bar{\mu}_0(N_s) e^{-\int_0^s \mu_0(u)du} \bar{\mu}_1(N_s, N_t) e^{-\int_s^t \mu_1(u)du} \lambda e^{-\lambda s} ds$$

$$+ \bar{\mu}_0(N_t) \exp\left\{ -\int_0^t \mu_0(u)du \right\} e^{-\lambda t}.$$

(ii) Let

$$F_t(B) = \frac{1}{L_N(t)}\left[\int_B I_{[0,t]}(u)L_N(t|\rho = u)\lambda e^{-\lambda u}du\right.$$
$$\left. + L_N(t|\rho > t)\int_t^\infty \lambda e^{-\lambda u}du \cdot \varepsilon_\infty(B)\right]$$

for any Borel set $B \subseteq \bar{\mathbb{R}}_+$, where ε_∞ is the unit mass at ∞. Similarly to the proof of Proposition 1 in Arjas et al.[1], which deals with the case of constant intensities μ_i, $i = 0, 1$, one can show that for time-varying intensities,

$$P(\rho \leq t|\mathcal{F}_{t^*}^N) = F_{t^*}([0,t])$$
$$= \int_0^t \frac{1}{L_N(t^*)}\bar{\mu}_0(N_s)e^{-\int_0^s \mu_0(u)du}\bar{\mu}_1(N_s, N_{t^*})e^{-\int_s^{t^*} \mu_1(u)du}\lambda e^{-\lambda s}ds$$

for $t \leq t^*$. In particular this yields

$$P(\rho \leq t^*|\mathcal{F}_{t^*}^N) = 1 - \frac{\bar{\mu}_0(N_{t^*})\exp\{-\int_0^{t^*}\mu_0(u)du - \lambda t^*\}}{L_N(t^*)}.$$

For $t > t^*$ we have

$$P(\rho \leq t|\mathcal{F}_{t^*}^N) = 1 - P(\rho > t|\rho > t^*, N_v, 0 \leq v \leq t^*)P(\rho > t^*|\mathcal{F}_{t^*}^N)$$
$$= 1 - P(\rho > t|\rho > t^*)P(\rho > t^*|\mathcal{F}_{t^*}^N)$$
$$= 1 - e^{-\lambda(t-t^*)}\frac{\bar{\mu}_0(N_{t^*})\exp\{-\int_0^{t^*}\mu_0(u)du - \lambda t^*\}}{L_N(t^*)}$$

and therefore $\hat{g}_s^{\mathbb{A}} = \frac{1}{L_N(t^*)}\bar{\mu}_0(N_{t^*})\lambda\exp\{-\int_0^{t^*}\mu_0(u)du - \lambda s\}$ for $s > t^*$.
(iii) The claim follows by substituting $t^* = t$ in (ii). $\qquad\square$

In the next theorem the $\mathbb{F}$-SSM representation of the indicator process $Y_t = I_{\{\rho \leq t\}}$ is given.

Theorem 4: Let

$$h_s = \begin{cases} (L_N(t^*))^{-1}\bar{\mu}_0(N_s)\bar{\mu}_1(N_s, N_{t^*})e^{-(\int_0^s \mu_0(u)du + \int_s^{t^*}\mu_1(u)du)}\lambda e^{-\lambda s}, & s \leq t^*, \\ (L_N(s))^{-1}\bar{\mu}_0(N_s)e^{-\int_0^s \mu_0(u)du}\lambda e^{-\lambda s}, & s > t^*. \end{cases}$$

$$(4)$$

Then the $\mathbb{F}$-SSM representation of Y is given by

$$Y_t = \int_0^t g_s^{\mathbb{F}}ds + m_t \quad \text{with} \quad g_s^{\mathbb{F}} = \frac{h_s}{P(\rho > s|\mathcal{G}_s)}I_{\{\rho > s\}},$$

where the $\mathbb{F}$-martingale $(m_t), t \in \mathbb{R}_+$, is bounded in L^2.

Proof: In Aven and Jensen[2] (Theorem 9, p. 51) it is shown that under suitable conditions, $g_s^{\mathbb{F}}$ can be derived as the limit $g_s^{\mathbb{F}} = \lim_{h\downarrow 0} D(s,h)$, where $D(s,h) = \frac{1}{h} E[Y_{s+h} - Y_s | \mathcal{F}_s]$. We have

$$
\begin{aligned}
D(s,h) &= \frac{1}{h} P(s < \rho \leq s + h | I_{\{\rho \leq u\}}, 0 \leq u \leq s, N_v, 0 \leq v \leq (s \vee t^*)) \\
&= \frac{1}{h} P(s < \rho \leq s + h | \rho > s, N_v, 0 \leq v \leq (s \vee t^*)) I_{\{\rho > s\}} \\
&= \frac{1}{h} \frac{P(s < \rho \leq s + h | N_v, 0 \leq v \leq (s \vee t^*))}{P(\rho > s | N_v, 0 \leq v \leq (s \vee t^*))} I_{\{\rho > s\}}.
\end{aligned}
$$

Furthermore, Lemma 3 (ii) implies

$$
\begin{aligned}
h_s &= \lim_{h\downarrow 0} \frac{1}{h} P(s < \rho \leq s + h | N_v, 0 \leq v \leq (s \vee t^*)) \\
&= \begin{cases} \frac{1}{L_N(t^*)} \bar{\mu}_0(N_s) \bar{\mu}_1(N_s, N_{t^*}) e^{-(\int_0^s \mu_0(u)du + \int_s^{t^*} \mu_1(u)du)} \lambda e^{-\lambda s}, & s \leq t^*, \\ \frac{1}{L_N(s)} \bar{\mu}_0(N_s) e^{-\int_0^s \mu_0(u)du} \lambda e^{-\lambda s}, & s > t^*. \end{cases}
\end{aligned}
$$

In Herberts[4] it is shown that the conditions concerning measurability of $g^{\mathbb{F}}$ and integrability of $D(s,h)$, which are stated in Theorem 9 in Aven and Jensen[2], are satisfied. This yields the SSM representation given in Theorem 4. The claim that $(m_t), t \in \mathbb{R}_+$, is bounded in L^2 is shown in Métivier[10] (p. 113). This completes the proof. $\qquad\square$

Remark 5:

(i) A formula for $P(\rho \leq t | \mathcal{G}_t)$ is given by Lemma 3 (ii) and (iii).

(ii) Note that the process $h = (h_s), s \in \mathbb{R}_+$, defined by (4) is adapted to $\mathbb{G}$. We will use this fact when considering conditional expectations. We have

$$
h_s = \begin{cases} \hat{g}_s^{\mathbb{A}} & \text{for } s \leq t^*, \\ \lambda(1 - \hat{Y}_s^{\mathbb{F}^N}) & \text{for } s > t^*, \end{cases} \tag{5}
$$

where $\hat{g}^{\mathbb{A}}$ is given in Lemma 3 (ii) and the formula for $s > t^*$ is implied by Lemma 3 (iii).

(iii) As one would expect, in the case of complete information Theorem 1 together with the SSM representation in Theorem 4 yields the optimal stopping time $\zeta^{\mathbb{F}} = \inf\{t \in \mathbb{R}_+ : g_t^{\mathbb{F}} \leq 0\} = \rho$.

Substituting the $\mathbb{F}$-SSM representation of Y into (3) we obtain the following $\mathbb{F}$-SSM representation of Z:

$$
Z_t = Z_0 + \int_0^t f_s ds + M_t, \tag{6}
$$

where

$$f_s = c_0 + (c_1 - c_0)Y_s + (k_0 - k_1)\frac{h_s}{P(\rho > s|\mathcal{G}_s)}I_{\{\rho > s\}}$$

and $M = (k_0 - k_1)m$ is bounded in L^2.

Using this decomposition of Z we will solve the optimization problem (MAX) for the different observation levels in the next subsections.

3.1. *Sequential Observation*

In this subsection we consider the problem of determining $\zeta^{\mathbb{F}^N} \in C^{\mathbb{F}^N}$ such that $EZ_{\zeta^{\mathbb{F}^N}} = \sup\{EZ_\tau : \tau \in C^{\mathbb{F}^N}\}$, where $\mathbb{F}^N = (\mathcal{F}_t^N), t \in \mathbb{R}_+$, is given by $\mathcal{F}_t^N = \sigma(N_s, 0 \leq s \leq t)$. Since ρ cannot be observed, Z is not adapted to $\mathbb{F}^N$ and therefore a projection to the observation filtration $\mathbb{F}^N$ is needed. Thus we consider the conditional expectation $\hat{Z}_t^{\mathbb{F}^N} = E[Z_t|\mathcal{F}_t^N]$. Since $EZ_\tau = E\hat{Z}_\tau^{\mathbb{F}^N}$ for all $\tau \in C^{\mathbb{F}^N}$ the optimization problem (MAX) is equivalent to maximizing $E\hat{Z}_\tau^{\mathbb{F}^N}$ in $C^{\mathbb{F}^N}$. This problem will be solved by means of Theorem 1 for which we need the $\mathbb{F}^N$-SSM representation of $\hat{Z}^{\mathbb{F}^N}$. This representation can be determined by applying Theorem 2 to equation (6) since $\mathbb{F}$ was chosen such that $\mathbb{F}^N$ is a subfiltration of $\mathbb{F}$. We obtain

$$\hat{Z}_t^{\mathbb{F}^N} = \hat{Z}_0^{\mathbb{F}^N} + \int_0^t \hat{f}_s^{\mathbb{F}^N}\, ds + \hat{M}_t^{\mathbb{F}^N},$$

where

$$\hat{f}_s^{\mathbb{F}^N} = E[f_s|\mathcal{F}_s^N]$$
$$= c_0 + (c_1 - c_0)\hat{Y}_s^{\mathbb{F}^N} + (k_0 - k_1)E\left[E\left[\frac{h_s}{P(\rho > s|\mathcal{G}_s)}I_{\{\rho > s\}}\,\Big|\,\mathcal{G}_s\right]\Big|\mathcal{F}_s^N\right]$$
$$= c_0 + (c_1 - c_0)\hat{Y}_s^{\mathbb{F}^N} + (k_0 - k_1)E[h_s|\mathcal{F}_s^N]$$

since h is adapted to $\mathbb{G}$. Some calculations carried out in detail in Herberts[4] show that $E[h_s|\mathcal{F}_s^N] = \lambda(1 - \hat{Y}_s^{\mathbb{F}^N})$. This yields

$$\hat{f}_s^{\mathbb{F}^N} = c_0 + \lambda(k_0 - k_1) - \hat{Y}_s^{\mathbb{F}^N}(c_0 - c_1 + \lambda(k_0 - k_1)). \tag{7}$$

Therefore the ILA-stopping rule $\zeta^{\mathbb{F}^N}$ is given by

$$\zeta^{\mathbb{F}^N} = \inf\{t \in \mathbb{R}_+ : \hat{f}_t^{\mathbb{F}^N} \leq 0\} = \inf\{t \in \mathbb{R}_+ : \hat{Y}_t^{\mathbb{F}^N} \geq z^*\} \text{ with}$$
$$z^* = \frac{c_0 + \lambda(k_0 - k_1)}{c_0 - c_1 + \lambda(k_0 - k_1)}.$$

For establishing the monotone case we will use a recursive formula for $\hat{Y}^{\mathbb{F}^N}$.

Lemma 6: $\hat{Y}^{\mathbb{F}^N}$ *satisfies*

$$\hat{Y}^{\mathbb{F}^N}_{T_0} = \hat{Y}^{\mathbb{F}^N}_0 = 0, \quad \hat{Y}^{\mathbb{F}^N}_{T_j} = \frac{\mu_1(T_j)\hat{Y}^{\mathbb{F}^N}_{T_j-}}{\mu_0(T_j) + (\mu_1(T_j) - \mu_0(T_j))\hat{Y}^{\mathbb{F}^N}_{T_j-}},$$

and for $T_j < t < T_{j+1}$:

$$\hat{Y}^{\mathbb{F}^N}_t = \hat{Y}^{\mathbb{F}^N}_{T_j} + \int_{T_j}^{t} (\lambda - (\mu_1(s) - \mu_0(s))\hat{Y}^{\mathbb{F}^N}_s)(1 - \hat{Y}^{\mathbb{F}^N}_s)ds. \tag{8}$$

Proof: The recursive formulas are derived analogously to the case of constant intensities μ_0 and μ_1 dealt with in Brémaud[3] (pp. 94-96). $\qquad\square$

Theorem 7: If

$$(\mu_1(t) - \mu_0(t))z^* \leq \lambda \ \forall t \in \mathbb{R}_+ \tag{9}$$

then $\zeta^{\mathbb{F}^N} = \inf\{t \in \mathbb{R}_+ : \hat{Y}^{\mathbb{F}^N}_t \geq z^*\}$ satisfies

$$EZ_{\zeta^{\mathbb{F}^N}} = \sup\{EZ_\tau : \tau \in C^{\mathbb{F}^N}\}.$$

Proof: First we establish the monotone case (1), then we show that $\hat{M}^{\mathbb{F}^N}$ is uniformly integrable. By Lemma 6 the jump size of $\hat{Y}^{\mathbb{F}^N}$ at T_j is given by

$$\hat{Y}^{\mathbb{F}^N}_{T_j} - \hat{Y}^{\mathbb{F}^N}_{T_j-} = \frac{(\mu_1(T_j) - \mu_0(T_j))\hat{Y}^{\mathbb{F}^N}_{T_j-}(1 - \hat{Y}^{\mathbb{F}^N}_{T_j-})}{\mu_0(T_j) + (\mu_1(T_j) - \mu_0(T_j))\hat{Y}^{\mathbb{F}^N}_{T_j-}} > 0,$$

therefore the jump size of $\hat{f}^{\mathbb{F}^N}$ is negative since $c_0 > c_1$ and $k_0 \geq k_1$. For $T_j < t < T_{j+1}$ we have

$$\frac{d}{dt}\hat{Y}^{\mathbb{F}^N}_t = (\lambda - (\mu_1(t) - \mu_0(t))\hat{Y}^{\mathbb{F}^N}_t)(1 - \hat{Y}^{\mathbb{F}^N}_t)$$

by (8), therefore $\hat{Y}^{\mathbb{F}^N}$ has increasing paths as long as

$$(\mu_1(t) - \mu_0(t))\hat{Y}^{\mathbb{F}^N}_t \leq \lambda.$$

Since we assume $(\mu_1(t) - \mu_0(t))z^* \leq \lambda$ for all t and because we furthermore have

$$\hat{f}^{\mathbb{F}^N}_t = c_0 + \lambda(k_0 - k_1) - \hat{Y}^{\mathbb{F}^N}_t(c_0 - c_1 + \lambda(k_0 - k_1)) \to c_1 < 0$$

as $t \to \infty$, the monotone case is satisfied for $\hat{f}^{\mathbb{F}^N}$.

The uniform integrability of $\hat{M}^{\mathbb{F}^N}$ is established by showing that $\hat{M}^{\mathbb{F}^N}$ is bounded in L^2, which is done by means of Theorem 2. We have $EZ_0^2 =$

$k_0^2 < \infty$, and the martingale part M in the $\mathbb{F}$-SSM representation (6) of Z is bounded in L^2. Instead of showing $E[(\int_0^\infty |f_s|ds)^2] < \infty$ it suffices to prove $E[(\int_0^\infty |\tilde{f}_s|ds)^2] < \infty$ for

$$\tilde{f}_s = f_s - c_1 = \left((c_0 - c_1) + (k_0 - k_1)\frac{h_s}{P(\rho > s|\mathcal{G}_s)} \right) I_{\{\rho > s\}}$$

since the martingale part in the SSM representation of

$$\hat{Z}_t^{\mathbb{F}^N} - c_1 t = E[Z_t - c_1 t | \mathcal{F}_t^N] = Z_0 + \int_0^t E[\tilde{f}_s | \mathcal{F}_s^N]ds + \hat{M}_t$$

is the same as the one of $\hat{Z}_t^{\mathbb{F}^N}$. The square integrability of $\int_0^\infty |\tilde{f}_s|ds$ can be shown by splitting the domain of integration into $[0, t^*]$, where $h_s = \hat{g}_s^{\mathbb{A}}$, and $[t^*, \infty)$, where $h_s/P(\rho > s|\mathcal{G}_s) = \lambda$. The details are given in Herberts[4].

The result of Theorem 7 is now a consequence of Theorem 1. $\square$

Remark 8: A similar result was obtained in Jensen and Hsu[7] for the case of constant intensities μ_i, $i = 0, 1$.

3.2. *Ex Post Analysis*

Here we consider the case that the counting process N has been observed up to the fixed time t^* and now the position of the change point has to be determined (in particular, it has to be decided if $\rho(\omega) \leq t^*$ or $\rho(\omega) > t^*$). The corresponding filtration $\mathbb{A} = (\mathcal{A}_t), t \in \mathbb{R}_+$, is given by $\mathcal{A}_t \equiv \mathcal{F}_{t^*}^N = \sigma(N_s, 0 \leq s \leq t^*)$ for all $t \in \mathbb{R}_+$.

The optimal stopping time $\zeta^{\mathbb{A}}$ is determined by means of the $\mathbb{A}$-SSM representation of $\hat{Z}_t^{\mathbb{A}} = E[Z_t | \mathcal{A}_t]$. Applying the Projection Theorem (Theorem 2) to the $\mathbb{F}$-SSM representation (6) of Z yields

$$\hat{Z}_t^{\mathbb{A}} = \hat{Z}_0^{\mathbb{A}} + \int_0^t \hat{f}_s^{\mathbb{A}}ds + \hat{M}_t^{\mathbb{A}}, \tag{10}$$

where $\hat{f}_s^{\mathbb{A}} = E[f_s|\mathcal{A}_s]$ and $\hat{M}^{\mathbb{A}}$ is an $\mathbb{A}$-martingale with $\hat{M}_0^{\mathbb{A}} = 0$. Since $\mathcal{A}_t \equiv \mathcal{A}_{t^*}$ for all $t \in \mathbb{R}_+$ we have

$$\hat{M}_t^{\mathbb{A}} = E[\hat{M}_t^{\mathbb{A}}|\mathcal{A}_t] = E[\hat{M}_t^{\mathbb{A}}|\mathcal{A}_0] = \hat{M}_0^{\mathbb{A}} = 0$$

for all $t \in \mathbb{R}_+$ and therefore

$$\hat{Z}_t^{\mathbb{A}} = \hat{Z}_0^{\mathbb{A}} + \int_0^t \hat{f}_s^{\mathbb{A}}ds.$$

For $\hat{f}^{\mathbb{A}}$ we obtain

$$\hat{f}_s^{\mathbb{A}} = E\left[c_0 + (c_1 - c_0)Y_s + (k_0 - k_1)\frac{h_s}{P(\rho > s|\mathcal{G}_s)}I_{\{\rho > s\}} \middle| \mathcal{A}_s \right]$$

$$= c_0 + (c_1 - c_0)\hat{Y}_s^{\mathbb{A}} + (k_0 - k_1)E\left[E\left[\frac{h_s}{P(\rho > s|\mathcal{G}_s)}I_{\{\rho > s\}} \middle| \mathcal{G}_s \right] \middle| \mathcal{F}_{t^*}^N \right]$$

$$= c_0 + (c_1 - c_0)\hat{Y}_s^{\mathbb{A}} + (k_0 - k_1)E[h_s|\mathcal{F}_{t^*}^N]$$

$$= c_0 + (c_1 - c_0)\hat{Y}_s^{\mathbb{A}} + (k_0 - k_1)\hat{g}_s^{\mathbb{A}} \text{ (see Herberts}[4]).$$

In the case of ex post analysis we can even maximize $\hat{Z}^{\mathbb{A}}$ pathwise instead of just $E\hat{Z}^{\mathbb{A}}$ since the martingale part in the $\mathbb{A}$-SSM representation of $\hat{Z}^{\mathbb{A}}$ vanishes. Furthermore, here we can do without the monotone case because at each time t one already has the full information about $(N_s), 0 \leq s \leq t^*$, and therefore $\hat{f}_s^{\mathbb{A}}$ is known for all $s \in \mathbb{R}_+$, i.e., $\hat{f}_s^{\mathbb{A}}$ is $\mathcal{A}_0$-measurable for all $s \in \mathbb{R}_+$. Thus the global maximum of $\hat{Z}^{\mathbb{A}}$ can be obtained pathwise analogously to the case of a deterministic function.

In the following we nevertheless examine the monotone case for $\hat{f}^{\mathbb{A}}$ since we will need it for the next subsection (the combination of ex post and sequential analysis) and also in order to be able to give an explicit solution of the optimization problem (MAX) and in order to compare the conditions needed in the ex post analysis with the ones of the sequential analysis. The next theorem states conditions under which $\hat{f}^{\mathbb{A}}$ satisfies the monotone case and the ILA-stopping rule solves the maximization problem:

Theorem 9: If

$$(k_0 - k_1)(\mu_1(t) - \mu_0(t) - \lambda) \leq c_0 - c_1 \ \forall t \in \mathbb{R}_+ \tag{11}$$

then $\zeta^{\mathbb{A}} = \inf\{t \in \mathbb{R}_+ : \hat{f}_t^{\mathbb{A}} \leq 0\}$ satisfies $EZ_{\zeta^{\mathbb{A}}} = \sup\{EZ_\tau : \tau \in C^{\mathbb{A}}\}$.

Proof: In order to apply Theorem 1 we have to show that $\hat{f}^{\mathbb{A}}$ satisfies the monotone case. The formula for $\hat{g}^{\mathbb{A}}$ given in Lemma 3 (ii) shows that $\hat{g}_s^{\mathbb{A}}$ is differentiable on $(T_k(\omega), T_{k+1}(\omega))$, $k \in \mathbb{N}$, and on (t^*, ∞) with

$$\frac{d}{ds}\hat{g}_s^{\mathbb{A}} = \begin{cases} (\mu_1(s) - \mu_0(s) - \lambda)\hat{g}_s^{\mathbb{A}} & \text{for } s \leq t^*, \\ -\lambda\hat{g}_s^{\mathbb{A}} & \text{for } s > t^*. \end{cases}$$

Since

$$\hat{Y}_t^{\mathbb{A}} = \int_0^t \hat{g}_s^{\mathbb{A}} ds,$$

this yields for $T_k(\omega) < s < T_{k+1}(\omega)$, $s \leq t^*$:

$$\frac{d}{ds}\hat{f}_s^{\mathbb{A}} = (c_1 - c_0 + (k_0 - k_1)(\mu_1(s) - \mu_0(s) - \lambda))\hat{g}_s^{\mathbb{A}}.$$

Hence, $\hat{f}^{\mathbb{A}}$ is decreasing between its jump points if and only if

$$(k_0 - k_1)(\mu_1(s) - \mu_0(s) - \lambda) \le c_0 - c_1$$

for all $s \in \mathbb{R}_+$. Because of the continuity of $\hat{Y}^{\mathbb{A}}$ we obtain the jump size

$$\hat{f}^{\mathbb{A}}_{T_k} - \hat{f}^{\mathbb{A}}_{T_k-} = (k_0 - k_1)\frac{1}{L_N(t^*)}\bar{\mu}_0(k-1)\bar{\mu}_1(k, N_{t^*})(\mu_0(T_k) - \mu_1(T_k))$$

$$\cdot \exp\left\{-\int_0^{T_k}\mu_0(u)du - \int_{T_k}^{t^*}\mu_1(u)du\right\}\lambda e^{-\lambda T_k}$$

which is nonpositive under our condition $\mu_0(t) \le \mu_1(t)$ for all $t \in \mathbb{R}_+$.

For $s > t^*$ we have

$$\frac{d}{ds}\hat{f}^{\mathbb{A}}_s = (c_1 - c_0 - \lambda(k_0 - k_1))\hat{g}^{\mathbb{A}}_s$$

which is nonpositive for $k_0 \ge k_1$ and $c_0 > c_1$. Furthermore, $\lim_{s\to\infty}\hat{f}^{\mathbb{A}}_s = c_1 < 0$. Therefore condition (11) guarantees that $\hat{f}^{\mathbb{A}}$ satisfies the monotone case, and Theorem 1 yields the result of Theorem 9. $\square$

Remark 10: Condition (11) is a weaker assumption than condition (9) needed for establishing the monotone case when sequentially observing the process N since

$$(\mu_1(t) - \mu_0(t))z^* \le \lambda \; \forall t \in \mathbb{R}_+$$

$$\Leftrightarrow \lambda(k_0 - k_1)(\mu_1(t) - \mu_0(t) - \lambda) + c_0(\mu_1(t) - \mu_0(t)) \le \lambda(c_0 - c_1) \; \forall t \in \mathbb{R}_+.$$

Since $c_0(\mu_1(t) - \mu_0(t)) \ge 0$ for all $t \in \mathbb{R}_+$ this yields (11).

3.3. *Combination of Ex Post and Sequential Analysis*

In this subsection we combine the last two approaches and consider the filtration $\mathbb{G}$ given by $\mathcal{G}_t = \sigma(N_s, 0 \le s \le t \vee t^*)$. Equation (6) yields the following $\mathbb{G}$-SSM representation of $\hat{Z}^{\mathbb{G}}_t = E[Z_t|\mathcal{G}_t]$:

$$\hat{Z}^{\mathbb{G}}_t = \hat{Z}^{\mathbb{G}}_0 + \int_0^t \hat{f}^{\mathbb{G}}_s ds + \hat{M}^{\mathbb{G}}_t, \tag{12}$$

where $\hat{M}^{\mathbb{G}}$ is a $\mathbb{G}$-martingale and $\hat{f}^{\mathbb{G}}$ is given by

$$\hat{f}^{\mathbb{G}}_s = E[f_s|\mathcal{G}_s]$$

$$= c_0 + (c_1 - c_0)\hat{Y}^{\mathbb{G}}_s + (k_0 - k_1)E\left[\frac{h_s}{P(\rho > s|\mathcal{G}_s)}I_{\{\rho>s\}}\Big|\mathcal{G}_s\right]$$

$$= c_0 + (c_1 - c_0)\hat{Y}^{\mathbb{G}}_s + (k_0 - k_1)h_s.$$

Using this representation, the optimal stopping time $\zeta^{\mathbb{G}}$ can be determined by means of Theorem 1. Here the monotone case has to be assumed again since at time t one cannot anticipate N_s and therefore $\hat{f}_s^{\mathbb{G}}$ for $s > t \vee t^*$.

Theorem 11: If

$$(\mu_1(t) - \mu_0(t)) \frac{c_0 + \lambda(k_0 - k_1)}{c_0 - c_1 + \lambda(k_0 - k_1)} \leq \lambda \ \forall t \in \mathbb{R}_+ \tag{13}$$

then $\zeta^{\mathbb{G}} = \inf\{t \in \mathbb{R}_+ : \hat{f}_t^{\mathbb{G}} \leq 0\}$ satisfies $EZ_{\zeta^{\mathbb{G}}} = \sup\{EZ_\tau : \tau \in C^{\mathbb{G}}\}$.

Remark 12: Here we have the same assumption as in the sequential case.

Proof: We have $\hat{f}_s^{\mathbb{G}} = \hat{f}_s^{\mathbb{A}}$ for $s \leq t^*$, $\hat{f}_s^{\mathbb{G}} = \hat{f}_s^{\mathbb{F}^N}$ for $s > t^*$ and $\hat{f}_{t^*}^{\mathbb{A}} = \hat{f}_{t^*}^{\mathbb{F}^N}$. Condition (13) equals condition (9) which guarantees that $\hat{f}^{\mathbb{F}^N}$ satisfies the monotone case. By Remark 10, condition (11), under which $\hat{f}^{\mathbb{A}}$ is decreasing, is implied by (13). Therefore, $\hat{f}^{\mathbb{G}}$ satisfies the monotone case if condition (13) holds. The uniform integrability of $\hat{M}^{\mathbb{G}}$ is implied by Theorem 2 since Z_0 and $\int_0^\infty |\tilde{f}_s| ds$ are square integrable and the martingale M in (6) is bounded in L^2 (see the proof of Theorem 7). Therefore the result of Theorem 11 is a consequence of Theorem 1. $\square$

The following examples illustrate the result of Theorem 11.

Example 13: For $k_0 = k_1$, the optimal stopping time is given by the $c_0/(c_0 - c_1)$-"quantile"

$$\zeta^{\mathbb{G}} = \inf\left\{t \in \mathbb{R}_+ : P(\rho \leq t|\mathcal{G}_t) \geq \frac{c_0}{c_0 - c_1}\right\}$$

of $\hat{Y}^{\mathbb{G}}$ if

$$(\mu_1(t) - \mu_0(t)) \frac{c_0}{c_0 - c_1} \leq \lambda \ \forall t \in \mathbb{R}_+.$$

In particular, in the symmetric case of $c_1 = -c_0$ the optimal stopping time is just the "median" of $\hat{Y}^{\mathbb{G}}$.

Example 14: If we choose $c_0 = 0, c_1 = -c, k_0 = 1$ and $k_1 = 0$ for some constant $c > 0$ then we obtain the gain function

$$Z_t = -I_{\{\rho > t\}} - c(t - \rho)^+$$

which was introduced by Shiryaev[11]. In this case we have $\hat{f}_s^{\mathbb{G}} = h_s - c\hat{Y}_s^{\mathbb{G}}$, and we obtain the following optimality result: If $\mu_1(t) - \mu_0(t) - \lambda \leq c$ for all $t \in \mathbb{R}_+$ then the ILA-stopping rule $\zeta^{\mathbb{G}} = \inf\{t \in \mathbb{R}_+ : c\hat{Y}_t^{\mathbb{G}} \geq h_t\}$ satisfies $EZ_{\zeta^{\mathbb{G}}} = \sup\{EZ_\tau : \tau \in C^{\mathbb{G}}\}$.

Example 15: If $\mu_0 \equiv \mu_1$ then condition (13) is always satisfied. In this case, the observation of N does not provide any information about the change point ρ, and $\hat{Y}^{\mathbb{G}}$ is just the prior distribution function of ρ and h its density. This implies

$$\hat{f}_t^{\mathbb{G}} = c_0 + (c_1 - c_0)(1 - e^{-\lambda t}) + (k_0 - k_1)\lambda e^{-\lambda t} = 0$$

$$\Leftrightarrow t = -\frac{1}{\lambda} \ln\left(-\frac{c_1}{c_0 - c_1 + \lambda(k_0 - k_1)}\right).$$

This is the same result one obtains when maximizing EZ_t for t deterministic, i.e., it corresponds to the case of no information about the failure times T_j. For $\mu_0 \equiv \mu_1$ and $\lambda \downarrow 0$ we obtain $\zeta^{\mathbb{G}} \to \infty$, and $\lambda \to \infty$ implies $\zeta^{\mathbb{G}} \to 0$.

The model dealt with in this chapter can also be generalized to a time-varying hazard rate $\lambda(\cdot)$ instead of the constant λ; see Herberts[4].

References

1. E. Arjas, P. Haara and I. Norros: Filtering the histories of a partially observed marked point process. *Stochastic Processes Appl.* **40**, No. 2, 225-250 (1992).
2. T. Aven and U. Jensen: *Stochastic Models in Reliability* (Springer, New York, 1999).
3. P. Brémaud: *Point Processes and Queues. Martingale Dynamics* (Springer, New York, 1981).
4. T. Herberts: Optimal stopping with estimated parameters in burn-in and detection models. *PhD Thesis, University of Ulm* (2002).
5. T. Herberts and U. Jensen: Rates of convergence of estimated change points with applications in reliability. *Calcutta Statist. Assoc. Bull.* **52**, 117-141 (2002)
6. T. Herberts and U. Jensen: Optimal detection of a change point in a point process for different observation schemes. *Research Report, University of Ulm* (2002).
7. U. Jensen and G. Hsu: Optimal stopping by means of point process observations with applications in reliability. *Mathematics of Operations Research* **18** (3), 645-657 (1993).
8. G. Kallianpur: *Stochastic Filtering Theory* (Springer, New York, 1980).
9. R.S. Liptser and A.N. Shiryaev: *Statistics of Random Processes II: Applications* (Springer, New York, 1978).
10. M. Métivier: *Semimartingales: A Course on Stochastic Processes* (De Gruyter, Berlin, 1982).
11. A.N. Shiryaev: On optimum methods in quickest detection problems. *Theory Probab. Appl.* **8**, 22-46 (1963).
12. J. van Schuppen: Filtering, prediction and smoothing observations, a martingale approach. *SIAM J. Appl. Math.* **32**, 552-570 (1977).

13

THE RELIABILITY OF QUADRATIC DYNAMIC SYSTEMS

Arvid Naess

Department of Mathematical Sciences
Norwegian University of Science and Technology
A. Getz vei 1, NO-7491 Trondheim, Norway
E-mail: arvidn@math.ntnu.no

A key quantity for the assessment of the reliability of a dynamic system subjected to a stochastic load process is the mean level crossing rate of the response. This chapter discusses a new method for calculating this quantity for a stochastic response process represented as a second order stochastic Volterra series. The derivation of this procedure consists of three stages. First the expression for the mean crossing rate is rewritten in terms of a joint characteristic function. Secondly, it is shown that a closed form expression for this joint characteristic function can be derived. Thirdly, it is then demonstrated how the method of steepest descent can be applied to the numerical calculation of the mean crossing rate. It is shown by an example that the numerical accuracy of this method is apparently very high.

1. Introduction

From the point of view of practical assessment of the response statistics of engineering structures subjected to stochastic load processes, a quantity of particular importance is the average rate of upcrossings of high levels by the response $Z(t)$. This is the key to e.g. estimation of extreme values.

In this chapter response processes will be considered that can be expressed as a second order stochastic Volterra series, that is, a stochastic Volterra series that has been truncated after the second order term. A substantial amount of work has been done to derive methods for efficient analysis of this model, see Naess[6] for references to such work.

The type of stochastic Volterra series model that will be studied can be expressed as a sum of a linear and a nonlinear, quadratic transformation of a

Gaussian process. Such a representation of the response process would apply to the standard model for expressing the total wave forces or horizontal excursion responses of e.g. a tension leg platform in a random sea way. Another example would be the response of a linear structure to a quadratic wind loading where the wind speed is modelled as a Gaussian process.

2. The Mean Crossing Rate

Let $(\Omega, \mathcal{F}, P)$ be a complete probability space, and let $Z(t)$ be a real (strictly) stationary stochastic process with continuously differentiable sample paths (a.s.). It is assumed throughout that the distribution function of $Z(0)$, denoted by $F_Z(z) = P(Z(0) \leq z)$ is absolutely continuous. For every fixed level $\zeta \in R$, let $N_Z^+(\zeta)$ denote the number of upcrossings of the level ζ by $Z(t)$ per unit of time,[4] and let $\nu_Z^+(\zeta) = E[N_Z^+(\zeta)]$ be the mean upcrossing rate. Under the assumed conditions on $Z(t)$, it can be proved[10] that if $E[|Z(0)|] < \infty$, then

$$\nu_Z^+(\zeta) = E[\dot{Z}^+ | Z = \zeta] \, f_Z(\zeta) \tag{1}$$

where the equality holds a.s. (Lebesgue) with respect to ζ, $\dot{Z}^+ = \max(\dot{Z}, 0)$ and $f_Z(\zeta)$ is the probability density function (PDF) of $Z(0)$.

Assuming that the distribution of $(\dot{Z}|Z = \zeta)$ is absolutely continuous with a PDF $f_{\dot{Z}|Z}(s|\zeta)$, it follows that for a.e. ζ

$$\nu_Z^+(\zeta) = \int_0^\infty s \, f_{\dot{Z}|Z}(s|\zeta) \, ds \, f_Z(\zeta) = \int_0^\infty s \, f_{Z\dot{Z}}(\zeta, s) \, ds \tag{2}$$

where $f_{Z\dot{Z}}(\cdot, \cdot)$ denotes the joint PDF of $Z(0)$ and $\dot{Z}(0) = dZ(t)/dt|_{t=0}$. Equation (2) is often referred to as the Rice formula.[4] $\nu_Z^+(\zeta)$ is assumed throughout to be finite, and it is referred to as the mean upcrossing rate of the level ζ. With additional assumptions on the joint densities of $Z(0), \dot{Z}(0)$ and $Z(0), h^{-1}(Z(h) - Z(0))$ for small values of h, it can be shown that equation (2) is valid for every ζ.[4,5] However, the required conditions to ensure equality in equation (2) for every value of ζ are not easy to verify, but most importantly, the stronger version is rarely needed. In reliability applications the critical levels for which the crossing rate is required can in general only be given with finite accuracy. This means that the crossing rate needs to be known for values of ζ belonging to small intervals who's length is determined by the level of the accuracy. Hence, the a.s. result is sufficient under such circumstances.

Denote the characteristic function of the joint variable $(Z, \dot{Z})$ by $M(\cdot, \cdot)$. Then $M(u, v) = E[\exp(iuZ + iv\dot{Z})]$. Assuming that $M(\cdot, \cdot)$ is an integrable function, that is, $M(\cdot,\cdot) \in L^1(\mathbf{R}^2)$, it follows that (a.s.)

$$f_{Z\dot{Z}}(z,s) = \frac{1}{(2\pi)^2} \int_{-\infty}^{\infty} \int_{-\infty}^{\infty} M(u,v) \exp\left(-iuz - ivs\right) du\, dv \qquad (3)$$

By substituting from equation (3) back into equation (2), the mean crossing rate is formally expressed in terms of the characteristic function, but this is not a very practical expression. It has been shown[7,8] that a combination of equations (2) and (3) makes it possible to derive a more useful formula for the mean crossing rate expressed in terms of the characteristic function $M(u,v)$. Naess[8] has shown that under suitable conditions on the characteristic function M, for a.e. ζ

$$\nu_Z^+(\zeta) = -\frac{1}{(2\pi)^2} \int_{-\infty}^{\infty} \left(\fint_{-\infty}^{\infty} \frac{1}{v} \frac{\partial M(u,v)}{\partial v}\, dv \right) e^{-iu\zeta}\, du \qquad (4)$$

where the inner integral wrt v is interpreted as a principal value integral in the following sense: $\fint_{-\infty}^{\infty} = \lim_{\epsilon \to 0+} \left\{ \int_{-\infty}^{-\epsilon} + \int_{\epsilon}^{\infty} \right\}$.

3. The Response Process and its Joint Characteristic Function

The second order stochastic Volterra series representation of the response process $Z(t)$, can be expressed as follows

$$Z(t) = \int_0^{\infty} h_1(\tau)X(t-\tau)\, d\tau + \int_0^{\infty} \int_0^{\infty} h_2(\tau_1,\tau_2)X(t-\tau_1)X(t-\tau_2)\, d\tau_1 d\tau_2 \tag{5}$$

In equation (5), $X(t)$ denotes a stationary, real Gaussian process. $X(t)$ could represent a random wave elevation process or a stochastic wind velocity field. We have chosen here to limit the exposition to the unidirectional case. How to deal with the multi-directional case is explained in detail by Naess.[6] Based on the results from this reference, it will be recognized that all results obtained in this chapter apply equally well to the multi-directional case.

The functions $h_1(\tau)$ and $h_2(\tau_1,\tau_2)$ characterize the physical system that is modeled. Thus, $h_1(\tau)$ is an ordinary impulse response function defining a linear dynamical system, while $h_2(\tau_1,\tau_2)$, which is referred to as the quadratic impulse response function, characterizes the second-order properties of the physical system. In contrast to the linear impulse response function h_1, h_2 does not have a direct physical interpretation, except, of course, that it completely specifies the second order contribution to the total response process when the input process $X(t)$ is known.

It has been shown [6,9] that for a wide class of second order stochastic Volterra series representing a system subjected to a stationary Gaussian process, the response $Z(t)$ can be represented as follows

$$Z(t) = \sum_{\alpha=1}^{n} \left(c_\alpha W_\alpha(t) + \lambda_\alpha W_\alpha(t)^2 \right) \tag{6}$$

where the c_α and λ_α are real constants, and the $W_\alpha(t)$ are continuously differentiable stationary (real) Gaussian processes. For each fixed t, the $W_\alpha(t)$ constitutes a set of independent $N(0,1)$ random variables.

The derivation of an explicit expression for the characteristic function of the joint variable $(Z, \dot{Z})$ for $Z = Z(t)$ and $\dot{Z} = dZ(t)/dt$, will be based on the representation given by equation (6). For this purpose we introduce the Gaussian vector variables $W = (W_1, \ldots, W_n)'$, where $W_\alpha = W_\alpha(t)$, $\alpha = 1, \ldots, n$, and $\dot{W} = (\dot{W}_1, \ldots, \dot{W}_n)'$ (When X is a vector or a matrix, X' denotes the transpose of X). Then the Gaussian vector $(W', \dot{W}')'$ has covariance matrix

$$\Sigma = \begin{pmatrix} \Sigma_{11} & \Sigma_{12} \\ \Sigma_{21} & \Sigma_{22} \end{pmatrix}, \tag{7}$$

where $\Sigma_{11} = \left(E[W_\alpha W_\beta] \right)$, $\Sigma_{12} = \left(r_{\alpha\beta} \right) = \left(E[W_\alpha \dot{W}_\beta] \right)$, $\Sigma_{21} = \left(E[\dot{W}_\alpha W_\beta] \right)$, and $\Sigma_{22} = \left(s_{\alpha\beta} \right) = \left(E[\dot{W}_\alpha \dot{W}_\beta] \right)$

In the present context, $\Sigma_{11} = I$, where I denotes the $n \times n$ identity matrix, and $\Sigma'_{12} = \Sigma_{21}$. It can also be shown that $r_{\alpha\beta} = E[W_\alpha \dot{W}_\beta] = -E[\dot{W}_\alpha W_\beta] = -r_{\beta\alpha}$, that is, $\Sigma'_{12} = -\Sigma_{12}$.

Naess[7] has shown that the characteristic function of the joint variable $(Z, \dot{Z})$ is given by the expression

$$M_{Z\dot{Z}}(u, v) = \frac{1}{\sqrt{\det(A)}} \exp\left(-\frac{1}{2} v^2 c' V c + \frac{1}{2} t' A^{-1} t \right) \tag{8}$$

where

$$t = (i u I + i v \Sigma_{12} - 2 v^2 D V) c \tag{9}$$

and

$$A = I - 2iuD - 2iv\left(D\Sigma_{21} + \Sigma_{12}D \right) + 4v^2 DVD \tag{10}$$

In the absence of the linear component, that is, when $Z(t) = \sum_{j=1}^{n} \lambda_j W_j(t)^2$, then

$$M_{Z\dot{Z}}(u, v) = \frac{1}{\sqrt{\det(A)}} \tag{11}$$

The marginal characteristic function $M_Z(u) = E[\exp\{i\,u\,Z\}]$ is now easily obtained by the relation $M_Z(u) = M_{Z\dot{Z}}(u, 0)$. In particular, it is seen that $A = \text{diag}(1 - 2iu\lambda_1, \ldots, 1 - 2iu\lambda_n)$. This gives $\det(A) = \prod_{j=1}^{n}(1 - 2\,i\,u\,\lambda_j)$ and $A^{-1} = \text{diag}(d_1, \ldots, d_n)$, where $d_j = (1 - 2\,i\,u\,\lambda_j)^{-1}$. Hence

$$M_Z(u) = \frac{1}{\sqrt{\prod_{j=1}^{n}(1 - 2\,i\,u\,\lambda_j)}} \exp\left(-\frac{1}{2} \sum_{j=1}^{n} \frac{u^2\,c_j^2}{1 - 2\,i\,u\,\lambda_j} \right) \tag{12}$$

which is in agreement with previously derived results.[3,1]

4. Numerical Calculation

The final step is the numerical calculation of the mean crossing rate by using equation (4) in combination with the expression for the joint characteristic function given by equation (8). However, a direct application of equation (4) for the numerical calculation turns out to pose severe problems due to the oscillatory term $e^{-iu\zeta}$. A way to alleviate this problem is to invoke the saddle point method, or the method of steepest descent. For this purpose equation (4) is rewritten as

$$\nu_Z^+(\zeta) = -\frac{1}{(2\pi)^2} \int_{-\infty}^{\infty} \frac{1}{v} \left(\int_{-\infty}^{\infty} \exp\left\{ -iu\zeta + \ln \frac{\partial M(u, v)}{\partial v} \right\} du \right) dv \tag{13}$$

which can be done for the case at hand. It is now observed that the function $g(\varpi) = -i\varpi\zeta + \ln \frac{\partial M(\varpi, v)}{\partial v}$ considered as a function of the complex variable ϖ is holomorphic in a region C, which includes the real line, of the complex plane $\mathbf{C}$. Assume that for each value of v and ζ, a saddle point $\varpi_s = \varpi_s(v, \zeta) = u_s - ib_s$ of the surface $\varpi \to |g(\varpi)|$ can be identified in C. Such a saddle point is also a saddle point for the surface $\varpi \to Re(g(\varpi))$. If the path of steepest descent on this last surface through the saddle point is approximately orthogonal to the imaginary axis, then, by shifting the path of integration from the real line to pass through the saddle point, a numerically stable and accurate estimate of the inner integral of equation (13) is usually obtained since along any path of steepest descent, the imaginary part of the function g remains constant, thus oscillations are avoided. If the path of steepest descent at the saddle point deviate from

     A. Naess

the direction orthogonal to the imaginary axis, then the integration line of the inner integral of equation (13) should be replaced by an integration path in C that follows the steepest descent path through the saddle point over a sufficient distance to achieve numerical accuracy, and which starts at $-\infty - i\,b_l$ and ends up at $\infty - i\,b_u$, where b_l and b_u are suitable constants. The direction of the steepest descent path from the saddle point on the surface $\varpi \to Re\big(g(\varpi)\big)$ is determined by the vector $-\overline{g'(\varpi)}$ for $\varpi \neq \varpi_s$.[2]

4.1. *Example*

To illustrate the accuracy of the numerical method, a simple model for the slow-drift response of a moored offshore structure is used. Specifically, the response process for this case may be written as ($c_1 = c_2 = 0$, $\lambda = \lambda_1 = \lambda_2$)

$$Z(t) = \lambda\left(W_1(t)^2 + W_2(t)^2 \right).$$ For this process, the upcrossing rate $\nu_Z^+(\zeta)$ is given by the relation

$$\nu_Z^+(\zeta) = \frac{\hat{\sigma}_1}{\sqrt{2\pi}} \exp\left(-\frac{\zeta}{2\lambda} + \frac{1}{2}\ln\left(\frac{\zeta}{\lambda}\right) \right) \tag{14}$$

where $\hat{\sigma}_1 = \sqrt{s_{11} - (r_{12})^2}$. This special case provides a suitable test for the accuracy of the numerical method.

Let $\tilde{\nu}_Z^+(\zeta)$ denote the average upcrossing rate of $Z(t)$ calculated by the numerical method. Table 1 compares the analytical with the numerical upcrossing rate for different levels, and it is seen that the agreement is very good indeed.

ζ	$\nu_{Z_2}^+(\zeta)$	$\tilde{\nu}_{Z_2}^+(\zeta)$
0.5	$1.37 \cdot 10^{-2}$	$1.37 \cdot 10^{-2}$
1.0	$4.31 \cdot 10^{-3}$	$4.30 \cdot 10^{-3}$
2.0	$2.995 \cdot 10^{-4}$	$2.995 \cdot 10^{-4}$
3.0	$1.802 \cdot 10^{-5}$	$1.793 \cdot 10^{-5}$
5.0	$5.618 \cdot 10^{-8}$	$5.618 \cdot 10^{-8}$
7.0	$1.605 \cdot 10^{-10}$	$1.592 \cdot 10^{-10}$

Acknowledgements

The author is grateful to U. B. Machado for carrying out the numerical calculations used in the example.

References

1. M. Grigoriu, *Applied Non-Gaussian Processes* (PTR Prentice Hall, Englewood Cliffs, 1995).
2. P. Henrici, *Applied and Computational Complex Analysis, Vol. II* (John Wiley and Sons, Inc., New York, 1974).
3. M. Kac and A. J. F. Siegert, On the theory of noise in radio receivers with square law detectors, *Journal of Applied Physics* **18**, 383 (1947).
4. R. M. Leadbetter, G. Lindgren and H. Rootzen, *Extremes and Related Properties of Random Sequences and Processes* (Springer-Verlag, New York, 1983).
5. M. B. Marcus, Level crossings of a stochastic process with absolutely continuous sample paths, *The Annals of Probability* **5**, 52 (1977).
6. A. Naess, Statistical analysis of nonlinear, second-order forces and motions of offshore structures in short-crested random seas, *Probabilistic Engineering Mechanics* **5**, 192 (1990).
7. A. Naess, Crossing rate statistics of quadratic transformations of Gaussian processes, *Probabilistic Engineering Mechanics* **16**, 209 (2001).
8. A. Naess, Croossing Rate Statistics of Second Order Stochastic Volterra Series of Gaussian Processes, *Preprint Statistics* **7**, Dept. of Mathematical Sciences, Norwegian University of Science and Technology (2002).
9. A. Naess, Representation of the Response Process of Quadratic Filters Subjected to Gaussian Noise, *Stochastic Analysis and Applications* **20**, 947 (2002).
10. U. Zähle, A general Rice formula, Palm measure, and horizontal window conditioning for random fields, *Stochastic Processes and their Applications* **17**, 265 (1984).

14

INSPECTION AND MAINTENANCE FOR STOCHASTICALLY DETERIORATING SYSTEMS

Martin Newby and Richard Dagg

School of Engineering & Mathematical Sciences,
City University, LONDON, UK
Email: m.j.newby@city.ac.uk

The chapter develops models that allow the determination of optimum inspection and maintenance policies. The models are based on Lévy processes and allow the use of the Markov property to construct renewal type integral equations for costs and dynamic programmes for determining optimum policies. The extension to models with imperfect observation or covariates is indicated. The models provide a single framework and depend only on the determination of an appropriate transition density to "plug in" to the equation. Some numerical results and a comparison of different policies are reported.

1. Motivations

Models of wear and degradation have been widely used in reliability and other applications[3]. Many of the models have been descriptive and entail parameter estimation and inference about model parameters without a decision mechanism. Wiener process (Brownian motion) models are tractable and able to reflect the empirical behaviour of a process while not necessarily according with the underlying mechanisms[4,15,16]. In this article the class of models is broadened and their applications extended. The central idea is the maintenance of the Markov property while relaxing the conditions on the process as far as possible. The practical applications came out of a number of reliability and maintenance problems[3].

Failure Prediction - The Paris-Erdogan fatigue crack growth model [5,12] is transformed by a power transformation of a Wiener process. Using the Paris-Erdogan model, the stochastic model for the growth of a crack under cyclical loading can be formulated as a simple transformation to a Wiener process. The underlying stochastic process X_t transforms via a Box-Cox transformation to a Wiener process Y_t with a finite escape time[5].

Maintenance Planning - Systems frequently show a deterioration in performance over time even though they are maintained. For example while a system is fully maintained it may lose efficiency over time until it fails either economically or technically.

2. Degradation Processes

Wear and degradation are natural concepts used to describe the state of a system as it progresses through time. They arise from aging of materials, accumulations of damage, or loss of capacity[6,7]. Many reliability models arise from shock models in which a discrete process describes the arrival of shocks, and the magnitude of the shocks may be continuous or discrete. More useful for this paper is the idea of accumulated damage. Van Noortwijk [13] points out that in many systems which are subject to shocks, the order in which the damage (i.e. the shocks) occur is immaterial. This suggests that the deterioration incurred in equal time intervals form a set of exchangeable random variables[2]. This also implies that the distribution of the degradation incurred is independent of the time scale, i.e. has stationary increments. Exchangeable and stationary increments are similar to the properties of independent and stationary increments suggesting Lévy processes[7] may provide effective models. Independent increments seem intuitively correct in many applications but stationarity is less appealing. Independence may hold[13], but in some cases the accumulated degradation depends on the age or the state of the system.

Time homogeneity and independent increments suggest a continuous time Markov process, the class of Lévy processes which include the compound Poisson process, the Wiener process, and the gamma process.

3. Lévy Process Restrictions

Lévy processes can be represented as a sum of a Wiener process and a jump process[10]. A process with continuous sample paths is a Wiener process. There is thus a conflict between the desire for continuous sample paths and monotonicity in the class of Lévy processes. A Wiener process with a large drift parameter and small variance is often used to approximate monotonicity. The discontinuous sample paths of jump processes are indistinguishable from continuous paths when not continuously observed making the restriction to continuous sample paths unnecessary. Using class of Lévy processes retains the Markov property, which simplifies the determination of optimum strategies[11]. Many processes are not monotone, for example, where the process represents the use of a capacity, or where there are limits on use; engines can run often at more than the design rating for a short period without causing failure, but will require extra compensatory maintenance at the next scheduled maintenance.

4. Structure of the Argument

The state of the system is represented by a stochastic process X_t and inspection is the only way to determine the state of the system. The processes are characterized by the transition density for the increments in X_t. Three cases are considered

perfect inspection in which the state X_t can be determined. The unconditional density $f_{X_t - X_s | X_s}$ determines the process;

imperfect inspection in which an indirect observation Y_t of the actual state X_t is made and the conditional densities $f_{X_t | Y_t}$ and $f_{Y_t | X_t}$ are known;

covariate models in which a vector of covariates Z_t is observed and allows the conditional density $f_{X_t | Z_t}$ to be determined.

The models developed all have a similar structure with the details of the model reflected in the transition densities $f_{X_t | .}$. Imperfect inspection and the covariate models have the same structure and the unobserved process X_t is removed by taking expectations.

5. Replacement Models with Perfect Inspection

The state space is partitioned into disjoint intervals $A_0, A_1, A_2, \ldots, A_n$. There is a minimum cost of inspection C_0. The policy is that if $X_t \in A_0$ no action is taken and if $X_t \in A_i$, the system is restored to new condition at cost C_i. The replacement system is identical to the original system. Failure occurs at the first moment the process hits the set A_n and is immediately observed with instantaneous replacement of the system. The cost of replacing a failed system is C_n, and costs increase with the state of the system: $C_i < C_j$ for $i < j$.

The treatment of non-monotonic processes may be extended by allowing the system to enter and leave state A_n between inspections. Action will then be taken at the next inspection when it is determined that the system has crossed the failure boundary. A general degradation model thus needs value of the process between inspections as well as at inspection. Using M_t, the maximum value of the process X_t in the period $[0, t)$, formally

$M_t = \sup\{X_u | 0 \le u \le t\}$, solves this difficulty. The actions taken after an inspection are completely determined by the state of the bivariate process (X_t, M_t). Where failure occurs immediately the boundary is crossed the supplementary variable $H^x_{A_n}$, the hitting time of the critical set starting from state x may be used[13].

5.1. *Expected Cost per Unit Time and Expected Cycle Length*

The argument is simplified by a non-standard notation. The cost per cycle for a system with initial level of degradation level x and inspection interval τ is $V(x, \tau)$, . Similarly, the length of a cycle under the same conditions is $L(x, \tau)$. The variables V and L are not functions, but are random variables, $V(x, \tau) = V|x, \tau$ and $L(x, \tau) = L|x, \tau$, conditioning on x and τ. Conditional probabilities are also indicated with a superscript, $P[A|X = x] \equiv P^x[A]$

5.2. *Expected Cost per Cycle*

For the Lévy processes, the results are independent of the current time, allowing it to be taken as $t = 0$, and all variables are conditional on $X_0 = x$. The cost function is the sum of the cost of continuing (i.e. not replacing the system) and the costs for each possible replacement event. The cost of both preventive and corrective actions is

$$Z(x, \tau) = \sum_{i=0}^{n-1} C_i \mathbf{1}_{\{X_\tau \in A_i, M_\tau \notin A_n\}} + C_n \mathbf{1}_{\{M_\tau \in A_n\}}$$

where $\mathbf{1}_A$ is the indicator function of the set A. The expected cost is

$$c(x, \tau) = E[Z(x, \tau)]$$
$$= \sum_{i=0}^{n-1} C_i P^x (X_\tau \in A_i, M_\tau \notin A_n) + C_n P^x (M_\tau \in A_n) \tag{1}$$

The total cost is

$$V(x, \tau)|X_\tau, M_\tau = [C_0 + V(X_\tau, \tau)] \mathbf{1}_{\{X_\tau \in A_0, M_\tau \notin A_n\}} + Z(x, \tau)$$

whose expected cost is

$$E\left[V(x,\tau)|X_\tau, M_\tau\right] = E\left[V(X_\tau,\tau)\mathbf{1}_{\{X_\tau \in A_0, M_\tau \notin A_n\}} + Z(x,\tau)\right]$$
$$= E\left[V(X_\tau,\tau)|X_\tau, M_\tau \mathbf{1}_{\{X_\tau \in A_0, M_\tau \notin A_n\}}\right] + c(x,\tau)$$

Define $f_\tau(y,m|x)$ to be the joint density of X_τ and M_τ, conditional upon $X_0 = x$; with this density the probabilities P^x are

$$P^x\left(X_\tau \in A_i, M_\tau \notin A_n\right) = \int\limits_{y \in A_i} \int\limits_{m \in S \backslash A_n} f_\tau(m,y|x)dmdy$$

and taking expectations

$$v(x,\tau) = \int\limits_{A_0} \int\limits_{S \backslash A_n} v(y,\tau)f_\tau(y,m|x)dmdy + c(x,\tau)$$

where S represents the whole state space.

To deal with the possibility of a negative level of degradation the integral is split into two parts and the problem removed by setting the associated costs to zero. The cost function for negative values is $v(y,\tau) = v(0,\tau)$ for $y < 0$. From the assumptions $A_0 = (-\infty, s_0)$, so write $A_0 = (-\infty, 0) \cup [0, s_0)$ to obtain

$$\int\limits_{A_0} \int\limits_{S \backslash A_n} v(y,\tau)f_\tau(y,m|x)dmdy$$
$$= \int\limits_{-\infty}^{0} \int_{S \backslash A_n} v(y,\tau)f_\tau(y,m|x)dmdy + \int\limits_{0}^{s_0} \int\limits_{S \backslash A_n} v(y,\tau)f_\tau(y,m|x)dmdy$$
$$= v(0,\tau)P^x\left(X_\tau < 0, M_\tau \notin A_n\right) + \int\limits_{0}^{s_0} \int\limits_{S \backslash A_n} v(y,\tau)f_\tau(y,m|x)dmdy$$

Call the probability of a negative degradation without failure

$$u(x,\tau) = P^x\left(X_\tau < 0, M_\tau \notin A_n\right)$$

The integral equation reduces to:

$$v(x,\tau) = c(x,\tau) + v(0,\tau)u(x,\tau) + \int_0^{s_0} v(y,\tau)K_\tau(y|x)dy \qquad (2)$$

with kernel

$$K_\tau(y|x) = \int_{S\setminus A_n} f_\tau(y, m|x)dm$$

The kernel is the probability density that the process is in state y at the next inspection and does not hit A_n before the next inspection, given the current state is x and the inspection interval is τ. For the Wiener process model kernel is defined by normal densities and can be computed in closed form[13].

The integral equation above has general form

$$v(x) = p(x) + q(x)v(0) + \int_0^s v(y)K(y, x)dy$$

a Fredholm equation of the second kind. When required effective numerical methods are available[8].

5.3. *Expected Length of a Cycle*

The expected time to inspection $l(x,\tau) = E[L(x,\tau)|X_0 = x]$ is obtained in the same way as the average cost, the only complication is that on absorption in A_n, a full period τ is not completed, and the hitting time of A_n is needed. Since A_n is an interval, the hitting time of the least point in that interval, s_n, is enough. Define $H^x_{s_n}$ to be the hitting time of this point starting from $X_0 = x$. Again the (random) time until replacement, starting in state x, conditional on X_τ, M_τ and $H^x_{s_n}$ is

$$L(x,\tau)|X_\tau, M_\tau, H^x_{s_n} = [\tau + L(X_\tau, \tau)]\mathbf{1}_{\{X_\tau \in A_0, M_\tau \notin A_n\}}$$
$$+\tau\mathbf{1}_{\{X_\tau \in A_1 \cup \cdots \cup A_{n-1}, M_\tau \notin A_n\}} + H_{s_n}\mathbf{1}_{\{M_\tau \in A_n\}}$$

Clearly the expected value of $\tau\mathbf{1}_{\{X_\tau \in A_1 \cup \cdots \cup A_{n-1}, H^x_{s_n} > \tau\}} + H_{s_n}\mathbf{1}_{\{H^x_{s_n} < \tau\}}$
is the mean length of the time between failures

$$E\left(\tau\mathbf{1}_{\{X_\tau \in A_1 \cup \cdots \cup A_{n-1}, H^x_{s_n} > \tau\}} + H_{s_n}\mathbf{1}_{\{H^x_{s_n} < \tau\}}\right) = \int_0^\tau [1 - G_\tau(h|x)]dh$$

Taking expectations with respect to the joint density of X_τ , M_τ , $H^x_{s_n}$

$$l(x,\tau) = \int_0^\tau [1 - G_\tau(h|x)]dh + \int_{A_0}\int_{S\setminus A_n} l(y,\tau)f_\tau(y, m|x)dmdy \qquad (3)$$

The distribution G_τ is the distribution functions of $H^x_{s_n}$, conditional on the current state x. The kernel K_τ is the same as above. This integral equation has the same form as the equation for the expected cost per cycle.

5.4. *Optimal Inspection Policies*

The underlying process is Markov, and each replacement is a renewal point. The renewal reward theorem[11] shows that for a renewal process the expected cost per unit time is the ratio of the expected cost per cycle and the expected length per cycle, with probability one. For a new system with inspection policy τ the ratio of cost to length is

$$\mathbf{C}(0,\tau) = \frac{v(0,\tau)}{l(0,\tau)}$$

For such a system the optimum policy $\tau*$ is

$$\tau* = \arg\min_{\tau>0} \{\mathbf{C}(0,\tau)\}$$

5.5. *Optimal Inspection: Discounted Cost Criterion*

The discounted cost criteria is simpler than the previous case, and discounting is closer to real applications. This criterion does not use the renewal reward theorem and extends easily to non-periodic inspections. The system is identical to that above, and the assumptions of section 5.

The policy, π, will be described more fully in particular cases; its realization in the n-th interval is π_n. The discount rate is δ . Define $V_\delta(x,\pi)$ is the discounted total cost for a system with current level of degradation x and policy π. Define t_i to be the inter-event time (an event being either an inspection or failure), then if $C(X_{t_{n-1}}, X_{t_n}; \pi_{n-1})$ is the cost of applying policy π in the n-th interval,

$$v_\delta(x,\tau) = E\left(V_\delta(x,\tau)|X_0 = x\right)$$
$$= E\left[\sum_{n=1}^{\infty} e^{-\delta(t_1+\cdots+t_n)} C(X_{t_{n-1}}, X_{t_n}; \pi_{n-1}))|X_0 = x\right]$$

5.6. *Periodic Inspection*

In the simple periodic case the policy is a set of fixed time instants $\pi = i\tau, i = 1, 2, \ldots$. Since the system is returned to the new state after each

action, the cost can be built up from the cost of starting the system up
after each inspection and repair together with the cost of replacement on
failure and we can write the discounted cost as

$$v_\delta(x,\tau) = E\left(V_\delta(x,\tau)|X_0 = x\right)$$

$$= E\left[\sum_{n=1}^{\infty} e^{-\delta(t_1+\cdots+t_n)}C(X_{t_{n-1}}, X_{t_n}; \tau)|X_0 = x\right]$$

$C(x,y;\tau)$ is the (random) cost incurred if the system is in state y at
time t_n when it was in state x at time t_{n-1} .

The cost function is the sum of the cost in the event of continuing
(no replacement) and the costs for each possible replacement event. The
argument is almost identical to the average cost criterion apart from the
discount factor.

$$V_\delta(x,\tau)|X_\tau, M_\tau, H_{s_n}^x = e^{-\delta\tau}\left[(C_0 + V_\delta(X_\tau, \tau))\mathbf{1}_{\{X_\tau \in A_0, M_\tau \notin A_n\}}\right.$$

$$+ \sum_{i=1}^{n-1}(C_i + V_\delta(0,\tau))\mathbf{1}_{\{X_\tau \in A_i, M_\tau \notin A_n\}}\Big]$$

$$+ e^{-\delta H_{s_n}^x}(C_n + V_\delta(0,\tau))\mathbf{1}_{\{H_{s_n}^x < \tau\}}$$

The density $f_\tau(y,m|x)$ retains its meaning from section 5.2. Taking
expectations of the above expression with respect to the joint density of
X_τ, M_τ, and $H_{s_n}^x$.

$$v_\delta(x,\tau) = E\left[V_\delta(x,\tau)|X_\tau, M_\tau, H_{s_n}^x\right]$$

$$v_\delta(x,\tau) = \int_{A_0}\int_{S\setminus A_n} e^{-\delta\tau}v(y,\tau)f_\tau(y,m|x)dmdy$$

$$+ e^{-\delta\tau}C_0 P^x\left(X_\tau \in A_0, M_\tau \notin A_n\right)$$

$$+ \sum_{i=1}^{n-1} e^{-\delta\tau}\left(C_i + v_\delta(0,\tau)\right)P^x\left(X_\tau \in A_i, M_\tau \notin A_n\right)$$

$$+ \left(C_n + v_\delta(0,\tau)\right)\int_0^\tau e^{-\delta h}g_\tau(h|x)dh$$

where $g_\tau(h|x)$ is the probability density function of $H_{s_n}^x$, the hitting time
of the critical set from x.

Negative degradation is dealt with as above, $v_\delta(x,\tau) = v_\delta(0,\tau)$ for $x < 0$. The integral term in the above expression becomes

$$\int_{A_0}\int_{S\setminus A_n} e^{-\delta\tau}\, v_\delta(y,\tau)f_\tau(y,m|x)dmdy =$$

$$e^{-\delta\tau}v_\delta(0,\tau)P^x\left(X_\tau < 0, M_\tau \notin A_n\right)$$

$$+\int_0^{s_0}\int_{S\setminus A_n} e^{-\delta\tau}v_\delta(y,\tau)f_\tau(y,m|x)dmdy$$

The details follow from the average cost case,

$$v_\delta(x,\tau) = c_\delta(x,\tau) + v_\delta(0,\tau)u_\delta(x,\tau) + \int_0^{s_0} v_\delta(y,\tau)K_\tau(y|x)dy$$

where

$$c_\delta(x,\tau) = \sum_{i=0}^{n-1}e^{-\delta\tau}C_iP^x\left(X_\tau \in A_i, M_\tau \notin A_n\right) + C_n\int_0^\tau e^{-\delta h}g_\tau(h|x)dh \quad (4)$$

$$u_\delta(x,\tau) = e^{-\delta\tau}P^x\left(X_\tau \notin A_0^+ \cup A_n, M_\tau \notin A_n\right) + \int_0^\tau e^{-\delta h}g_\tau(h|x)dh \quad (5)$$

$$K_\tau(y|x) = \int_{S\setminus A_n} e^{-\delta\tau}f_\tau(y,m|x)dm \quad (6)$$

A_0^+ the set of positive states which are in A_0, i.e. $A_0^+ = A_0 \setminus (-\infty, 0)$.

The optimal policy is given by

$$\tau_\delta* = \arg\inf_{\tau>0}\left\{v(0,\tau)\right\}$$

The functions c, u and K are more complicated than for the average cost case. However, the integral equation has the same form as in the previous case.

5.7. *Discounted Monotone Process Model – periodic inspection*

Since X_t and M_t can be identified the integral equation in standard form becomes

$$v_\delta(x,\tau) = c_\delta(x,\tau) + v_\delta(0,\tau)u_\delta(x,\tau) + \int_0^{s_0} v_\delta(y,\tau)K_\tau(y|x)dy$$

with

$$c_\delta(x,\tau) = \sum_{i=0}^{n-1} e^{-\delta\tau}C_i P^x\left(X_\tau \in A_i\right) + C_n \int_0^\tau e^{-\delta h}g_\tau(h|x)dh \qquad (7)$$

The process is monotonic (increasing) so the level of degradation cannot be negative, $u_\delta(x,\tau)$ is given by

$$u_\delta(x,\tau) = e^{-\delta\tau}P^x\left(X_\tau \notin A_0 \cup A_n\right) + \int_0^\tau e^{-\delta h}g_\tau(h|x)dh \qquad (8)$$

and the transition kernel is

$$K_\tau(y|x) = e^{-\delta\tau}f_\tau(y|x) \qquad (9)$$

5.8. *Discounted Non-Periodic Inspection*

Non-periodic inspection is more general and often more useful than the previously considered case of periodic inspection, since it generally results in policies with lower costs. This is because each inspection interval is in some way determined by the state of the system at the previous inspection. The policy π is not adaptive, but a sequence of fixed instants $\pi = \{\tau_1, \tau_2, \ldots, \tau_n\}$. The discounted cost function is

$$v_\delta(x,\pi) = E\left[V_\delta(x,\tau_0)|X_0 = x\right]$$
$$= E\left[\sum_{n=1}^{\infty} e^{-\delta(t_1+\cdots+t_n)}C(X_{t_{n-1}}, X_{t_n}; \tau_{n-1})|X_0 = x\right]$$

the notation is as in section 5.6 and τ_i replaces τ in the cost equation. We are only concerned with deterministic stationary policies. The Markov property implies the current state determines the future development of the system.

Define the value function for the δ-optimal policy

$$v_\delta(x) = \inf_\pi v_\pi(x)$$

The optimality equation is given by

$$v_\delta(x) = \inf_{\tau>0} \left\{ c(x,\tau) + \int_{-\infty}^{\infty} \int_0^{\infty} e^{-\delta(\tau \wedge h)} v_\delta(y) f_\tau(y, h|x) dh dy \right\}$$

where c represents the discounted costs incurred at the next inspection, after an inspection interval τ is chosen. The dynamic programming solution follows[11]. In the particular case here the optimality equation is given by

$$v_\delta(x) = \inf_{\tau>0} \left\{ c_\delta(x,\tau) + v_\delta(0)u_\delta(x,\tau) + \int_0^{s_0} v_\delta(y) K_\tau^\delta(y|x) dy \right\}$$

where c, u and K are given by equations (1, 2, 3) above.

A standard argument[11] shows that

$$(T_\delta v)(x) = \inf_{\tau>0} \left\{ c_\delta(x,\tau) + v(0)u_\delta(x,\tau) + \int_0^{s_0} v(y) K_\tau^\delta(y|x) dy \right\}$$

is a contraction mapping. The optimum policy is obtained iterating the policy $\pi_k = \left\{ \tau_0^{(k)}, \ldots, \tau_n^{(k)} \right\}$ in an improvement algorithm[9].

5.9. *Monotone Process Model – Non-Periodic Inspection*

Applying the same reasoning as above to the monotone degradation case the optimality equation is

$$v_\delta(x) = \inf_{\tau>0} \left\{ c_\delta(x,\tau) + v_\delta(0)u_\delta(x,\tau) + \int_0^{s_0} v_\delta(y) K_\tau(y|x) dy \right\}$$

with c_δ, u_δ and K_δ defined exactly as in equations (7),(8) and (9)

6. Example: Gamma Process

Gamma process models have been widely used[1,13,14]. The arguments are simplified because the process is monotone. The increments of the degradation process have a gamma distribution

$$X_t - X_s \sim Gamma\left(\alpha(t - s), \beta\right)$$

with transition density function

$$f_\tau(y|x) = \frac{\beta^{\alpha\tau}(y-x)^{\alpha\tau-1}e^{-\beta(y-x)}}{\Gamma(\alpha\tau)}$$

Monotonicity makes it is simple to find the distribution of H_c^x, the time to reach a level c, as an incomplete gamma ratio

$$G_c(h|x) = P^x(H_c^x \le h) = P^x(X_h > c) = \frac{\Gamma(\alpha h; \beta(c-x))}{\Gamma(\alpha h)}$$

In a simple version the system fails if the degradation reaches a critical level L and is replaced; if inspection reveals degradation greater than $W < L$, make a planned replacement. Inspection costs C_I, preventive replacement costs $C_I + C_R$, and replacement on failure costs C_F. This model has

$$A_0 = [0, W), A_1 = [W, L), A_2 = [L, \infty), C_0 = C_I, C_1 = C_I + C_R, C_2 = C_F$$

Three examples, periodic inspection (average and discounted costs), and non-periodic discounted costs for a gamma process are given.

6.1. *Periodic Inspection : Average Cost per Unit Time*

Since the gamma process is monotonic the equations (1), (2), and (3) of sections 5.2 and 5.3 are simplified. The probability of negative degradation is zero, so the function $u(x, \tau) \equiv 0$ and the maximum M_τ disappears from the model.

$$c(x, \tau) = \sum_{i=0}^{n} C_i P^x(X_\tau \in A_i)$$

$$P^x(X_\tau \in A_i) = \int_{s_i}^{s_{i+1}} f_\tau(y|x)dy$$

and as before, $f_\tau(y|x)$ is the probability density function of X_τ given $X_0 = x$, and $G_L(h|x)$ is the cumulative distribution function of H_L^x, the hitting time of L starting from $X_0 = x < L$. These are as follows:

$$f_\tau(y|x) = \frac{\beta^{\alpha\tau}(y-x)^{\alpha\tau-1}e^{-\beta(y-x)}}{\Gamma(\alpha\tau)}$$

In this particular case

$$c(x,\tau) = C_I + C_R G_W(\tau|x) + (C_F - C_R - C_I)G_c(\tau|x)$$

The average cost and length become

$$v(x,\tau) = C_I + C_R G_W(\tau|x) + (C_F - C_R - C_I)G_L(\tau|x) + \int_x^W v(y,\tau)f_\tau(y|x)dy$$

$$l(x,\tau) = \tau - \int_0^\tau G_L(h|x)dh + \int_x^W l(y,\tau)\,f_\tau(y|x)dy$$

The integral equations are solved numerically[8] to compute values of $v(x)$ and $l(x)$ for various values of x to determine the average cost function **C**.

6.2. *Periodic Inspection: Discounted Total Cost*

In the case of periodic inspection under the discounted total cost criterion we must solve the integral equation

$$v_\delta(x,\tau) = c_\delta(x,\tau) + v_\delta(0,\tau)u_\delta(x,\tau) + \int_0^{s_0} v_\delta(y,\tau)K_\tau(y|x)dy$$

The explicit forms of c_δ, u_δ and K_δ follow from equations (7), (8) and (9) of section 5.7.

$$c_\delta(x,\tau) = e^{-\delta\tau}\{C_I + C_R G_W(\tau|x) + (C_F - C_R - C_I)G_L(\tau|x)\}$$
$$+\delta C_F \int_0^\tau e^{-\delta h}G_L(h|x)dh$$

$$u_\delta(x,\tau) = e^{-\delta\tau}G_W(\tau|x)) + \delta \int_0^\tau e^{-\delta h}G_L(h|x)dh$$

$$K_\tau(y|x) = e^{-\delta\tau}f_\tau(y|x)$$

with $u_\delta(x,\tau)$ following from an integration by parts.

6.3. *Non-Periodic Inspection: Discounted Total Cost*

In the case of non-periodic inspection the dynamic programming equation the results of 6.2 above are extended exactly as section 5.8 is an extension of the argument in 5.7.

6.4. *Results*

In all three cases, the parameters are chosen so that varying their relative values demonstrates the effects of the whole set of parameters. The assumptions are: i) the degradation process is Gamma with $\alpha/\beta = 1$, α controls the variability of the underlying process; ii) the failure limit of the process $L = 1$ and the replacement limit W is varied; iii) The cost of inspection $C_I = 1$, the replacement and failure costs C_R and C_F are varied.

The most striking results are for the shape of the cost function $v(\tau)$ under the periodic inspection policy. Two patterns, a single local minimum with a clear optimum and a monotone decreasing function implying no inspection are not surprising. The third pattern demonstrated multiple local minima which appear to imply both frequent inspection and infrequent failure or infrequent inspection and frequent failure are equally good policies.

Table 1.   Discounted Total Cost as Percentage of Periodic Optimal DTC

repair limit,W	C_R	C_F	variance low $a = 20$			medium $a = 10$			high $a = 5$		
			10	30	100	10	30	100	10	30	100
	5		100	99.5	94.5	100	98	97.5	100	99	98
0.3	20			100	99.5		100	98.5		100	100
	50				100			100			100
	5		98	85	81.5	98	88	84.5	99	92	88.5
0.6	20			99	94		99.5	94.5		100	95
	50				98			98.5			99
	5		91.5	65.2	74	96	76	83.5	98.5	84.5	89.5
0.9	20			91	85		94.5	89		98	92.5
	50				96.5			95			97

The numerical results show little difference between the periodic solutions, but in Table 1 the discounted non-periodic solution does show some large potential savings.

7. Models with Covariates

The degradation process itself is not observed, instead a surrogate is observed and is the basis for decision making, in all other respects, the approach is unchanged. The process $X = \{X_t : t \geq 0\}$ denotes the actual degradation process of the system, and the process $Y = \{Y_t : t \geq 0\}$ the corresponding covariate process. Initially no assumptions are made about the relationship between the processes.

The state-space of the observed covariate process S_c is partitioned into intervals $A_0, A_1, A_2, \ldots, A_n$; $A_0 = [0, s_0)$, $A_k = [s_k, s_{k+1})$, $k = 0, 1, \ldots, n-1$, $s_k < s_{k+1}$ and $s_{n+1} = \infty$. The actual degradation process has state space S_d partitioned into the set of failed states $B = [b, \infty)$ and its complement $B^c = [0, b)$. Inspection reveals the true state of the covariate process Y_t; a new system has initial covariate level $Y_0 = \lambda \in A_0$ and degradation level 0 and the rules are as in section 5. However, the system fails immediately X_t enters the set B which is observed immediately with immediate replacement of the system at cost C_F. Since the decisions are made using the process Y_t, the effect of the underlying process X_t is removed by taking expectations. The transition densities now relate to the observed process but are conditioned on the state of the underlying process.

The transition density used is, in the general case, $f_{Y_\tau, M_\tau}(u, m)$, the joint density of Y_τ, M_τ (maximum of X_t) conditional on $X_0 = x$.

$$
\begin{aligned}
f_{Y_\tau, M_\tau}(u, m) &= \int_0^\infty f_{Y_\tau, M_\tau | X_\tau}(u, m | z) f_{X_\tau}(z) dz \\
&= \int_0^\infty f_{Y_\tau | X_\tau}(u | z) f_{M_\tau | Y_\tau, X_\tau}(m | u, z) f_{X_\tau}(z) dz \\
&= \int_0^\infty f_{Y_\tau | X_\tau}(u | z) f_{M_\tau, X_\tau}(m, z) dz
\end{aligned}
$$

For a Wiener process the density can be obtained in terms of the standard normal density function. In the monotone case a similar argument gives $f_\tau(u, z | x)$, the joint density of Y_τ, and X_τ, conditional on $X_0 = x$.

The simplest case of periodic inspection is developed, the discounted versions also follow the same line and imitate the arguments above.

The expected costs and cycle length conditional on the state X_t are derived exactly as above, the argument consists of little more than writing y in place of x and replacing the densities and kernels as appropriate. Then the conditioning on the unobservable process is removed by taking expectations. The equations for the cost and length of a cycle are reduced to the standard form

$$
v(y, \tau) = c(y, \tau) + u(y, \tau) v(0, \tau) + \int_0^{s_0} v(u, \tau) K_\tau(u | y) du
$$

$$l(y,\tau) = d(y,\tau) + l(0,\tau)u(y,\tau) + \int_0^{s_0} l(u,\tau)K_\tau(u|x)du$$

with d, c, u and K defined by

$$c(y,\tau) = \int_0^b \left\{ \sum_{i=0}^n C_i P^x \left(Y_\tau \in A_i, M_\tau \notin B\right) + C_F P^x \left(M_\tau \in B\right) \right\} f(x|y)dx$$

$$d(y,t) = \int_0^b \left\{ \tau P^{x,y}(Y_\tau \notin A_0, M_\tau \notin B) + \int_0^\tau hg_B(h|x)dh \right\} f(x|y)dx$$

$$u(y,\tau) = \int_0^b P^x \left(Y_\tau < 0, M_\tau \notin B\right) f(x|y)dx$$

$$K_\tau^{(}u|y) = \int_0^b \int_0^b f_\tau(u,m|x)f(x|y)dmdx$$

Thus, as in the case of perfect observations, the average cost per unit time is

$$\mathbf{C}(\lambda,\tau) = \frac{\mathrm{v}(\lambda,\tau)}{\mathrm{l}(\lambda,\tau)}$$

which can be minimized in the usual way.

7.1. *Example : Imperfect Inspection of a Gamma Process*

The imperfect inspection of Wiener process is relatively straightforward[15]. Here a gamma process, as above, is subject to a statistically independent additive error, $\varepsilon_t \sim N(0, \nu^2)$. With these assumption it is clear that $Y_t|X_t \sim N(X_t, \nu^2)$.

The distribution of the true level of degradation given the observed degradation level requires more care. The system has not failed, so

$$f(x|y)dx = \Pr(X_t = x|Y_t = y, 0 \le X_t \le c)dx = \frac{\phi\left(\frac{x-y}{v}\right)dx}{\left[\Phi\left(\frac{y}{\nu}\right) - \Phi\left(\frac{y-c}{\nu}\right)\right]}$$

Conversely, the joint density of the future observed and true levels of degradation is

$$f_{Y_t, X_t | X_0}(u, z | x) = f_{Y_t | X_0, X_t}(u | x, z) f_{X_t | X_0}(z | x) = f_{Y_t | X_t}(u | z) f_{X_t | X_0}(z | x)$$

The given distributions lead to

$$f_\tau(u, z | x) = \frac{1}{\nu \sqrt{2\pi}} \exp\left\{ -\frac{1}{2\nu^2}(u - z)^2 \right\} \frac{\beta^{\alpha\tau}(z - x)^{\alpha\tau - 1} \exp\left\{ -\beta(z - x) \right\}}{\Gamma(\alpha\tau)}$$

The failure time distribution is again $G_c(h | x) = Q(\alpha h, \beta(c - x))$.

Using these distributions, the elements of the integral equations are determined and the optimization

$$\tau* = \arg \min_\tau \mathbf{C}(\lambda, \tau)$$

can be completed.

8. Summary

The approach developed in this chapter extends the standard versions of age replacement and condition monitoring in a systematic way based on the relatively tractable class of Lévy processes. The structure of the argument in all cases, both for perfect and imperfect inspection reduces the problem to a standard form of integral equation. This creates a fixed framework for analysis which depends only on the determination of appropriate transition densities and transition kernels. In simpler cases, such as gamma or Wiener processes, the densities and kernels can be determined explicitly in closed form. All integral equations are of the Fredholm type with effective numerical methods of solution.

References

1. M.A. Abdel–Hameed, A Gamma Wear Process, *IEEE Transactions on Reliability*, **R-24**, 152-154 (1975).
2. J. M. Bernardo and A. F. M. Smith, *Bayesian Theory* (John Wiley and Sons, New York, 1992).
3. R. Dagg, *Optimal Inspection and Maintenance for Stochastically Deteriorating Systems* (PhD thesis, City University, London, 2000).
4. M. L. T. Lee and G. A. Whitmore, Stochastic processes directed by randomised time, *Journal of Applied Probability*, **30**,302-314 (1993).
5. M. J. Newby, Analysing Crack Growth, *Proceedings of the Institution of Mechanical Engineers. Part G – Aeronautical Engineering*, **212**, 157-166 (1998).

6. K. S. Park, Optimal Continuous-Wear Limit replacement under Periodic Inspections, *IEEE Transactions on Reliability*, **37**, 97-102 (1988a).
7. K. S. Park, Optimal Wear-limit Replacement with Wear dependent Failures, *IEEE Transactions on Reliability*, **37**, **37**, 293-294 (1988b).
8. W. H.Press, S. A. Teukolsky, W. T. Vetterling and B. P. Flannery, *Numerical Recipes in C*, 2nd *Edition* (Cambridge University Press, Cambridge, UK 1992).
9. M. L. Puterman, *Markov Decision Processes: Discrete Stochastic Dynamic Programming* (John Wiley and Sons, New York. 1994).
10. L. C. Rogers and D. Williams, *Diffusions, Markov Processes and Martingales, Volume 1: Foundations* 2nd *Edition)* (John Wiley and Sons, Chichester 1994).
11. S. M. Ross, *Applied Probability Models with Optimization Applications*, (Holden-Day Inc., San Francisco. 1970).
12. K. Sobczyk, Stochastic Models for Fatigue Damage of Materials, *Advances in Applied Probability*, **19**, 652-673 (1987).
13. J. M. Van Noortwijk, *Optimal Maintenance Decisions for Hydraulic Structures under Isotropic Deterioration* (PhD Thesis, Technical University of Delft, Netherlands 1996).
14. L. M. Wenocur, A Reliability Model Based on the Gamma Process and its Analytic Theory, *Advances in Applied Probability* **21**, 899-918 (1988).
15. G. A. Whitmore, Estimating Degradation By a Wiener Diffusion Process Subject to Measurement Error, *Lifetime Data Analysis*,**1**,307-319 (1995).
16. G. A. Whitmore and F. Schenkelberg, Modelling Accelerated degradation data using Wiener diffusion with a time scale transformation, *Lifetime Data Analysis*, **3**, 27-45,(1997).

Part V

MODELLING IN SURVIVAL AND RELIABILITY ANALYSIS

15

A LOOK BEHIND SURVIVAL DATA:
UNDERLYING PROCESSES AND QUASI-STATIONARITY

Odd O. Aalen

Section of Medical Statistics
P.O.Box 1122 Blindern, N-0317 Oslo, Norway.
E-mail: o.o.aalen@basalmed.uio.no

Håkon K. Gjessing

Norwegian Institute of Public Health
P.O.Box 4404, Nydalen, N-0403 Oslo, Norway
E-mail: hakon.gjessing@fhi.no

In survival and event history analysis the focus is usually on the mere occurrence of events. Not much emphasis is placed on understanding the processes leading up to these events. The simple reason for this is that these processes are usually unobserved. However, one may consider the structure of possible underlying processes and draw some general conclusions from this. One important concept being of use here is quasi-stationary distributions. These arise as limiting distributions on transient spaces where probability mass is continuously being lost to some set of absorbing states. Due to this leaking of probability mass, the limiting distribution is just stationary in a conditional sense, that is, conditioned on non-absorption. Quasi-stationarity is a research theme in stochastic process theory, with several established results, although not too much work has been done. Quasi-stationary distributions act as attractors on the set of individual underlying processes, and can be a tool for understanding the shape of the hazard rate.

We shall explain the use of this concept in survival analysis. Stochastic models based on Markov chains, diffusion processes and Lévy processes will be mentioned.

1. Introduction

In survival and event history analysis one studies the times until occurence of events. The typical analysis just focuses on the time until an event, with no consideration of the underlying process leading up to the event. The reason for this is that, usually, one will not have much, if any, information concerning this process. It is our opinion, however, that one should not disregard the underlying process. It might be fruitful and illuminating to model such a process, even though in specific cases it might be speculative. In this respect there is some analogy to frailty theory, where the explicit consideration of frailty helps the understanding of a phenomenon even in situations where frailty cannot be explicitly estimated.

Certainly, one would think that events are usually manifestations of some process happening beneath the surface. There are exceptions to this, like the occurence of pure accidents where no deeper reasons can be sought. However, in most cases there must be an underlying process. One can then raise the question what kind of survival models follow from specific under-lying processes.

We shall assume that all individuals move around on a state space according to some stochastic process. The risk of the event occurring varies over the state space, in a continuous or discontinuous fashion. A simple special case would be when the event corresponds to the process hitting a special state or set of states.

When the underlying process is modelled as a Markov process, then often a phenomenon termed *quasi-stationarity* arises. This means that distribution on the state space of surviving individuals (that is, those for which the event has not yet happened) converges to some limiting distribution. An implication of this is that the hazard rate of the event converges towards a constant when time increases. The phenomenon of quasi-stationarity may be of interest in itself, but it turns out that quasi-stationarity may also be a tool for understanding the various shapes that hazard rates can assume. The hazard rate is a fundamental concept in survival and event history analysis, being for instance the basis of both the Cox model and the counting process approach. In spite of apparent simplicity, however, it is really a complex concept that is poorly understood.

The interesting thing about the quasi-stationary distribution is that it acts as an *attractor* on surviving individuals. In general there will be two forces operating on the individuals, namely the diffusion in the state space and the occurence of events, which takes individuals out of the state space. The joint effect of this is embodied in an attraction to the quasi-stationary distribution. This might give valuable information on how, for instance, the hazard rate changes over time.

In Aalen and Gjessing[2] these ideas are spelt out in detail for a number of examples and special models. The aim of the present paper is to review these issues with some new cases and examples. We will look at Markov chains, Wiener and Ornstein-Uhlenbeck processes, as well as Lévy processes. We shall use both first-passage times models as well as "killing" models, that is models where each individual has his own stochastic hazard rate.

2. Phase-type models

The simplest type of stochastic process used to describe the underlying developments leading up to an event, would be a finite-state Markov chain. It might contain a transient space and an absorbing state, as indicated in Figure 1. The time to absorption starting somewhere in the transient state space is said to follow a phase-type distribution. Often the transient space might form a single class, but this does not have to be the case. As pointed

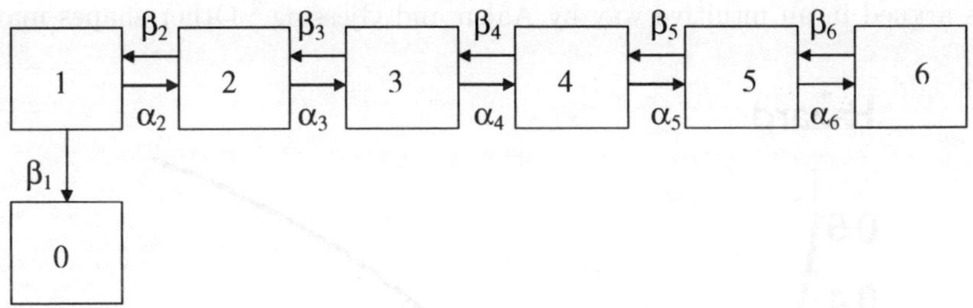

Fig. 1.   Markov chain with transition intensities

out in Aalen[1] the shape of the hazard rate of a phase-type distribution depends on how far away the starting point is from the absorbing state. This is demonstrated in Figure 2: It is evident that starting far away from the absorbing state yields an increasing hazard, starting in the middle of the state space yields a unimodal hazard, and starting very close to the absorbing state yields a decreasing hazard. Hence the three major hazard shapes arise naturally dependent on the distance from the starting state to the absorbing state. This seems to be a general phenomenon occuring, at least approximately, in many models.[2] One question that naturally arises is what is meant by a large or a small distance from the absorbing state; how do we calibrate the concept of distance. It turns out that the quasi-stationary distribution is helpful here. In the present phase-type model with all α's and β's equal to 1, the quasi-stationary distribution on states 1 to

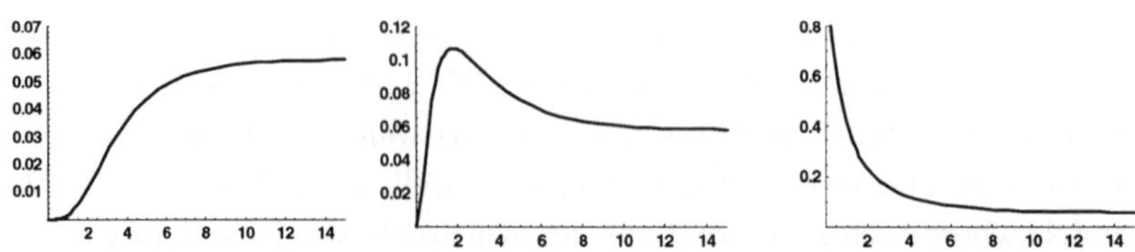

Fig. 2. Hazard rates for time to absorption when the starting is in state 6 (left panel), state 3 (middle panel) and state 1 (right panel). All α's and β's are equal to 1.

6 equals $\{0.058, 0.113, 0.161, 0.200, 0.227, 0.241\}$. Starting out with this distribution yields a constant hazard rate equal to 0.058, and it is seen that the curves in Figure 2 converge to this value. From any starting distribution the distribution of survivors on the state space will converge towards the quasi-stationary one. When evaluating distance from the absorbing state in this context, one must compare it to the quasi-stationary distribution as argued in an intuitive way by Aalen and Gjessing.[2] Other shapes may

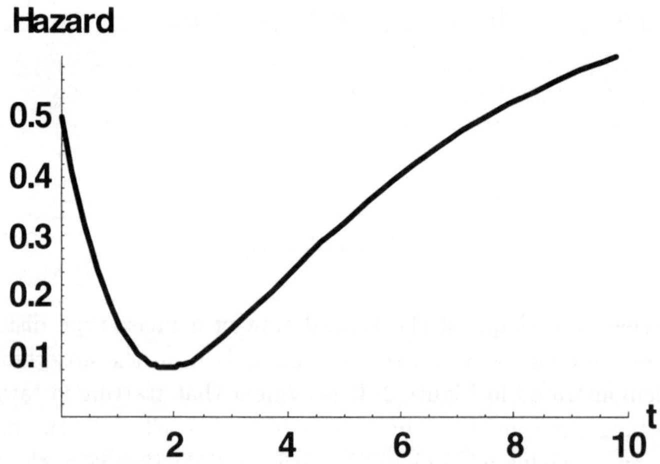

Fig. 3. Hazard rate for time to absorption starting in state 6 with a direct transition to state 0 of rate γ. Here $\gamma = 0.5$, all α's equal 0 and all β's equal 1.

easily arise in the same setting. The bathtub shape is of interest in reliability theory. This may arise in the model of Figure 2 if, for instance, we add a direct transition from state 6 to state 0 with transition rate γ. An illustration is given in Figure 3 for a set of parameter values.

3. Diffusion processes

A natural next step is to consider Markov processes with a continuous state space, in particular, diffusions with an absorbing boundary. It is well known that quasi-stationarity exist for a number of such processes, and also distributions for first passage times are known in some cases. To be more precise, we shall consider a diffusion process on the positive half line with zero as the absorbing state. Let $\sigma^2(x)$ and $\mu(x)$ be the variance and drift diffusion coefficients, respectively, in state x, and let $\phi_t(x)$ be the density on the state space at time t, conditional on non-absorption up to time t.

For diffusion processes there exists a direct connection between the distribution of survivors on the transient state space and the hazard rate of the time to absorption. More precisely, it can be shown that the hazard rate of absorption at time t is given as:[2]

$$\omega_t = \frac{\sigma^2(0)}{2}\phi_t'(0),$$

This is illustrated in Figure 4. It can be used to argue for the validity of

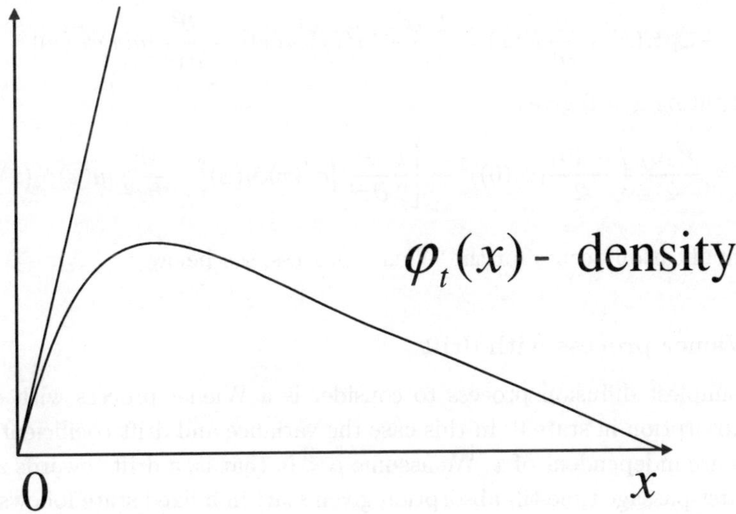

Fig. 4. The hazard rate of time to absorption is proportional to the derivative at 0 of the distribution of survivors

connections similar to those found in the phase-type case: If the starting distribution is close to the absorbing point compared to the quasi-stationary one, then $\phi_t(x)$ will have a flux away from the absorbing state, and one would expect the hazard rate ω_t to decrease. If the starting distribution is far away compared to the quasi-stationary one, it will move towards the absorbing state, yielding an increasing ω_t. Unimodal hazard rates would also be expected to occur in many situations, for instance the following one: Assume that the process starts in a point which is quite close to the absorbing one. Then the hazard rate will increase from 0 while $\phi_t(x)$ spreads out. Soon $\phi_t(x)$ will start losing mass on the left due to absorption while still diffusing further towards the right, yielding a decreasing hazard. The point is further illustrated in Aalen and Gjessing (Figure 6).[2]

Following the arguments in the latter reference, we can also go a step further and derive the condition on $\phi_t(x)$ that yields a local increase or decrease in the hazard rate ω_t. From Kolmogorov's forward equation one can derive:

$$-\omega_t\phi_t(x) + \frac{\partial}{\partial t}\phi_t(x) = \frac{1}{2}\frac{\partial^2}{\partial x^2}\left[\sigma^2(x)\phi_t(x)\right] - \frac{\partial}{\partial x}\left[\mu(x)\phi_t(x)\right]$$

Differentiation with respect to x yields:

$$-\omega_t\phi_t'(x) + \frac{\partial}{\partial t}\phi_t'(x) = \frac{1}{2}\frac{\partial^3}{\partial x^3}\left[\sigma^2(x)\phi_t(x)\right] - \frac{\partial^2}{\partial x^2}\left[\mu(x)\phi_t(x)\right]$$

Substituting $x = 0$ gives:

$$\frac{\partial}{\partial t}\omega_t = \frac{\sigma^2(0)}{2}\left(\frac{\sigma^2(0)}{2}(\phi_t'(0))^2 + \left[\frac{1}{2}\frac{\partial^3}{\partial x^3}\left[\sigma^2(x)\phi_t(x)\right] - \frac{\partial^2}{\partial x^2}\left[\mu(x)\phi_t(x)\right]\right]_{x=0}\right) \tag{1}$$

A simplification occurs for the Wiener process, see below.

4. Wiener process with drift

The simplest diffusion process to consider is a Wiener process with drift and absorption in state 0. In this case the variance and drift coefficients σ^2 and μ are independent of x. We assume $\mu < 0$, that is, a drift towards zero. The first-passage time till absorption given start in a fixed state follows the well known inverse Gaussian distribution.

It is known that in this case there is a whole class of quasi-stationary distributions. One of those is the limiting distribution of survivors given a fixed initial state. This is given by the following gamma distribution:

$$\phi_{\mathrm{can}}(x) = \frac{\mu^2}{\sigma^4}\,x\exp\left(-\frac{\mu x}{\sigma^2}\right)$$

and yields a constant hazard of absorption equal to $\theta_0 = (\mu/\sigma)^2/2$. It is demonstrated by Aalen and Gjessing how the distance of the starting point from the absorbing boundary influences the shape of the hazard rate.

The more general quasi-stationary distribution is given by:

$$\phi(x;\theta) = \frac{\theta}{\eta}\left(\exp\left(-\frac{\mu-\eta}{\sigma^2}x\right) - \exp\left(-\frac{\mu+\eta}{\sigma^2}x\right)\right)$$

for $0 < \theta < \theta_0$, where we define $\eta = \sqrt{\mu^2 - 2\sigma^2\theta}$. Here the constant hazard of absorption equals θ. An illustration of the quasi-stationary distributions is shown in Figure 5. The canonical one is that with the least heavy tail. It is seen that the other quasi-stationary distributions may be quite heavy-tailed. As mentioned in the introduction it is interesting to view quasi-

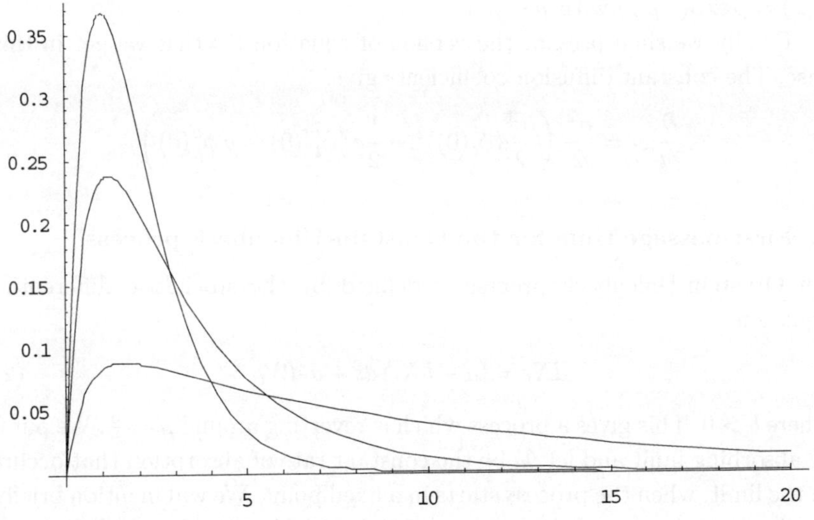

Fig. 5.   Quasi-stationary distributions for the Wiener process with drift absorbed in zero. The distribution with the smallest tail is the quasi-stationary one. (Parameters: $\mu = 1$, $\sigma^2 = 1$. The parameter θ assumes the three values 0.5, 0.3 and 0.1)

stationary distributions as attractors, telling us something about how the distribution of individuals on the state space will change over time. In the present case, the various quasi-stationary distributions have different domains of attraction, and the question arising is which initial distributions yield the various quasi-stationary distributions in the limit. A quite general answer to this is given by Martinez et al.[8] Let $F(x)$ be the cumulative distribution function of the initial distribution, and $f(x)$ the density if it is well defined. Then the following results can be derived from Martinez et al:

(1) if $\lim_{x \uparrow \infty} -(1/x) \ln f(x) = \rho < \mu/\sigma^2$ then the limiting distribution is $\phi(x; \theta)$ where $\theta = \mu\rho - \frac{1}{2}\rho^2\sigma^2$.
(2) if $\underline{\lim}_{x \to \infty} -(1/x) \ln(1-F(x)) \geqslant \mu/\sigma^2$ then the limiting distribution is $\phi_{\text{can}}(x)$. An alternative sufficient condition is:

$$E_f[X \exp(\frac{\mu}{\sigma^2} X)] < \infty$$

All distributions which are limited to a finite interval on the positive real line, i.e. have bounded support, satisfy the conditions in (2) and hence will converge to the canonical quasi-stationary limit. Distributions that trivially fulfil the condition in result 1 are the exponential distributions $f(x) = \rho \exp(-\rho x)$ with $\rho < \mu/\sigma^2$.

Finally we shall present the version of equation (1) that we get in this case. The constant diffusion coefficients give:

$$\frac{\partial}{\partial t}\omega_t = \frac{\sigma^2}{2}\left(\frac{\sigma^2}{2}(\phi_t^{'}(0))^2 + \frac{1}{2}\sigma^2\phi_t'''(0) - \mu\,\phi_t''(0)\right)$$

5. First-passage time for the Ornstein-Uhlenbeck process

An Ornstein-Uhlenbeck process is defined by the stochastic differential equation

$$dX_t = (a - bX_t)\,dt + \sigma\,dW_t \tag{2}$$

where $b > 0$. This gives a process which is reverting around $\mu = \frac{a}{b}$. We put 0 as absorbing limit and let θ_0 be the constant rate of absorption that occurs in the limit, when the process starts in a fixed point. We will mention briefly results from an unpublished paper;[3] see also Ricciardi and Sato.[10] One can show that θ_0 is the smallest value of θ that yields $H_{\theta/b}(-a/(\sigma\sqrt{b})) = 0$, where $H_\nu(\cdot)$ is the Hermite function. The canonical quasi-stationary distribution is given by:

$$\psi_0(x) = \frac{\sqrt{b}}{\sigma} e^{\frac{x(2a-bx)}{\sigma^2}} \frac{H_{\theta_0/b}\left(\frac{-a+bx}{\sigma\sqrt{b}}\right)}{H_{-1+\theta_0/b}\left(-\frac{a}{\sigma\sqrt{b}}\right)} \tag{3}$$

The canonical distribution is a simple modification of a normal distribution to handle the limitation to the positive real axis. In fact, except for a constant factor $\psi_0(x)$ is a parabolic cylinder function, see e.g. Abramowitz and Stegun.[4] By formulas 19.8.1 and 19.13.1 of this reference it follows that the Hermite function part increases asymptotically as $x^{\theta_0/b}$, and therefore the tail of the distribution is not much more heavy than that of a normal distribution.

Consider the symmetric case $a = 0$. Then θ_0 is the smallest zero of $H_{\theta/b}(0)$ considered as as function of θ. This zero occurs for $\theta/b = 1$, hence $\theta_0 = b$. From equation (3) one gets the canonical quasi-stationary distribution:

$$\psi_0(x) = \frac{\sqrt{b}}{\sigma} e^{\frac{-b\,x^2}{\sigma^2}} H_1\left(\frac{b\,x}{\sigma\sqrt{b}}\right) = 2\frac{b}{\sigma^2} x\, e^{\frac{-b\,x^2}{\sigma^2}}$$

6. Modelling the hazard rate as the square of an Ornstein-Uhlenbeck process

The present section is inspired by work of Woodbury and Manton,[11] Myers[9] and Yashin and Vaupel.[12] Previously we have studied the hazard rate as the time to barrier crossing in an Ornstein-Uhlenbeck process. We shall now use the same process in a different setting, namely for modelling the hazard rate directly. We assume that the hazard of an event depends on where the process is in the state space (such models are called "killing models" in the probabilistic literature).

Let X_t be defined by the stochastic differential equation (2). We shall here present just an example, and it is then convenient to assume a model defined by the individual hazard rate $Z_t = X_t^2$. Those for which the event occurs leave the population. The distribution of X_t for survivors then converges to a limit, which is the quasi-stationary dstribution.

Usually, the Ornstein-Uhlenbeck process is only defined for $b \geq 0$, however in the present case we can have any value of b, positive or negative.

The distribution of X_t at time 0 is supposed to be a normal distribution with mean m_0 and variance γ_0. The distribution of X_t given survival can be shown to be normally distributed at any time t, with mean m_t and variance γ_t. The population hazard rate for this model is given by

$$\mu(t) = m_t^2 + \gamma_t$$

where m_t and γ_t are the solutions of the following differential equations:

$$m_t' = a - b\,m_t - 2m_t\gamma_t \tag{4}$$

$$\gamma_t' = -2b\,\gamma_t + \sigma^2 - 2\gamma_t^2 \tag{5}$$

see e.g. Yashin and Vaupel.[12]

When t increases the mean and variance of the normal distribution of survivors converge to a limit. This may be derived from equations (4) and (5) by putting the left hand side equal to zero, yielding:

$$m_\infty = \frac{a}{\sqrt{2\sigma^2 + b^2}}, \qquad \gamma_\infty = \frac{1}{2}(-b + \sqrt{2\sigma^2 + b^2})$$

Hence, the normal distribution with this mean and variance is a quasi-stationary distribution. When looking at the shapes of the hazard rate, it

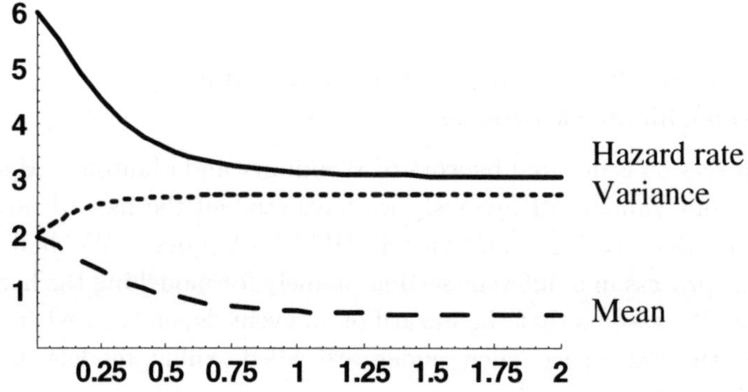

Fig. 6. Showing the hazard rate as well as the mean and variance of the normal distribution as a function of time. (Parameters: $a = 2$, $b = 2$, $\sigma^2 = 2$, $m_0 = 2$, $\gamma_0 = 2$)

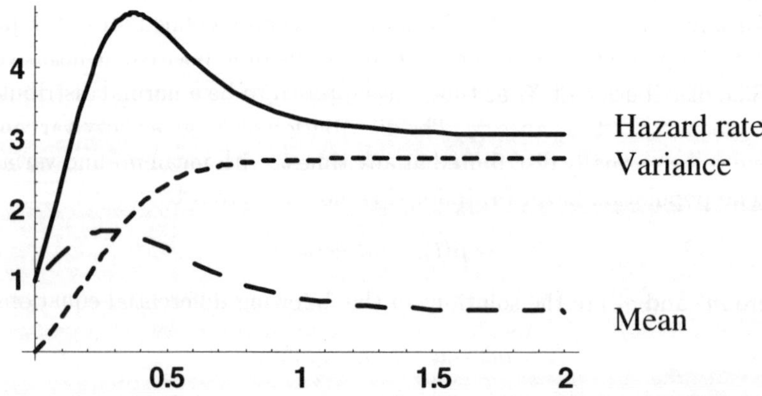

Fig. 7. Showing the hazard rate as well as the mean and variance of the normal distribution as a function of time. (Parameters: $a = 2$, $b = 2$, $\sigma^2 = 2$, $m_0 = 1$, $\gamma_0 = 0$)

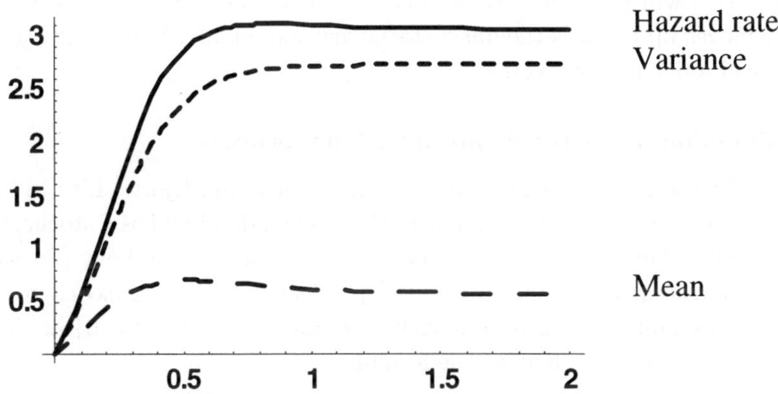

Fig. 8. Showing the hazard rate as well as the mean and variance of the normal distribution as a function of time. (Parameters: $a = 2$, $b = 2$, $\sigma^2 = 2$, $m_0 = 0$, $\gamma_0 = 0$)

appears once more that the basic shapes in Figure 2 typically appear. We have no general results concerning this, but experience with plotting seems to confirm this. A few examples are shown in the figures where equations (4) and (5) are solved numerically. In Figure 6, one sees that the mean of the normal distribution decreases, while the variance increases only slightly, both converging to stationarity. The individual hazard rates would then typically decrease too, and the figure shows a decreasing population hazard rate. Hence, the example clearly shows how the hazard rate depends on changes in the underlying distribution. Similar results are shown in Figure 7. Here the population hazard increases initially because both the mean and the variance increases. Later on the mean and hence the hazard decrease. The reason for this development is probably as follows: Since the variance starts out at zero it will immediately start increasing due to diffusion. Due to the drift the mean will also increase. However, after a while the right part of the distribution will lose mass due to high hazard and the mean and the hazard will start to decrease. A final example with mostly increasing mean and hazard rate is shown in Figure 8. Here both the mean and variance start in zero, and so the diffusion of the Ornstein-Uhlenbeck process will result in increasing mean and variance.

It should be noted from equation (5) that the development of the variance γ_t does not depend on the mean m_t. Hence, starting out with $\gamma_t = \gamma_\infty$, will give a distribution with constant spread, and it will be only the position on the line that changes. In this case, starting below the quasi-stationary distribution will yield an increasing hazard as the distribution on the state space moves up towards the quasi-stationary one. Similarly, starting above will yield a decreasing hazard.

7. Modelling the hazard rate by a Lévy process

We shall review some results from Gjessing, Aalen and Hjort.[7] Like in the previous section, we shall assume here that each individual has a stochastic hazard rate. This time the rate is defined by a nonnegative Lévy process, that is, a process with nonnegative, independent, time-homogeneous increments. The Laplace transform of such a process $Z = \{Z(t) : t \geq 0\}$ at time t is given by the Lévy–Khintchine formula

$$L(c;t) = E \exp\{-cZ(t)\} = \exp\{-t\Phi(c)\},$$

where $c \geq 0$ is the argument of the Laplace transform. The function $\Phi(c)$ is called the Laplace exponent of the Lévy process. The family of Lévy processes contains a number of important special cases, like compound Poisson processes, gamma processes, stable processes etc. All nonnegative Lévy processes are limits of compound Poisson processes.

The individual hazard rates are defined by the following process:

$$h(t) = \int_0^t a(u, t - u)\, dZ(R(u)).$$

Here, we consider a nonnegative deterministic rate function $r(t)$ with integral $R(t) = \int_0^t r(u)\, du$, and where $Z(R(t))$ is the time-transformed Lévy process. The function $a(u, s)$ is a weight function determining the influence of previous jumps in $Z(t)$. The weight function has two time scales, namely the time scale of the stochastic process (first argument), and time as measured in distance from the current time (second argument).

It can be proved that the population hazard rate is:

$$\mu(t) = \int_0^t \Phi'(b(u, t))\, a(u, t - u)\, r(u)\, du$$

where $b(u, t) \stackrel{\text{def}}{=} \int_u^t \lambda(s)\, a(u, s - u)\, ds$.

To consider a simple special case, let $a(t, v) \equiv a(v)$ depend only on the second argument, and let $r(t) \equiv 1$. Define $A(v) = \int_0^v a(u)\, du$. Then $b(u, t) = A(t - u)$, and the population survival function and hazard rate are then given by

$$S(t) = \exp(-\int_0^t \Phi(A(v))\, dv) \quad \text{and} \quad \mu(t) = \Phi(A(t)). \tag{6}$$

We shall further assume that Φ is bounded, this is equivalent to the process being compound Poisson. Then $\mu(t)$ converges to a limit when $t \to \infty$, and hence one would expect there to be a quasi-stationary limit.

In order to exhibit the quasi-stationary distribution we shall assume that the weight function is constant, that is $a(v) = a$. Furthermore, we consider a compound Poisson process with gamma distributed jumps. In that case $\Phi(c) = \rho\{1 - (\nu/(\nu + c))^n\}$ and the Lévy measure is

$$\Pi(dx) = \rho\frac{\nu^n}{\Gamma(n)}x^{n-1}e^{-\nu x}\,dx.$$

Note that the Lévy measure is the product of the rate ρ of the underlying Poisson process and the probability density of the jump size.

It follows from equation (6) that

$$\mu(t) = \rho\{1 - (\nu/(\nu + a\,t)^n\}$$

When $t \longrightarrow \infty$, one sees that $\mu(t) \longrightarrow \rho$. If $n > 1$ it can be proved that there exists a quasi-stationary distribution which is the value at time 1 of a compound Poisson process with Lévy measure

$$\tilde{\Pi}(dx) = \rho\frac{\nu^n}{a^n\Gamma(n)}x^{n-2}e^{-\nu x/a}\,dx,$$

Hence the jump times of this compound Poisson process occurs with a rate $\rho\nu/(a(n - 1))$, and the jump sizes are gamma distributed with scale parameter ν/a and shape parameter $n - 1$.

8. Other models

Bains[5] suggests a model where survival the depends on the functioning of a highly interconnected network of individual elements with constant failure rates. When connection is lost in the network, it is regarded as failing. Interestingly, the hazard rate of this increases with time, but then shows asymptotic constancy. Hence, there is likely to be quasi-stationarity occurring, although it might be difficult to exhibit the quasi-stationary distribution due to the complexity of the network. See also the paper by Gavrilov and Gavrilova[6] which discusses similar issues to Bains.

Typically, quasi-stationarity occurs in situations where there is a transient states space where all states are connected, that is, they are forming a single class. This is the case for the typical phase-type distribution, as well as the diffusion process models. However, there are exceptions from this, the Levy-process being one such example. What creates quasi-stationarity in, for instance, the compound Poisson process example above, is the fixed rate of leaving state zero, that is the state when $h(t) = 0$. This can be considered a "bottle neck" for the further rise of the hazard. Such bottle-necks may determine the dominant eigenvalue for the process. They could easily be present in phase-type and other models and create a quasi-stationary distribution from the bottle-neck and onwards even though there may not be reversibility in the state space.

References

1. O. O. Aalen, Phase type distributions in survival analysis, *Scandinavian Journal of Statistics*, **22** , 447-463 (1995).
2. O. O. Aalen and H. Gjessing, Understanding the shape of the hazard rate, *Statistical Science*,**16**, 1-22 (2001).
3. O. O. Aalen and H. Gjessing, Survival models based on the Ornstein-Uhlenbeck process, Unpublished manuscript (2002).
4. M. Abramowitz and I. A. Stegun, *Handbook of mathematical functions* (Dover Publications, New York, 1964).
5. W. Bains, Statistical mechanic prediction of non-Gompertzian ageing in extremely aged populations, *Mechanisms of Ageing and Development*, **112**, 89-97 (2000).
6. , L. A. Gavrilov and N. S. Gavrilova, The reliability theory of aging and longevity, *Journal of Theoretical Biology*,**213**, 527-545 (2001).
7. H. Gjessing, O. O. Aalen and N. L. Hjort, Frailty models based on Lévy processes, Submitted, (2002).
8. S. Martinez, P. Picco and J. San Martin, Domain of attraction of the quasi-stationary distributions for the Brownian motion with drift, *Advances of Applied Probability*, **30**, 385-408 (1998).
9. L. E. Myers, Survival functions induced by stochastic covariate processes, *Journal of Applied Probability*,**18**, 523-29 (1981).
10. L. M. Ricciardi and S. Sato, First-passage-time density and moments for the Ornstein-Uhlenbeck process, *Journal of Applied Probability*, **25**, 43-57 (1988).
11. M. A. Woodbury and K. G. Manton, A random walk model of human mortality and aging, *Theoretical Population Biology*, **11**, 37-48 (1977).
12. A. I. Yashin and J.W. Vaupel, Measurement and estimation in heterogeneous populations, in *Lecture Notes in Biomathematics 65: Immunology and Epidemiology*, Eds. G. W. Hoffman and T. Hraba (Springer, Berlin, 1986), pp. 198-206.

16

ON SOME PROPERTIES OF DEPENDENCE AND AGING FOR RESIDUAL LIFETIMES IN THE EXCHANGEABLE CASE

Bruno Bassan and Fabio Spizzichino

Dipartimento di Matematica, Università "La Sapienza"
P. A. Moro 5, I–00185 Roma, Italy
E-mail: bassan@mat.uniroma1.it, fabio.spizzichino@uniroma1.it

For a pair of exchangeable lifetimes we analyze several properties of the joint distribution of the residual lifetimes, conditional on histories of survivals. Attention is focused on properties of dependence, univariate aging and bivariate aging. In particular, we deal with notions of PQD, NBU, bivariate NBU and their negative counterparts. The role of copulæ in this analysis is highlighted.

The case of time-transformed exponential lifetimes, where the copulæ involved are Archimedean, is analyzed in detail. Several examples and remarks illustrate the main points of this note and may give hints for future research.

1. Introduction

There is a general interest, in the relevant literature on survival analysis, toward the interplay between dependence properties among individuals and features of the distribution of their residual lifetimes as long as their age increases. Here, we consider some specific issues in the case of similar, dependent individuals.

Let X and Y be the lifetimes of two individuals; here we assume X and Y to be exchangeable and study how their dependence and aging features evolve as time elapses and the individuals continue to survive. In other words we are interested in analyzing, for any *age* $r > 0$, dependence and aging properties of the joint conditional law of the residual lifetimes $X - r, Y - r$, given the survival data $\{X > r, Y > r\}$.

In this note we concentrate attention on some technical, rather specific, aspects of this general subject.

We deal with three main features: dependence, marginal aging, i.e. the

235

one-dimensional aging properties of the marginal laws, and what we may call multivariate aging. The latter notion can be given several interpretations. We consider here multivariate aging in the sense studied by Bassan and Spizzichino [4], which can be roughly summarized as follows: a joint law has positive multivariate aging if, conditionally on a same history of survivals and failures, the marginal law of the residual lifetime of a younger item dominates in some stochastic order sense the corresponding law of an older item.

In Bassan and Spizzichino [6] it was pointed out that multivariate aging properties of a joint law can be translated into dependence properties of a function B which has all the properties of a copula, except possibly for the fact that it need not be 2-increasing. Recall that a copula, in the simplest case, is a joint distribution function on the unit square with uniform marginals. For an overview of copulæ and of dependence concepts such as TP_2, PQD etc., see e.g. Joe[9], Drouet-Mari and Kotz[8] and Nelsen[11]. Furthermore, we refer to Marshall and Olkin[10] for the definition and properties of Schur-concavity, whereas we refer to Barlow and Mendel[2] and Spizzichino[13] for a discussion about the role of Schur-concavity in the analysis of multivariate aging. Finally, we refer to Shaked and Shantikumar[12] for stochastic orders and for details about basic univariate aging notions such as NBU, IFR etc..

Let $\mathcal{G}$ be the class of one-dimensional, continuous survival functions $\bar{G}$ which are positive and strictly decreasing on $\mathbb{R}_+$ and such that $\bar{G}(0) = 1$. We denote by $\mathcal{F}^{(2)}$ or simply $\mathcal{F}$ the class of 2-dimensional exchangeable survival functions on $\mathbb{R}_+^2$ whose one-dimensional marginals are in $\mathcal{G}$.

Consider now exchangeable random lifetimes X, Y, each with survival function $\bar{G} \in \mathcal{G}$. We consider their joint survival function $\bar{F}(x,y) = \mathbb{P}(X > x, Y > y)$, their marginal survival function $\bar{G}(x) = \mathbb{P}(X > x) = \bar{F}(x,0)$, the *survival copula*

$$K(u,v) = \bar{F}\left(\bar{G}^{-1}(u), \bar{G}^{-1}(v)\right), \qquad 0 \leq u,v \leq 1, \tag{1}$$

and the *multivariate aging function*

$$B(u,v) = \exp\left\{-\bar{G}^{-1}\left(\bar{F}(-\log u, -\log v)\right)\right\}, \qquad 0 \leq u,v \leq 1. \tag{2}$$

Recall that the survival copula describes the dependence structure of the joint law, and that each of the pairs $(B, \bar{G}), (K, \bar{G})$ completely specifies $\bar{F}$, as the following relations show:

$$\bar{F}(x,y) = K\left(\bar{G}(x), \bar{G}(y)\right) = \bar{G}\left(-\log B\left(e^{-x}, e^{-y}\right)\right). \tag{3}$$

The role of the function B in describing multivariate aging has been studied in Bassan and Spizzichino[6][7]. As we mentioned before, it shares many properties of copulæ, but it need not be 2-increasing. More specifically, B, like a copula, is increasing in each variable and satisfies

$$B(0, v) = B(u, 0) = 0; \qquad B(u, 1) = u; \qquad B(1, v) = v, \qquad \forall u, v \in [0, 1],$$

but the 2-variation $B(u + h, v + k) + B(u, v) - B(u + h, v) - B(u, v + k)$ need not be positive. In the absolutely continuous case, this translates into the fact that the mixed partial derivative need not be positive. We may occasionally refer to functions with the above properties as *semi-copulæ*.

As mentioned above, our purpose in this note is the analysis of dependence and aging properties of residual lifetimes of two individuals, conditionally on their survival up to the age $r > 0$.

For this reason we concentrate attention on the joint survival function of the residual lifetimes at epoch r, namely

$$\bar{F}_r(x, y) = \mathbb{P}(X \geq x + r, Y \geq y + r | X \geq r, Y \geq r) = \frac{\bar{F}(x + r, y + r)}{\bar{F}(r, r)}. \quad (4)$$

Let $K_r, B_r, \bar{G}_r$ denote the survival copula, multivariate aging function and marginal survival function of $\bar{F}_r$, respectively.

In the following, we shall analyze several aspects of these objects and, more specifically, we shall obtain some sufficient or necessary conditions for their positive (or negative) dependence, positive (or negative) unidimensional aging, and positive (or negative) bivariate aging.

We shall also state some simple result about the way these properties evolve when the age r increases.

In order to maintain the length of the paper within reasonable limits, we choose to illustrate basic ideas by concentrating attention only on certain specific notions of dependence and aging. For what concerns concepts of dependence, we shall consider the concepts of Positive and Negative Quadrant Dependence (PQD and NQD). For what concerns notions of univariate and bivariate aging, we shall consider New Better than Used and New Worse than Used (NBU and NWU), Bivariate New Better than Used and New Worse than Used (BNBU and BNWU), that, in a sense, are the notions of aging that correspond to quadrant dependence; the meaning of such correspondence will be briefly sketched in the sequel and can be more completely explained in terms of the method presented in Bassan and Spizzichino[7].

More precisely, this note will be organized as follows. In Section 2 we recall some properties of dependence and aging; in particular, we focus our attention on notions that can be described in terms of quadrant dependence properties of suitable copulæ. In Section 3 we deal with relations and compatibility results among features of dependence and aging. In Section 4 we consider residual lifetimes, and we study how their survival functions are affected by dependence and aging properties of the original survival function. Finally, in Section 5 we focus our attention to the special case of time-transformed exponential models, for which some additional results and examples can be provided.

For brevity reasons we also limited the analysis to the case of bidimensional joint distributions; our arguments can be extended to the case of $n > 2$ components, in a more or less straightforward way.

2. Quadrant dependence and aging

In this section we concentrate attention on several properties of the concepts of "quadrant dependence"; we shall see the related role in the description of dependence and of aging for the joint distribution of the lifetimes X, Y. We start with a few well-known definitions and corresponding properties of the survival copula K.

Definition 1: We say that a distribution function F is *positively quadrant dependent* (PQD) if $F(x, y) \geq G(x)G(y), \forall x, y \in \mathbb{R}$.

It is well known that F is PQD iff $\bar{F}(x, y) \geq \bar{G}(x)\bar{G}(y), \forall x, y \in \mathbb{R}$. This condition will be also referred to as: $\bar{F}$ is PQD. When the inequalities are reversed, we say that F (or $\bar{F}$) is NQD.

In the study of quadrant dependence properties it is also of interest to consider the notion of TP_2. The latter is a notion of dependence stronger than PQD. In fact, we shall see below that the TP_2 property of $\bar{F}$ implies that all the joint survival functions $\bar{F}_r$ of the residual lifetimes are PQD.

Proposition 2: *Let $\bar{F} \in \mathcal{F}$. The following are equivalent.*

(1) $\bar{F}$ is PQD (NQD) if and only if K is PQD (NQD);
(2) $\bar{F}$ is TP_2 (RR_2) if and only if K is TP_2 (RR_2).

Definition 3: Let $\bar{F}, \bar{H} \in \mathcal{F}$, and assume that they have the same univariate marginals. We say that H is *more concordant*, or more PQD, than F if $\bar{H}(\mathbf{x}) \geq \bar{F}(\mathbf{x}), \forall \mathbf{x} \in \mathbb{R}^2_+$.

The following proposition relates the notion of more concordant laws with the notion of pointwise comparison of survival copulæ. Together with

Theorem 16, it can provide the tools for the analysis of the evolution of the dependence of residual lifetimes, as time elapses. We shall carry out this analysis for Time Transformed Exponential models in Section 5.1.

Proposition 4: *Let $\bar{H}$ and $\bar{F}$ admit the same one-dimensional marginals. Then $\bar{H}$ is more concordant than $\bar{F}$ iff $K_{\bar{H}} \geq K_{\bar{F}}$.*

We now turn our attention to the function B and to the meaning of PQD properties of it. Let us first briefly recall the following definitions of univariate aging.

Definition 5: A positive random variable with survival function $\bar{G}$ is said to be NBU (*New Better than Used*) if $\bar{G}(x + y) \leq \tilde{G}(x)\tilde{G}(y)$, $\forall x, y \geq 0$. It is said to be NWU if the opposite inequality holds.

Observe that the notion of NWU is actually equivalent to the PQD property of a suitable copula: $\bar{G}$ is NWU iff the copula $\bar{G}\left(\bar{G}^{-1}(x) + \bar{G}^{-1}(y)\right)$ is PQD. See Averous and Dortet-Bernadet[1].

In general, our interest for dependence properties of B lies in that they allow us to describe bivariate aging properties of $\bar{F}$. See Bassan and Spizzichino[6][7]).

In particular, in the sequel we shall use the following definition: a bivariate exchangeable law is said to be *Bivariate New Better than Used* (BNBU) if its function B is PQD, i.e. satisfies $B(u, v) \geq uv, \forall u, v \in (0, 1)$. Indeed, this condition is equivalent to the following inequality:

$$\mathbb{P}\left(X > t|Y > r\right) \geq \mathbb{P}\left(Y > t + r|Y > r\right). \tag{5}$$

This relation can be given the following interpretation: conditionally on a same history according to which one item is new and another item is aged at least r, the new item is to be preferred. The notion of BNWU is defined by reversing the above inequalities.

It will be also useful to consider the condition of Schur-concavity of $\bar{F}$. The latter is a condition stronger of aging than BNBU; in particular, as we shall see below, $\bar{F}$ is Schur-concave iff $\bar{F}_r$ is BNBU for every $r \geq 0$.

The following proposition shows the relation between the function B and the Schur properties of the joint survival function. For the proof, see Bassan and Spizzichino[7].

Proposition 6: *$\bar{F} \in \mathcal{F}$ is Schur-concave (convex) if and only if*

$$B(us, v) \geq (\leq) B(u, vs), \quad u \geq v, 0 < s < 1.$$

In particular, $\bar{F}$ is Schur constant if and only if $B(u_1, u_2) = u_1 u_2$.

3. Relations among dependence, multivariate aging and marginal aging

In this section we shall address the problem of the mutual relations among the concepts of dependence and aging introduced in the previous section. Let us start with some general considerations.

Remark 7: Positive multivariate aging does not imply positive marginal aging. We present here an example based on the notions of bivariate and univariate aging of specific interest in this note, namely, we provide an example of a pair (T_1, T_2) which is BNBU but not marginally NBU.

Let T_1, T_2 be i.i.d. conditionally on a positive random variable Θ, with

$$\bar{G}\left(x \mid \theta\right) = \mathbb{P}\left(T_1 > x \mid \Theta = \theta\right) = \exp\left\{-\theta^2\right\}, \qquad \mathbb{P}\left(\Theta > \theta\right) = \exp\left\{-\beta\theta\right\}.$$

Thus, the conditional law of T_1 is a Weibull$(\theta, 2)$, which is IFR, so that $\bar{F}\left(x, y \mid \theta\right) = \bar{G}\left(x \mid \theta\right)\bar{G}\left(y \mid \theta\right)$ is Schur-concave. Since Schur-concavity is preserved under mixtures, the unconditional joint survival function

$$\bar{F}\left(x, y\right) = \int_0^\infty \bar{G}\left(x \mid \theta\right)\bar{G}\left(y \mid \theta\right)e^{-\beta\theta}d\theta = \frac{\beta}{\beta + x^2 + y^2}$$

is also Schur-concave. Thus, $B(u, v) \geq uv$, and we have positive multivariate aging. On the other hand, the (unconditional) marginal survival function $\bar{G}\left(x\right) = \beta\left[\beta + x^2\right]^{-1}$ need not be NBU, so that we are not necessarily faced with positive univariate aging. In fact, for example, take $\beta = 1$, and observe that $\bar{G}\left(4\right) = 1/17 > \bar{G}\left(2\right)\bar{G}\left(2\right) = 1/25$.

Remark 8: Positive dependence may coexist with negative multivariate aging. Again, we shall present an example involving the notions of dependence (PQD or NQD) and bivariate aging (BNBU or BNWU) of interest here. In fact, we shall exhibit a pair (T_1, T_2) whose survival function is PQD and BNWU.

Let T_1, T_2 be i.i.d. conditionally on a r.v Θ, with a DFR conditional marginal law. Then both the conditional and the unconditional joint survival function are Schur-convex, and hence we are faced with negative bivariate aging. On the other hand, T_1, T_2 are positively dependent, since they all positively depend on a same r.v. Θ.

Here is an example: Let $\mathcal{L}(T_1 \mid \Theta = \theta)$ be a Weibull$(\theta, \frac{1}{2})$, i.e. $\bar{G}(x \mid \theta) = \exp\{-\theta\sqrt{x}\}$, and let $\mathcal{L}(\Theta)$ be a standard exponential. Then

$$\bar{F}(x, y) = \int_0^\infty e^{-\theta\sqrt{x}}e^{-\theta\sqrt{y}}e^{-\theta} = \frac{1}{1 + \sqrt{x} + \sqrt{y}}.$$

Some simple calculations show that $K(u,v) = \left(u^{-1} + v^{-1} - 1\right)^{-1} \geq uv$ and $B(u,v) = uv\exp\left\{-2\sqrt{\log u \log v}\right\} \leq uv$, so that $\bar{F}$ is PQD and BNWU. Observe that B is not a copula, in this case.

Remark 9: Negative dependence may coexist with positive multivariate aging. See Remark 18 below.

Although, as we have seen in the previous remarks, there is more freedom than one may expect in the relations among dependence and aging, not all combinations are allowed. In fact, in Bassan and Spizzichino[7] this issue is addressed in some generality. In what follows we give a sample of the type of results that can be obtained

Proposition 10: *Let $F \in \mathcal{F}$. The following implications hold.*

(1) If $\bar{F}$ is BNBU and $\bar{G}$ is NWU, then K is PQD.
(2) If K is NQD and $\bar{G}$ is NWU, then $\bar{F}$ is BNWU.
(3) If $\bar{F}$ is BNBU and K is NQD, then $\bar{G}$ is NBU.
(4) If $\bar{F}$ is BNWU and $\bar{G}$ is NBU, then K is NQD.
(5) If K is PQD and $\bar{G}$ is NBU, then $\bar{F}$ is BNBU.
(6) If F is BNWU and K is PQD, then $\bar{G}$ is NWU.

Proof: Let us consider the first claim. It is evident from (5) that $\bar{F}$ is BNBU if and only if $\bar{F}(x,y) \geq \bar{F}(0,x+y) = \bar{G}(x+y)$. Hence, if $\bar{G}$ is NWU, we have $\bar{F}(x,y) \geq \bar{G}(x+y) \geq \bar{G}(x)\bar{G}(y)$, and the claim follows.

As far as the second point is concerned, we have $\bar{F}(x,y) \leq \bar{G}(x)\bar{G}(y) \leq \bar{G}(x+y)$.

Finally, for the third claim, we have $\bar{G}(x+y) \leq \bar{F}(x,y) \leq \bar{G}(x)\bar{G}(y)$. Clearly, the first inequality is due to the BNBU property and the second to NQD. The other points are proved similarly. $\square$

Notice that the above proof is straightforward. In Bassan and Spizzichino[7] a more general method is provided that allows us to prove in a unified way a class of compatibility results like those in Proposition 10.

This method is based on the use of a suitable axiomatic definition of correspondence among concepts of dependence, univariate aging and bivariate aging. For example, the notions of aging axiomatically corresponding to PQD are NBU and BNBU.

4. Residual lifetimes

In the previous section we considered the functions $K, B,$ and $\overline{G}$, corresponding to a joint survival function $\overline{F}$, and analyzed the relation existing among dependence and aging properties of them. In this section we switch our interest toward the survival functions $\overline{F}_r$ defined in (4); namely, we are interested in the law of the residual lifetimes of two exchangeable components, knowing that they are still functioning (alive) at time r.

For $r \geq 0$, let then $\overline{F}_r$ be given by (4), and let $K_r, B_r,$ and $\overline{G}_r$ (or K_{F_r}, etc.) denote the survival copula, multivariate aging function and marginal survival function of $\overline{F}_r$, respectively. Explicit formulas for $K_r, B_r,$ and $\overline{G}_r$ can be obtained very easily, but are rather involved and not too useful for our purposes.

Here, we aim rather to analyze how properties of $K, B,$ and $\overline{G}$ reflect both on dependence and aging properties of $\overline{F}_r$ for fixed $r > 0$ and on the evolution of $K_r, B_r,$ and $\overline{G}_r$, for increasing values of r.

First, we show that that the evolution of the multivariate aging functions of the residual lifetimes, B_r, depends only on the "initial" function B.

Proposition 11: *The family $\{B_r(u, v) \mid r \geq 0\}$ is determined only by* $B(u, v)$.

The proof is a simple consequence of the following lemma.

Lemma 12: *Let $\overline{F} \in \mathcal{F}$ be strictly decreasing in each of the variables. Then, for all $r > 0$ and $u, v \in (0, 1)$, $B_r(u, v)$ is the unique solution γ of the equation*

$$B\left(ue^{-r}, ve^{-r}\right) = B\left(\gamma e^{-r}, e^{-r}\right) \tag{6}$$

Proof: We have

$$\overline{F}_r(-\log u, -\log v) = \frac{\overline{F}(r - \log u, r - \log v)}{\overline{F}(r, r)} = \frac{\overline{G}\left(-\log\left(B\left(ue^{-r}, ve^{-r}\right)\right)\right)}{\overline{F}(r, r)}.$$

On the other hand $\overline{F}_r(-\log u, -\log v) = \overline{G}_r\left(-\log\left(B_r(u, v)\right)\right)$. Now,

$$\overline{G}_r\left(-\log z\right) = \frac{\overline{F}(r - \log z, r)}{\overline{F}(r, r)} = \frac{\overline{G}\left(-\log\left(B\left(ze^{-r}, e^{-r}\right)\right)\right)}{\overline{F}(r, r)}, \quad \forall z \in \mathbb{R}_+,$$

so that

$$\overline{F}_r(-\log u, -\log v) = \frac{\overline{G}\left(-\log\left(B\left(B_r(u, v)e^{-r}, e^{-r}\right)\right)\right)}{\overline{F}(r, r)}$$

Thus, we see that $B_r(u,v)$ is a solution of (6). Uniqueness follows from the facts that $\bar{F}(r,r) > 0$, that $\bar{G}$ is strictly decreasing, and that $\bar{F}$ is strictly decreasing. $\qquad\Box$

Remark 13: A statement similar to Proposition 11 does not hold for the survival copulæ; namely, K_r depends on K and also on B, in general. However, in Proposition 14 below, it will be shown in particular that certain properties of dependence of K_r are determined by properties of K alone. Some more specific results in this vein will be proved in Section 5 for time-transformed exponential models.

Proposition 14: *Let $\bar{F} \in \mathcal{F}$. Then:*

(1) If $\bar{F}$ is TP_2 (respectively: RR_2), then $\bar{F}_r$ is PQD (respectively: NQD), for every $r \geq 0$.

(2) $\bar{F}$ is Schur-concave (respectively: Schur-convex) if and only if $\bar{F}_R$ is BNBU (respectively: BNWU), for every $r \geq 0$.

(3) If the application $x \mapsto \bar{F}(x,y)$ is log-concave (resp.: log-convex) for every $y \geq 0$, then $\bar{G}_r$ is NBU (resp.: NWU), for every $r \geq 0$.

Proof: If $\bar{F}$ is TP_2, then $\bar{F}(x+r,y+r)\bar{F}(r,r) \geq \bar{F}(x+r,r)\bar{F}(r,y+r)$, which is equivalent to

$$\bar{F}_r(x,y) = \frac{\bar{F}(x+r,y+r)}{\bar{F}(r,r)} \geq \frac{\bar{F}(x+r,r)}{\bar{F}(r,r)}\frac{\bar{F}(r,y+r)}{\bar{F}(r,r)} = \bar{G}_r(x)\bar{G}_r(y),$$

and the first claim is proved. The case of $\bar{F}$ RR_2 is dealt with in a similar way.

Next, let $\bar{F}$ be Schur-concave. Then $\bar{F}(x,y) \geq \bar{F}(x+y,0) = \bar{G}(x+y)$, and this is equivalent, to $B_F(u,v) \geq uv$, in view of (3). It is immediately seen that the Schur-concavity of $\bar{F}$ is equivalent to the Schur-concavity of all the members of the family $\{\bar{F}_r \mid r \geq 0\}$. Hence, if $\bar{F}$ is Schur-concave, then $B_r(u,v) \geq uv$, $\forall r \geq 0$. The claim in the case of Schur-convexity is proved analogously.

Conversely, suppose that $B_r(u,v) \geq uv$, $\forall r \geq 0$. Then, for every $x,y,r \geq 0$, $\bar{F}_r(x,y) \geq \bar{F}_r(x+y,0)$, and this, recalling that $\bar{F}$ is everywhere strictly positive, is equivalent to

$$\bar{F}(x+r,y+r) \geq \bar{F}(x+y+r,r), \qquad \forall x,y,r \geq 0. \tag{7}$$

Now, take two vectors ordered according to the majorization order: $(a',b') \preceq (a'',b'')$. Without loss of generality, we assume that $a' < b'$ and $a'' < b''$.

Then, we have: $a' \leq a'', b' \geq b''$ and $a' + b' = a'' + b''$. Now, set $r = a', x = a'' - a'$ and $y = b' - a''$. Hence, recalling (7), we can write

$$\bar{F}(a', b') = \bar{F}(r, x + y + r) \leq \bar{F}(x + r, y + r) = \bar{F}(a'', b''),$$

and this shows that $\bar{F}$ is Schur-concave.

Finally, let $x \mapsto \bar{F}(x, y)$ be log-concave for every $y \geq 0$. This implies, in particular, the following:

$$\frac{\bar{F}(x + y + r, r)}{\bar{F}(x + r, r)} \leq \frac{\bar{F}(y + r, r)}{\bar{F}(r, r)}, \qquad \forall r, x, y \geq 0.$$

This, in turn, is equivalent to $\bar{G}_r(x + y) \leq \bar{G}_r(x)\bar{G}_r(y), \qquad \forall r, x, y \geq 0$, i.e. $\bar{G}_r$ is NBU. The claim in the NWU case is proved similarly. $\qquad\square$

Combining together Propositions 10 and 14, we get some other implications among dependence and aging properties of $\bar{F}$ and $\{\bar{F}_r | r \geq 0\}$.

Proposition 15: *Let $\bar{F} \in \mathcal{F}$. Then:*

(1) If $\bar{F}$ is Schur-concave and $\bar{G}_r$ is NWU for every $r > 0$, then $\bar{F}_r$ is PQD, for every $r > 0$.

(2) If $\bar{F}$ is RR_2 and $\bar{G}_r$ is NWU, for every $r > 0$, then $\bar{F}$ is Schur-convex.

(3) If $\bar{F}$ is Schur-concave and RR_2, then $\bar{G}_r$ is NBU, for every $r > 0$.

(4) If $\bar{F}$ is Schur-convex and $\bar{G}_r$ is NBU for every $r > 0$, then $\bar{F}_r$ is NQD, for every $r > 0$.

(5) If $\bar{F}$ is TP_2 and $\bar{G}_r$ is NBU, for every $r > 0$, then $\bar{F}$ is Schur-concave.

(6) If $\bar{F}$ is Schur-convex and TP_2, then $\bar{G}_r$ is NWU, for every $r > 0$.

5. Time-transformed exponential models and their residual lifetimes

We consider now the class of the so called *time-transformed exponentials*, namely of those laws whose survival function $\bar{F}$ can be written in the form

$$\bar{F}(x, y) = W(R(x) + R(y)) \tag{8}$$

for some strictly decreasing, convex survival function W and some strictly increasing continuous functions $R : \mathbb{R}_+ \to \mathbb{R}_+$ such that $R(0) = 0$ and $\lim_{x \to \infty} R(x) = \infty$. If (8) holds, we write $\bar{F} \sim TTE(W, R)$.

Relevant particular cases of TTE models are the following: (1) The independent laws (take $W(z) = \exp\{-z\}$). (2) The Schur-constant laws (take $R(z) = z$). (3) The proportional hazard models; in this case, one takes $W(x) \int_0^\infty \exp\{-\theta x\} \, d\Pi(\theta)$, i.e. $W(x)$ is the Laplace transform of a probability distribution Π on $(0, \infty)$ and X, Y are conditionally i.i.d., with conditional marginal survival function $\bar{G}(x|\theta) = e^{-\theta R(x)}$.

As we shall see, these distributions are strictly related to the so-called Archimedean copulæ.

5.1. *TTE models and Archimedean copulæ.*

Consider functions $Q : [0, 1]^2 \to [0, 1]$ such that there exists a decreasing function $\phi : (0, 1] \to [0, \infty]$, such that $\phi(1) = 0$ and

$$Q(u, v) = \phi^{-1}(\phi(u) + \phi(v)),\qquad(9)$$

If $\phi(0) < \infty$, the inverse is defined by setting $\phi^{-1}(x) = 0$ for $\phi(0) < x < \infty$.

It is known that a function Q of the form (9) is a copula if and only if ϕ is convex. In such a case, we call Q an *Archimedean copula* and ϕ its *generator*. If ϕ is not convex, we have a semi-copula, as defined in Section 1.

The following theorem (see Nelsen[11]) allows us to order Archimedean copulæ.

Theorem 16: Let $C_i(u, v) = \phi_i^{-1}(\phi_i(u) + \phi_i(v))$, $i = 1, 2$. Then C_1 is less concordant than C_2 iff $\phi_1 \circ \phi_2^{-1}$ is subadditive, or, equivalently, if $\phi_2 \circ \phi_1^{-1}$ is superadditive.

In the following proposition we shall see, in particular, that if $\bar{F}$ is $TTE(W, R)$, then its survival copula K is Archimedean and depends on W but not on R; furthermore, its multivariate aging function B is an Archimedean (semi-)copula and depends on R but not on W. The proof is simple and will be omitted.

Proposition 17: *Let $\bar{F}$ be $TTE(W, R)$. Then $K_{\bar{F}}$ is an Archimedean copula with generator W^{-1}, and $B_{\bar{F}}$ is an Archimedean (semi-)copula with generator $R \circ (-\log)$*

Thus, in TTE models dependence and bivariate aging are, in a sense, "separated": dependence is based on W and bivariate aging on R. This gives a useful tool to generate examples, as the following remark shows.

Remark 18: As mentioned before, negative dependence may coexist with positive multivariate aging. For example, let

$$\bar{F}(x, y) = \exp\left\{1 - e^{x^2 + y^2}\right\}$$

This law is $TTE(W, R)$, with $W(x) = \exp\{1 - e^x\}$ and $R(x) = x^2$. Now,

$$K_F(u, v) = W(W^{-1}(u) + W^{-1}(v))$$
$$= uv \exp\{-(\log u)(\log v)\} \le uv.$$

On the other hand, $B_F(u, v) = \exp\left\{-\sqrt{(-\log u)^2 + (-\log v)^2}\right\} \ge uv$

5.2. *Evolution of the dependence of residual lifetimes for time transformed exponentials*

We turn now our attention to residual lifetimes. In the next proposition, we show that the joint law of the residual lifetimes is again TTE.

Proposition 19: *Let $\bar{F} \in \mathcal{F}$ be $TTE(W, R)$. Then $\bar{F}_r$ is $TTE(W_r, R_r)$, with*

$$W_r(x) = \frac{W(2R(r) + x)}{W(2R(r))}, \qquad R_r(x) = R(r + x) - R(r).$$

Thus, the survival copula K_r of $\bar{F}_r$ is again an Archimedean copula, with generator $W_r^{-1}(u) = W^{-1}(uW(2R(r))) - 2R(r)$, Furthermore, the aging function B_r is an Archimedean (semi-)copula with generator $R(r - \log x) - R(r)$.

Proof: The result follows from a simple computation. $\square$

Remark 20: If $\bar{F}$ follows a proportional hazard models, one can easily see that the same applies for $\bar{F}_r$, $\forall r > 0$. As we mentioned before, W is a Laplace transform, and hence K is TP_2 (see Joe[9]). By the above argument, it follows that the same holds for W_r and K_r. Compare this result with the first claim of Proposition 14.

Clearly, the results of Section 4 can be specialized to the case of TTE models, yielding several interesting examples. From now on, we prefer to consider the ordering of concordance between the survival copulæ of residual lifetimes at different ages.

Proposition 21: *Let $\bar{F} \in \mathcal{F}$ be $TTE(W, R)$. Then K_{r+a} is less concordant than K_r if and only if*

$$W_{r+a}^{-1} \circ W_r(x) = W^{-1}\left(\frac{W(2R(r) + x)}{W(2R(r))} W(2R(r + a))\right) - 2R(r + a)$$

is subadditive.

Proof: It is a simple application of Theorem 16. $\qquad\square$

Proposition 19 shows that the aging function B_r depends only on R, in accordance with Proposition 11. On the contrary, notice that the survival copula K_r depends on R as well as on W; thus, the dependence of the residual lifetimes is affected both by dependence and aging properties of the original survival function (see also remark 13). However, as far as the evolution of dependence is concerned, the fact that concordance decreases with time does not depend on R. The following proposition deals with this issue.

Proposition 22: *Let $\bar{F} \sim TTE\,(W, R)$ be such that $K_{t_2} \leq K_{t_1} \quad \forall t_1 \leq t_2$, i.e. such that the joint law of the residual lifetimes is less and less concordant as time elapses. Then the same is true for all TTE laws with the same W and arbitrary $\tilde{R}$.*

Proof: Let $\tilde{F} \sim TTE\left(W, \tilde{R}\right)$, and let $\tilde{W}_r = \frac{W\left(2\tilde{R}(r)+x\right)}{W\left(2\tilde{R}(r)\right)}$. The concordance of the residual lifetimes of $\tilde{F}$ decreases with time iff

$$\tilde{W}_{t_2}^{-1} \circ \tilde{W}_{t_1}(x) = W^{-1}\left(\frac{W\left(2\tilde{R}\left(t_1\right)+x\right)}{W\left(2\tilde{R}\left(t_1\right)\right)} W\left(2\tilde{R}\left(t_2\right)\right)\right) - 2\tilde{R}\left(t_2\right) \quad (10)$$

is subadditive, $\forall t_1 \leq t_2$. Set $y_i := R^{-1}\left(\tilde{R}\left(t_i\right)\right), i = 1, 2$. Then $y_1 \leq y_2$, and we see that $\tilde{W}_{t_2}^{-1} \circ \tilde{W}_{t_1} = W_{y_2}^{-1} \circ W_{y_1}$, so that the conclusion follows. $\qquad\square$

Corollary 23: *Let $\bar{F} \sim TTE(W, R)$. Then $K_{t_2} \leq K_{t_1} \quad \forall t_1 \leq t_2$ if and only if*

$$\psi_{a,r}(x) := W^{-1}\left(\frac{W(2r + x)}{W(2r)} W(2(r + a))\right) - 2(r + a) \quad (11)$$

is subadditive for every $r, a \geq 0$.

Proof: The conclusion follows from Proposition 22, by letting $R(x) = x$ $\square$

Corollary 24: *Let $\bar{F} \sim TTE(W, R)$. If $h_c(x) := W^{-1}\left(cW(x)\right)$ is concave for every $c < 1$, then $K_{t_2} \leq K_{t_1} \quad \forall t_1 \leq t_2$.*

Proof: Notice that $\psi_{a,r}(0) = 0$. Hence, to show that it is subadditive, it is sufficient to show that it is concave. Clearly,

$$\psi_{a,r}(x) = h_{\hat{c}}(2r + x) - 2(r + a) \quad \text{with} \quad \hat{c} = \frac{W(2r + a)}{W(2r)}.$$

Since W is decreasing, $\hat{c} < 1$. Hence $h_{\hat{c}}$ is concave, and therefore so is ψ_a. $\square$

Conditions for the increase of concordance, namely for $K_{t_2} \geq K_{t_1}$ $\forall t_1 \leq t_2$, can be obtained as in Proposition 22 and its corollaries, in terms of convex and superadditive functions.

Notice in particular that $K_r = K$, $\forall r > 0$ iff $W_{r+a}^{-1} \circ W_r$ is linear in x, $\forall a, r > 0$. This will be the case in Example 25 below. Afterwords, we shall provide examples in which we see how surviving may increase or decrease positive dependence. The last example deals with negatively dependent copulæ.

Example 25: Let $\bar{F} \in \mathcal{F}^{(2)}$ be $TTE(W, R)$, with an arbitrary $R \in \mathcal{R}$ and with W given by

$$W(x) = \left(\frac{\beta}{\beta + x}\right)^{\alpha}, \tag{12}$$

where $\alpha, \beta > 0$. This is a proportional hazard model with a Gamma(α, β) as mixing distribution. We have $W^{-1}(u) = \beta \left(u^{-\frac{1}{\alpha}} - 1\right)$, so that

$$K_{\bar{F}}(u, v) = W\left(W^{-1}(u) + W^{-1}(v)\right) = \left(u^{-\frac{1}{\alpha}} + v^{-\frac{1}{\alpha}} - 1\right)^{-\alpha},$$

which does not depend on β.

Now, let us consider the residual lifetimes. We have

$$W_r(x) = \frac{W\left(2R(r) + x\right)}{W\left(2R(r)\right)} = \frac{\left(\frac{\beta}{\beta + 2R(r) + x}\right)^{\alpha}}{\left(\frac{\beta}{\beta + 2R(r)}\right)^{\alpha}} = \left(\frac{\beta + 2R(r)}{\beta + 2R(r) + x}\right)^{\alpha},$$

which is of the same type of (12), with $\beta' = \beta + 2R(r)$. Hence, the above argument shows that

$$K_{\bar{F}_r}(u, v) = W_r\left(W_r^{-1}(u) + W_r^{-1}(v)\right) = W\left(W^{-1}(u) + W^{-1}(v)\right) = K_{\bar{F}}(u, v).$$

As we mentioned before, we could have proved the result by observing that $W_{r+a}^{-1} \circ W_r$ is linear in x.

The fact that dependence does not change as time elapses comes as no surprise in this model. In fact, we can reason as follows. For the sake of simplicity, fix $R(x) = x$; we are then dealing with two lifetimes X and Y, that are conditionally i.i.d. exponentially distributed given a parameter Θ, with $\Theta \sim$ Gamma(α, β). The Gamma family is conjugated to this statistical model, in the usual Bayesian sense. Thus, conditionally on the event $\{X > r, Y > r\}$, the residual lifetimes $X - r, Y - r$ are still conditionally i.i.d exponentially distributed given Θ, where Θ follows now a Gamma$(\alpha, \beta + 2r)$.

Example 26: Let $\bar{F}$ be $TTE(W, R)$, with $W(x) = [1 + \log(1 + x)]^{-1}$ and $R \in \mathcal{R}$ arbitrary. Then, for every $r, a > 0$, $K_{r+a} \geq K_r$. All these copulæ are PQD.

Example 27: Let $\bar{F}$ be $TTE(W, R)$, with $W(x) = \exp\{-\sqrt{x}\}$ and $R \in \mathcal{R}$ arbitrary. Then, for every $r, a > 0$, $K_{r+a} \leq K_r$. All these copulæ are PQD, and the limiting copula is the independent copula.

Example 28: Let $\bar{F}(x, y) = 2(1 + e^{x+y})^{-1}$. This law is $TTE(W, R)$, with $W(x) = 2\left(1 + e^x\right)^{-1}$ and $R(x) = x$. One may check that for every $r, a > 0$, $K_{r+a} \geq K_r$. All these copulæ are NQD, and the limiting copula is the independent copula.

References

1. J. Averous and J.L. Dortet-Bernadet, Dependence for Archimedean copulas and aging properties of their generating functions, Preprint (2000).
2. R.E. Barlow and M.B. Mendel, Similarity as a characteristic of Wear-out, in *Reliability and Decision Making*, Eds. R.E. Barlow, C.A. Clarotti, F. Spizzichino (Chapman and Hall, London, 1993).
3. R.E. Barlow and F. Spizzichino, Schur-concave survival functions and survival analysis, *Journal of Computational and Applied Mathematics*, **46**, 437–447 (1993).
4. B. Bassan and F. Spizzichino, Stochastic comparison for residual lifetimes and Bayesian notions of multivariate aging, *Advances in Applied Probability*, **31**, 1078–1094 (1999).
5. B. Bassan and F. Spizzichino, On a multivariate notion of New Better than Used, in *Proceedings of the Conference "Mathematical Methods of Reliability", Bordeaux*, 167–169 (2000).
6. B. Bassan and F. Spizzichino, Dependence and multivariate aging: the role of level sets of the survival function, in *System and Bayesian Reliability*, Eds. Y. Hayakawa, T. Irony, M. Xie (World Scientific, River Edge, 2001) pp. 229–242.
7. B. Bassan and F. Spizzichino, Relations among univariate aging, bivariate aging and dependence for exchangeable lifetimes, Preprint (2002).
8. D. Drouet Mari and S. Kotz, *Correlation and Dependence* (Imperial College Press, London, 2001).
9. H. Joe, *Multivariate Models and Dependence Concepts* (Chapman & Hall, London, 1997).
10. A.W. Marshall and I. Olkin, *Inequalities: Theory of Majorization and Its Applications* (Academic Press, New York, 1979).
11. R.B. Nelsen, *An Introduction to Copulas* (Springer, New York, 1999).
12. M. Shaked and J.G. Shanthikumar, *Stochastic Orders and their Applications* (Academic Press, London, 1994).
13. F. Spizzichino, *Subjective Probability Models for Lifetimes* (CRC Press, Boca Raton, 2001).

17

EFFICIENT COMPUTATIONAL TECHNIQUES FOR POWER SYSTEMS RELIABILITY ASSESSMENT

Dumitru Cezar Ionescu, Paul Ulmeanu and Adrian Constantinescu

Department of Reliability, Faculty of Power Engineering,
"POLITEHNICA" University,
313 Spl. Independentei, 77206 Bucharest, Romania
E-mail: dc_ionescu@rectorat.pub.ro, paul@fiab.pub.ro,
adrianconst@netscape.net

Ioan Rotaru

National Company "NuclearElectrica Inc."
33-35 Magheru Blvd. , 7000 Bucharest, Romania
E-mail: irotaru@snn.rdsnet.ro

The chapter presents the utilization of the Generalized Stochastic Petri Nets to model the reliability of the power systems. It is known that a dependability analysis for this kind of systems is often very difficult due to the multiple dependencies on the specific maintenance policies, on the great number of operation conditions and the possibility to consider non-exponential distribution of the operation periods and the time for maintenance activities. The main idea to use the GSPN formalism for the reliability modelling was to benefit of the advantages offered to generate more exactly the system's states taking into account its dynamic evolution and to use the Monte Carlo simulation of the model in the special situations imposed by the power systems complexity. In the first part is presented the main ideas for building-up a GSPN model, and the possibilities of its evaluation using stochastic processes or Monte Carlo simulation. A test network is modelled and evaluated. In the second part a study case is presented in order to establish the configuration for the Medium Voltage Distribution Systems (main auxiliary services) for a nuclear power plant (Cernavoda Nuclear Power Plant, CANDU reactor).

1. Introduction

Modelling and performance assessment of power systems play an important role in the optimal design and efficient operation of such systems. The reliability modelling of the power systems is already a classical issue, as the reliability analysis is a obligatory request and the modelling techniques are characterized by advanced performance. The actual concerns are mostly related to the efficient algorithms field, respectively to the time consuming and the accuracy of the methods used. Usually, there are two main ways to model the power systems reliability:

- the space states method based on stochastic processes (markovian or semi-markovian);
- the simplified methods based on logical formalisms (block diagrams, structure functions, fault-trees and so on).

The main difficulties concerning the power systems reliability modelling are:

- the dynamic modelling of the power system evolution in accord to the dependencies generated by operation conditions and maintenance tasks ;
- the evaluation of restoration procedures, taking into account component failures, systems' structure and the protections.

2. Reliability modelling of the power systems using GSPN formalism

The main idea of using the GSPN formalism to model the power system reliability is to build a Petri net considering the following two aspects:

- the timed behavior of the system, taking into account the component's evolution (operation, corrective or preventive maintenance, standby and so on) ;
- the system's performance evaluation in accord with its structure and components' behavior.

The proposed GSPN model is structured in two parts. The first part, labelled "System Evolution" (SE), contains several Petri sub-nets assigned to the power systems components in order to simulate their evolution using timed transitions. The second part, labelled "System Performance Evaluation" (SPE), is used to check if the load-points of the power system (e.g. a distribution network) are connected to the source node. SPE simulates the power flow from one system node to other system node, through the transmission lines, using immediate transitions.

After a power system's event (failure, repair, so on), modelled by a SE timed transition, an evaluation of the system's performance is performed using the SPE. The connection between the SE and SPE is made by two places labelled CHANGE and MODE. To exemplify the modelling approach, a test network is analyzed in order to compare the results obtained with classical methods and to compare different evaluation techniques. The test network and its Petri net are presented in the Figures 1 and 2.

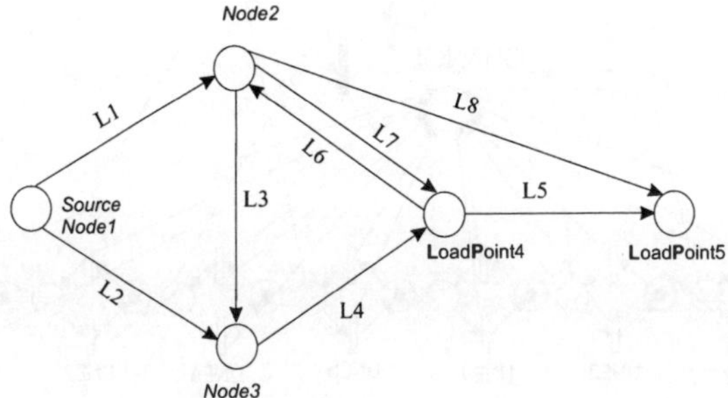

Fig. 1.   The simplified diagram for the test network (S source, i network node, L transmission line)

The GSPN model is a defined by [1,2,5] .:

$$M_{GSPN} = (P, T, \Pi, I, O, H, W, PAR, PRED, MP) \tag{1}$$

where:

- P is the set of places, T is the set of transitions, I, O, H are the input, output and inhibition functions respectively;
- $MP : P \rightarrow N$ is the marking function, PAR and $PRED$ functions who associate natural numbers or predicates restrictions to the places' markings;
- $\Pi : T \rightarrow N$ is the priority function that maps transitions onto natural numbers, representing the priority level;
- $W : T \rightarrow R$ is the function that indicates for the transitions firing; the "rate" for the timed transitions, and the "weight" for the immediate ones.

- the places
$P(pc\_1\_1, pc\_1\_2, pc\_2\_1, pc\_2\_2, \ldots, pc\_8\_1, pc\_8\_2, CHANGE,$
$MODE, node\_1(SOURCE), node\_2, \ldots, node\_5)$

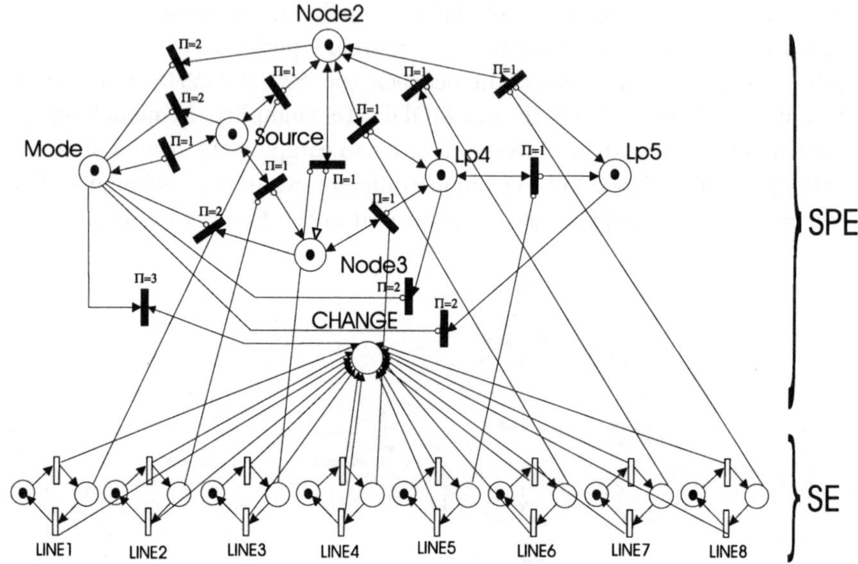

Fig. 2.   The Petri net modelling the test network

where pc_i_1 and pc_i_2 are the places of components (transmission lines), indicating their operation and maintenance states; $CHANGE$ and $MODE$ are the connection places between the two parts of the Petri net; $node\_i$ are the places for the nodes of the network used to indicate the power flow in the network (two nodes are connected with a transmission line).

- the transitions

$T(tl\_1\_1, tl\_1\_2, tl\_2\_1, tl\_2\_2, \ldots, tl\_8\_1, tl\_8\_2,$
$t\_CHANGE\_MODE, t\_MODE\_SOURCE,$
$t\_reset\_nod\_1\_MODE, t\_reset\_nod\_2\_MODE, \ldots,$
$t\_reset\_nod\_5\_MODE, t\_SOURCE\_node\_2, t\_SOURCE\_node\_3,$
$t\_node\_2\_node\_3, t\_node\_2\_node\_4, t\_node\_3\_node\_4, t\_node\_4\_node\_2,$
$t\_node\_4\_node\_5)$

where tl_i_1 and tl_i_2 are the timed transition between the components' states (from operation to repair and vice-versa) , t_CHANGE_MODE is a sink transition (has only input arcs), t_MODE_SOURCE is a source transition (has not input arcs) and $t\_reset\_nod\_i$, the reset transitions used to delete the tokens from the places $node\_i$. The $t\_reset\_nod\_i$ are a sink transitions too (its have not output places).

- the priorities

The priority function is : $\Pi(0, 0, \ldots, 0, 0, 3, 1, 2, \ldots, 2)$

where all timed transitions have zero priority and all immediate transitions have different levels of priority in order to assure the following logic: at the beginning of the start of the simulation of the electrical energy circulation, there have to be deleted the tokens from all $node\_i$ places. After that, the immediate transitions must be fired from the place $SOURCE$ to the places $node\_i$.

- the marking

The initial marking is:

$M_0(p) = [1, 0, 1, 0, \ldots, 1, 0, 0, 1, 1, 1, 1, \ldots, 1]$

where $M(pc\_i\_1) = 1$, $M(pc\_i\_2) = 0$, $i = \overline{1,8}$, $M(CHANGE) = 0$, $M(MODE) = 1$, $M(node\_i) = 1 (i = \overline{1,5})$

- the parameters and conditions (PAR and $PRED$ functions)

The imposed conditions are ensuring that the Petri net is $1 - bounded$.

- the I, O, H sets

I, O and H are input, output and inhibitor arcs. All input and output arcs have the multiplicity 1. In the Figure 2 is shown how the inhibitor arcs were used:

- to control the firing of the SPE immediate transitions. If the component i is in operation (modelled by the marking $M(pc\_i\_2) = 0$) in SE sub-net, a token is added in the $node\_i$ places in SPE sub-net. If the component i is in failure state, the marking becomes $M(pc\_i\_2) = 1$ and the transitions $t\_node\_i\_node\_j$ can not be fired. The power flow can't pass the transmission lines between the nodes i and j.

- to limit the number of the tokens in a $node\_i$ at 1 (if a token was added to a place $node\_i$,it will be no more possible to add another token because the transition is disabled using an inhibitor arc).

- the W function

With this function there can be modelled the law of the transitions firing (the dynamic evolution). For the timed transitions, this function indicates the parameters of the time period elapsed from the enabling moment of the transition to the firing moment (law for the time operation, time repair and so on). The timed transitions in SE are exponential with the rates $1/MTTF/(1/MTTR)$. For the immediate transitions in SPE, this function indicates the weight of the transition in order to solve a priority in a conflict situation. For the modelled network it was not necessary to use the weights because for the SE sub-net there are no conflicting transitions and for the SPE sub-net the conflicts are structural solved.

The system's states belong the reachability set of the Petri model $RS(M_0)$, where M_0 is the initial marking.

Considering the GSPN formalism, there are two possibilities to model the behavior of the system:

- using a markovian (semi-markovian) approach
- using a Monte Carlo simulation

For the both approaches, there is necessary to identify the $RS(M0)$ set and to build the reachability graph. For this, it is presented hereby the particularities of marking evolution:

- in the initial state M_0, the system is assuming to be in an operation state. That means all the network nodes are energized and no component is in failure state: $M_0(1, 0, 1, 0, \ldots, 1, 0, 0, 1, 1, 1, 1, 1, 1)$.
  In this marking there are enabled all SE timed transitions $tl\_i\_1, i = \overline{1,8}$. Therefore M_0 is a tangible marking.
- let be fired one of the timed transitions (tl_i_1). The new marking is: $M_{0i}(1, 0, , 0, 1, \ldots, 1, 0, 1, 1, 1, 1, 1, 1, 1)(pc\_i\_1 = 0, pc\_i\_2 = 1)$.
  The failure of the ith component generates two consequences:

  - the places pc_i_2 and $CHANGE$ receive a token each one;
  - the SPE immediate transitions are not enabled because of the action of the inhibitor arc from pc_i_2 place (where is a token).

  In this marking there are enabled the timed transitions tl_i_2 , $tl\_i\_1, i = \overline{2,8}$ and the immediate transition t_CHANGE_MODE. The marking M_{0i} is vanishing.
- because the immediate transition t_CHANGE_MODE has priority $(\Pi(t\_CHANGE\_MODE) = 3)$ it is firing first.

  The new marking is:
  $M_{0i}^1(1, 0, \ldots, 0, 1, \ldots, 1, 0, 0, 0, 1, 1, 1, 1, 1)$.
  where   the   places $CHANGE$ and $MODE$ were emptied (t_CHANGE_MODE is a sink transition). In this marking there are enabled all the reset immediate transitions, $t\_reset\_nod\_i\_MODE, i = \overline{1,5}$ (sink transitions).

- because all reset transitions have the same priority level 2, they will be fired sequently and there will be generated a series of three vanishing markings and a tangible one (the last), this being:
$M_{oi}^{(2)\text{-}4}(1,0,\ldots,0,1,\ldots,1,0,0,0,0,0,0,0,0)$.
The tokens from the nodes were deleted because the transitions $t\_reset\_nod\_i\_MODE$ are sink transitions. In this marking there is only one enabled transition, t_MODE_SOURCE.
- the firing of the t_MODE_SOURCE transition adds a token in the $SOURCE$ place and the new marking is: $M_{oi}^{(3)}(1,0,\ldots,0,1,\ldots,1,0,0,1,1,0,0,0,0)$. A new series of markings is generated by firing the immediate enabled transitions between nodes. The series of the markings are:
$M_{oi}^{(4)\text{-}1}(1,0,\ldots,0,1,\ldots,0,0,1,1,1,0,0,0,0)$ $\qquad \ldots$
$M_{oi}^{(4)\text{-}4}(1,0,\ldots,0,1,\ldots,1,0,0,1,1,1,1,1,1)$
The final marking is a tangible one and the others are vanishing.
- in the final marking $M_{oi}^{(4)\text{-}4}(1,0,\ldots,0,1,\ldots,1,0,0,1,1,1,1,1,1)$ there are enabled only the timed transitions $tl\_i\_2, tl\_i\_1, i = \overline{2,8}$. It means there is possible the repairing of the failed component or a new failure occurrence, the next marking evolution being similary with the evolution presented before (a series of vanishing markings finished with a tangible one). This initial part of the reachability graph is presented in the Figure 4 and the generalized reduced one in the Figure 5.

In the Figure 6 there is comparatively presented the Loss of the Load Probability obtained by markovian assessment and Monte Carlo simulation [3]. For failures and repairs there were used the same rates values ($\lambda = \mu = 0.01h^{-1}$). This Figure shows that for 10000 histories the Monte Carlo simulation gives a very good approximation. The Monte Carlo simulation performed was an forced one , generating sample paths through the state space of the system.[3]

3. A Study Case Presentation

The $GSPN$ model presented with some particular developments (nonexponential timed transitions or deterministic ones) is applied for a more complex power systems. There is considered a Medium Voltage Distribution System ($MVDS$) from a Nuclear Power Plant (NPP) [4] in order to compare four configurations for the main auxiliary services.

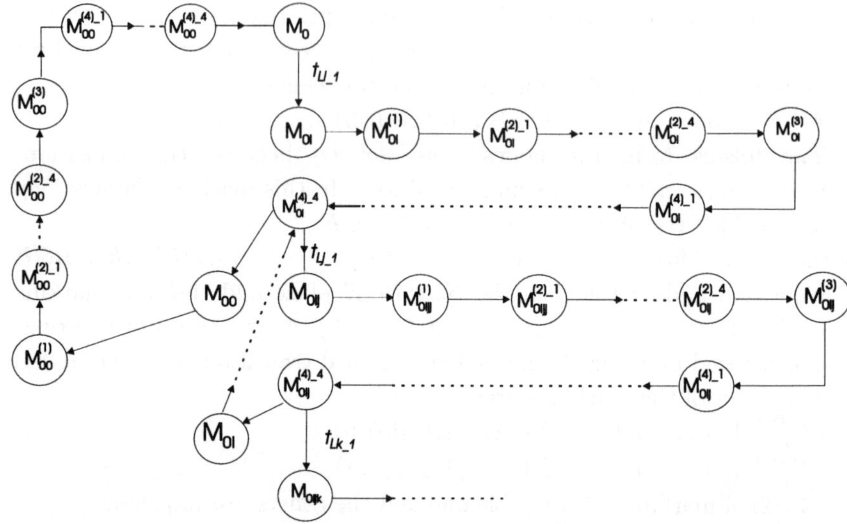

Fig. 3.   The reachability graph between the first and the second tangible markings

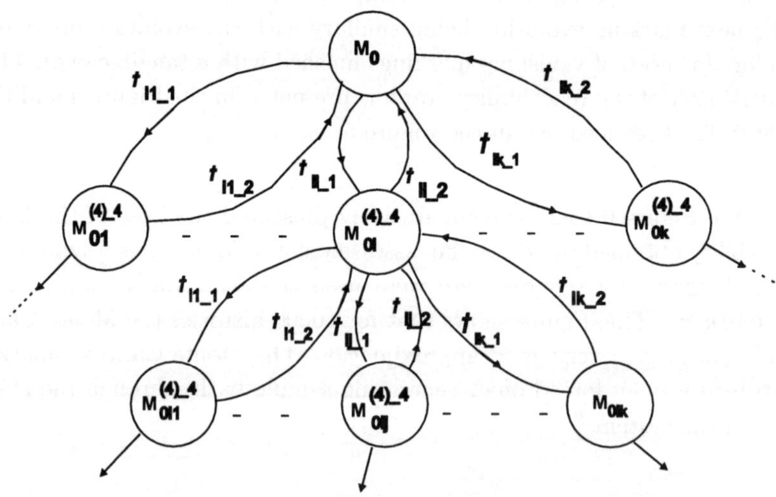

Fig. 4.   The generalized reduced reachability graph

3.1. *Target MDVS Configurations*

In the Figure 6 (a) it is presented a main line $MVDS$ configuration and in Figure 6 (b) and Figure 7 all simplified networks (similarly with the initial analyzed network).

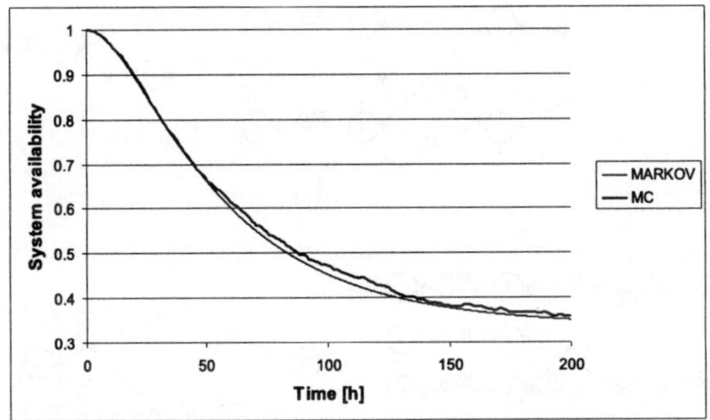

Fig. 5.   Comparative results Markov vs. Monte Carlo simulation

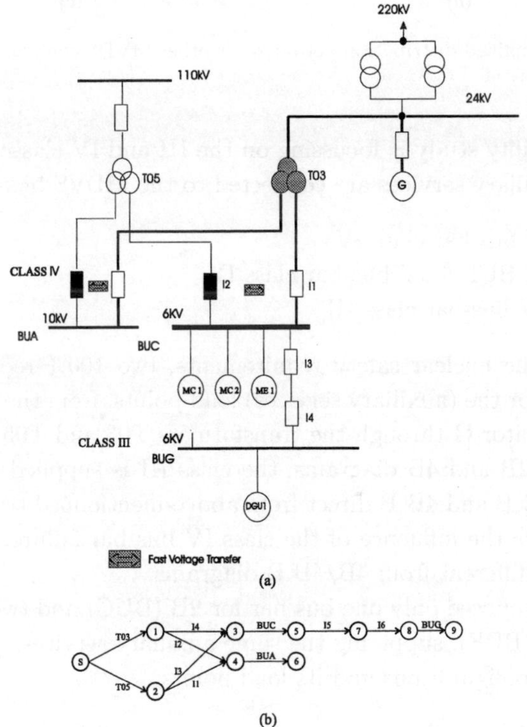

Fig. 6.   The MVDS 2B (half part) diagram (a. main on-line diagram; b. equivalent
distribution network)

Following the reliability nuclear requirements, the NPP auxiliary services are divided in four classes (I, II, III, IV).

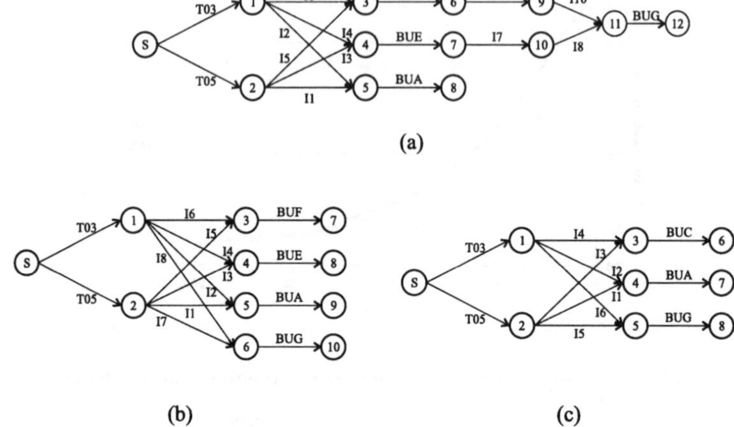

Fig. 7.   The simplified distribution networks for other MVDS configurations (a. 4B; b. 4B_B; c. 2B_B)

Our reliability study is focussing on the III and IV classes.

These auxiliary services are connected to the MDVS bus-bars labelled:

- BUA 10 kV bus-bar class IV
- BUC, BUE, BUF 6 kV bus-bar class IV
- BUG 0.4 kV bus-bar class III

To meet the nuclear safety requirements, two 100% redundant paths are provided for the (auxiliary services) load points, from the off-site GRID or main generator G through the transformers T03 and T05 (see Figure 7 (a)). For the 2B and 4B diagrams, the class III is supplied from class IV and for the 2B_B and 4B_B direct from above-mentionned transformers. In order to reduce the influence of the class IV bus bar failure, the 2B/2B_B diagrams are different from 4B/4B_B diagrams.

There is proposed only one bus bar for 2B (BUC) and two bus bars for 4B (BUE and BUF), supplying the same auxiliary services.

The four configurations and its load points:

- 2B, Figure 6 (a) and (b) - the load nodes are 5, 6 for class IV and 9 for class III
- 2B_B Figure 7 (a) - the load nodes are 6, 7, for class IV and 8 for class III
- 4B Figure 7 (b) - the load nodes are 6,7,8 for class IV and 12 for class III
- 4B_B Figure 7 (c) - the load nodes are 7,8,9 for class IV and 10 for class

III.

The above-mentioned bus-bars, the circuit breakers (labelled I), and the transformers (labelled T) are the components considered in the reliability analysis (Figure 6(a)).

3.2. *Modelling*

Taken into consideration the components' particularities, SE sub-nets are built, as follows.

The available states for a circuit breaker are: circuit breaker in closed position, circuit breaker in open position and circuit breaker in repairing state, after a active failure or a failure to trip (see Figure 8).

The available states for a bus/transformer are: bus/transformer in operational state, bus/transformer in passive state, following a failure of a neighboring component and bus/transformer in repairing state (see Figure 9). In both cases, the immediate transitions are conditioned by the state of neighboring components. The transitions T1 and T4 are used to simulate the failure to trip probability of the circuit breaker (following the demands from "open" to "closed" position). The transitions T2 and T3 have the same meaning for commutation in the opposite sense.

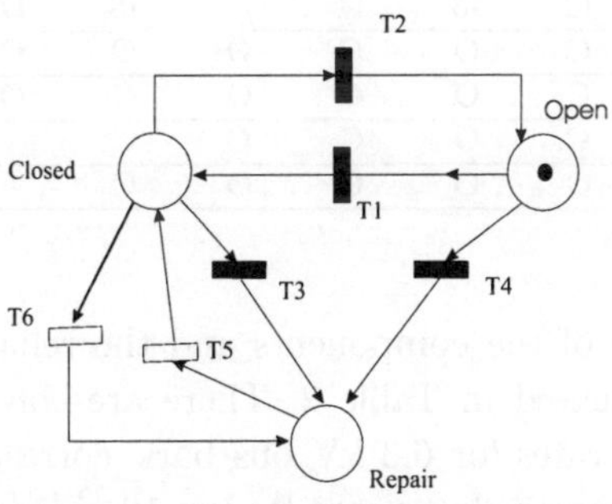

Fig. 8.   Petri sub-net for circuit breaker

The SPE sub-nets are developed for each MDVS proposed configuration, following the model presented in the section 2.

3.3. *Reference operating condition for reliability analysis*

The reference condition assumed for the reliability analysis is the full load operation because is the normal operating condition. The main interest of

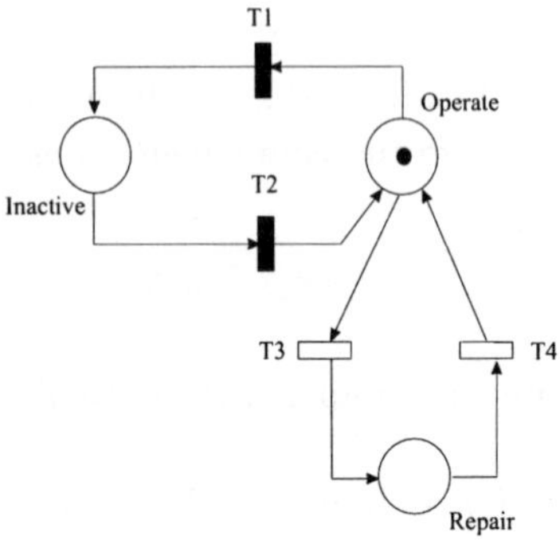

Fig. 9.   Petri sub-net for bus-bar / transformer

the comparatione of the configurations is to supply the final bus bar BUG class III. The reference conditions for the circuit breakers are presented in Table 1, where O - normally open and C - normally closed (see Figure 6 (a)).

Table 1.   Normal states for the MDVS breakers

Main One Line	Circuit Breakers									
Diagram	I1	I2	I3	I4	I5	I6	I7	I8	I9	I10
Cod 4B	0	C	O	C	O	C	C	C	C	O
Cod 4B_B	O	C	O	C	O	C	O	C	-	-
Cod 2B	O	C	O	C	C	C	-	-	-	-
Cod 2B_B	O	C	O	C	O	C	-	-	-	-

The failure modes of the components and the reliability data used in this analysis are presented in Table 2. There are considered two sets of figures for the failure rates for 6.3 kV bus bars, corresponding to case A, $\lambda = 10^{-1}$ [1/year], respectively to case B, $\lambda = 10^{-2}$ [1/year].

Table 2.   Component reliability indices

Component	Failure Mode	λ [1/year]	q	$MTTR$ [hrs]
10.5 kV	Active failure	0.01	-	96
6.3 kV class IV buses	Active failure	0.01 (0.1)	-	96
6.3 kV class IV buses	Active failure	0.02 (0.2)	-	96
Circuit Breakers	Active failure	0.007	-	107
	Failure to trip	-	0.01	-
Transformers	Active failure	0.02	-	768

In order to check the reliability performances of MDVS, two specific scenarios are considered, namely: first scenario - normal source power of MVDS, either from off-site grid G, either from on-site main generator; second scenario - only on-site power source from main turbine generator, during a 30 minutes unavailability off-site grid G.

Figure 10 (first scenario) shows in the case A the LOLP of BUG bus bar as a function of time, the off-grid is *available*, but the BUG bus bar is considered in ideal conditions (no failure).

Figures 11 and 12 (second scenario) show in the cases A and B the Loss of Load Probability (LOLP) of BUG bus bar as a function of time, the first 30 minutes indicated in time axis the off-grid G is considered *unavailable*.

Figure 13 (second scenario) shows in the case B the LOLP of BUG bus bar as a function of time, the first 30 minutes indicated in time axis the off-grid G is considered *unavailable*.

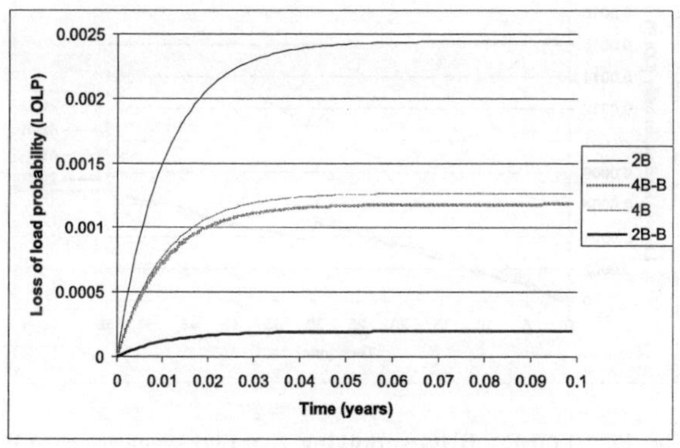

Fig. 10.   LOLP - BUG supplying / first scenario, case A

3.4. *Study Case Conclusion*

Following the Loss of Load Probability (LOLP) criteria, for the both scenarios and both cases (A,B), the merit order is the same: 2B_B, 4B_B, 4B, 2B, as it is presented in the Figures 10, 11, 12, and 13.

4. Conclusion

The main goal of this chapter was to analyze the advantages offered by Petri nets formalism to model the power systems in order to evaluate its

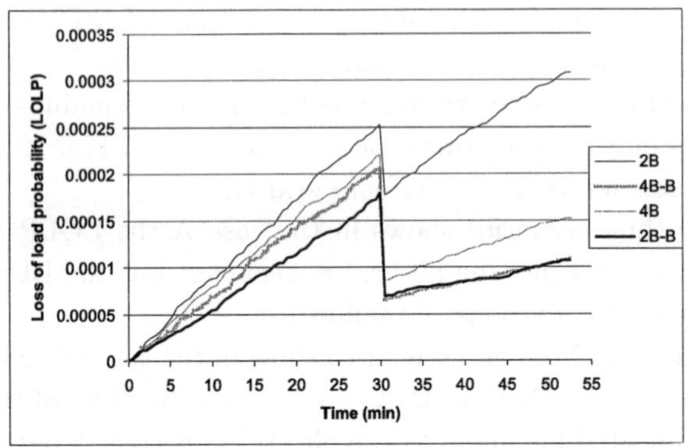

Fig. 11.   LOLP - BUG supplying / second scenario, case B

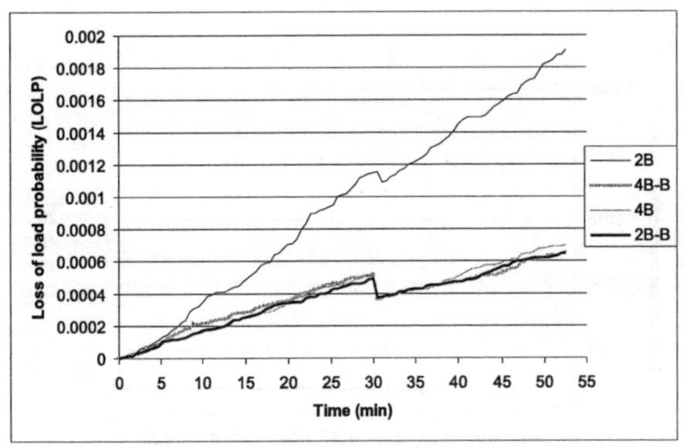

Fig. 12.   LOLP - BUG supplying / second scenario, case B

reliability and to manage the state space explosion. It is clear that GSPN models offers a good way to solve very practical aspects of the power systems as dependencies, conflicts, different kinds of delays laws. A computer code based on object-oriented programming and dynamic memory allocation are develping at the Reliability Department, Faculty of Power Engineering, University POLITEHNICA of Bucharest. [6,7]

References

1. K. Jensen and G. Rozenberg, *High-level Petri Nets: Theory and Application* (Springer-Verlag, Berlin, 1991).
2. M. Ajmone Marsan, G. Balbao, G. Conte, S. Donatelli and G. Franceschinis,

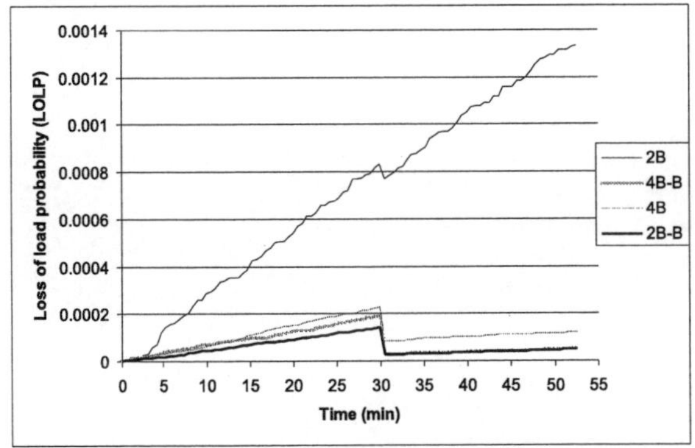

Fig. 13.   LOLP - BUG supplying / second scenario, case B, BUG as ideal bus-bar

Modelling with generalized stochastic Petri nets (John Wiley & Sons Ltd., Chichester, 1995).

3. M. Marseguerra and E. Zio, *Basics of the Monte Carlo Method with Application to System Reliability* (LiLoLe-Verlag GmbH, Publ. Co. Ltd., 2000).

4. *BOP Safety Related Performance Requierements*, Design Guide 79-036950-DG003 Rev. 3 (1979).

5. R. Billinton and R. N. Allan, *Reliability Evaluation of Engineering Systems: Concepts and Techniques* (Plenum Press, New York, 1983).

6. A. P. Ulmeanu and D. C. Ionescu, The Computer-Assisted Analysis of the Semi-Markovian Stochastic Petri Nets and an Application, in *Statistical and Probabilistic Models in Reliability*, Eds. D.C. Ionescu and N. Limnios (Birkhauser, Boston, 1999).

7. D. C. Ionescu, A. P. Ulmeanu, A. Constantinescu, and I. Rotaru Reliability Modeling of Medium Voltage Distribution Systems of Nuclear Power Plants using Generalized Stochastic Petri Nets, *The Second Euro-Japanese Workshop on Stochastic Risk Modelling for Finance, Insurance, Production and Reliability* Chamonix, France, 155–164 (18-20 September 2002).

18

PREDICTING DAMAGE

Nozer D. Singpurwalla

Department of Statistics, The George Washington University
Washington, DC, 20052, USA
E-mail: nozer@research.circ.gwu.edu

Chung Wai Kong

Sales & Revenue Management Department, Singapore Airlines
07-C Airline House, 25 Airline Road, Singapore 819829
E-mail: chungwai_kong@singaporeair.com.sg

Andrew W. Swift

Mathematical Sciences Department, Worcester Polytechnic Institute
100 Institute Road, Worcester, MA, 01609, USA
E-mail: swift@wpi.edu

This chapter is motivated by a real problem, involving the prediction of damage to an item under different conditions. Problems of this type are generic. What is noteworthy here is that information about damage comes from three sources: a physics based model, actual test data, and informed testimonies from specialists. The problem then is to fuse or integrate this information in a coherent manner. Our proposed approach is Bayesian and involves some subtle manoeuvres of the likelihood function.

1. Introduction

The problem that we propose to address in this chapter involves the testing of several copies of an item for failure or survival under different test conditions. The result of each test is binary, taken to be zero for survival, and one for failure. It is often the case that the test conditions cannot be precisely controlled; thus it is possible to test only one copy of the item at any test condition.

Our aim is to use a combination of observed test data and expert testimony to predict the item's survivability under environments at which it has not, or cannot, be tested. Our prediction will come in the form of an assessment of the item's probability of survival. A key feature of our work is the systematic pooling of observed data and expert testimonies that are conditioned on unobserved physical parameters.

Our work here is motivated by that of McDonald[5], in which he considers a mine which is designed to be placed in shallow waters to deter enemy vessels. However, the enemy is aware of such mines and aims to destroy them via an underwater explosive charge that is fired at some distance.

2. A Normative Approach to Problem Formulation

Consider an environment in which an item is to be tested. The environment is defined by two covariates C_1 and C_2, which can be controlled, but not precisely. The outcome of the test is a Bernoulli random variable X, where $X = 1$ if the item fails the test, and $X = 0$, otherwise. An engineer E is required to predict X when C_1 and C_2 are set to W and R respectively. E tests n (presumed identical) copies of the item with covariates set at (W_i, R_i), $i = 1, \ldots, n$ in order to observe the outcome X_i. It is supposed that because C_1 and C_2 cannot be precisely controlled, that only one item can be tested at any (W_i, R_i). Upon completion of the tests, E records each X_i as x_i, where x_i is either zero or one. Hence, the available data is the vector $\{(W_i, R_i, x_i) ; i = 1, \ldots, n\}$; which we will denote by $\underline{d}$. Thus, E's problem is to specify $P_E (X = x; (W, R), \underline{d})$, for $x = 0$ and 1.

In order for E to arrive upon $P_E (X = x; (W, R), \underline{d})$, suppose that a panel of k experts $E_1, \ldots, E_k$, who are knowledgeable about the underlying dynamics of the problem are consulted. However, the experts can provide an assessment of X if they are appraised of an abstract physical quantity Q, where Q depends on W and R. However, the nature of the relationship between Q, W, and R is not precisely known, so we may write $Q = f(W, R) + \epsilon$, where f is some function, and ϵ is an error. In our scenario, Q is kinetic energy.

2.1. *The Nature of Expert Testimonies*

In the previous section, we discussed a panel of experts who will each provide an assessment of X in the light of the kinetic energy Q. Thus, each of these experts E_i, $i = 1, \ldots, k$, will specify $P_{E_i} (X = x; Q)$ for $x = 0$ and 1, their probability that the item fails or passes the test respectively. For definitiveness, we will focus on the case $x = 1$, and for ease we will denote $P_{E_i}(X = 1; Q)$ as $p_i(Q)$. Due to the nature of the scenario, it is reasonable to suppose that $p_i(Q)$ is non-decreasing in Q, and ranges from 0 when $Q = 0$, to 1 when Q is infinite.

In reality, it may not be possible to elicit the entire function $p_i(Q)$ from E_i. Rather, E_i may only deliver $p_i(Q)$ for a finite set of values of Q, $Q_1 < Q_2 < \cdots < Q_n$. In such a case we may interpolate $p_i(Q)$ between the elicited values of Q, however, it is important that E_i's responses be non-decreasing in Q.

Thus, from each of the k experts, we elicit a dose-response curve $p_i(Q)$, $i = 1, \ldots, k$, wherein our scenario the dose is the abstract physical parameter Q.

This leads to several issues that should be addressed before we continue. Firstly, since $p_i(Q)$ is non-decreasing in Q, the sequence of responses $\{p_i(Q_1), p_i(Q_2), \ldots, p_i(Q_n), \ldots, \}$ will be dependent for each i, $i = 1, \ldots, k$. Also, as the panel of experts are each providing an assessment on the same X, it is reasonable to expect that there will also be dependence between the experts' responses. That is, that for any fixed value of Q, say Q^* the sequence of responses $\{p_1(Q^*), \ldots, p_k(Q^*)\}$ is also dependent. Both of these dependencies have both been previously considered individually. The within expert dependence, namely, the dependence in the sequence $\{p_i(Q_1), p_i(Q_2), \ldots, p_i(Q_n), \ldots, \}$ in the pioneering papers of Ramsey[6] and of Ferguson[3], and the between expert dependence, the dependence in the sequence $\{p_1(Q^*), \ldots, p_k(Q^*)\}$, by (among others) Lindley[4]. However, a simultaneous consideration of both sources of dependence has not been attempted before.

Recall that E needs to assess $X = x$, when C_1 is set to W and C_2 is set to R, and that E has access to data $\underline{d}$, and the testimonies of the experts $E_1, \ldots, E_k$, when $x = 1$ (assuming that the experts were supplied with a Q which corresponds to W and R). Recall that the relationship between (W, R) and Q is not deterministic, but rather that $Q = f(W, R) + \epsilon$.

2.1.1. *A Model for Q*

In order to tackle the problem at hand, we must first specify a distribution for Q. Due to the nature of the problem, a suitable choice is a lognormal distribution with parameters μ and σ, where μ and σ are related to W and

R via the relationships:

$$\mu(W, R) = \mu = \ln m$$
$$\sigma(W, R) = \sigma = .2\mu,$$
$$\text{where } m = e^{-\frac{1}{2}\sigma^2} K_E W^{1/3} \left(\frac{W^{1/3}}{R}\right)^{\alpha_E} \tag{1}$$

and K_E, α_E are known constants. As a shorthand, we denote the above as $(Q; \mu, \sigma) \sim \Lambda(\mu, \sigma)$. For a motivation for this choice of a model for Q, see Cole[1]. Since $\sigma = .2\mu$, our lognormal distribution is denoted as $(Q; \mu) \sim \Lambda(\mu, .2\mu)$.

Suppose now that E is convinced that for specified values of W and R, Q is always greater than some threshold, say γ, so that it is $(Q - \gamma)$ that has a lognormal distribution, that is, that $((Q - \gamma); \mu) \sim \Lambda(\mu, .2\mu)$.

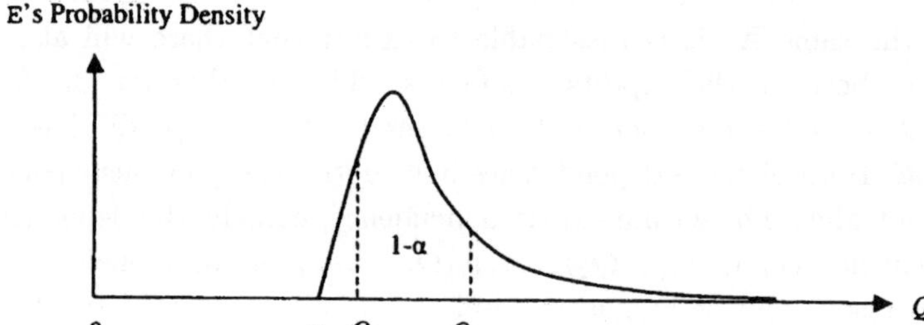

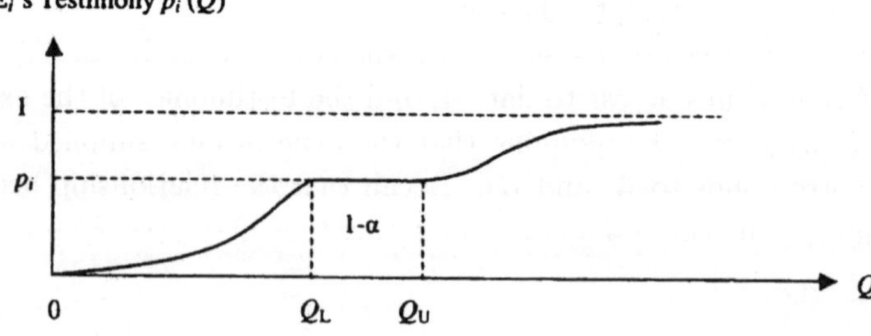

Fig. 1.

Suppose now, that for each of the k experts, $P_{E_i}(X = 1; Q)$ is constant over the range of values of Q for which its lognormal density is appreciable;

see Figure 1. Specifically, for some small α, say $\alpha = 0.05$, and a specified $\gamma, \gamma \in [0, \infty)$, E is able to specify two numbers Q_L and Q_U, with $Q_L < Q_U$, such that

$$P_E(Q_L \leq (Q - \gamma) \leq Q_U = 1 - \alpha.$$

Whilst Q_L and Q_U need not be unique, they should be chosen by E to make $(Q_U - Q_L)$ a minimum. Thus, under the above, we have that, for each i, $i = 1, \ldots, k$,

$$P_{E_i}(X = 1; Q_L < (Q - \gamma) < Q_U) = p_i,$$

and for $(Q - \gamma)$ outside the region $[Q_L, Q_U]$, $P_{E_i}(X = 1; Q)$ is simply $p_i(Q)$.

Under the above scheme, E elicits the expert testimonies $\underline{p} = (p_1, \ldots, p_k)$, for any $(Q - \gamma) \in [Q_L, Q_U]$ ensuring that each p_i is a constant within the above range. It is important to note that the p_i's need not be independent; indeed, the case when they are dependent is a more realistic one and should be considered.

2.2. *Assessment Based on Data and Expert Testimony*

From the previous section, it follows that E's problem is to assess $P_E(X = 1; (W, R), \underline{d}, \underline{p}(Q))$, where $\underline{d} = \{(W_i, R_i, x_i); i = 1, \ldots, n\}$, with $x_i = 1$ or 0, $\underline{p}(Q) = (p_1(Q), \ldots, p_k(Q))$, and Q has a lognormal distribution with parameters $\mu(W, R)$ and $\sigma(W, R)$. We treat $\underline{d}$ and $\underline{p}(Q)$ as separate sources of information, one hard data, the other expert judgement, both of which must be incorporated to address the stated problem. The question is, how should one proceed.

The first thing to note is that $\underline{p}(Q)$ was elicited with reference to Q, and hence one can extend the conversation to Q. Using the law of total probability and that fact that if $Q \in [Q_L, Q_U]$ then $\underline{p}(Q) = \underline{p}$, one obtains,

$$P_E(X = 1; (W, R), \underline{d}, \underline{p}(Q))$$

$$= \int_{Q \in [Q_L, Q_U]} P_E(X = 1 | Q; (W, R), \underline{d}, \underline{p}) \, P_E(Q; (W, R), \underline{d}, \underline{p}) \, dQ$$

$$+ \int_{Q \notin [Q_L, Q_U]} P_E(X = 1 | Q; (W, R), \underline{d}, \underline{p}(Q)) \, P_E(Q; (W, R), \underline{d}, \underline{p}(Q)) \, dQ$$

$$(2)$$

Clearly, $P_E(Q; (W, R), \underline{d}, \underline{p})$ (and similarly $P_E(Q; (W, R), \underline{d}, \underline{p}(Q))$) does not depend on $\underline{d}$ or $\underline{p}$ and its dependence on W and R can be encapsulated by μ (and maybe also γ). Thus $P_E(Q; (W, R), \underline{d}, \underline{p})$

and $P_E\left(Q;(W,R),\underline{d},\underline{p}(Q)\right)$ can be replaced by $P_E\left(Q;\mu\right)$. However, $P_E\left(Q \notin [Q_L, Q_U]\right) = \alpha$, where α is a small number, and thus the value of the second integral above is negligible and can be disregarded. Thus, equation 2 can be rewritten as

$$
\begin{aligned}
&P_E\left(X = 1; (W,R), \underline{d}, \underline{p}(Q)\right) \\
&\approx P_E\left(X = 1; (W,R), \underline{d}, \underline{p}\right) \\
&= \int_{Q \in [Q_L, Q_U]} P_E\left(X = 1 | Q; (W,R), \underline{d}, \underline{p}\right) P_E(Q; \mu) dQ
\end{aligned}
\tag{3}
$$

where Q has a lognormal distribution with parameters μ and $.2\mu$.

Invoking Bayes' Law on the above yields the following relationship:

$$
\begin{aligned}
&P_E\left(X = 1 | Q; (W,R), \underline{d}, \underline{p}\right) \\
&\propto L_E\left(X = 1 | Q; (W,R), \underline{d}, \underline{p}\right) P_E\left(X = 1 | Q; (W,R), \underline{d}\right),
\end{aligned}
\tag{4}
$$

where the first term on the right hand side is E's likelihood that $X = 1$, were E to know Q, and in the light of (W,R), $\underline{d}$, and $\underline{p}$. The second term is E's prior probability that $X = 1$, were E to know Q, and in the light of (W,R), and $\underline{d}$ alone.

We now suppose that the expert testimonies $\underline{p}$ are based on Q alone, where Q is of course between Q_L and Q_U. Then, to specify the likelihood, all that matters is $\underline{p}$ alone, thus

$$
L_E\left(X = 1 | Q; (W,R), \underline{d}, \underline{p}\right) = L_E\left(X = 1; \underline{p}\right)
$$

and so equation 4 simplifies to

$$
\begin{aligned}
&P_E\left(X = 1 | Q; (W,R), \underline{d}, \underline{p}\right) \\
&\propto L_E\left(X = 1; \underline{p}\right) P_E\left(X = 1 | Q; (W,R), \underline{d}\right)
\end{aligned}
\tag{5}
$$

In the sections that follow we describe an approach for evaluating each of the two terms on the right hand side of equation 5 whilst also addressing the dependence in the expert testimonies.

2.2.1. *Assessment of the Likelihood given Expert Testimonies*

In order to specify $L_E\left(X = 1; \underline{p}\right)$, E's likelihood of the event $(X = 1)$ in the light of expert testimonies $\underline{p}$, we first examine the case of a single expert E_i, whose testimony p_i has been declared to E. Recall that p_i is E_i's probability of the event $(X = 1)$ when $Q \in [Q_L, Q_U]$. To incorporate E's judgement about the expertise of the expert into the analysis, we follow a strategy proposed by Lindley[4], and consider the log-odds q_i, where $q_i = \log\left(p_i/\left(1 - p_i\right)\right)$. With log-odds, specifying $L_E\left(X = 1; p_i\right)$ is equivalent to specifying $L_E\left(X = 1; q_i\right)$. Using the approach of Lindley[4], we suppose that were q_i to be unknown, then its distribution conditional on the event $(X = 1)$, is normal with mean μ_1 and variance σ^2. That is, that $\left(q_i | X = 1; \mu_1, \sigma^2\right) \sim N(\mu_1, \sigma^2)$. Similarly, $\left(q_i | X = 0; \mu_0, \sigma^2\right) \sim N(\mu_1, \sigma^2)$. Using these normal distributions as models for the likelihoods of the events $(X = 1)$ and $(X = 0)$, it is easy to see that the likelihood ratio is of the form

$$\log \frac{L_E\left(X = 1; q_i\right)}{L_E\left(X = 0; q_i\right)} = \frac{\mu_1 - \mu_0}{\sigma^2}\left(q_i - \frac{\mu_1 + \mu_0}{2}\right). \tag{6}$$

The likelihood ratio is of interest to us as it provides us with an interpretation of the quantities μ_1, μ_0, and σ^2, and hence a method for specifying them. For example, the quantity $(\mu_1 + \mu_0)/2$ acts like a bias term for correcting q_i; the correction is made by E. The corrected term is then multiplied by $(\mu_1 + \mu_0)/\sigma^2$, where the value of $(\mu_1 + \mu_0)/\sigma^2$ depends upon E's opinion of the confidence of E_i. If E believes that E_i has a tendency to be underconfident or ultra cautious then E can inflate E_i's testimony by choosing $(\mu_1 + \mu_0)/\sigma^2 > 1$. Similarly, if E believes E_i is overconfident, then E can deflate E_i's testimony by choosing $(\mu_1 + \mu_0)/\sigma^2 < 1$. If E agrees with E_i then E should choose $(\mu_1 + \mu_0)/\sigma^2 = 1$. If E wishes to adopt E_i's testimony without any modification or correction, then $(\mu_1 + \mu_0)$ should be zero, and $(\mu_1 - \mu_0) = \sigma^2$; i.e. $\mu_1 = -\mu_0 = \sigma^2/2$.

In order to specify μ_1 and μ_0, E should consider E_i's credibility with regards to optimistic or pessimistic tendencies. For example, were $(X = 0)$ true, then a highly credible expert may declare p_i as being .1 whereas an expert who tends to be an optimist may declare p_i as being .4. Similarly, if $(X = 1)$ were to be true, then a credible expert may declare p_i to be .9, whilst an expert that tends to be a pessimist my declare p_i to be .6. In specifying μ_1 and μ_0 all combinations of the above scenarios can be accommodated. In particular, if an expert is credible with respect to both $(X = 0)$ and $(X = 1)$ then E would set $\mu_1 = \mu_0 = \mu$.

Once μ_1, μ_0, and σ^2 have been specified, say as μ_1^*, μ_0^*, and $\widehat{\sigma}^2$, we can turn our attention to $L_E\left(X = 1; q_i\right)$, however this requires careful understanding of the meaning of the likelihood function, as follows.

As $\left(q_i | X = 1; \mu_1, \sigma^2\right) \sim N(\mu_1, \sigma^2)$, the normal distribution provides us with a model for the likelihood function of μ_1, for a fixed value of q_i and for a fixed value of σ^2, say $\widehat{\sigma}^2$, see Figure 2. Let us denote this likelihood as $L_E\left(\mu_1; q_i, \widehat{\sigma}^2\right)$.

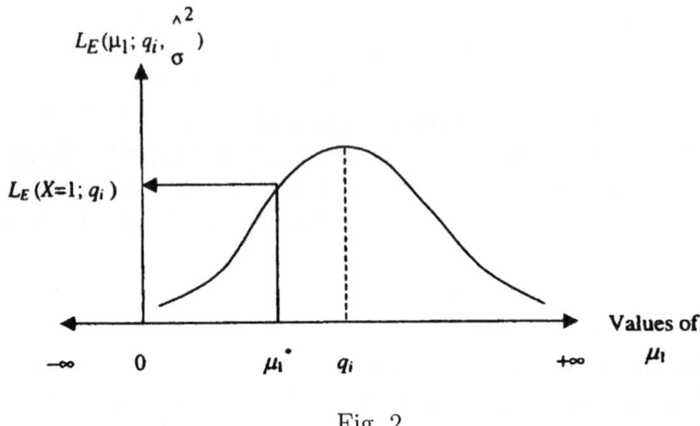

Fig. 2.

Clearly, this likelihood function will attain its maximum at q_i. However, E has previously chosen μ_1^* to correspond to the case $(X = 1)$, and hence the likelihood of $(X = 1)$ for a fixed q_i and $\hat{\sigma}^2$ is the value of $L_E\left(\mu_1^*; q_i, \hat{\sigma}^2\right)$. Thus we have that

$$L_E\left(X = 1; p_i\right) = L_E\left(X = 1; q_i\right) = L_E\left(\mu_1^*; q_i, \hat{\sigma}^2\right)$$

The extension of the above to the case of multiple experts is relatively straightforward. Firstly, we note that we are now looking for $L_E\left(X = 1; \underline{q}\right)$ rather than $L_E\left(X = 1; q_i\right)$, where we suppose that $\left(\underline{q}|X = 1, \underline{\mu}, \Sigma\right)$ has a multivariate normal distribution with a mean vector $\underline{\mu}$ and correlation matrix Σ. The diagonal elements of Σ are 1, and its off-diagonal elements represent the correlations between the expert testimonies q_i, $i = 1, \ldots, k$, as perceived by E.

To specify the likelihood $L_E\left(X = 1; \underline{p}\right)$, E examines $L_E\left(\underline{\mu}; \underline{q}, \Sigma^*\right)$, and with $\underline{\mu}$ specified (by E) as $\underline{\mu}^*$, obtains $L_E\left(X = 1; \underline{q}\right)$ as $L_E\left(\underline{\mu}^*; \underline{q}, \Sigma^*\right)$, where Σ^* is E's specification of Σ. Thus, we have that

$$L_E\left(X = 1; \underline{p}\right) = L_E\left(X = 1; \underline{q}\right) = L_E\left(\underline{\mu}^*; \underline{q}, \Sigma^*\right)$$

2.2.2. *Assessment of Uncertainty Based on Data Alone*

To evaluate $P_E\left(X = 1|Q; (W, R), \underline{d}\right)$, the second term on the right hand side of equation 5, it is required to introduce a model that links the data $\underline{d}$ to the uncertainty about X under the condition that Q is known. This is a standard procedure that is used in both accelerated testing and dose response studies. Whilst there are several models that exist under the heading of logits and probits (see Cox[2]), that could be deemed suitable for use here, the one we find attractive is the "Weibitt Model" in which $P(X = 1|Q, \alpha, \beta) = 1 - \exp\left(-\alpha Q^\beta\right)$, where α and β are unknown parameters. Estimation of α and β is based on the data $\underline{d}$, as described below.

Suppose that E, starting with $P_E\left(X = 1|Q; (W, R), \underline{d}\right)$, extends the conversation to α and β. Thus,

$$P_E\left(X = 1|Q; (W, R), \underline{d}\right)$$

$$= \int_{(\alpha,\beta)} P_E\left(X = 1|Q, \alpha, \beta; (W, R), \underline{d}\right) \pi_E\left(\alpha, \beta; (W, R), \underline{d}\right) d\alpha d\beta$$

where $\pi_E\left(\alpha, \beta; (W, R), \underline{d}\right)$ represents E's uncertainty about (α, β) in the light of (W, R) and $\underline{d}$.

However, the Weibitt model does not entail (W, R) and $\underline{d}$; thus, we claim that X is independent of (W, R) and $\underline{d}$, given Q. Similarly, as (W, R) can be arbitrarily chosen, it follows that inference about (α, β) cannot be based on (W, R). Thus, the above expression simplifies to

$$P_E\left(X = 1|Q; (W, R), \underline{d}\right)$$

$$= \int_{(\alpha,\beta)} P_E\left(X = 1|Q, \alpha, \beta\right) \pi_E\left(\alpha, \beta; \underline{d}\right) d\alpha d\beta$$

$$= \int_{(\alpha,\beta)} \left(1 - e^{-\alpha Q^\beta}\right) \pi_E\left(\alpha, \beta; \underline{d}\right) d\alpha d\beta. \tag{7}$$

In order to evaluate this integral we will need to know $\pi_E\left(\alpha, \beta; \underline{d}\right)$. Recalling that $\underline{d} = (W_i, R_i, x_i)_{i=1}^n$, our task is therefore to evaluate $\pi_E\left(\alpha, \beta; (W_i, R_i, x_i)_{i=1}^n\right)$. As a first step, let us (temporarily) assume that the x_i's are not known. Under this assumption, the above can be written as $\pi_E\left(\alpha, \beta|X_1, \ldots, X_n; (W_i, R_i)_{i=1}^n\right)$ which by Bayes' Law can be expressed as

$$\pi_E\left(\alpha, \beta|X_1, \ldots, X_n; (W_i, R_i)_{i=1}^n\right)$$
$$\propto P_E\left(X_1, \ldots, X_n|\alpha, \beta; (W_i, R_i)_{i=1}^n\right) P_E\left(\alpha, \beta; (W_i, R_i)_{i=1}^n\right), \tag{8}$$

where $P_E\left(\alpha, \beta; (W_i, R_i)_{i=1}^n\right)$ is E's prior on α and β, in the light of $(W_i, R_i)_{i=1}^n$. Clearly, $(W_i, R_i)_{i=1}^n$ do not have a bearing on α and β and thus $P_E\left(\alpha, \beta; (W_i, R_i)_{i=1}^n\right)$ is simply $P_E\left(\alpha, \beta\right)$. If we now assume that, in the light of $(W_i, R_i)_{i=1}^n$, the X_i's are (conditionally) independent, then equation 8 simplifies to

$$\pi_E\left(\alpha, \beta | X_1, \ldots, X_n; (W_i, R_i)_{i=1}^n\right)$$

$$\propto \prod_{i=1}^n P_E\left(X_i = x_i | \alpha, \beta; (W_i, R_i)_{i=1}^n\right) P_E\left(\alpha, \beta\right). \qquad (9)$$

Now, let us consider an individual term in the product in the above equation, namely $P_E\left(X_i = x_i | \alpha, \beta; (W_i, R_i)_{i=1}^n\right)$. Clearly, for $j \neq i$, (W_j, R_j) has no bearing on X_i and thus $P_E\left(X_i = x_i | \alpha, \beta; (W_i, R_i)_{i=1}^n\right)$ is simply $P_E\left(X_i = x_i | \alpha, \beta; (W_i, R_i)\right)$, and by extending the conversation to Q_i we get

$$P_E\left(X_i = x_i | \alpha, \beta; (W_i, R_i)\right)$$

$$= \int_{Q_i} P_E\left(X_i = x_i | \alpha, \beta, Q_i; (W_i, R_i)\right) P_E\left(Q_i; (W_i, R_i)\right) dQ_i$$

$$= \int_{Q_i} P_E\left(X_i = x_i | \alpha, \beta, Q_i\right) P_E\left(Q_i; (W_i, R_i)\right) dQ_i. \qquad (10)$$

Recall that the above equation was developed under the assumption that the X_i's have not been observed, however, in reality X_i has been observed as x_i; thus the first term in the integral is no longer a probability, but rather a likelihood for α and β, were Q_i to be known, and for the fixed value x_i. We denote this likelihood as $L_E\left(\alpha, \beta | Q_i; x_i\right)$. Singpurwalla[7] states that (as a subjectivist) E is free to specify this likelihood in any manner that seams appropriate, however, we shall suppose that E follows convention and specifies the likelihood based on the probability model for X_i. Specifically,

$$L_E\left(\alpha, \beta | Q_i; x_i\right) = \left(e^{-\alpha Q_i^\beta}\right)^{1-x_i} \left(1 - e^{-\alpha Q_i^\beta}\right)^{x_i}.$$

Thus, using this likelihood in equation 10 yields

$$P_E\left(X_i = x_i | \alpha, \beta; (W_i, R_i)\right)$$

$$= \int_{Q_i} \left(e^{-\alpha Q_i^\beta}\right)^{1-x_i} \left(1 - e^{-\alpha Q_i^\beta}\right)^{x_i} P_E\left(Q_i; W_i, R_i\right) dQ_i,$$

where $Q_i \sim \Lambda\left(\mu_i, .2\mu_i\right)$, with $\mu_i\left(W_i, R_i\right) \overset{def}{=} \mu_i = -\frac{1}{2}\sigma^2 + \ln\left[K_E W_i^{1/3} \left(\frac{W_i^{1/3}}{R_i}\right)^{\alpha_E}\right]$.

With this likelihood in place, and with $x_1, \ldots, x_n$ replacing $X_1, \ldots, X_n$, equation 9 becomes

$$\pi_E\left(\alpha, \beta; (W_i, R_i, x_i)_{i=1}^n\right)$$
$$= \pi_E\left(\alpha, \beta; \underline{d}\right)$$
$$\propto \prod_{i=1}^n \int_{Q_i} \left(e^{-\alpha Q_i^\beta}\right)^{1-x_i} \left(1 - e^{-\alpha Q_i^\beta}\right)^{x_i} P_E\left(Q_i; W_i, R_i\right) dQ_i P_E\left(\alpha, \beta\right).$$

$$(11)$$

However, recall that $\pi_E\left(\alpha, \beta; \underline{d}\right)$ is the missing ingredient in equation 7, E's assessment of the uncertainty about $X = 1$, in the light of (W, R) and data alone, but assuming that Q were known. Thus, inserting equation 11 into equation 7 gives us

$$P_E\left(X = 1 | Q; (W, R), \underline{d}\right)$$
$$\propto \int_{(\alpha, \beta)} \left(1 - e^{-\alpha Q^\beta}\right) [\prod_{i=1}^n \int_{Q_i} \left(e^{-\alpha Q_i^\beta}\right)^{1-x_i} \left(1 - e^{-\alpha Q_i^\beta}\right)^{x_i}$$
$$\times P_E\left(Q_i; W_i, R_i\right) dQ_i] P_E\left(\alpha, \beta\right) d\alpha d\beta$$

With regard to above, it still remains for us to elicit $P_E(\alpha, \beta)$, E's joint prior on α and β. One approach is to examine several plots of $1 - \exp\left(-\alpha Q^\beta\right)$ versus Q for different values of α and β, to arrive upon a range of values for α and β which makes these plots have the desired S-shaped property. Investigation shows that S-shapedness occurs when $\alpha \in [5, 10]$ and $\beta \in [5, 10]$, and as there is no justification for preferring any particular values of α and β within these intervals, we shall assume that $P_E\left(\alpha, \beta\right)$ is uniform over the region $5 \leq \alpha \leq 10$ and $5 \leq \beta \leq 10$. Thus $P_E\left(\alpha, \beta\right) = 1/25$ and

$$P_E\left(X = 1 | Q; (W, R), \underline{d}\right)$$
$$\propto \frac{1}{25} \int_5^{10} \int_5^{10} \left(1 - e^{-\alpha Q^\beta}\right) [\prod_{i=1}^n \int_{Q_i} \left(e^{-\alpha Q_i^\beta}\right)^{1-x_i} \left(1 - e^{-\alpha Q_i^\beta}\right)^{x_i}$$
$$\times P_E\left(Q_i; W_i, R_i\right) dQ_i] d\alpha d\beta \qquad (12)$$

Thus we now specified both terms on the right hand side of equation 5. Specifically, if we ignore to constant $1/25$ in equation 12 above, then we have

$$P_E\left(X = 1|Q;(W,R),\underline{d},\underline{p}\right) \propto L_E\left(X = 1;\underline{p}\right) \int_5^{10}\int_5^{10}\left(1 - e^{-\alpha Q^\beta}\right)$$

$$\times \left[\prod_{i=1}^{n}\int_0^\infty \left(e^{-\alpha Q_i^\beta}\right)^{1-x_i}\left(1 - e^{-\alpha Q_i^\beta}\right)^{x_i} \cdot P_E\left(Q_i; W_i, R_i\right)dQ_i\right]d\alpha d\beta \quad (13)$$

2.2.3. *Integration of All Information*

Equation 13 is one step away from our desired goal of E's assessment of $(X = 1)$ as it entails a conditioning on Q. To get our desired result, we average out over Q (see equation 3) to write out E's assessment of $(X = 1)$ in the light of $(W, R), \underline{d}$, and $\underline{p}$, as

$$P_E\left(X = 1; (W, R), \underline{d}, \underline{p}\right)$$

$$\propto \int_{Q\in[Q_L, Q_U]} L_E\left(X = 1;\underline{p}\right)\left\{\int_5^{10}\int_5^{10}\left(1 - e^{-\alpha Q^\beta}\right)\right.$$

$$\times \left[\prod_{i=1}^{n}\int_0^\infty\left(e^{-\alpha Q_i^\beta}\right)^{1-x_i}\left(1 - e^{-\alpha Q_i^\beta}\right)^{x_i}\right.$$

$$\times \left. P_E\left(Q_i; W_i, R_i\right)dQ_i\right]d\alpha d\beta\right\}P_E\left(Q; (W, R)\right)dQ \quad (14)$$

where

$$Q \sim \Lambda\left(\mu, .2\mu\right) \text{ with } \mu = -\frac{1}{2}\sigma^2 + \ln\left[K_E W^{1/3}\left(\frac{W^{1/3}}{R}\right)^{\alpha_E}\right],$$

$$Q_i \sim \Lambda\left(\mu_i, .2\mu_i\right) \text{ with } \mu = -\frac{1}{2}\sigma^2 + \ln\left[K_E W_i^{1/3}\left(\frac{W_i^{1/3}}{R_i}\right)^{\alpha_E}\right],$$

$$L_E\left(X = 1;\underline{p}\right) = L_E\left(\underline{\mu}^*; \underline{q}, \Sigma\right), \text{ and}$$

Q_L and Q_U are those values of Q for which

$$P_E\left(Q_L \le Q \le Q_U\right) = 1 - \alpha$$

for some specified small value of α, $\alpha > 0$.

Similarly, we have that

$$P_E\left(X=0;(W,R),\underline{d},\underline{p}\right)$$

$$\propto \int_{Q\in[Q_L,Q_U]} L_E\left(X=0;\underline{p}\right)\{\int_5^{10}\int_5^{10}\left(e^{-\alpha Q^\beta}\right)$$

$$\times [\prod_{i=1}^n \int_{Q_i}\left(e^{-\alpha Q_i^\beta}\right)^{1-x_i}\left(1-e^{-\alpha Q_i^\beta}\right)^{x_i}$$

$$\times P_E\left(Q_i;W_i,R_i\right)dQ_i]d\alpha d\beta\}P_E\left(Q;(W,R)\right)dQ \tag{15}$$

where in order to obtain $L_E\left(X=0;\underline{p}\right)$ an analogue to $\underline{\mu}^*$, say $\underline{\mu}_0^*$, needs to be specified.

One final thing must be considered, namely that equations 14 and 15 are not normalized. However, this is a relatively simple task. Let S be the sum of the right hand terms of equations 14 and 15, so that $P_E\left(X=1;(W,R),\underline{d},\underline{p}\right)$ equals the right hand side of equation 14 divided by S, and similarly for $P_E\left(X=0;(W,R),\underline{d},\underline{p}\right)$.

3. Computation

In Section 2.2.2 we described the likelihood $L_E\left(\alpha,\beta|Q_i;x_i\right)$ using the Weibitt model where for $x_i=1$,

$$L_E\left(\alpha,\beta|Q_i;x_i\right)=\left(e^{-\alpha Q_i^\beta}\right)^{1-x_i}\left(1-e^{-\alpha Q_i^\beta}\right)^{x_i}.$$

However, this likelihood provides little discrimination for the different values of α and β and hence a scaled Weibitt model introducing a parameter η is preferred. Thus,

$$L_E\left(\alpha,\beta|Q_i;x_i\right)=\left(\exp\left(-\alpha\left(\frac{Q_i}{\eta}\right)^\beta\right)\right)^{(1-x_i)}\left(1-\exp\left(-\alpha\left(\frac{Q_i}{\eta}\right)^\beta\right)\right)^{x_i}$$

This scaled likelihood is able to provide better discrimination over a range of values of Q_i. In our application, η was taken to be 400.

Incorporating the above scaling into equation 14 and then making the following substitutions

$$g_0\left(x,\alpha,\beta,(W,R)\right)\equiv\frac{\exp\left(-\alpha\left(\frac{Q_i}{400}\right)^\beta\right)}{x}\exp\left(-\frac{(\log x-\mu(W,R))^2}{2\sigma(W,R)^2}\right),$$

$$g_1\left(x,\alpha,\beta,(W,R)\right)\equiv\frac{1-\exp\left(-\alpha\left(\frac{Q_i}{400}\right)^\beta\right)}{x}\exp\left(-\frac{(\log x-\mu(W,R))^2}{2\sigma(W,R)^2}\right)$$

yields the following integral of interest

$$P_E\left(X=1;(W,R),\underline{d},\underline{p}\right) \propto L_E\left(X=1;\underline{p}\right)\int_\alpha\int_\beta\left[\int_Q g_1\left(Q,\alpha,\beta,(W,R)\right)dQ\right]$$

$$\times\left[\prod_{i=1}^n\int_{Q_i} g_{x_i}\left(Q_i,\alpha,\beta,(W_i,R_i)\right)dQ_i\right]d\beta d\alpha.$$

Furthermore, if we define the function $h_1\left(\alpha,\beta\right)$ as

$$h_1\left(\alpha,\beta\right)=\int_Q g_1\left(Q,\alpha,\beta,(W,R)\right)dQ\left[\prod_{i=1}^n\int_{Q_i} g_{x_i}\left(Q_i,\alpha,\beta,(W_i,R_i)\right)dQ_i\right],$$

then our expression for $P_E\left(X=1;(W,R),\underline{d},\underline{p}\right)$ simplifies to

$$P_E\left(X=1;(W,R),\underline{d},\underline{p}\right)\propto L_E\left(X=1;\underline{p}\right)\int_\alpha\int_\beta h_1\left(\alpha,\beta\right)d\beta d\alpha.$$

Evaluating $h_1\left(\alpha,\beta\right)$ entails calculating the value of $(n+1)$ integrals, and thus evaluating $P_E\left(X=1;(W,R),\underline{d},\underline{p}\right)$ requires the evaluation of $(n+3)$ integrals. The evaluation of these integrals has to be done by numeric techniques. See Singpurwalla, Cui and Kong[8] for a methodology for calculating these integrals as well as a worked example.

4. Conclusions

In this chapter we have proposed a Bayesian approach for addressing a commonly occurring problem in the engineering and biological sciences. Our approach allows for fusing of two sources of information, data and expert testimonies, but under the caveat that expert testimonies are conditioned on parameters that have an interpretation within the underlying science.

Whilst the posed problem and our proposed approach appears to look simple and deceptively straightforward, we feel that the reality is different. Whilst each piece of the approach may appear relatively easy, putting them together requires some subtle manoeuvres.

Acknowledgments

This work was supported by the Office of Naval Research under Grant N00014-99-1-0875 and by the Army Research Office under A MURI Grant DAAD 19-01-1-0502.

References

1. R. H. Cole, *Underwater Explosions*. (Dover Publications Inc., New York, 1965).
2. D. R. Cox, *The Analysis of Binary Data*. (Methuen & Co Ltd., London, 1970).
3. T. S. Ferguson, A Bayesian analysis of some nonparametric problems, *Annals of Statistics*. **1**, 209-230 (1973).
4. D. V. Lindley, The Use of Probability Statements, in *Accelerated Life Testing and Expert's Opinions in Reliability*, Eds. C. A. Clarotti and D. V. Lindley (North-Holland, New York, 1988).
5. W. W. McDonald, Personal Communication (2000).
6. F. L. Ramsey, A Bayesian approach to bioassay, *Biometrics,* **28**, 841-858 (1972).
7. N. D. Singpurwalla, Some cracks in the empire of chance (flaws in the foundations of reliability), *International Statistical Review* (with Discussion), **10**, 53-78 (2002).
8. N. D. Singpurwalla, Y. Cui and C. W. Kong, Information Fusion for Damage Prediction, in *Case Studies in Reliability and Maintenance*, Eds. W.R. Blishke and D.N.P. Murthy (John Wiley and Sons Inc., New York, 2003) pp. 251-265.

19

WEIBULL-RELATED DISTRIBUTIONS FOR THE MODELLING OF BATHTUB SHAPED FAILURE RATE FUNCTIONS

M. Xie

Department of Industrial and Systems Engineering
National University of Singapore, Kent Ridge Crescent, Singapore
E-mail: mxie@nus.edu.sg

C.D. Lai

Statistics, Institute of Information Sciences and Technology
Massey University, Palmerston North, New Zealand
E-mail: c.lai@massey.ac.nz

D.N.P. Murthy

Department of Mechanical Engineering
University of Queensland, Brisbane, Australia
E-mail: p.murthy@mailbox.uq.edu.au

Weibull distribution is very useful in reliability and survival analysis because of its ability in modelling increasing and decreasing failure rate functions. However, when the failure rate function is of a bathtub shape, a single Weibull distribution will not be sufficient. Bathtub failure rate is a common phenomenon in reliability and it is an important ageing property. The system lifetime cycle can usually be divided into three distinct stages: early life, useful life and wear out period. Models for bathtub shaped failure rate functions are needed while it is also useful for a model to be generalisation of the Weibull distribution. In this paper, some models extending the traditional two-parameter Weibull distribution are presented and discussed. We also review some related models that can be used for the modelling of bathtub-shaped failure rate function.

1. Introduction

For a lifetime distribution function $F(t)$, the failure rate function is defined as $r(t) = f(t)/(1-F(t))$ where $f(t)$ is the probability density function. The failure rate function is an important aging characteristic of a distribution in reliability analysis. It can be increasing (IFR) or decreasing (DFR) and these classes of lifetime distributions have been studied by Barlow and Proschan[4]. For some interesting reviews and discussions on the failure rate function and its applications in reliability modelling and analysis, see e.g., Lee and Thompson[20], Bartoszewicz[5], Deshpande et al.[8], Boland et al.[6], and Guess et al.[10].

Weibull distribution (Weibull[31]), which is a generalization of the exponential distribution, has been widely used because of its flexibility (Murthy et al.[28]). The two-parameter Weibull distribution has the following form

$$F\left(t\right) = 1 - \exp\left\{-\left(t/\alpha\right)^{\beta}\right\}, \alpha > 0, \beta \geq 0 \qquad (1)$$

where α is the scale parameter and β is the shape parameter. The failure rate function is given by

$$r\left(t\right) = \frac{f\left(t\right)}{R\left(t\right)} = \frac{\beta}{\alpha}\left(t/\alpha\right)^{\beta-1} \qquad (2)$$

It can be seen that for $\beta < 1$, the failure rate function is decreasing while for $\beta > 1$, it is increasing. When $\beta = 1$, we have the well-known exponential distribution which has a constant failure rate.

An important advantage of the Weibull distribution is its graphical interpretation. We simply have the relationship

$$\ln\ln\frac{1}{1-F(t)} = \beta\ln t - \beta\ln\alpha \qquad (3)$$

and this is useful from a practical point of view in a sense that Weibull probability paper can be easily constructed. Weibull probability plot serves as a simple model validation procedure in addition to simple parameter estimation. If the plotted points cannot be fitted with a straight line, the model should be rejected. If the model is considered acceptable, then the slope and intercept of the fitted straight line can be used to estimate the shape and scale parameters with Equation (3).

However, for many complex systems, the failure rate function exhibits a bathtub shape. It decreases at the beginning, and then stabilizes, and towards the end of the lifecycle, it enters the wear-out region and the failure rate starts to increase. A distribution function is said to be BFR (bathtub shaped failure rate) if its failure rate function decreases at the beginning and then remains almost constant for a period before it increases. In other words, the failure rate function has a bathtub shape. Although bathtub-shaped failure rate functions are common in practice, there are few lifetime distribution models that possess this property. The traditional Weibull distribution widely used in reliability analysis can only model monotonely increasing or decreasing failure rate functions. An earlier review of bathtub shaped reliability models was given in Rajarshi and Rajarshi[30] and some later bathtub shape models can be found in Haupt and Schabe[12] and Mudholkar and Srivastava[24]. For a recent review of bathtub-shape failure distribution, see Lai et al.[18].

It can be noted that although Weibull distribution is flexible, a single Weibull model cannot be used to model all three phases of a bathtub curve at the same time. Usually a Weibull probability plot can be generated. For data sets with bathtub shaped failure rate function, the initial slope is smaller than one and the final slope is greater than one.

In Figure 1, the first part of the data can be fitted by a straight line with a slope less than 1 which corresponds to a shape parameter less than one, and hence having decreasing failure rate. On the other hand, the last part tends to have a slope greater than one corresponding to an increasing failure rate. Hence for the complete data set, we have a decreasing and then increasing failure rate which signifies a bathtub-shaped failure rate function.

For a detailed analysis of Weibull probability plot and and plots of other Weibull related distributions, see Murthy et al.[28].

In this chapter, we review some models that extend the two parameter Weibull distribution with the possibility of having a failure rate function of bathtub shape. A recently proposed model by the authors is first introduced. Another model called the exponentiated Weibull introduced in Mudholkar and Srivastava[24] is then discussed. An interesting approach of introducing an additional parameter by Marshall and Olkin[21] leading to an extended Weibull distribution is also presented. An additive Weibull and some other models will also be briefly discussed.

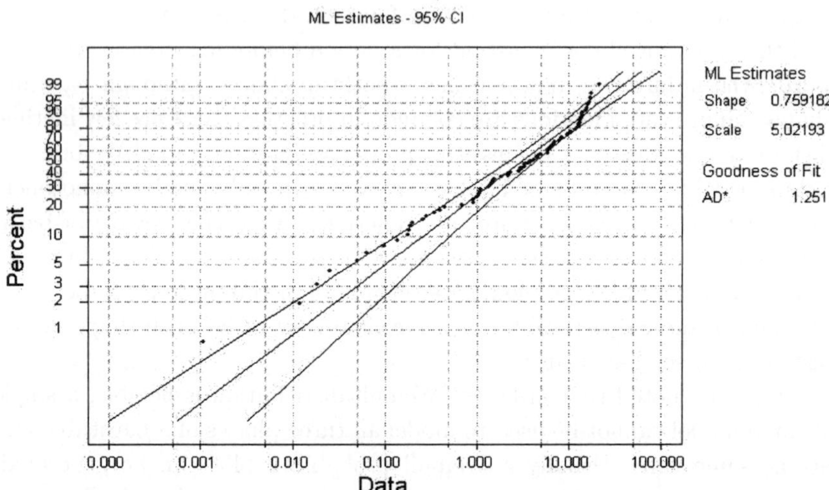

Fig. 1.   A typical Weibull probability plot on a set of data that exhibits a decreasing failure rate and then constant followed by an increasing failure rate at the end.

2. A modified Weibull model

Lai et al.[19] introduced a new three-parameter model that generalizes the basic Weibull distribution. The model has a bathtub-shaped failure rate function. The distribution function is given by

$$G(t) = 1 - \exp(-at^b e^{\lambda t}), \quad t \geq 0, \tag{4}$$

where the parameters $a > 0$, $b \geq 0$ and $\lambda \geq 0$ are to be estimated when a set of data is available. This model reduces to a two-parameter Weibull distribution when $\lambda = 0$ and hence it is another generalisation of the Weibull distribution.

The failure rate function can be shown to be

$$r(t) = a(b + \lambda t)t^{b-1}e^{\lambda t} \tag{5}$$

The shape of the failure rate function depends only on b because of the factor t^{b-1} and the remaining two parameters have no influence. In fact, for $b \geq 1$, $r(t)$ is increasing in t, implying an increasing failure rate (IFR). For $0 < b < 1$, $r(t)$ initially decreases and then increases in t, implying a bathtub shape for the failure rate function, and the change point t^* can be shown to be

$$t^* = \frac{\sqrt{b} - b}{\lambda} \qquad (6)$$

An interesting feature is that this is independent of a. Note that t^* increases as λ decreases.

It is clear from (5) that the modified Weibull distribution indeed reduces to a two parameter Weibull distribution when $\lambda=0$. Hence it is clearly a generalization of the two parameter Weibull distribution. Furthermore, when $b = 0$ the model reduces to

$$R(t) = \exp(-ae^{\lambda t}) \qquad (7)$$

which is a Type I extreme-value distribution. Sometimes this is also referred to as a log-Weibull distribution.

With the Weibull transformation $y = \ln\{-\ln R(t)\}$ and $x = \ln(t)$, we have that

$$y = \ln a + bx + \lambda e^x$$

and graphical estimation procedures similar to the Weibull probability plot can be obtained. See Lai et al.[19]. In fact, by introducing $z = e^x$ and using the Weibull transformation, multiple regression analysis can be used to obtain an estimate of the model parameters.

3. Exponentiated Weibull model

The exponentiated Weibull distribution was first studied by Mudholkar and Srivastava[24] and is given by the distribution function

$$G(t) = [F(t)]^{\nu} = \{1 - \exp[-(t/\alpha)^{\beta}]\}^{\nu} \qquad (8)$$

where $F(t)$ is the standard two-parameter Weibull distribution and $\nu > 0$.

When $\nu = 1$, this distribution reduces to the standard two-parameter Weibull distribution. When ν is an integer, the model is a special case of the multiplicative Weibull distribution. Such a model can be used to model a parallel system comprised of ν identical and independently distributed components whose lifetime distribution is Weibull.

The failure rate function of the exponentiated Weibull distribution can be shown to be of the following form

$$r(t) = \nu \left\{ \frac{[F(t)]^{\nu} - [F(t)]^{\nu+1}}{F(t) - [F(t)]^{(\nu+1)}} \right\} \left[\left(\frac{\beta}{\alpha}\right) \left(\frac{t}{\alpha}\right)^{\beta-1} \right] \tag{9}$$

It is can be shown that for small t, we have

$$r(t) \approx \left(\frac{\beta\nu}{\alpha}\right) \left(\frac{t}{\alpha}\right)^{\beta\nu-1}$$

In other words, for small t, $r(t)$ can be approximated by the failure rate of a two-parameter Weibull distribution with shape parameter $(\beta\nu)$ and scale parameter α. For large t (limiting case when $t \to \infty$), we have

$$r(t) \approx \left(\frac{\beta}{\alpha}\right) \left(\frac{t}{\alpha}\right)^{(\beta-1)}$$

In other words, for large t, $r(t)$ can be approximated by the failure rate of a two-parameter Weibull distribution with shape parameter β and scale parameter α. For further results, see Mudholkar et al.[25] and Jiang and Murthy[16] who studied the shapes of the failure rate function for the exponentiated Weibull distribution and its characterisation in the parameter space.

4. A Weibull extension model

A Weibull extension model is recently proposed in Xie et al.[35]. The cumulative distribution function is:

$$F(t) = 1 - \exp\left\{ -\lambda\alpha \left[e^{(t/\alpha)^{\beta}} - 1 \right] \right\}. \tag{10}$$

The failure rate function of the Weibull extension model has the following form:

$$r(t) = \lambda\beta \, (t/\alpha)^{\beta-1} \exp\left[(t/\alpha)^{\beta} \right], \quad \alpha, \beta, \lambda > 0, \quad t \geq 0. \tag{11}$$

It can be shown from Equation (11) that when $\beta \geq 1$, the failure rate function is an increasing function; and when $0 < \beta < 1$, the failure rate function has bathtub shape property (Xie et al.[35]. The change point of the bathtub curve in this case is

$$t^* = \alpha(1/\beta - 1)^{1/\beta} \tag{12}$$

This new model is asymptotically related to the traditional two-parameter Weibull distribution. As the scale parameter α approaches infinity, we have that

$$F(t) = 1 - \exp\left\{\lambda\alpha\left[1 - e^{(t/\alpha)^\beta}\right]\right\}$$
$$= 1 - \exp\left\{\lambda\alpha\left[1 - 1 - (t/\alpha)^\beta + o((t/\alpha)^\beta)\right]\right\} \approx 1 - \exp\{-\lambda\alpha^{1-\beta}t^\beta\}$$

which is a standard two-parameter Weibull distribution with a shape parameter of β (when λ tends to ∞ with α in such a manner that $\alpha^{\beta-1}/\lambda$ is held constant).

The Weibull extension model is based on a model given in Chen[7] which is a two-parameter distribution without a scale parameter. On the other hand, it also has the ability of modeling a bathtub shape or increasing failure rate function. A useful property of this distribution (Chen[7]) is that, the confidence intervals for the parameter β and the joint confidence regions for the parameters β and γ have a closed form.

5. An additive Weibull model

The additive Weibull model (Xie and Lai[34]) is a special case of competing risk model. It simply combines two Weibull distributions; one with a decreasing failure rate and another with an increasing failure rate. It has the cumulative hazard function given in the following form:

$$H(t) = (at)^b + (ct)^d, t, a, c \geq 0, b > 1, d < 1. \tag{13}$$

Based on this form of cumulative hazard function, the reliability function is simply given by

$$R(t) = \exp\left\{-(at)^b - (ct)^d\right\}. \tag{14}$$

The corresponding failure rate function can be shown to be

$$r(t) = ab(at)^{b-1} + cd(ct)^{d-1}, t \geq 0 \tag{15}$$

For a component that has a bathtub-shaped failure rate, the initial failures are usually caused by design faults and initial problems, which lead to a decreasing failure rate with a Weibull shape parameter less than one. The last part of the bathtub shaped failure rate is usually caused by material fatigue or component aging, and this corresponds to an increasing failure rate with a Weibull shape parameter greater than one. This model incorporates both types of failures.

It follows from (14) that this model can also be interpreted as the lifetime of a system which consists of two independent Weibull components that are arranged in a series. Let T_1 denote the lifetime of component 1, which is Weibull with parameters a and b and let T_2 denote the lifetime of component 2 which is Weibull with parameters c and d. If the system lifetime is T, then $T = \text{Min}\ (T_1, T_2)$ has the distribution as in an additive Weibull model. The interpretation is the same as for a "competing risk" problem in a survival analysis in which individuals die from any one of the two competing causes of death; and the additive Weibull is a special case with the underlying distributions to Weibull.

It is easy to see that the failure rate function has a bathtub-shape. The second term in (15) which is the dominating one for small t, is decreasing. For large t, the first term in (15) dominates and it is an increasing function. That the failure rate curve must be bathtub-shaped can also be shown as follows. The derivative of $r'(t)$ is given by

$$r'(t) = t^{d-2}\left[a^b b(b-1)t^{b-d} - c^d d(1-d)\right] \tag{16}$$

and then, $r'(t) = 0$ if and only if

$$t = t_0 = \left[\frac{c^d d(1-d)}{a^b b(b-1)}\right]^{1/(b-d)} \tag{17}$$

As $b > 1$ and $d < 1$, it is obvious that $t_0 > 0$. Also, $r'(t) > 0$ when $t > t_0$ and $r'(t) < 0$ when $t < t_0$. It is now obvious that our $r(t)$ has a bathtub shape.

This model should be able to fit to various types of data partly because this general additive Weibull model has four parameters $a, b, c,$ and d. Standard procedures such as the maximum likelihood method can be used to estimate the model parameters, and for large data sets, a graphical method is also available (Xie and Lai[34]). However, the estimation becomes more complex as the number of parameters increases. As a special case of our general model with $a = c$ and $d=1/b$, a two parameter model is obtained. When $b > 1$, and therefore $d = 1/b < 1$, the reduced model retains the bathtub shape.

The reduced model has only two parameters, a and b. Its cumulative hazard function is given by

$$H(t) = (at)^b + (at)^{1/b}, \ t \geq 0; \ a > 0, b > 1. \tag{18}$$

The corresponding failure rate function is

$$r(t) = ab(at)^{b-1} + a(at)^{1/b-1}/b \tag{19}$$

It can be seen that for this two-parameter additive Weibull, the failure rate function is of a bathtub shape.

6. Some related models

In this section we present some related or common models for bathtub shaped failure rate function.

Hjorth model:

The model by Hjorth[13] has the following failure rate function,

$$r(t) = \frac{\alpha}{1 + \beta t} + 2\gamma t, \ 0 < \gamma \leq \frac{\alpha\beta}{2}. \tag{20}$$

In fact, this is a well-known model for the bathtub shaped failure distribution. However, this model does not have a direct relationship with Weibull distribution.

A more general model is

$$r(t) = \frac{\alpha}{1 + \beta t} + \gamma\delta t^{\delta-1}, \alpha, \beta, \gamma > 0; \delta > 2, \tag{21}$$

and it is clear that $r(0) = \alpha, \ r(t) \to \infty$ as $t \to \infty$. For this model, when $\alpha = 0$, it reduces to a two-parameter Weibull distribution.

A quadratic failure rate function:

Bain and Engelhardt[3] considered quadratic failure rate

$$r(t) = \alpha + \beta t + \gamma t^2; \ \alpha \geq 0, \ \beta < 0, \ \gamma > 0 \tag{22}$$

which has a bathtub shape. Note that $r(0) = \alpha$, $r(t) \to \infty$ as $t \to \infty$. This model can be considered as an additive model with three independent components arranged in series, each with constant, linearly increasing, quadratic failure rate function, respectively.

Sectional model with two Weibull distributions:

Murthy and Jiang[27] consider two sectional models involving two Weibull distributions. The first model has a failure rate function given by

$$r(t) = \begin{cases} (\beta_1/\eta_1)(t/\eta_1)^{\beta_1-1}, & 0 \leq t \leq t_0, \\ (\beta_2/\eta_2)\left(\frac{t-\gamma}{\eta}\right)^{\beta_2-1}, & t_0 < t < \infty \end{cases} \tag{23}$$

satisfying the following two conditions:

$$t_0 = \left[\beta\eta_1^{\beta_1}/\eta_2^{\beta_2}\right]^{1/(\beta_1-\beta_2)}, \gamma = (1-\beta)t_0 \tag{24}$$

so that $r(t)$ is continuous at t_0.

The second model is the same as the first except that $\gamma = 0$. Now the shift parameter γ has no influence on the shape type of the failure rate. Thus the two models are essentially the same. For $\beta_1 < \beta_2$, $r(t)$ would have a bathtub shape when $\beta_1 < 1$ and $\beta_2 > 1$.

Jiang and Murthy[14] extended their results to give two sectional models involving three Weibull distributions. Four types of bathtub shapes are possible for both models. It can be noted that this can be seen to be the same as piecewise distributions involving many parameters. Usually a large data set is needed to estimate the shape and scale parameters in each of the Weibull distributions, and the change points of the distributions.

Generalized Weibull family:

Modholkar et al.[26] considered a generalized Weibull family having

$$r(t) = \frac{(t/\sigma)^{(1/\alpha)-1}}{\alpha\sigma\left[1 - \lambda(t/\sigma)^{1/\alpha}\right]}, \quad \alpha, \sigma > 0; \ \lambda \text{ real.} \tag{25}$$

The range of this generalized Weibull random variable is $(0, \infty)$ for $\lambda \leq 0$ and $(0, \sigma/\lambda^\alpha)$ for $\lambda > 0$. This failure rate function $r(t)$ has a bathtub shape for $\alpha > 1$ and $\lambda > 0$, and

$$r(t) \to \infty \text{ as } t \to 0 \text{ or } \sigma \ / \ \lambda^\alpha. \tag{26}$$

The distribution function is

$$F(t) = \left[1 - \left(1 - \lambda(t/\sigma)^{1/\alpha}\right)^{1/\lambda}\right] \qquad (27)$$

Generalized gamma distribution:
Glaser[9], McDonald and Richards[22] and Agarwal and Kalla[2] considered a generalized gamma distribution having probability density:

$$f(t) = ct^{\alpha v - 1}\exp[-(t/\beta)^{\gamma}], \quad \alpha > 0, \ \beta > 0, \ v > 1, \ v\alpha < 1. \qquad (28)$$

Note that this is also a type of generalisation of the Weibull distribution.
Marshall-Olkin's extended Weibull Distribution:
A three-parameter model generalising the basic Weibull distribution is presented in Marshall and Olkin[21]. The model was obtained via an interesting method of introducing a parameter into a family of distributions. Given a probability distribution function $F(t)$ or survival probability function $\overline{F}(t)$, a new survival function can be obtained as

$$\overline{G}(t,\nu) = \frac{\nu\overline{F}(t)}{1 - (1-\nu)\overline{F}(t)} = \frac{\nu\overline{F}(t)}{F(t) + \nu\overline{F}(t)}, \quad 0 < \nu < \infty. \qquad (29)$$

When $\nu = 1$, $\overline{G} = \overline{F}$ and hence the original distribution is a special case of the generalised distribution.

When $F(t)$ is a two-parameter Weibull distribution function, a new three-parameter distribution is obtained. It is termed an extended Weibull distribution in Marshall and Olkin[21]. The Marshall-Olkin extended Weibull distribution is

$$G(t,v,\alpha,\beta) = 1 - \frac{v\exp[-(t/\alpha)^{\beta}]}{1 - (1-v)\exp[-(t/\alpha)^{\beta}]}, \quad 0 < v,\alpha,\beta < \infty, \qquad (30)$$

The corresponding failure rate function is given by

$$r(t,v,\alpha,\beta) = \frac{(v/\alpha)(t/\alpha)^{\beta-1}}{1 - (1-v)\exp[-(t/\alpha)^{\beta}]}. \qquad (31)$$

Marshall and Olkin[21] carried out a partial study of the failure rate function. It can be shown that the failure rate function is increasing when $\alpha \geq 1$, $\beta \geq 1$ and decreasing when $\alpha \leq 1$, $\beta \leq 1$. If $\beta > 1$, then the failure rate function is initially increasing and eventually increasing, but there may be one interval where it is decreasing. Similarly, when $\beta < 1$, the failure rate function is initially decreasing and eventually decreasing, but there may be one interval where it is increasing.

7. Parameter estimation and model validation

As for any statistical distribution, there are a number of model parameters that have to be estimated. Techniques such as the method of moments, maximum likelihood estimation, percentile estimation, Bayesian estimation can be applied to all models. They are straightforward although usually tedious.

As mentioned before, Weibull distribution has been widely used in reliability analysis partly because of the existence of the so called Weibull probability plotting procedure. It will be useful to develop methods that can extend this procedure even for more complicated model. For models with three or more parameters, fitting the plot to a single straight line will not be sufficient. As Xie and Lai[34] and Lai et al.[19] pointed out, it is possible to produce two plots with different transformations, so that two slopes and intercepts can be used. The same idea is presented in Jiang and Murthy[16] who studied the asymptotes. In fact, for all models described in Sections 2-5, a graphical approach can be used for parameter estimation.

It can also be pointed out that a graphical plotting serves as a useful tool for model validation. In reliability analysis, it is important to make use of the fitted distribution in decision making. An assumption in such an analysis is that the distribution is the correct one. It is very useful if simple model validation procedure can be developed. Graphical methods play an important role in this respect, although statistical approaches are also available.

There are tests that are available for testing the exponentiality against aging alternatives (Lai[17]). Methods for testing against bathtub-shaped failure rate functions have also been studied. Interested readers can refer to Aarset[1], Xie[33], Lai[17], Guess et al.[10], among others.

8. Discussions

There are a number of ways to obtain a distribution with a bathtub shaped failure rate function. In this article, some simple extensions of the Weibull distribution are discussed. Although the sectional and additive Weibull distributions are suitable for generating bathtub-shaped failure rate distributions, mixtures of two Weibull distributions can never be a BFR distribution. See, for example, Glaser[9], Pamme and Kunitz[29], and Jiang and

Murthy[15].

Most of the models contain more parameters than the two-parameter Weibull distribution. The traditional method of maximum likelihood estimation can be used for parameter estimation with complete or censored data. It should be pointed out that an important advantage of the Weibull distribution is its graphical interpretation, and it is of practical interest to develop appropriate graphical methods for model validation and parameter estimation. Hence when the exponentiality is not an appropriate assumption due to aging, Weibull distribution can be a good alternative.

This chapter focuses on the failure rate function, with which the bathtub distribution is associated. However, an interesting reliability characteristic is the mean residual life which is related to the bathtub shaped failure rate (Mi^{23}). Gupta and Akman[11] shows that a bathtub shaped failure rate function does not imply that the mean residual life is of an upside-down bathtub shape. A sufficient condition for this to be true, on the other hand, is $r(0) \geq 1/\mu$ where μ is the mean of the distribution.

A practical problem is the model selection. Our focus in this Chapter is on the description of the models and their basic properties. These models have the ability to fit data sets which exhibit a bathtub shaped failure rate. However, when the data set is small, most of the models would still fit the data set reasonably well and further investigation is needed on model selection. Indeed, several of the models are new and have not been widely studied and they open up new directions for further research in several aspects.

Acknowledgements

This research is partly supported by a research grant from the National University of Singapore for the project "Reliability Analysis of Engineering Systems" (RP3992679).

References

1. M. V. Aarset, How to identify a bathtub hazard rate, *IEEE Transactions on Reliability*, **R-36**, 106-108 (1987).
2. S. K. Agarwal. and S. T. Kalla, A generalized gamma distribution and its application in reliability, . *Communications in Statistics A:-Theory and Methods*, **25**, 201-210 (1996).
3. J. J. Bain and M. Englehardt, *Statistical Analysis of Reliability and Life-Testing. Models* (Marcel Dekker, New York, 1991).

4. R. E. Barlow and F. Proschan, *Statistical Theory of Reliability and Life Testing: Probability Models* (To Begin With, Silver Spring, MD, 1981).

5. J. Bartoszewicz, Dispersive ordering and monotone failure rate distributions, *Advances in Applied Probability,* **17**, 472-474 (1985).

6. P. J. Boland, E. El-Neweihi and F. Proschan, Applications of the hazard rate ordering in reliability and order statistics, *Journal of Applied Probability,* **31**, 180-192 (1994).

7. Z. Chen, A new two-parameter distribution with bathtub shape or increasing failure rate function, *Statistics & Probability Letters,* **49**, 155-161 (2000).

8. J. V. Deshpande, S. C. Kochar and H. Singh, Aspects of positive ageing, *Journal of Applied Probability,* **23**, 748-758 (1986).

9. R. E. Glaser, Bathtub and related failure rate characterizations, *Journal of the American Statistical Association,* **75**, 667-672 (1980).

10. F. Guess, K. H. Nam and D. H. Park, Failure rate and mean residual life with trend changes, *Asia-Pacific Journal of Operations Research,* **15**, 239-244 (1998).

11. R. C. Gupta and H. O. Akman, Mean residual life functions for certain types of non-monotonic ageing, *Communications in Statistics - C: Stochastic Models,* **11**, 219-225 (1995).

12. E. Haupt and H. Schabe, A new model for a lifetime distribution with bathtub shaped failure rate, *Microelectronics and Reliability,* **32**, 633-639 (1992).

13. U. Hjorth, A reliability distribution with increasing, decreasing, constant and bathtub-shaped failure rates, *Technometrics,* **22**, 99-107 (1980).

14. R. Jiang and D. N. P. Murthy, Two sectional models involving three Weibull distributions, *Quality and Reliability Engineering International,* **13**, 83-96 (1997).

15. R. Jiang and D. N. P. Murthy, Mixture of Weibull distributions - parametric characterization of failure rate function, *Applied Stochastic Models and Data Analysis,* **14**, 47-65 (1998).

16. R. Jiang and D. N. P. Murthy, The exponentiated Weibull family: A graphical approach, *IEEE Transactions on Reliability,* **48**, 68-72 (1999).

17. C. D. Lai, Tests of univariate and bivariate stochastic ageing, *IEEE Transactions on Reliability,* **43**, 233-241 (1994).

18. C. D. Lai, M. Xie and D. N. P. Murthy, Bathtub-shaped failure rate life distributions. Ch.3, Vol.10, pp.69-104, *Handbook of Statistics* (Elsevier, London, 2001).

19. C. D. Lai, M. Xie and D. N. P. Murthy, Modified Weibull model, *IEEE Transactions on Reliability,* To appear (2003).

20. L. Lee and W. A. Thompson Jr., Failure rate - a unified approach, *Journal of Applied Probability,* **13**, 176-182 (1976).

21. A. W. Marshall and I. Olkin, A new method for adding a parameter to a family of distributions with application to the exponential and Weibull families, *Biometrika,* **84**, 641-652 (1997).

22. J. B. McDonald and D. O. Richards, Model selection: Some generalized distributions, *Communications in Statistics - A: Theory and Methods,* **16**,

1049-1047 (1987).

23. J. Mi, Bathtub failure rate and upside-down bathtub mean residual life. *IEEE Transactions on Reliability*, **44**, 388-391 (1995).

24. G. S. Mudholkar and D. K. Srivastava, Exponentiated Weibull family for analyzing bathtub failure-rate data, *IEEE Transactions on Reliability*, **42**, 299-302 (1993).

25. G. S. Mudholkar, D. K. Srivastava and M. Freimer, The Exponentiated Weibull family: a reanalysis of the bus-motor-failure data, *Technometrics*, **37**, 436-445 (1995).

26. G. S. Mudholkar, D. K. Srivastava and G. D. Kollia, A generalization of the Weibull distribution with application to analysis of survival data, *Journal of the American Statistical Association*, **91**, 1575-1583 (1996).

27. D. N. P. Murthy and R. Jiang, Parametric study of sectional models involving two Weibull distributions, *Reliability Engineering and System Safety*, **56**, 151-159 (1997).

28. D. N. P. Murthy, M. Xie and R. Jiang, *Weibull Distribution* (Wiley, New York, 2003).

29. H. Pamme and H. Kunitz, Detection and modeling of aging properties in lifetime data, in *Advances in Reliability*, Ed. A. P. Basu (Elsevier Science Publishers, 1993) pp. 291-302.

30. S. Rajarshi and M. B. Rajarshi, Bathtub distributions: a review, . *Communications in Statistics – A: Theory and Methods*, **17**, 2597-2621 (1988).

31. W. Weibull, A statistical distribution of wide applicability, *Journal of Applied Mechanics*, **18**, 293-297 (1951).

32. K. L. Wong, The physical basis for the roller-coaster hazard rate curve for electronics, *Quality & Reliability Engineering International*, **7**, 489-495 (1991).

33. M. Xie, Some total time on test quantiles useful for testing constant against bathtub-shaped failure rate distributions, *Scandinavian Journal of Statistics*, **16**, 137-144 (1989).

34. M. Xie and C. D. Lai, Reliability analysis using an additive Weibull model with bathtub-shaped failure rate, *Reliability Engineering and System Safety*, **52**, 87-93 (1995).

35. M. Xie, T. N. Goh and Y. Tang, A modified Weibull extension with bathtub-shaped failure rate function, *Reliability Engineering and Systems Safety*, **76**, 279-285 (2002).

Part VI

STATISTICAL METHODS FOR DEGRADATION DATA

20

ESTIMATION FROM SIMULTANEOUS DEGRADATION AND FAILURE TIME DATA

V. Bagdonavičius, A. Bikelis and V. Kazakevičius

Department of Mathematical Statistics, University of Vilnius, Vilnius, Lithuania

M. Nikulin

*UFR Sciences et Modelisation, Victor Segalen University Bordeaux 2,
BP 69, 33076 Bordeaux Cedex, France
E-mail: nikou@sm.u-bordeaux2.fr*

The paper considers degradation and failure time models with multiple failure modes. Dependence of traumatic failure intensities on the degradation level are included into the models. Nonparametric estimators of various reliability characteristics are proposed and their asymptotic properties are established.

1. Introduction

Functioning of a unit is characterized by its degradation process and by the random moments of its potential failures which may be of various modes. We call a failure of a unit *natural* if the degradation attains some critical level. Other failures are called *traumatic*. These can be related with production defects, caused by mechanical damages or by fatigue of components etc.

The intensities of the traumatic failures of different modes depend on degradation. As a rule these intensities are increasing functions of degradation values. In this paper we consider the models, where other factors such as, e.g., functioning time do not affect the traumatic failures intensities.

The first problem considered in this paper is statistical estimation of the effect of degradation (wear) on the intensity of traumatic failures. The solution of this problem gives the basis for estimation of various reliability characteristics and for prediction of residual reliability of any concrete unit

301

given degradation value at a given moment.

Asymptotic properties of estimators are investigated.

2. The Model

Suppose that the life time of a unit is determined by the following random objects: the degradation process of the unit $Z(t)$ and the moments of its potential traumatic failures of q different modes $T^1, \ldots, T^q$. For example, $Z(t)$ may be the value of tire wear at the moment when a tire has run t km (in this case "time" is the tire run), the size of fatigue crack, the size of failure-causing conducting filament of chlorine-copper compound in a printed-circuit board, luminosity of light emitting diode at the moment t (see Meeker and Escobar[5]), etc. Usually $q = 1$, sometimes several failure modes are observed. For example, traumatic failures of different modes may be tire punctures, bursting, dividing of the protector into layers, etc.

Denote by T^0 the moment, when the degradation attains some critical value z_0. We suppose that this also leads to the termination of functioning of the unit and call T^0 the moment of potential *natural* failure. Thus the moment of the observed unit's failure is

$$T = \min\left(T^0, T^1, \ldots, T^q\right). \tag{2.1}$$

We suppose that the degradation process is modeled by the general path model (see Meeker and Escobar[5], Bagdonavičius and Nikulin[2]).

$$Z(t) = g(t, A);$$

here $A = (A_1, \ldots, A_r)$ is a random vector with positive components and the distribution function π, and g is a specified continuously differentiable in t function, which increases from 0 to $+\infty$ when t increases from 0 to $+\infty$. For example, $g(t, a) = t/a$ (linear degradation, $r = 1$) or $g(t, a) = (t/a_1)^{a_2}$ (convex or concave degradation, $r = 2$). A classical example of linear degradation is tire wear. Another example is the increase in a resistance over time[6]. Three other above mentioned examples of degradation processes can be modeled by path models with convex or concave degradation.

Denote by h the function inverse to g with respect to the first argument. Evidently, it is continuously differentiable and increasing in t. Moreover,

$$T^0 = h(z_0, A). \tag{2.2}$$

We also suppose that the random variables T^k $(k = 1, \ldots, q)$ are conditionally independent (given $A = a$) and have the intensities depending only on the degradation level. The latter means that

$$\mathbf{P}\{T^k > t \mid A = a\} = \exp\left\{-\int_0^t \lambda_k\big(g(s,a)\big)ds\right\}, \qquad (2.3)$$

where the λ_k are some positive functions. If we set

$$\Lambda_k(z) = \int_0^z \lambda_k(y)dy,$$

then (2.3) can be rewritten in the following form

$$\mathbf{P}\{T^k > t \mid A = a\} = \exp\left\{-\int_0^{g(t,a)} h'(z,a)d\Lambda_k(z)\right\}.$$

This model has some remote relations with frailty models, interpreting A as the frailty variable, see Hougaard[4]. The difference from the classical frailty models is that observing the degradation process it is possible to estimate the distribution of the frailty variable.

Equality (2.3) implies that

$$\mathbf{P}\{t < T^{(k)} \le t + \Delta \mid T^{(k)} > t,\ Z(t) = z\} = \lambda_k(z)\Delta + o(\Delta), \quad \text{as } \Delta \to 0.$$

So the intensity $\lambda_k(z)$ is proportional to the conditional probability to have a traumatic failure of the kth mode in a small time interval given that at the beginning of this interval the unit is functioning and its wear is z.

There is a number of papers dealing with estimation of degradation process and non-traumatic failure distribution parameters using degradation data and the path models (see Meeker and Escobar[5] for references). In this paper we generalize the problem. The purpose is to estimate and predict not only reliability characteristics related with the degradation process itself but also various characteristics related with traumatic failures using degradation and failure time data.

3. Main Reliability Characteristics of a Unit

3.1. *Unconditional reliability characteristics*

Recall that the life time of a unit is defined by (2.1), where T^0 is given in (2.2). Set

$$\Lambda_\cdot(z) = \sum_{k=1}^q \Lambda_k(z) \quad \text{and} \quad H(z,a) = \int_0^z h'(y,a)d\Lambda_\cdot(y).$$

Then the survival function and the mean of the random variable T equal, respectively,

$$S(t) = \int_{g(t,a)<z_0} e^{-H(g(t,a),a)} d\pi(a) \tag{3.1}$$

and

$$e = \int \left\{ h(z_0,a) - \int_0^{z_0} h'(z,a)[h(z_0,a) - h(z,a)]e^{-H(z,a)} d\Lambda.(z) \right\} d\pi(a). \tag{3.2}$$

Denote by V the indicator of the failure mode:

$$V = \begin{cases} 0 & \text{if } T = T^0; \\ 1 & \text{if } T = T^1; \\ \vdots & \\ q & \text{if } T = T^q. \end{cases} \tag{3.3}$$

Important reliability characteristics of a unit are the probability $P^{(k)}(t) = \mathbf{P}\{T \leqslant t,\ V = k\}$ of a failure of the kth mode in the interval $[0; t]$ and its limit value $P^{(k)} = P^{(k)}(\infty)$. For $k = 1, \ldots, q$ we have

$$P^{(k)}(t) = \int d\pi(a) \int_0^{g(t,a)\wedge z_0} h'(z,a)e^{-H(z,a)} d\Lambda_k(z), \tag{3.4}$$

where $a \wedge b = \min(a,b)$. In particular,

$$P^{(k)} = \int d\pi(a) \int_0^{z_0} h'(z,a)e^{-H(z,a)} d\Lambda_k(z). \tag{3.5}$$

Summing over $k = 1, \ldots, q$ we obtain the probability $P^{(tr)}(t)$ of the traumatic failure in the interval $[0; t]$ and its limit value $P^{(tr)} = P^{(tr)}(\infty)$:

$$P^{(tr)}(t) = 1 - \int e^{-H(g(t,a)\wedge z_0,a)} d\pi(a) \tag{3.6}$$

and

$$P^{(tr)} = 1 - \int e^{-H(z_0,a)} d\pi(a). \tag{3.7}$$

Finally, by (3.6) and (3.1),

$$P^{(0)}(t) = \int_{g(t,a)\geqslant z_0} e^{-H(z_0,a)} d\pi(a) \tag{3.8}$$

and

$$P^{(0)} = \int e^{-H(z_0,a)} d\pi(a). \tag{3.9}$$

3.2. *Residual reliability characteristics*

Suppose that the degradation curve of a concrete unit was observed until the moment t. For example, the wear z of a specified tire is known at the "moment" t. The following (conditional) characteristics are important to estimate: the probability to fail in the interval $(t; t + \Delta]$, the probability of a failure of the kth mode and the probability of a traumatic failure in the same interval, the mean residual life time of the unit. We denote them, respectively, by $Q(\Delta)$, $Q^{(k)}(\Delta)$, $Q^{(tr)}(\Delta)$ and ε.

Set

$$t_0 = h(z_0, a), \quad z = g(t, a) \quad \text{and} \quad z_1 = g(t + \Delta, a). \tag{3.10}$$

If $k = 1, \ldots, q$ then

$$Q^{(k)}(\Delta) = \int_z^{z_0 \wedge z_1} h'(y, a) e^{-[H(y,a) - H(z,a)]} d\Lambda_k(y). \tag{3.11}$$

Summing over $k = 1, \ldots, q$ yields

$$Q^{(tr)}(\Delta) = 1 - e^{-[H(z_0 \wedge z_1, a) - H(z,a)]}. \tag{3.12}$$

It is evident that $Q^{(0)}(\Delta) = 0$, $Q(\Delta) = Q^{(tr)}(\Delta)$ if $t + \Delta < t_0$ and $Q^0(\Delta) = 1 - Q^{(tr)}(\Delta)$, $Q(\Delta) = 1$ if $t + \Delta \geqslant t_0$. Therefore

$$Q^{(0)}(\Delta) = \begin{cases} 0 & \text{for } \Delta < t_0 - t; \\ e^{-[H(z_0,a) - H(z,a)]} & \text{for } \Delta \geqslant t_0 - t; \end{cases} \tag{3.13}$$

and

$$Q(\Delta) = \begin{cases} 1 - e^{-[H(z_1,a) - H(z,a)]} & \text{for } \Delta < t_0 - t; \\ 1 & \text{for } \Delta \geqslant t_0 - t. \end{cases} \tag{3.14}$$

Finally, the mean residual life time of a unit equals

$$\varepsilon = \int_0^\infty \Delta dQ(\Delta) = \int_z^{z_0} h'(y, a)[h(y, a) - t] e^{-[H(y,a) - H(z,a)]} d\Lambda.(y)$$

$$+ (t_0 - t) e^{-[H(z_0,a) - H(z,a)]}. \tag{3.15}$$

4. Non-parametric Estimation of the Cumulative Intensities

Most of reliability characteristics are functionals of the cumulative intensities Λ_k and of the distribution function π. Therefore we begin with non-parametric estimation of these functions.

The data. Suppose that n units are on test and the failure moments T_i, the failure modes V_i and the degradation curves $Z_i(t) = (g(t, A_i) \mid t \leqslant T_i)$ are observed.

Examples of degradation curves are found in Meeker and Escobar[5], fig. 13.1, 13.3, 13.4., 13.7.

If the degradation curves are known then the vectors of degradation parameters $A_i = (A_{i1}, \ldots, A_{ir})$ are also known. We suppose that the measuring errors of degradation are negligible, and the values of A_i can be found using $m_i \geq r$ observed degradation values $Z_{ij} = Z_i(t_{ij})$ of the ith unit minimizing the sum $\sum_{j=1}^{m_i} (g(t_{ij}, A_i) - Z_{ij})^2$ with respect to A_i. Therefore the data can be defined as a collection of vectors

$$(A_1, T_1, V_1), \ldots, (A_n, T_n, V_n),$$

which are independent copies of the vector (A, T, V) defined in Section 2.

The failure of any mode is observed if it occurs earlier than failures of other modes. So we have competing risk situation. Note that traumatic failure moments depend on degradation (which determines non-traumatic failure).

Set $Z_i = g(T_i, A_i)$. It is the degradation value of the ith unit at its failure moment.

The best estimator of the distribution function π is the empirical distribution function

$$\hat{\pi}(a) = \frac{1}{n} \sum_{i=1}^{n} \mathbf{1}_{\{A_i \leqslant a\}} = \frac{1}{n} \sum_{i=1}^{n} \mathbf{1}_{\{A_{i1} \leqslant a_1, \ldots, A_{ir} \leqslant a_r\}}.$$

It is well-known that this estimator is uniformly consistent, i.e. almost surely

$$\sup_a \mid \hat{\pi}(a) - \pi(a) \mid \to 0,$$

as $n \to \infty$. Moreover, if the function π is absolutely continuous then the random function $\sqrt{n}(\hat{\pi} - \pi)$ tends in distribution in the Skorokhod space $D^r[0; \infty]$ to a zero mean Gaussian field W_0 with the covariance function

$$\mathbf{E}[W_0(a) W_0(a')] = \pi(a \wedge a') - \pi(a)\,\pi(a');$$

where $a \wedge a' = (a_1 \wedge a'_1, \ldots, a_r \wedge a'_r)$.

Estimation of the cumulative intensities Λ_k is based on counting processes

$$N_k(z) = \sum_{i=1}^{n} \mathbf{1}_{\{Z_i \leqslant z, V_i = k\}}. \tag{4.1}$$

Obviously, $N_k(z)$ is the number of units having a failure of the kth mode before or at the moment when the degradation attains the level z.

Theorem 1: Let $\mathcal{F}_z$ be the σ-algebra generated by the random variables $A_1, \ldots, A_n$ and $N_1(y), \ldots, N_q(y)$ with $y \leq z$. Then the process $N_k(z)$ can be written as the sum

$$N_k(z) = \int_0^z \lambda_k(y) Y(y) dy + M_k(z),$$

where

$$Y(z) = \sum_{Z_i \geqslant z} h'(z, A_i) \tag{4.2}$$

and M_k is a martingale with respect to the filtration $(\mathcal{F}_z \mid 0 \leqslant z \leqslant z_0)$. Moreover, the predictable covariation of the processes M_k and M_l is given by

$$< M_k, M_l > (z) = \delta_{kl} \int_0^z \lambda_k(y) Y(y) dy,$$

where $\delta_{kl} = \mathbf{1}_{\{k=l\}}$ stands for the Kronecker symbol.

The proofs of this and of the following theorems are given in Section 6.

The theorem implies that the optimal estimator of the cumulative intensity $\Lambda_k(z)$ is of Nelson-Aalen type (see Andersen et al.[1], pp. 177-178):

$$\hat{\Lambda}^{(k)}(z) = \int_0^z Y^{-1}(y)\, dN_k(y) = \sum_{Z_i \leqslant z, V_i = k} Y^{-1}(Z_i). \tag{4.3}$$

To get asymptotic properties of $\hat{\Lambda}_k$ we need two additional assumptions. Set

$$c_0(a) = \sup_{z \leqslant z_0} h'(z, a), \quad c_1(a) = \sup_{z_1 < z_2 \leqslant z_0} \frac{|h'(z_2, a) - h'(z_1, a)|}{z_2 - z_1}$$

and

$$b(z) = \mathbf{E}[h'(z, A_i)\mathbf{1}_{\{Z_i \geqslant z\}}] = \int h'(z, a)e^{-H(z,a)} d\pi(a).$$

Assumption 1. $\mathbf{E}[c_0(A) + c_1(A)] < \infty.$
Assumption 2. $\inf_{z \leqslant z_0} b(z) > 0.$

By Assumption 1, $b(z) \leqslant \mathbf{E}c_0(A) < \infty$ for all $z \leqslant z_0$. Therefore, almost surely,

$$\sup_{z \leqslant z_0} |n^{-1} Y(z) - b(z)| \to 0,$$

as $n \to \infty$ (this can be proved analogously as the Glivenko-Cantelli theorem). Then Theorems IV.1.1 and IV.1.2 in [1] (see remark below the proof of the second theorem) imply the following result.

Theorem 2: Under Assumptions 1-2, the estimator $\hat{\Lambda}_k$ is uniformly consistent, i.e.

$$\sup_{z \leqslant z_0} |\hat{\Lambda}_k(z) - \Lambda_k(z)| \xrightarrow{P} 0,$$

as $n \to \infty$.

Theorem 3: Set $\Lambda = (\Lambda_1, \ldots, \Lambda_q)$ and $\hat{\Lambda} = (\hat{\Lambda}_1, \ldots, \hat{\Lambda}_q)$. If Assumptions 1-2 hold, then the random vector function $\sqrt{n}(\hat{\Lambda} - \Lambda)$ tends in distribution in the space $D^q[0; z_0]$ to $W = (W_1, \ldots, W_q)$, the vector of independent zero mean Gaussian processes with the covariance function

$$\mathbf{E}[W_k(z)W_k(z')] = \sigma_k^2(z \wedge z'),$$

where

$$\sigma_k^2(z) = \int_0^z \frac{\lambda_k(y)}{b(y)} dy. \tag{4.4}$$

The next theorem shows that the estimators $\hat{\pi}$ and $\hat{\Lambda}$ are asymptotically independent.

Theorem 4: If Assumptions 1-2 hold and the function π is absolutely continuous then the random vector function $\sqrt{n}(\hat{\pi} - \pi, \hat{\Lambda} - \Lambda)$ tends in distribution in the space $D^r[0; \infty] \times D^q[0; z_0]$ to a random vector function with independent components.

5. Non-parametric Estimation of the Main Reliability Characteristics

5.1. *Unconditional reliability characteristics*

Estimating the survival function $S(t)$, the mean failure time e and the probabilities $P^{(k)}(t)$, $P^{(k)}$, $P^{(tr)}(t)$, $P^{(tr)}$ we replace the functions π and Λ_k in (3.1)-(3.2), (3.4)-(3.9) by $\hat{\pi}$ and $\hat{\Lambda}_k$:

$$\hat{S}(t) = \frac{1}{n} \sum_{g(t,A_i)<z_0} e^{-\hat{H}(g(t,A_i),A_i)},$$

$$\hat{e} = \frac{1}{n} \sum_{i=1}^n \left\{ h(z_0, A_i) - \sum_{V_j \neq 0} h'(Z_j, A_i) \frac{h(z_0, A_i) - h(Z_j, A_i)}{Y(Z_j)} e^{-\hat{H}(Z_j,A_i)} \right\},$$

for $k = 1, \ldots, q$,

$$\hat{P}^{(k)}(t) = \frac{1}{n} \sum_{i=1}^{n} \sum_{V_j=k,\, Z_j \leqslant g(t,A_i)} \frac{h'(Z_j, A_i)}{Y(Z_j)} e^{-\hat{H}(Z_j, A_i)},$$

$$\hat{P}^{(k)} = \frac{1}{n} \sum_{i=1}^{n} \sum_{V_j=k} \frac{h'(Z_j, A_i)}{Y(Z_j)} e^{-\hat{H}(Z_j, A_i)},$$

$$\hat{P}^{(tr)}(t) = 1 - \frac{1}{n} \sum_{i=1}^{n} e^{-\hat{H}(g(t,A_i), A_i)},$$

$$\hat{P}^{(tr)} = 1 - \frac{1}{n} \sum_{i=1}^{n} e^{-\hat{H}(z_0, A_i)},$$

$$\hat{P}^{(0)}(t) = 1 - \frac{1}{n} \sum_{g(t,A_i) \geqslant z_0} e^{-\hat{H}(z_0, A_i)},$$

$$\hat{P}^{(0)} = 1 - \hat{P}^{(tr)};$$

here

$$\hat{H}(z, a) = \sum_{V_l \neq 0,\, Z_l \leqslant z} \frac{h'(Z_l, a)}{Y(Z_l)}$$

and $Y(z)$ is defined by (4.2).

Set

$$\sigma^2(z) = \sum_{k=1}^{q} \sigma_k^2(z), \quad \sigma_{-k}^2(z) = \sigma^2(z) - \sigma_k^2(z),$$

where $\sigma_k^2(z)$ are given in (4.4).

Theorem 5: Let θ be any of the unconditional reliability characteristics considered above. If Assumption 2 holds, $\mathbf{E}[(1 + c_0(A))^2(1 + c_0(A) + c_1(A))] < \infty$ and the distribution function π is absolutely continuous then the distribution of the random variable $\sqrt{n}(\hat{\theta} - \theta)$ tends to the normal law with the mean 0 and the variance $V(\hat{\theta})$, where:

$$V(\hat{S}(t)) = \int_0^{z_0} \left\{ \iint_{y<g(t,a)<z_0} h'(y,a) e^{-H(g(t,a),a)} d\pi(a) \right\}^2 d\sigma^2(z)$$

$$+ \int_{g(t,a)<z_0} e^{-2H(g(t,a),a)} d\pi(a) - S^2(t),$$

$$V(\hat{e}) = \int_0^{z_0} \Big\{ \int h'(z,a)\Big(\int_z^{z_0} h'(u,a)[h(z_0,a) - h(u,a)]e^{-H(u,a)}d\Lambda.(u)$$

$$- [h(z_0,a) - h(z,a)]e^{-H(z,a)}\Big)d\pi(a)\Big\}^2 d\sigma^2(z)$$

$$+ \int \Big\{ h(z_0,a) - \int_0^{z_0} h'(z,a)[h(z_0,a) - h(z,a)]e^{-H(z,a)}d\Lambda.(z)\Big\}^2 d\pi(a) - e^2,$$

$$V(\hat{P}^{(k)}(t)) = \int_0^{z_0} \Big\{ \int_{g(t,a)>z} h'(z,a)d\pi(a) \int_z^{z_0 \wedge g(t,a)} h'(u,a)e^{-H(u,a)}d\Lambda_k(u)\Big\}^2 d\sigma^2_{-k}(z)$$

$$+ \int_0^{z_0} \Big\{ \int_{g(t,a)>z} h'(z,a)\Big(e^{-H(z,a)} - \int_z^{z_0 \wedge g(t,a)} h'(u,a)e^{-H(u,a)}d\Lambda_k(u)\Big)d\pi(a)\Big\}^2 d\sigma^2_k(z)$$

$$+ \int \Big\{ \int_0^{z_0 \wedge g(t,a)} h'(z,a)e^{-H(z,a)}d\Lambda_k(z)\Big\}^2 d\pi(a) - [P^{(k)}(t)]^2,$$

$$V(\hat{P}^{(tr)}(t)) = \int_0^{z_0} \Big\{ \int_{g(t,a)>z} h'(z,a)e^{-H(z_0 \wedge g(t,a),a)}d\pi(a)\Big\}^2 d\sigma^2(z)$$

$$+ \int e^{-2H(z_0 \wedge g(t,a),a)}d\pi(a) - [P^{(tr)}(t)]^2,$$

$$V(\hat{P}^{(0)}(t)) = \int_0^{z_0} \Big\{ \int_{g(t,a)\geqslant z} h'(z,a)e^{-H(z_0,a)}d\pi(a)\Big\}^2 d\sigma^2(z)$$

$$+ \int_{g(t,a)\geqslant z_0} e^{-2H(z_0,a)}d\pi(a) - [P^{(0)}(t)]^2.$$

Approximate $(1-\alpha)$-confidence interval for θ has the form

$$\hat{\theta} \pm z_{1-\alpha/2}\sqrt{\hat{V}(\hat{\theta})/n}, \tag{5.1}$$

where $\hat{V}(\hat{\theta})$ is obtained by replacing the unknown quantities $\pi(a)$, $\Lambda_k(z)$ and $d\sigma_k^2(z)$ in the expression of $V(\hat{\theta})$ by $\hat{\pi}(a)$, $\hat{\Lambda}_k(z)$ and $d\hat{\sigma}_k^2(z) = nY^{-1}(z)d\hat{\Lambda}_k(z)$, respectively.

5.2. *Residual reliability characteristics*

Estimators for the conditional mean ε and the conditional failure probabilities $Q^{(k)}(\Delta)$, $Q^{(tr)}(\Delta)$ are obtained by replacing the cumulative intensities Λ_k in (3.11)–(3.15) by their nonparametric estimators $\hat{\Lambda}_k$: for $k = 1,\dots,q$,

$$\hat{Q}^{(k)}(\Delta) = \sum_{z<Z_i\leqslant z_1, V_i=k} \frac{h'(Z_i,a)}{Y(Z_i)}e^{-[\hat{H}(Z_i,a)-\hat{H}(z,a)]},$$

$$\hat{Q}^{(tr)}(\Delta) = 1 - e^{-[\hat{H}(z_1,a)-\hat{H}(z,a)]},$$

$$\hat{\varepsilon} = \sum_{Z_i>z,V_i\neq0} h'(Z_i,a)\frac{h(Z_i,a)-t}{Y(Z_i)}\,e^{-[\hat{H}(Z_i,a)-\hat{H}(z,a)]} + (t_0-t)e^{-[\hat{H}(z_0,a)-\hat{H}(z,a)]},$$

where t_0, z and z_1 are defined by (3.10). Moreover,

$$\hat{Q}^{(0)}(\Delta) = \begin{cases} 0 & \text{for } \Delta < t_0 - t; \\ 1 - \hat{Q}^{(tr)}(\Delta) & \text{for } \Delta \geqslant t_0 - t; \end{cases}$$

$$\hat{Q}(\Delta) = \begin{cases} \hat{Q}^{(tr)}(\Delta) & \text{for } \Delta < t_0 - t; \\ 1 & \text{for } \Delta \geqslant t_0 - t; \end{cases}$$

Theorem 6: Let θ denotes any of the residual characteristics considered above. If Assumptions 1-2 hold then the distribution of the random variable $\sqrt{n}(\hat{\theta} - \theta)$ tends to the normal law with the mean 0 and the variance $V(\hat{\theta})$. The asymptotic variances are: for $k = 1,\ldots,q$,

$$V(\hat{Q}^{(k)}(\Delta)) = \int_z^{z_0\wedge z_1}\left\{h'(y,a)\int_y^{z_0\wedge z_1}h'(u,a)e^{-[H(u,a)-H(z,a)]}d\Lambda_k(u)\right\}^2 d\sigma^2_{-k}(y)$$

$$+\int_z^{z_0\wedge z_1}[h'(y,a)]^2\left\{e^{-[H(y,a)-H(z,a)]}-\int_y^{z_0\wedge z_1}h'(u,a)e^{-[H(u,a)-H(z,a)]}d\Lambda_k(u)\right\}^2 d\sigma^2_k(y),$$

$$V(\hat{Q}^{(tr)}(\Delta)) = (1 - Q^{(tr)}(\Delta))^2\int_z^{z_0\wedge z_1}[h'(y,a)]^2 d\sigma^2(y),$$

$$V(\hat{\varepsilon}) = \int_z^{z_0}[h'(y,a)]^2\left\{\int_y^{z_0}h'(u,a)[h(u,a)-h(y,a)]e^{-[H(u,a)-H(z,a)]}d\Lambda.(u)\right.$$

$$\left. - [h(z_0,a)-h(y,a)]e^{-[H(z_0,a)-H(z,a)]}\right\}^2 d\sigma^2(y).$$

Approximate $(1-\alpha)$-confidence intervals for θ have the form (5.1) where $\hat{V}(\hat{\theta})$ is obtained from the formulas for $V(\hat{\theta})$ by replacing $\Lambda_k(z)$ and $d\sigma^2_k(z)$ by $\hat{\Lambda}_k(z)$ and $d\hat{\sigma}^2_k(z) = nY^{-1}(z)d\hat{\Lambda}_k(z)$, respectively.

6. Proofs

Proof of Theorem 1. Let $0 \leq y < z \leq z_0$. We need to prove that

$$\mathbf{E}[N_k(z) - N_k(y) \mid \mathcal{F}_y] = \mathbf{E}[\int_y^z \lambda_k(u)Y(u)du \mid \mathcal{F}_y].$$

312   *V. Bagdonavičius, A. Bikelis, V. Kazakevičius and M. Nikulin*

By additivity of the mathematical expectation, it suffices to consider the case $n = 1$.

If $A_1 = a$ and $Z_1 \leqslant y$ then $N_k(z) = N_k(y)$. If $A_1 = a$ and $Z_1 > y$ then the random variable $N_k(z)$ takes two values: 1 with the probability p and 0 with the probability $1 - p$, where

$$p = \mathbf{P}\{T_1^k \leqslant h(z,a),\ T_1^1,\ldots,T_1^{k-1},T_1^{k+1},\cdots,T_1^q \mid T_1 > h(y,a), A_1 = a\}$$

$$= e^{H(y,a)} \int_{h(y,a)}^{h(z,a)} \lambda_k(g(t,a))e^{-H(g(t,a),a)}dt$$

$$= \int_y^z h'(u,a)e^{-[H(u,a)-H(y,a)]}d\Lambda_k(u).$$

Thus,

$$\mathbf{E}[N_k(z) - N_k(y) \mid \mathcal{F}_y] = \mathbf{1}_{\{Z_1>y\}} \int_y^z h'(u,a)e^{-[H(u,a)-H(y,a)]}d\Lambda_k(u).$$

Analogously, for $u > y$,

$$\mathbf{E}[\mathbf{1}_{\{Z_1>u\}} \mid \mathcal{F}_y] = \mathbf{1}_{\{Z_1>y\}}e^{-[H(u,a)-H(y,a)]}.$$

Hence

$$\mathbf{E}[\int_y^z Y(u)d\Lambda_k(u) \mid \mathcal{F}_y] = \int_y^z h'(u, A_1)\mathbf{E}[\mathbf{1}_{\{Z_1>u\}} \mid \mathcal{F}_y]d\Lambda_k(u)$$

$$= \mathbf{1}_{\{Z_1>y\}} \int_y^z h'(u, A_1)e^{-[H(u,a)-H(y,a)]}d\Lambda_k(u).$$

$\square$

Proofs of other results are based on the following simple fact about weak convergence of distributions in metric spaces.

Consider two separable metric spaces Y and Z, a sequence of Y-valued random variables (η_n) and a Z-valued random variable ζ (all defined on the same probability space). Let $\mathcal{L}(\eta_n \mid z)$ denote the regular variant of the conditional distribution of η_n given $\zeta = z$. If, almost surely, $\mathcal{L}(\eta_n \mid \zeta)$ weakly converges to the distribution of some Y-valued random variable η, then we write $\eta_n \mid \zeta \xrightarrow{d} \eta$.

Lemma 7: *Suppose X is another separable metric space and (ξ_n) is a sequence of $\sigma(\zeta)$-measurable X-valued random variables. If ξ and η are independent random variables, $\xi_n \xrightarrow{d} \xi$ and $\eta_n \mid \zeta \xrightarrow{d} \eta$, then $(\xi_n,\eta_n) \xrightarrow{d} (\xi,\eta)$.*

Proof: We must prove that

$$\mathbf{E}[\varphi(\xi_n)\psi(\eta_n)] \to \mathbf{E}[\varphi(\xi)]\mathbf{E}[\psi(\eta)] \tag{6.1}$$

for all bounded continuous functions $\varphi : X \to \mathbb{R}$ and $\psi : Y \to \mathbb{R}$. By conditions of the lemma,

$$\mathbf{E}[\psi(\eta_n) \mid \zeta] \to \mathbf{E}[\psi(\eta)]$$

almost surely. Therefore, by Theorem 4.4 of Billingsley[3],

$$(\xi_n, \mathbf{E}[\psi(\eta_n) \mid \zeta]) \xrightarrow{d} (\xi, \mathbf{E}[\psi(\eta)])$$

and, by continuity of the function $(x, c) \mapsto \varphi(x)c$,

$$\varphi(\xi_n)\,\mathbf{E}[\psi(\eta_n) \mid \zeta] \xrightarrow{d} \varphi(\xi)\,\mathbf{E}[\psi(\eta)].$$

Since the random variable on the left-hand side is bounded, Theorem 5.4 of Billingsley[3] yields

$$\mathbf{E}\big[\varphi(\xi_n)\mathbf{E}[\psi(\eta_n) \mid \zeta]\big] \to \mathbf{E}[\varphi(\xi)]\,\mathbf{E}[\psi(\eta)].$$

This implies (6.1) because $\varphi(\xi_n)$ is $\sigma(\zeta)$-measurable. $\qquad\square$

Proof of Theorem 4. We can apply Lemma 7 with $\xi_n = \sqrt{n}\,(\hat{\pi} - \pi) \in D^r[0;\infty]$, $\eta_n = \sqrt{n}\,(\hat{\Lambda} - \Lambda) \in D^q[0;\infty]$ and $\zeta = \bar{A} = (A_i \mid i \geqslant 1) \in (\mathbb{R}_+^r)^\infty$, if we show that almost surely

$$\mathcal{L}(\sqrt{n}\,(\hat{\Lambda} - \Lambda) \mid \bar{A}) \to \mathcal{L}(W). \tag{6.2}$$

Fix an arbitrary $\bar{a} = (a_i \mid i \geqslant 1) \in (\mathbb{R}_+^r)^\infty$ and denote by $\mathbf{P}_{\bar{a}}$ the conditional probability $\mathbf{P}(\cdot \mid \bar{A} = \bar{a})$. Repeating the proof of Theorem 1 shows that

$$N_k(z) - \int_0^z \lambda_k(y)Y(y)dy$$

is a martingale with respect to the filtration $(\mathcal{F}_z \mid 0 \leqslant z \leqslant z_0)$ and the probability $\mathbf{P}_{\bar{a}}$. Moreover, by the weak law of large numbers,

$$\sup_{z \leqslant z_0} \Big| \frac{1}{n}Y(z) - \frac{1}{n}\sum_{i=1}^n h'(z, a_i)e^{-H(z,a_i)} \Big|$$

tends to 0 in probability as $n \to \infty$. Hence we can repeat the proof of Theorem IV.1.2 of [1] to get (6.2) for all $\bar{a}$ such that

$$\sup_{z \leqslant z_0} \Big| \frac{1}{n}\sum_{i=1}^n h'(z, a_i)e^{-H(z,a_i)} - b(z) \Big| \to 0. \tag{6.3}$$

314   V. Bagdonavičius, A. Bikelis, V. Kazakevičius and M. Nikulin

It remains to notice that (6.3) is verified for π^∞-almost all sequences $\bar{a}$: this can be proved analogously as the Glivenko-Cantelli theorem with the use of Assumption 1. $\square$

Let D_+ be the set of all non-decreasing functions from $D[0; z_0]$, equal to 0 at the point 0, and C be the set of continuous functions from $D[0; z_0]$, equal to 0 at the point 0. All unconditional reliability characteristics considered in Subsection 5.1 have the form

$$\theta = \int \varphi(\Lambda, a) d\pi(a),$$

where $\varphi : D_+^q \times \mathbb{R}_+^r \to \mathbb{R}$ is a non-negative Borel function. Asymptotic normality of the estimator $\hat{\theta}$ follows from the following lemma.

Lemma 8: *Suppose that for any sequence $(\Lambda^n) \subset D_+^q$ such that $\Delta\Lambda^n = \sqrt{n}\,(\Lambda^n - \Lambda) \to \Delta\Lambda \in C^q$ we have:*
    (i) $\int \varphi(\Lambda^n, a) d\pi(a) \to \int \varphi(\Lambda, a) d\pi(a) < \infty$;
    (ii) for all a, $\sqrt{n}\,(\varphi(\Lambda^n, a) - \varphi(\Lambda, a)) \to \varphi'(\Delta\Lambda, a)$,
    (iii) for all a, $\sqrt{n}\,|\varphi(\Lambda^n, a) - \varphi(\Lambda, a)| \leqslant \bar{\varphi}(a)$;
here $\bar{\varphi}$ is a continuous functions such that

$$\int \bar{\varphi}(a) d\pi(a) < \infty. \tag{6.4}$$

Then $\sqrt{n}\,(\hat{\theta} - \theta) \xrightarrow{d} \xi + \eta$, where

$$\eta = \int \varphi'(W, a) d\pi(a)$$

and ξ is a Gaussian zero mean random variable, independent of W, with the variance

$$\mathbf{Var}(\xi) = \int \varphi^2(\Lambda, a) d\pi(a) - \theta^2.$$

Proof: Notice that $\sqrt{n}\,(\hat{\theta} - \theta) = \xi_n + \eta_n$, where

$$\xi_n = \sqrt{n} \int \varphi(\hat{\Lambda}, a) d[\hat{\pi}(a) - \pi(a)] = \frac{1}{\sqrt{n}} \sum_{i=1}^{n} [\varphi(\hat{\Lambda}, A_i) - \int \varphi(\hat{\Lambda}, a) d\pi(a)]$$

and

$$\eta_n = \sqrt{n} \int [\varphi(\hat{\Lambda}, a) - \varphi(\Lambda, a)] d\pi(a).$$

By (i) and the central limit theorem, $\xi_n \xrightarrow{d} \xi$. Therefore the lemma follows from Lemma 7 if we prove that, for π^∞-almost all $\bar{a}$,

$$\mathcal{L}(\eta_n \mid \bar{a}) \to \mathcal{L}(\eta). \tag{6.5}$$

Fix an arbitrary $\bar{a}$ such that

$$\sup_a |\hat{\pi}(a) - \pi(a)| \to 0. \tag{6.6}$$

By the Glivenko-Cantelli theorem, (6.6) holds for π^∞-almost all $\bar{a}$. The proof of Theorem 1 shows that $\sqrt{n}\,(\hat{\Lambda} - \Lambda) \mid \bar{A} \xrightarrow{d} W$. Therefore (6.5) follows from Theorem 5.5 of Billingsley[3] if we prove that $\Lambda + \Delta\Lambda^n/\sqrt{n} \in D_+^q$ and $\Delta\Lambda^n \to \Delta\Lambda \in C^q$ imply

$$f_n(\Delta\Lambda^n) \to f(\Delta\Lambda); \tag{6.7}$$

here

$$f_n(\Delta\Lambda) = \sqrt{n} \int_0^\infty [\varphi(\Lambda + \Delta\Lambda/\sqrt{n}, a) - \varphi(\Lambda, a)]d\pi(a)$$

(for $\Lambda + \Delta\Lambda/\sqrt{n} \in D_+^q$) and

$$f(\Delta\Lambda) = \int \varphi'(\Delta\Lambda, a)d\pi(a)$$

(for $\Delta\Lambda \in C^q$). It remains to notice that (6.7) follows from (ii), (iii), (6.4) and the dominated convergence theorem. $\qquad\qquad\square$

Proof of Theorem 5. 1. Estimator $\hat{S}(t)$. In this case

$$\varphi(\Lambda, a) = e^{-H(g(t,a),a)} \mathbf{1}_{\{g(t,a) < z_0\}}(a).$$

By Assumption 1, the variation of the function $h'(\cdot, a)$ in the interval $[0; z_0]$ is finite (more exactly, it is less than $c_1(a)z_0$). Therefore

$$H(z, a) = h'(z, a)\Lambda.(z) - \int_0^z \Lambda.(y)dh'(y, a). \tag{6.8}$$

Suppose $\Lambda^n \in D_+^q$, $\Delta\Lambda^n = \sqrt{n}(\Lambda_n - \Lambda) \to \Delta\Lambda \in C^q$ and set

$$H^n(z, a) = \int_0^z h'(y, a)d\Lambda^n_.(y). \tag{6.9}$$

Then, by (6.8),

$$\sqrt{n}\,[H^n(z, a) - H(z, a)] \to h'(z, a)\Delta\Lambda.(z) - \int_0^z \Delta\Lambda.(y)dh'(y, a) \tag{6.10}$$

and

$$\sqrt{n}\,|H^n(z,a) - H(z,a)| \leqslant [c_0(a) + c_1(a)z_0]\sup_{y\leqslant z_0}|\Delta\Lambda^n_{\cdot}(y)|,$$

where $\Delta\Lambda_{\cdot} = \sum_{k=1}^s \Delta\Lambda_k$ and $\Delta\Lambda^n_{\cdot} = \sum_{k=1}^s \Delta\Lambda^n_k$.

We denote the right-hand side of (6.10) by $\int_0^z h'(y,a)d\Delta\Lambda_{\cdot}(y)$, although the function $\Delta\Lambda_{\cdot}$ is not of finite variation, in general. This is justified by the fact that replacing $\Delta\Lambda_{\cdot}$ by the random process $W_{\cdot} = \sum_{k=1}^q W_k$ gives the stochastic integral $\int_0^z h'(y,a)dW_{\cdot}(y)$.

Hence

$$\sqrt{n}\left(\varphi(\Lambda^n,a) - \varphi(\Lambda,a)\right) \to -\mathbf{1}_{\{g(t,a)<z_0\}}e^{-H(g(t,a),a)}\int_0^{g(t,a)} h'(z,a)d\Delta\Lambda_{\cdot}(z)$$

and, by the inequality $|e^{-u_1} - e^{-u_2}| \leqslant |u_1 - u_2|$,

$$\sqrt{n}\,|\varphi(\Lambda^n,a) - \varphi(\Lambda,a)| \leqslant [c_0(a) + c_1(a)z_0]\sup_{y\leqslant z_0}|\Delta\Lambda^n_{\cdot}(y)|.$$

Now Lemma 8 implies that $\sqrt{n}\left(\hat{S}(t) - S(t)\right) \xrightarrow{d} \xi + \eta$, where ξ and η are independent Gaussian zero mean random variables,

$$\eta = -\int_{g(t,a)<z_0} e^{-H(g(t,a),a)}d\pi(a)\int_0^{g(t,a)} h'(z,a)dW_{\cdot}(z)$$

$$= -\int_0^{z_0}\left\{\int_{z<g(t,a)<z_0)} h'(z,a)d\pi(a)\right\}dW_{\cdot}(z)$$

and

$$\mathbf{Var}(\xi) = \int_{g(t,a)<z_0} e^{-2H(g(t,a),a)}d\pi(a) - S^2(t).$$

It remains to notice that, by the stochastic integral properties,

$$\mathbf{Var}(\eta) = \int_0^{z_0}\left\{\int_{z<g(t,a)<z_0)} h'(z,a)d\pi(a)\right\}^2 d\sigma^2(z).$$

2. *Estimator $\hat{e}$.* In this case

$$\varphi(\Lambda,a) = h(z_0,a) - \int_0^{z_0} h'(z,a)[h(z_0,a) - h(z,a)]e^{-H(z,a)}d\Lambda_{\cdot}(z).$$

Suppose $\Lambda^n \in D^q_+$, $\Delta\Lambda^n = \sqrt{n}(\Lambda_n - \Lambda) \to \Delta\Lambda \in C^q$ and define $H^n(z,a)$ by (6.9). Then $\sqrt{n}\,[\varphi(\Lambda^n,a) - \varphi(\Lambda,a)]$ equals

$$-\sqrt{n}\int_0^{z_0} h'(z,a)[h(z_0,a) - h(z,a)][e^{-H^n(z,a)} - e^{-H(z,a)}]d\Lambda^n(z)$$

$$-\int_0^{z_0} h'(z,a)[h(z_0,a) - h(z,a)]e^{-H(z,a)}d\Delta\Lambda^n(z). \quad (6.11)$$

Similarly as above we get

$$\sqrt{n}\, h'(z,a)[h(z_0,a) - h(z,a)][e^{-H^n(z,a)} - e^{-H(z,a)}]$$
$$\leqslant c_0^2(a)z_0[c_0(a) + c_1(a)z_0]\, \sup_{y\leqslant z_0}|\Delta\Lambda^n(y)|.$$

Therefore the first term in (6.11) tends to

$$\int_0^{z_0}\Big\{h'(z,a)[h(z_0,a) - h(z,a)]e^{-H(z,a)}\int_0^z h'(y,a)d\Delta\Lambda.(y)\Big\}d\Lambda.(z)$$
$$= \int_0^{z_0}\Big\{h'(y,a)\int_y^{z_0} h'(z,a)[h(z_0,a) - h(z,a)]e^{-H(z,a)}d\Lambda.(z)\Big\}d\Delta\Lambda.(y)$$

and is dominated by

$$c_0^2(a)z_0[c_0(a) + c_1(a)z_0]\, \sup_{z\leqslant z_0}|\Delta\Lambda^n_.(z)|\Lambda^n_.(z_0).$$

The supremum and the variation of the function

$$h'(z,a)[h(z_0,a) - h(z,a)]e^{-H(z,a)}$$

in the interval $[0; z_0]$ are less than $c_0^2(a)z_0$ and $c_0(a)z_0[c_0(a) + c_1(a)z_0]$, respectively. Therefore the second term in (6.11) tends to

$$-\int_0^{z_0} h'(z,a)[h(z_0,a) - h(z,a)]e^{-H(z,a)}d\Delta\Lambda.(z)$$

and is dominated by

$$[2c_0^2(a)z_0 + c_0(a)c_1(a)z_0^2]\, \sup_{z\leqslant z_0}|\Delta\Lambda^n_.(z)|.$$

By Lemma 8, $\sqrt{n}\,(\hat{e} - e) \xrightarrow{d} \xi + \eta$ with

$$\mathbf{Var}(\eta) = \int_0^{z_0}\Big\{\Big[\int h'(y,a)\Big(\int_y^{z_0} h'(z,a)[h(z_0,a) - h(z,a)]e^{-H(z,a)}d\Lambda.(z)$$
$$- [h(z_0,a) - h(y,a)]e^{-H(y,a)}\Big)d\pi(a)\Big]^2 d\sigma^2(z)$$

and

$$\mathbf{Var}(\xi) = \int\Big\{h(z_0,a) - \int_0^{z_0} h'(z,a)[h(z_0,a) - h(z,a)]e^{-H(z,a)}d\Lambda.(z)\Big\}^2 d\pi(a)$$
$$- e^2.$$

3. *Estimator* $\hat{\Gamma}^{(k)}(t)$, $\hat{\Gamma}^{(k)}$, $\hat{\Gamma}^{(tr)}(t)$, $\hat{\Gamma}^{(tr)}$. Asymptotic normality of these estimators is proved analogously. □

Proof of Theorem 6. The theorem is proved by the ordinary functional delta-method. □

References

1. P. K. Andersen, Ø. Borgan, R. D. Gill and N. Keiding, *Statistical Models Based on Counting Processes* (Springer-Verlag, New York, 1993).
2. V. Bagdonavičius and M. Nikulin, *Accelerated Life Models* (Chapman&Hall/CRC, Boca Raton, 2002).
3. P. Billingsley, *Convergence of Probability Measures* (Wiley, New York, 1968).
4. P. Hougaard, *Analysis of Multivariate Survival Data* (Springer-Verlag, New York, 2000).
5. W. Q. Meeker and L. Escobar, *Statistical Methods for Reliability Data* (Wiley, New York, 1998).
6. K. Suzuki, K. Maki and S. Yokogawa, An analysis of degradation data of carbon film and properties of the estimators, in *Statistical Science and Data Analysis*, Eds. K. Matusita, M. Puri and T. Hayakawa (VSP, Utrecht, Netherlands, 1993).

21

ACCELERATED DESTRUCTIVE DEGRADATION TESTS: DATA, MODELS, AND ANALYSIS

Luis A. Escobar

Dept. of Experimental Statistics, Louisiana State University
Baton Rouge, LA 70803, USA
E-mail: luis@lsu.edu

William Q. Meeker

Dept. of Statistics, Iowa State University
Ames, IA 50011, USA
E-mail: wqmeeker@iastate.edu

Danny L. Kugler and Laura L. Kramer

Imaging & Printing Group, Hewlett-Packard
Corvallis, OR 97330, USA
E-mail: danny_ kugler@hp.com, laura_ kramer@hp.com

Degradation data analysis is a powerful tool for reliability assessment. Useful reliability information is available from degradation data when there are few or even no failures. For some applications the degradation measurement process destroys or changes the physical/mechanical characteristics of test units. In such applications, only one meaningful measurement can be can be taken on each test unit. This is known as "destructive degradation." Degradation tests are often accelerated by testing at higher than usual levels of accelerating variables like temperature.

This chapter describes an important class of models for accelerated destructive degradation data. We use likelihood-based methods for inference on both the degradation and the induced failure-time distributions. The methods are illustrated with the results of an accelerated destructive degradation test for an adhesive bond.

1. Introduction

1.1. *Motivation*

Today's manufacturers face strong pressure to develop newer, higher technology products in record time. In addition, there are competitive pressures to improve productivity, product field reliability, and overall quality. This implies the increased need for up-front testing of materials, components and systems. Traditional life tests (where time to failure is the response) may result in few or no failures, even when accelerated (e.g., by testing at higher-than-usual levels of temperature or voltage). Accelerated degradation tests can be useful for such up-front testing.

1.2. *Advantages and Difficulties of Degradation Data*

Degradation is the natural response for some tests. With degradation data, it is possible to make useful reliability inferences, even with no failures. In addition, modeling degradation data provides more justification and credibility for extrapolative acceleration models. This is because modeling is closer to the physics-of-failure mechanisms.

It may, however, be difficult, costly, or impossible to obtain degradation measures from some components or materials. Often taking degradation measures will require destructive measurements, which is the motivation for the current work. Also, the analysis of destructive degradation data generally requires the use of special software. For more information on degradation data (with special emphasis on repeated measures degradation), see Chapters 13 and 21 of Meeker and Escobar[3].

1.3. *Adhesive Bond B ADDT Data*

The objective of the experiment was to assess the strength of an adhesive bond over time. In particular, there is interest in estimating the proportion of devices with a strength below 40 Newtons after 5 years of operation (approximately 260 weeks) at room temperature of 25°C. The test needed to be completed in 16 weeks, and thus acceleration would be needed, as little or no degradation could be expected at 25°C during the length of the test. The test is destructive; strength can be measured only once on each unit.

The strength data are shown in Figure 1 (the data have been modified by a change in scale, in effect changing the units of strength, in order to protect proprietary information). There were six strength measurements at

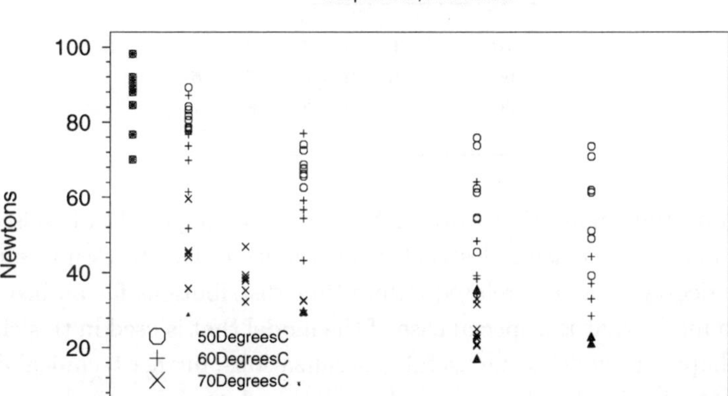

Fig. 1. Adhesive Bond B ADDT data scatter plot.

three different levels of temperature that had suspiciously low values. Fitting a model to accommodate these values gave estimates that the engineers involved in the problem knew to be inconsistent with actual failure probabilities. It was believed all of these lower readings could be attributed to the fabrication of the test units and that this problem could be avoided in actual production (a test would have to be conducted to verify that this was so). Thus, for purposes of modeling and analysis, it was decided to mark these observations as right-censored values, indicating that the actual level of strength is unknown, but certainly larger than the recorded value. In our plots, such observations are marked with a ▲.

The experiment included 8 units that were measured at the start of the experiment, with no aging. A total of 80 additional units were aged and measured according to the temperature and time schedule shown in Table 1.

1.4. *Related Literature*

Chapter 7 of Tobias and Trindade[12], Chapters 13 and 21 of Meeker and Escobar[3], and Meeker, Escobar, and Lu[5] present statistical methods for estimating a failure-time distribution from repeated measures degradation.

Table 1.   Adhesive Bond B test plan.

| Temp | Weeks Aged | | | | | |
°C	0	2	4	6	12	16
70		6	6	4	9	0
60		6	0	6	6	6
50		8	0	8	8	7
—	8					

The important pioneering work of Nelson[8] and Chapter 11 of Nelson[9] describe methods for using destructive degradation data to estimate performance degradation and related failure time distributions for an insulation, using a model that is a special case of the model that is used in this chapter. This chapter provides some useful generalizations, further technical details, a new application for destructive degradation data. In particular, we show how to deal with censored data, multiple accelerating variables, describe model identification and diagnostic tools, provide more details on the distribution of failure times, and provide discussion of acceleration factors.

1.5. *Overview*

This remainder of this chapter is organized as follows. Section 2 describes the degradation model that we use for destructive degradation. Section 3 outlines methods for ML estimation with right-censored data, both for individual test conditions and the full acceleration model. Section 4 gives formulas for the distribution of degradation. Section 5 gives formulas for the failure time distribution induced by the degradation model. Section 6 shows how to compute and interpret acceleration factors. Section 7 contains some concluding remarks and areas for future research.

2. Model

2.1. *Model for a Degradation Path*

The model for the actual degradation path of a unit at time t_i and a particular accelerating variable condition AccVar_j (e.g., temperature) is

$$\mathcal{D}_{ij} = \mathcal{D}(\tau_i, x_j, \beta)$$

where $\tau_i = h_t(t_i)$ and $x_j = h_a(\mathrm{AccVar}_j)$ are known monotone increasing transformations of t_i and AccVar_j, respectively. The form of the function $\mathcal{D}$ and the appropriate transformations may be suggested by physical-chemical theory, (see, for example Meeker and LuValle[6] and Meeker, Escobar, and Lu[5], past experience, or the data. When there is no possibility of confusion, τ_i and x_j are called the time and the AccVar condition, respectively. Rates in the model are with respect to transformed time $\tau = h_t(t)$. In our application, the path parameters β are fixed but unknown.

2.2. *Model for Degradation Sample Paths*

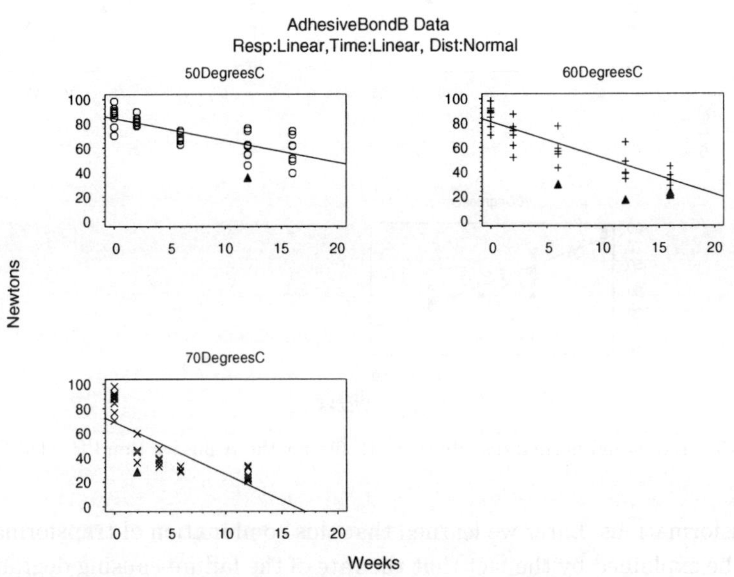

Fig. 2.   Adhesive Bond B ADDT data scatter plot at individual levels of temperature.

For unit k at time τ_i and accelerating variable condition x_j the sample path model is

$$y_{ijk} = h_d(\mathcal{D}_{ij}) + \epsilon_{ijk} = \mu_{ij} + \epsilon_{ijk}$$

where $\mu_{ij} = h_d(\mathcal{D}_{ij})$ and y_{ijk} are, respectively, monotone increasing transformations of $\mathcal{D}_{ij}$ and the observed degradation. ϵ_{ijk} is a residual deviation which describes unit-to-unit variability with $(\epsilon_{ijk}/\sigma) \sim \Phi(z)$, where $\Phi(z)$ is a completely specified distribution, for example $\Phi(z) = \Phi_{\mathrm{nor}}(z)$ provides a normal model and $\Phi(z) = \Phi_{\mathrm{sev}}(z)$ provides a smallest extreme value model.

Figure 2 suggests that some kind of transformation should be used to linearize the degradation paths. Several transformations were investigated on the Adhesive Bond B data and other similar data sets. In all cases the combination of a log transformation on degradation response and square root on time provided the best fit. Figure 3 shows the effect of using these

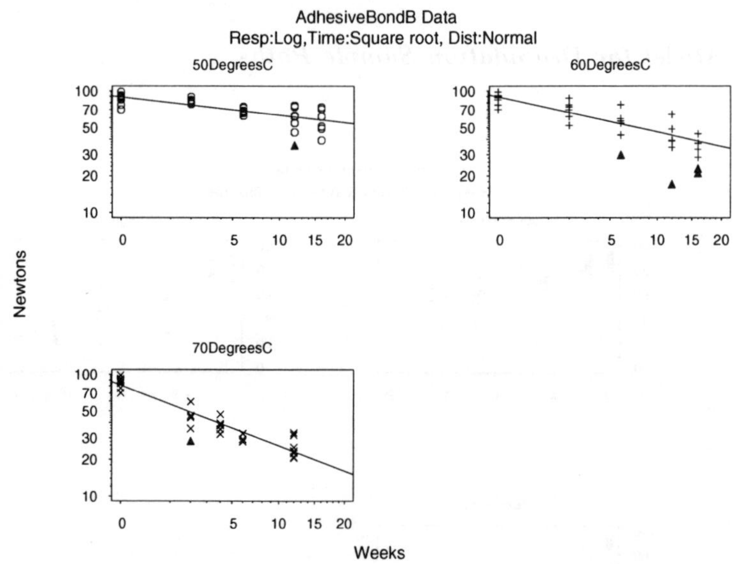

Fig. 3.   Individual normal distribution ML fits for the Adhesive Bond B ADDT data.

transformations. Later we learned that this combination of transformations can be explained by the fact that the rate of the failure-causing degradation mechanism is controlled by a process that can be described by the 2nd Law of Diffusion (otherwise known as Fick's Law).

2.3. *Model for Acceleration*

For the Adhesive Bond B application, the degradation rate (on the transformed time scale) is assumed to be described by the Arrhenius relationship. That is,

$$\mathcal{D}(\tau, x, \boldsymbol{\beta}) = \exp[\beta_0 + \beta_1 \exp(\beta_2 x)\tau]$$

where $\tau = \sqrt{\text{Weeks}}$ and $x = -11605/(°C_j + 273.15)$ is Arrhenius-transformed temperature. This relationship between degradation and temperature is depicted in Figure 4. On the log-degradation scale, the individual

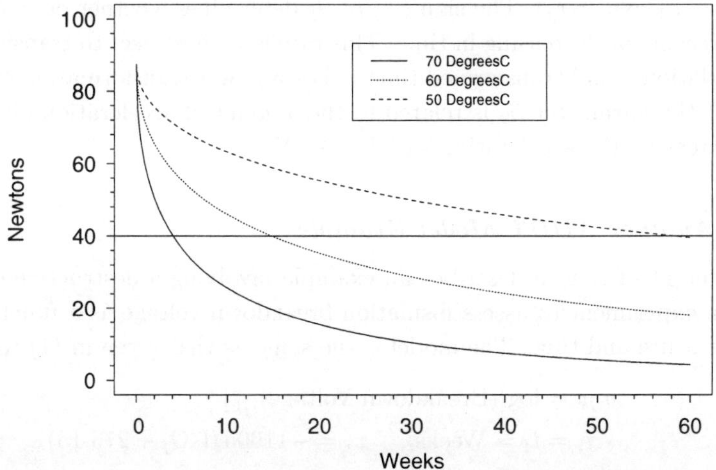

Fig. 4.   Degradation path model for three different temperatures.

regressions are linear in square root time τ. In particular, $\log[\mathcal{D}(\tau, x, \beta)] = \beta_0 + \beta_1 \exp(\beta_2 x)\tau$. These degradation sample-path models are, in general, of the form

$$y_{ijk} = \mu_{ij} + \epsilon_{ijk}$$
$$= \beta_0 + \beta_1 \exp(\beta_2 x_j)\tau_i + \epsilon_{ijk} \tag{1}$$

where y_{ijk}, τ_i, and x_j may be monotone transformations of the measured degradation, t_i, and the AccVar variable, respectively. This degradation model is linear in the sense that for a specified AccVar condition x_j, the degradation is linear in τ_i. For multiple AccVar situations (e.g., temperature and humidity), the term $\beta_2 x_j$ can be replaced by a linear combination of accelerating variables $\beta_2' x_j$. In either case, however, the regression model with the acceleration terms is nonlinear in the unknown parameter(s) β_2. Thus, even if there is no censoring, ordinary least squares cannot be used to estimate, simultaneously, all of the parameters of the model.

2.4. *Interpretation of the Parameters*

The interpretation of the degradation model parameters can be described as follows. For the linear degradation model in (1), β_0 is the degradation level when $\tau_i = 0$. For example, if $\tau_i = \sqrt{t_i}$ (or some other power transformation of time), then β_0 is degradation at time $t = 0$. If $\tau_i = \log(t_i)$ then β_0 is degradation at time $t = 1$. The degradation rate at AccVar level x_j is $v(x_j) = \beta_1 \exp(\beta_2 x_j)$. The sign ($\pm$) of β_1 determines whether degradation is increasing or decreasing in time. This rate is with respect to transformed degradation y and transformed time τ. For a power transformation of time $\tau = t^\kappa$ the parameter β_2 is related to the amount of acceleration obtained by increasing the accelerating variable AccVar.

2.5. *Another ADDT Model Example*

Chapter 11 of Nelson describes an example involving a destructive degradation experiment to assess insulation breakdown voltage as a function of temperature and time. The model is the same as that given in (1) with

$$y_{ijk} = \log[(\text{Breakdown Voltage})_{ijk}]$$
$$\tau_i = t_i = \text{Weeks}_i, \quad x_j = -11605/(^\circ C_j + 273.15)$$
$$(\epsilon_{ijk}/\sigma) \sim \Phi_{\text{nor}}(z)$$

and $\Phi_{\text{nor}}(z)$ is a standardized normal cdf.

3. Maximum Likelihood Estimation

3.1. *Estimation at Individual Conditions*

For the data at a fixed condition x_j of the AccVar with exact failure times and right-censored observations, the likelihood is

$$L_j(\boldsymbol{\theta}|\text{DATA}) = \prod_i \prod_{k=1}^{n_{ij}} \left[\frac{1}{\sigma}\phi\left(\frac{y_{ijk} - \mu_{ij}}{\sigma}\right)\right]^{\delta_{ijk}} \times \left[1 - \Phi\left(\frac{y_{ijk} - \mu_{ij}}{\sigma}\right)\right]^{1-\delta_{ijk}}$$

$$(2)$$

where $\mu_{ij} = \mu(\tau_i, x_j, \boldsymbol{\beta}) = \beta_0 + \beta_1 \exp(\beta_2 x_j)\tau_i$, δ_{ijk} indicates whether observation y_{ijk} is a failure ($\delta_{ijk} = 1$) or a right censored observation ($\delta_{ijk} = 0$), $\boldsymbol{\theta} = (\beta_0, \beta_1, \beta_2, \sigma)$ is the vector of unknown parameters, and n_{ij} is the number of observations at (τ_i, x_j). The logarithm of (2) can be maximized by using standard numerical function maximization methods. For fixed x_j, the identifiable parameters are the standard deviation of the error term, σ, the intercept β_0, and the slope of the line $v^{[j]} = \beta_1 \exp(\beta_2 x_j)$. Then for each specified condition of the AccVar x_j, three individual ML estimates are obtained, say $\widehat{\beta}_0^{[j]}$, $\widehat{v}^{[j]}$, and $\widehat{\sigma}^{[j]}$. The parameter $v^{[j]}$ can be interpreted as the degradation rate of μ_{ij} with respect to transformed time τ_i.

Figure 3 shows the data and the individual ML regression line estimates for the three different temperature levels in the Adhesive Bond B example using model (1) with a normally distributed residual component, i.e., $\Phi(z) = \Phi_{\mathrm{nor}}(z)$, y_{ijk} in the log-degradation scale, and the time τ_i in the square root scale. Figure 5 shows the same results, all on one plot. The

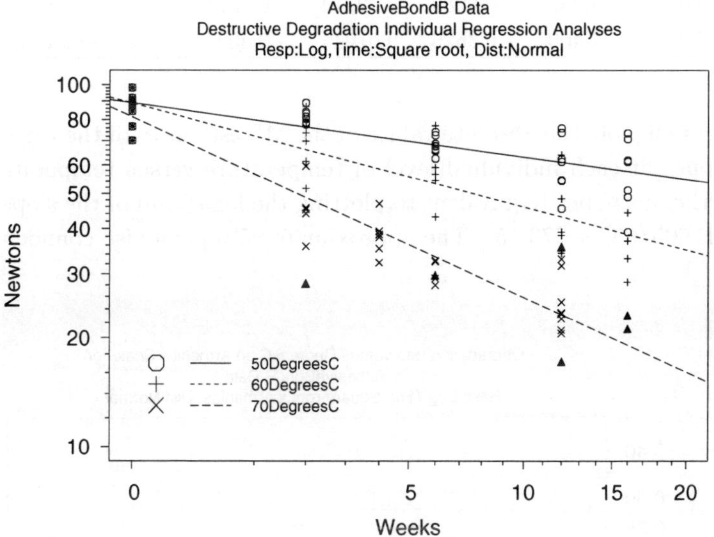

Fig. 5. Overlay of individual normal distribution fits for the Adhesive Bond B ADDT data.

parameter estimates for the simple regression model fit to the censored Adhesive Bond B data are given in Table 2. The standard errors were obtained by using local information (see, for example, Appendix Section B.6.4 in Meeker and Escobar[3]).

3.2. *Arrhenius Plot of Degradation Rates*

The ML estimates $\widehat{v}^{[j]}$ (slopes of the individual lines) can be used to identify the relationship between degradation rate and the AccVar. When the degradation is decreasing, use absolute values of the degradation rate. Because $\log(|\,v^{[j]}\,|) = \log(|\,\beta_1\,|) + \beta_2 x_j$ a plot of $\log(|\,\widehat{v}^{[j]}\,|)$ versus x_j should be approximately linear if the model relating the degradation rate and the AccVar level x_j is adequate.

Table 2. Normal distribution individual parameter ML estimates and 95% confidence intervals for the slope at each level of temperature.

AccVar$_j$	ML Estimates			95% Approximate Confidence Interval for $v^{[j]}$	
	$\widehat{\beta}_0^{[j]}$	$\widehat{v}^{[j]}$	$\widehat{\text{se}}_{\widehat{v}^{[j]}}$	Lower	Upper
50°C	4.490	−0.1088	0.01494	−0.1424	−0.08309
60°C	4.489	−0.2089	0.02214	−0.2571	−0.16969
70°C	4.400	−0.3626	0.01944	−0.4028	−0.32643

Figure 6 plots the absolute values of the ML estimates of the regression-line slopes at each individual level of temperature versus temperature on log-Arrhenius paper (equivalent to plotting the logarithm of the slopes versus $-11605/(°C + 273.15)$. The approximate 95% pointwise confidence in-

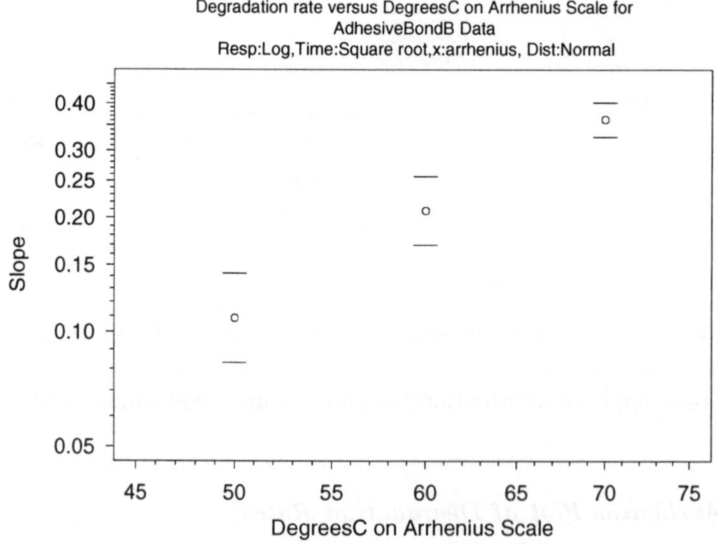

Fig. 6. Arrhenius plot of individual degradation rates normal distribution ML estimates.

tervals for the slopes aid in the interpretation of the plot, relative to the statistical importance of deviations from linearity. The nearly linear relationship in Figure 6 suggests good agreement with the Arrhenius model for temperature acceleration, at least within the range of the data.

3.3. *Likelihood for the Acceleration Model Using All Data*

For a sample of n units consisting of exact failure times and right-censored observations, the likelihood can be expressed as

$$L(\boldsymbol{\theta}|\text{DATA}) = \prod_j L_j(\boldsymbol{\theta}|\text{DATA}) \tag{3}$$

$$= \prod_{ijk} \left[\frac{1}{\sigma} \phi\left(\frac{y_{ijk} - \mu_{ij}}{\sigma} \right) \right]^{\delta_{ijk}} \times \left[1 - \Phi\left(\frac{y_{ijk} - \mu_{ij}}{\sigma} \right) \right]^{1-\delta_{ijk}}$$

where $\boldsymbol{\theta} = (\beta_0, \beta_1, \beta_2, \sigma)$, $\mu_{ij} = \beta_0 + \beta_1 \exp(\beta_2 x_j)\tau_i$, $x_j = -11605/(°C_j + 273.15)$, and δ_{ijk} indicates whether observation ijk is a failure ($\delta_{ijk} = 1$) or a right censored observation ($\delta_{ijk} = 0$).

For the Adhesive Bond B data, Figure 7 shows the ML estimates from the combined data for each of the three levels of temperature used in the experiment plus the use condition of 25°C. Note that all of the lines cross at the common intercept at time 0. The parameter estimates for the accel-

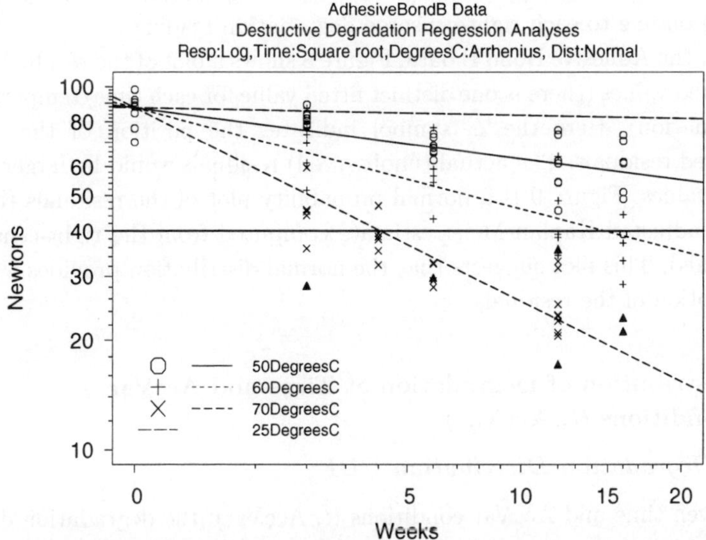

Fig. 7. Normal distribution Arrhenius model fit to the Adhesive Bond B ADDT data

eration model fit to the Adhesive Bond B data are given in Table 3. Again, standard errors are based on local information.

Table 3. ML estimates for the acceleration model fit to the Adhesive Bond B data.

Parameter	ML Estimate	Standard Error	95% Approximate Confidence Interval	
			Lower	Upper
β_0	4.471	0.03864	4.396	4.547
β_1	-8.641×10^8	1.595×10^9	-3.989×10^9	2.261×10^9
β_2	0.6364	0.05488	0.5375	0.7536
σ	0.1580	0.01233	0.1356	0.1841

3.4. *Residual Analysis*

Analysis of residuals to detect model departures is just as important for the destructive degradation models as it is for other regression models. The censored observations can make interpretation of such plots more complicated. We follow the approach in Nelson[7], yielding a right-censored residual corresponding to each right-censored degradation reading.

For the Adhesive Bond B data, Figure 8 shows a plot of the residuals versus fitted values (there is one distinct fitted value for each time/temperature combination). Here the $\triangle$ symbol indicates the position of the right-censored residuals. The actual (unobserved) residuals would be larger than these values. Figure 9 is a normal probability plot of the residuals (based on an adjusted Kaplan-Meier estimate computed from the right-censored residuals). This plot suggests that the normal distribution provides a good description of the residuals.

4. Distribution of Degradation at Time and AccVar Conditions (t, AccVar)

4.1. *Degradation Distribution CDF*

For given time and AccVar conditions (t, AccVar), the degradation distribution is

$$F_Y(y; \tau, x) = P(Y \leq y; \tau, x) = \Phi\left[\frac{y - \mu(\tau, x, \beta)}{\sigma}\right]$$

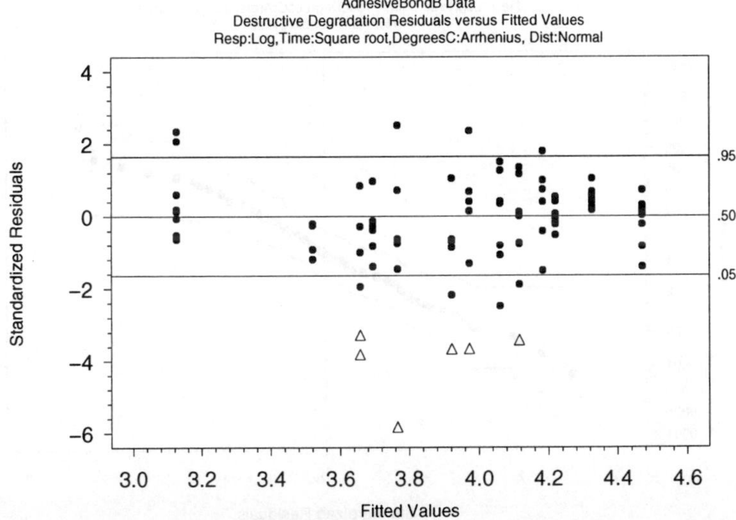

Fig. 8.   Adhesive Bond B ADDT data residuals versus fitted values

where $y = h_d(\text{degradation})$, $\mu(\tau, x, \beta) = \beta_0 + \beta_1 \exp(\beta_2 x)\tau$. The ML estimate of the degradation distribution for given (t, AccVar) is

$$\widehat{F}_Y(y; \tau, x) = \Phi\left(\frac{y - \widehat{\mu}}{\widehat{\sigma}}\right)$$

where $\widehat{\mu} = \widehat{\beta}_0 + \widehat{\beta}_1 \exp(\widehat{\beta}_2 x)\tau$, $\tau = h_t(t)$, $x = h_a(\text{AccVar})$, and $\widehat{\beta}$'s are ML estimates.

For the Adhesive Bond B data, the normal distribution ML estimate of $F_Y(y; \tau, x)$ at time and temperature (Weeks, $^\circ$C) is

$$\widehat{F}_Y(y; \tau, x) = \Phi_{\text{nor}}\left(\frac{y - \widehat{\mu}}{\widehat{\sigma}}\right)$$

where $\widehat{\mu} = \widehat{\beta}_0 + \widehat{\beta}_1 \exp(\widehat{\beta}_2 x)\tau$, $\tau = \sqrt{\text{Weeks}}$, $x = -11605/(^\circ\text{C} + 273.15)$. The ML estimates $\widehat{\beta}_0$, $\widehat{\beta}_1$, $\widehat{\beta}_2$, and $\widehat{\sigma}$ are given in Table 3.

4.2. *Degradation Distribution Quantiles*

The p quantile of the degradation distribution is $y_p = \mu(t, x, \beta) + \sigma\Phi^{-1}(p)$. The ML estimate of the p quantile on the transformed scale (log Newtons

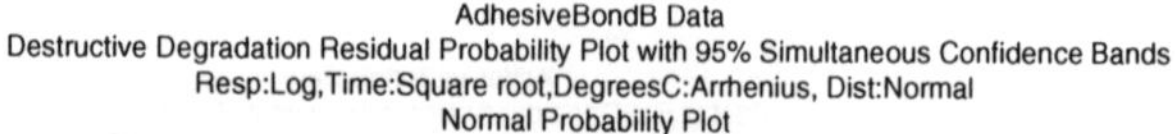
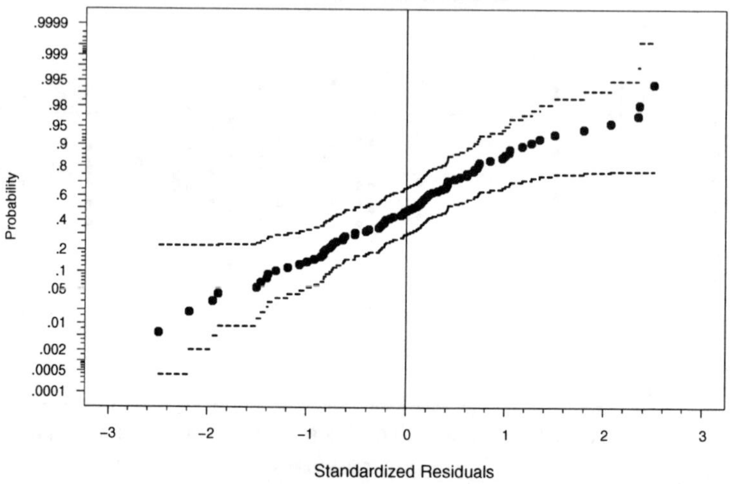

Fig. 9.   Adhesive Bond B ADDT data normal distribution residual probability plot

for the Adhesive Bond B example) is

$$\widehat{y}_p = \widehat{\mu} + \widehat{\sigma}\Phi^{-1}_{\text{nor}}(p).$$

5. Induced Failure Time Distribution at Fixed Values of $(\text{AccVar}, \mathcal{D}_f)$ for Decreasing Linear Degradation

5.1. *Failure Time CDF*

Observe that $T \leq t$ [i.e., $h_t(T) \leq \tau$] is equivalent to observed degradation being less than $\mathcal{D}_f$ (i.e., $Y \leq \mu_f$,) where $\mu_f = h_d(\mathcal{D}_f)$. Then

$$\begin{aligned}
F_T(t; x, \boldsymbol{\beta}) &= \Pr(T \leq t) \\
&= F_Y(\mu_f; x, \boldsymbol{\beta}) = \Phi\left[\frac{\mu_f - \mu(\tau, x, \boldsymbol{\beta})}{\sigma}\right] \\
&= \Phi\left(\frac{\tau - \nu}{\varsigma}\right), \text{ for } t \geq 0
\end{aligned} \qquad (4)$$

where $\tau = h_t(t)$,

$$\nu = \frac{(\beta_0 - \mu_f)\exp(-\beta_2 x)}{|\beta_1|} \quad \text{and} \quad \varsigma = \frac{\sigma \exp(-\beta_2 x)}{|\beta_1|}.$$

The failure time distribution in (4) is a mixed distribution with a *spike* of probability, $\Pr(T = 0) = \Phi\left[(\beta_0 - \mu_{\mathrm{f}})/\sigma\right]$ at $t = 0$. For $t > 0$ the cdf is continuous and it agrees with the cdf of a log-location-scale variable with standardized cdf $\Phi(z)$, with location parameter ν and scale parameter ς.

For the Adhesive Bond B application, $\mathcal{D}_{\mathrm{f}} = 40$. Figure 10 provides a visualization of the failure-time distribution induced by the degradation model at 25°C, based on the ML estimates of the Adhesive Bond B example. The figure shows clearly the reason for the spike of probability at time zero.

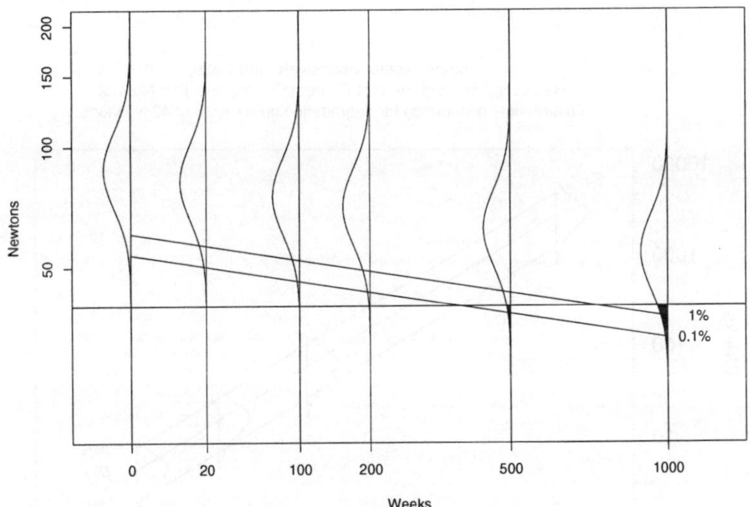

Fig. 10.   ML estimate showing proportion failing as a function of time at condition 25°C

5.2. *Failure Time Distribution Quantiles*

The p quantile of the failure time distribution can be expressed as follows. Let $p \geq \Phi\left[(\beta_0 - \mu_{\mathrm{f}})/\sigma\right]$ and

$$h_t(t_p) = \tau_p = \nu + \varsigma\Phi^{-1}(p) \tag{5}$$

where

$$\nu = \frac{(\beta_0 - \mu_{\mathrm{f}})\exp(-\beta_2 x)}{|\beta_1|} \quad \text{and} \quad \varsigma = \frac{\sigma\exp(-\beta_2 x)}{|\beta_1|}.$$

Then the p quantile of the failure time distribution is $t_p = h_t^{-1}\left[\nu + \varsigma\Phi^{-1}(p)\right]$. Substituting the expressions for ν and ς into (5), taking the logarithm, and simplifying gives

$$\log[h_t(t_p)] = \log(\tau_p) = -\beta_2 x + \log\left[\frac{(\beta_0 - \mu_f) + \sigma\Phi^{-1}(p)}{\mid \beta_1 \mid}\right].$$

This shows that the log of the transformed failure-time distribution quantiles are linear in the transformed AccVar condition x. If $p < \Phi\left[(\beta_0 - \mu_f)/\sigma\right]$, the p quantile of the failure time distribution is 0.

For the Adhesive Bond B example, Figure 11 is a model plot showing ML estimates of the failure-time distribution quantiles as a function of temperature on log-Arrhenius scales.

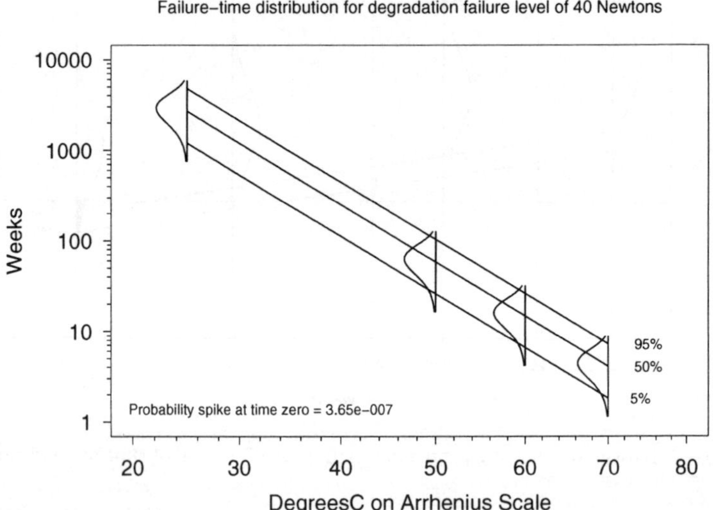

Fig. 11. Adhesive Bond B data: model plot ML estimates of failure time distribution as function of temperature

6. Acceleration Factors

Those who conduct accelerated tests often need to quote an "acceleration factor" to indicate the amount of time being saved by acceleration. Here we consider acceleration factors for time power transformations (i.e.,

$\tau = h_t(t) = t^\kappa$, where $\kappa > 0$). To obtain the effect of acceleration due to using higher than usual values of the AccVar x, let $\tau(x)$ and $\tau(x_U)$ be the (transformed) times to reach the critical degradation $\mathcal{D}_f$ when the (transformed) accelerating variable take values x and x_U, respectively. Solving for $\tau(x)$ and τ_{x_U} the equation

$$\mathcal{D}_f = \mathcal{D}[\tau(x), x, \boldsymbol{\beta}] = \mathcal{D}[\tau(x_U), x_U, \boldsymbol{\beta}]$$

gives

$$\frac{\tau(x_U)}{\tau(x)} = \frac{h_t[t(x_U)]}{h_t[t(x)]} = \exp[\beta_2(x - x_U)].$$

Using $\tau(x) = h_t[t(x)] = [t(x)]^\kappa$ and solving for $t(x_U)/t(x)$ yields

$$\mathcal{AF}(x) = \frac{t(x_U)}{t(x)} = \exp\left[\frac{\beta_2}{\kappa}(x - x_U)\right].$$

7. Concluding Remarks and Extensions

Destructive degradation tests provide methods for assessing reliability even with few or no failures. The methodology described here can be extended in a number of different directions to handle various related problems that arise in practice. In particular,

- Our example had only one accelerating variable, which is the most common type of accelerated test. As explained in Section 2.3, however, the statistical extension to multiple accelerating variables is straightforward (and indeed has been implemented in the SPLIDA software, as described in Meeker and Escobar[4]. The most recent version of SPLIDA is always available at www.public.iastate.edu/~splida). Possible interactions between accelerating variables can, however, complicate modeling and extrapolation to use conditions.

- Our modeling assumed that there is no measurement error. If there is a substantial amount of measurement error, then the estimate of σ will be biased high, providing similarly biased estimates of the failure-time distribution. If the magnitude of the measurement error standard deviation is known, the methods could be generalized to deal with this issue.

- In an extreme case, the degradation response may be given in terms of ordered categories (e.g., no degradation, light, medium, heavy, failed). Methods for handling such data are described in Agresti[1] and Johnson and Albert[2].

- In many applications, knowledge of the failure mechanism will provide useful prior information that can be used to improve, substantially, the precision of estimates of life at use conditions. Bayesian methods can be used to handle such situations.
- The model used in this paper assumes that the relationship between transformed degradation and transformed time is linear at each level of temperature. It is easy to find examples in which this assumption will not hold (e.g., the models used in Meeker and LuValle[6]. The extension of the methods presented here to such nonlinear models is in principle, straightforward. There can, however, be identifiability and convergence problems in the ML methods.
- The predictions of reliability developed in this paper assume that the use environment is constant. Nelson[9,11] describes statistical methods, based on a cumulative damage model, for estimating the life distribution under variable life conditions.
- Nelson[10] describes statistical methods for handling random initiation times that underlie some degradation processes.
- There are a number of open issues concerning the development of statistically efficient test planning that also meet practical constraints and provide useful amounts of power to detect departures fro the assumed model.

References

1. A. Agresti, *Analysis of Ordinal Categorical Data* (John Wiley & Sons, New York, 1984).
2. V. E. Johnson and J. H. Albert, *Ordinal Data Modeling* (Springer, New York, 1999).
3. W. Q. Meeker and L. A. Escobar, *Statistical Methods for Reliability Data* (John Wiley & Sons, New York, 1998).
4. W. Q. Meeker and L. A. Escobar, SPLIDA User's Manual (Available at www.public.iastate.edu/~splida/, 2002).
5. W. Q. Meeker, L. A. Escobar, and C. J. Lu, Accelerated degradation tests: modeling and analysis, *Technometrics*, **40**, 89–99 (1998).
6. W. Q. Meeker and M. J. LuValle, An accelerated life test model based on reliability kinetics, *Technometrics*, **37**, 133–146 (1995).
7. W. Nelson, Analysis of residuals from censored data, *Technometrics*, **15**, 697–715 (1973).
8. W. Nelson, Analysis of performance degradation data from accelerated tests, *IEEE Transactions on Reliability*, **R-30**, 3, 149–155 (1981).
9. W. Nelson, *Accelerated Testing: Statistical Models, Test Plans, and Data Analyses* (John Wiley & Sons, New York, 1990).

10. W. Nelson, Defect initiation and growth—a general statistical model & data analysis, paper presented at the 2nd annual Spring Research Conference, sponsored by the Institute of Mathematical Statistics and the Physical and Engineering Section of the American Statistical Association, Waterloo, Ontario, Canada, June 1995.

11. W. Nelson, Prediction of Field Reliability of Units, Each under Differing Dynamic Stresses, from Accelerated Test Data, in *Advances in Reliability*, Eds. C. R. Rao and N. Balakrishnan (Elsevier Science, Amsterdam, 2001).

12. P. A. Tobias and D. C. Trindade, *Applied Reliability,* Second Edition (Van Nostrand Reinhold, New York, 1995).

Part VII

STATISTICAL METHODS FOR MAINTAINED SYSTEMS

22

MODELLING HETEROGENEITY IN NUCLEAR POWER PLANT VALVE FAILURE DATA

Madhuchhanda Bhattacharjee and Elja Arjas

Rolf Nevanlinna Institute
University of Helsinki
Finland
E-mail: mab@rni.helsinki.fi, ela@rni.helsinki.fi

Urho Pulkkinen

VTT Industrial Systems
VTT, Finland
E-mail: urho.pulkkinen@vtt.fi

We develop an inferential procedure based on hierarchical Bayesian models which can take into account different kinds of heterogeneity in reliability/survival data. This is illustrated by an analysis of failure data from several motor operated closing valves at two nuclear power plants.

1. Introduction

Heterogeneity in data has been frequently encountered and incorporated in modelling and inference problems in areas like design of experiments, survival analysis, duration analysis, and many more. As far as reliability data are concerned this issue has been addressed only to a very limited extent from both these aspects. For example, burn-in problems have been considered for engineering systems, but possibly do not explain heterogeneous behaviour when longitudinal data are collected from such systems. Some relevant works in this context are by Pörn[8], Bedford et al.[2], Block and Savits [3], Lin and Singpurwalla[7], Hofer and Peshchke[5], Hougaard[6], etc.

This chapter develops inference procedures which can take into account different kinds of heterogeneity in reliability/survival data and enable inferences to be drawn not only on population parameters relating to the survival process, but also on underlying processes that lead to such observed heterogeneity between individuals.

We illustrate the proposed model by analysing a real life data set, wherein, although the observations consist of repeated failure times from similar systems, several factors can cause these systems to behave differently. In the current set-up, a large number of such variables can be suspected to be causing heterogeneity, but how exactly these affect the failure rate may be very difficult to establish.

2. Data

Our data pertain to several motor operated closing valves in different safety systems at two boiling water reactor plants in Finland, viz. TVO-I and TVO-II nuclear power units. Previous studies carried out on plants and their safety systems have shown significant variations in the number of valve failures and also indicated deviations in the reliability of some of the valves and their operating conditions (Pulkkinen and Simola[9]). The valves being parts of safety systems, such findings cast doubt about their performance when actually on demand in a real situation.

The data are based on 9' years operating experiences since the beginning of 1981, of 104 such motor operated valves, with 52 valves in each of the plants. These are parts of four different safety systems. Distribution of the valves according to Safety system and Plant is given in Table 1.

A motor operated valve, including the actuator, is an electromechanical system and consists of several parts, and hence can be quite complicated in structure. The valves have different design capacities, vary in diameter and actuator size and are manufactured by different manufacturers. The valves can experience different types of malfunctioning either when in operation or when pre-scheduled tests are carried out on them as a part of preventive maintenance (Simola and Laakso[10]).

Here we consider the failures of the type "External Leakage". An external leakage is recorded from a valve usually when there is a leakage from one of its subcomponents, such as a "Bonnet" or "Packing". Valves are continuously monitored for such failures and are rectified/repaired without delay. Here we assume that repair times are negligible.

Table 1. Distribution of valves according to safety system, plant and failure.

Safety System ($\equiv$ Valve Type)	Plant I		Plant II		All	
	Total	Failed	Total	Failed	Total	Failed
1. Shut-down cooling system	4	3	4	2	8	5
2. Containment vessel spray system	24	6	24	4	48	10
3. Core spray system	16	0	16	0	32	0
4. Auxiliary feed water system	8	1	8	0	16	1
All	52	10	52	6	104	16

As is expected, a large majority of systems did not fail during the 9 year follow up, as these are parts of safety systems and are built to perform successfully for long time periods. From the 104 valves 37 such external leakages were observed with the failures pertaining to 16 valves only.

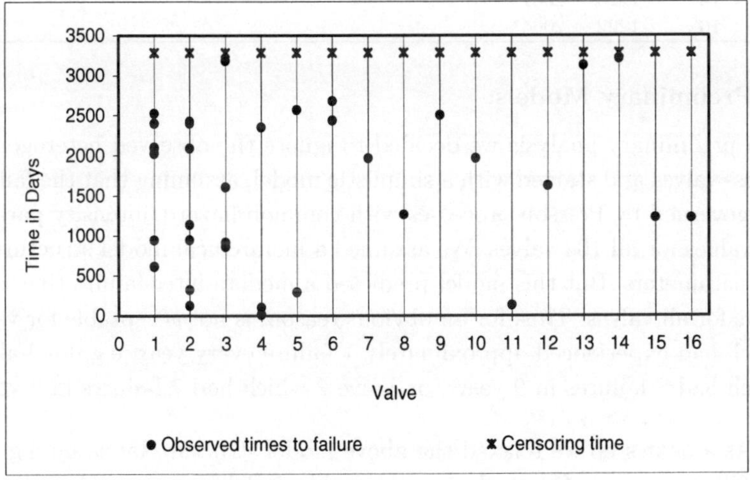

Fig. 1. Plots of cumulative times to failure of 20 valves.

The cumulative times to failures and the inter-failure times from these 16 valves are presented in Figure 1 and Table 2, respectively, with the valves being arranged according to the number of failures observed from them. The remaining 88 valves reporting no failures have not been included in this table, although each of them produces one censored time to first failure, censored at 3286-th day.

Table 2.   Inter-event times of external leakages from the valves.

Valve	Inter-failure times in days								
Number	1-st	2-nd	3-rd	4-th	5-th	6-th	7-th	8-th	9-th
1	610	4	329	1081	63	17	295	126	761*
2	126	197	620	189	955	312	27	860*	
3	860	55	691	1575	105*				
4	10	9	85	2248	934*				
5	293	2274	719*						
6	2434	242	610*						
7	1963	1323*							
8	1262	2024*							
9	2501	785*							
10	1963	1323*							
11	132	3154*							
12	1623	1663*							
13	3127	159*							
14	3211	75*							
15	1225	2061*							
16	1222	2064*							

3. Preliminary Models:

In a preliminary analysis we decided to ignore the observed heterogeneity across valves and started with a simplistic model, assuming that the failures are governed by Poisson processes with common hazard/intensity parameter values for all 104 valves. We assumed a hierarchical model structure for the parameters. But this model predicted a median inter-failure time of 17 years for all valves. This, for an obvious reason, is not acceptable for valves which had experienced approximately 1 failure every year, e.g. for Valve 1 which had 8 failures in 9 years, or Valve 2 which had 7 failures in 9 years.

As a next step we relaxed the above model structure by assuming that the failure process of any valve is governed by a Poisson process but the processes are not necessarily identical in distribution for all valves. We assumed an appropriate hierarchical structure for these processes. This simple extension immediately captured the heterogeneous behavior of the valves and valve specific predicted inter-failure times now had median values ranging approximately from 2 years to 26 years. This is consistent with the observed behavior of the respective valves, as some of them had failed often whereas most others did not fail during the observed period.

We further tried an extension of this model by dropping the Poisson assumption. Instead, we considered a point process model where failure intensity for each valve is piecewise constant and changes randomly at its own failure epochs, the intensities being drawn independently from a Gamma distribution. This was done observing that also the inter-event times from any specific valve varied widely, and under the homogeneous Poisson process assumption they are assumed to be coming from the same exponential distribution. But even though this was a simple and intuitive extension it lost the valve specific prediction ability as due to the implicit identical assumption the model did not contain any valve specific features. The large number of long (censored or uncensored) inter-failure times dominated the prediction and the predicted median inter-arrival time was estimated to be approximately 14 years for all valves. Note that, by integrating out the piecewise constant parameters, this model reduces to modelling the valves by identical renewal processes with Pareto distributed inter-failure times.

These three preliminary models (M1, M2 and M3, say) can be summarized as follows:

A. Inter-failure times: For $i = 1, \ldots, 104$ and $j \geq 1$,

M1. i.i.d. Homogeneous Poisson Process: $\qquad X_{i,j} \mid \theta \sim Exp(\theta)$

M2. Non-i.i.d. homogeneous Poisson Process: $\quad X_{i,j} \mid \theta_i \sim Exp(\theta_i)$

M3. Point process with piecewise constant

$$\text{intensity:} \qquad X_{i,j} \mid \theta_{ij} \sim Exp(\theta_{ij})$$

B. Hazard rate(s): For $i = 1, \ldots, 104$ and $j \geq 1$

M1. i.i.d. Homogeneous Poisson Process: $\qquad \theta \mid \beta \sim Gamma(1, \beta),$

M2. Non-i.i.d. homogeneous Poisson Process: $\quad \theta_i \mid \beta \sim Gamma(1, \beta),$

M3. Point process with piecewise constant

$$\text{intensity:} \qquad \theta_{ij} \mid \beta \sim Gamma(1, \beta),$$

C. Hyper parameter: (Under all three models M1-M3)

$$\beta \sim Gamma(5, 0.003)$$

Table 3. Preliminary analysis of valve failure data

Valve no.	Observed no. of failures	Predicted median failure time (apprx.) in years		
		i.i.d. HPP	Non-i.i.d. HPP	PP with piecewise const. intensity
1	8	17	2	14
2	7	17	2.3	14
10	1	17	10.5	14
17	0	17	25.5	14

Reasonably vague distributions were chosen for the hyper-parameters. A partial summary of the preliminary analysis is given in Table 3. From the preliminary analyzes we could see that, even in a situation in which there is considerable heterogeneity in the inter-failure times both between valves and within each valve, merely adding more latent variables may not be a good strategy for modelling. Here, in the third model, it resulted in a loss of the predictive performance.

4. Model and Analysis

It is apparent from the data that the valves exhibited varying behaviour during the period of study with respect to external leakages, with the majority experiencing no failures whereas some recording as many as 8. Some of the possible sources of heterogeneity have been described above.

The standard procedure is to express such heterogeneity in a mathematical form, by incorporating the variables (possibly) causing such deviations as "covariates". A popular method of incorporating the effects of a covariate on the probabilistic behaviour of a system is to use them while modelling the failure intensity/hazard rate of the system, possibly in a form known as the Cox regression model (Cox[4]). In the current set-up, we have a large number of such variables which can be suspected to be causing the valves to behave in different ways. Also, how exactly these affect the failure rate may be difficult to establish through any explicit mathematical formulation.

The inter-failure times of the valves are observed to be highly varying, with very small as well as very large values being observed from the same valve. No specific pattern like systematic deterioration, was observed. Several preliminary models for the valves, like i.i.d. and non-i.i.d. renewal processes, non-homogeneous Poisson processes with piecewise constant intensities, with the intensities changing/jumping at failure epochs, were attempted for initial analyses. Evident heterogeneity rejected models treating the valves in an i.i.d. manner. Also models without valve specific control on the models for the inter-failure times performed poorly from a prediction point of view. From these we noted that even if heterogeneity is allowed between valves, within valve heterogeneity, arising possibly due to repairs and other environmental factors, should be modelled carefully. An appropriate model should be able capture valve specific behaviour also.

Following Arjas and Bhattacharjee[1], we model the probabilistic behaviour of these systems in a hierarchical Bayesian model. The first layer of this model consists of the inter-failure times. As we have already pointed out above, due to evident heterogeneity in the valves the inter-failure times need to be modelled separately for each valve. Also for any fixed valve the inter-failure times exhibit a high amount of variation without any clear pattern over time in them. Hence for each valve we consider a renewal process with a Pareto inter-failure time distribution. This should be able to explain absence of time trend and to accommodate both very small and very large inter-failure times.

Next, we observe that there are at least two different patterns in the behaviour of the valves, with most not failing during the entire followup and a few failing even several times. This effectively means that each valve comes either from an, *a priori* unknown, group of "good" valves with possibly long failure times, or from a (smaller) group of "bad" valves with not so long inter-failure times. The two groups/classes are described by assigning to each valve a Bernoulli distributed latent variable identifying the group for that valve. Assuming that the inter-failure times are Pareto distributed, the two groups are characterised by drawing the shape parameters from two ordered Gamma distributions.

Our lack of further prior knowledge about these latent variables is expressed by assigning a Uniform distribution to the parameter of the Bernoulli distribution.

This model structure can be described briefly as follows:

A. Inter-failure times : $\quad X_{i,j} \mid \beta_0, \beta_1, C_i \sim Pareto(\beta_{C_i})$,
where $i = 1, \ldots, 104$ and $j \geq 1$,

B. Latent variables : $\quad C_i \mid p \sim Bernoulli(p), i = 1, \ldots, 104$

C. Group-specific parameters : $\quad \beta_1 \sim Gamma(.,.)$ and $\beta_0 = \beta_1 + \eta_1$,
where $\eta_1 \sim Gamma(.,.)$,

D. Mixing probability : $\quad p \sim Uniform(0, 1)$.

Several values of the hyper-parameters were chosen in order to carry out sensitivity analysis. Parameters (5, 0.001) and (5, 0.005) were found to be appropriate choices for the distributions of η_1 and β_1 respectively, giving rise to reasonably uninformative priors for β's . The proportions of the "good" and "bad" valves in the population may be estimated from the posterior distribution of the mixing hyper-parameter, namely, p. The posterior distribution of C_i describes, for the i-th valve, the chances of coming from the "good" or the "bad" group, which may be highly useful for many purposes, like reliability assessment and maintenance.

The structure of the model makes it possible to evaluate posterior characteristics on population level (Table 4) and also at generic valve level (Table 5). As is expected the proportion of "bad" valves is only 7 %, consistent with the fact that these are parts of safety systems. The benefit of considering the quality of a valve is evident when the predicted lifetime of a generic valve is considered. With no added information on quality the predicted life time is more than 30 years, whereas with the information that the valve is of poor quality, the prediction is drastically reduced to about 1 year only.

Table 4. Results of analysis of valve failure data: Population level characteristics

Characteristic	Posterior expectation
Proportion of "bad" valves	0.07
Proportion of "good" valves	0.93
β Parameter for "bad" valves	494
β Parameter for "good" valves	13650

Table 5. Results of analysis of valve failure data: Generic-valve characteristics.

Characteristic	Posterior estimate
Predictive probability for a generic valve to be a "good" valve	0.93
Predictive median time to first failure for a generic valve	31.67 years
Predictive median time to first failure for a "bad" valve	1.27 years
Predictive median time to first failure for a "good" valve	37 years

For planning maintenance activities it would certainly be useful to have an estimate of the expected number of failures in future. Given the data for 9 consecutive years we predicted the expected number of failures from all valves during the 10-th year and during 10-th & 11-th year (see Table 6). Unsurprisingly, the expected number of failures is approximately proportional to the duration of prediction. A closer inspection reveals that it is highly unlikely that more than many valves would experience more than one failure during the considered prediction periods.

Table 6.   Results of analysis of valve failure data: Maintenance related characteristics.

Characteristic	Predictive expectation
No. of failures from all 104 valves during the 10-th year	4.00
No. of failures from all 104 valves during the 10 & 11-th year	7.93
No. of valves to fail at least once during 10-th year	3.47
No. of valves to fail $\geq$ 2 times during 10-th year	0.41
No. of valves to fail $\geq$ 3 times during 10-th year	0.09
:	:
No. of valves to fail at least once during 10 & 11-th year	6.32
No. of valves to fail $\geq$ 2 times during 10 & 11-th year	1.06
No. of valves to fail $\geq$ 3 times during 10 & 11-th year	0.37
:	:

If maintenance programs are designed for specific valves or valve types, predictions for individual valves are needed. Examples of such predictions are given in Tables 7 and 8. Note that for any valve chances of failing beyond the study period would depend not only on its quality but also on when it had failed last during the study period.

The concept of "good" and "bad" valves was extended within each category of valves (i.e. valves belonging to a specific safety system). It was observed that indeed the estimates of the population parameters for different safety systems differ significantly. The posterior estimates of some of the model parameters are presented in Table 8.

We observe from Table 8 that the parameter estimates for Valve-type /Safety-system 1 differ noticeably from the others. This is in agreement with the information provided by Table 1, showing that a relatively higher proportion of valves from the first safety system failed during the study. This difference may also be due to the fact that unlike the other three groups, valves in Safety system 1 are continuously in operation, and this probably makes them more failure prone than the others.

Also the number of predicted failures for the 10-th and 11-th years are seen to be varying across valve type, which is consistent with the observed

Table 7.   Results of analysis of valve failure data: Unit level characteristics.

Characteristic	Posterior/predictive estimates for valve no.								
	1	2	3	4	5	6	7	8	9
Chances of being a "good" valve	0.00	0.00	0.02	0.01	0.72	0.73	0.97	0.96	0.97
Median future inter failure time	461	471	493	465	6760	7000	12770	12510	12910
Expected no. of failures during the 10-th year	0.31	0.29	0.54	0.28	0.10	0.11	0.03	0.03	0.03
Expected no. of failures during the 10 & 11-th year	0.64	0.60	0.97	0.56	0.21	0.22	0.06	0.06	0.07
Median (residual) time to next failure at the end of 9-th year	1248	1341	603	1468	7815	7907	14140	14450	13860

Characteristic	Posterior/predictive estimates for valve no.								
	10	11	12	13	14	15	16	17	
Chances of being a "good" valve	0.96	0.83	0.96	0.96	0.95	0.96	0.96	0.99	...
Median future inter failure time	12270	9119	12380	12330	12500	12480	12160	13080	...
Expected no. of failures during the 10-th year	0.03	0.04	0.03	0.04	0.05	0.03	0.03	0.02	...
Expected no. of failures during the 10 & 11-th year	0.06	0.08	0.06	0.09	0.09	0.06	0.06	0.04	...
Median (residual) time to next failure at the end of 9-th year	14170	13440	14100	12500	12550	14390	14270	16740	...

Table 8. Results of analysis of valve failure data: Valve type specific analysis

Characteristic	Posterior expectations for valve type			
	1	2	3	4
Proportion of "bad" valves	0.67	0.05	0.04	0.09
Proportion of "good" valves	0.33	0.95	0.96	0.91
β Parameter for "bad" valves	467	1273	1229	1193
β Parameter for "good" valves	5778	9559	12310	9230
Expected no. of failures during the 10-th year	1.93	1.56	0.81	0.54
Expected no. of failures during the 10 & 11-th year	3.80	3.10	1.60	1.08

data. Note that although the predicted number of failures for valve types 1 and 2 appear to be comparable, the number of valves in these two safety systems are quite different, being 8 and 48 respectively. Similar comments can be made about the other two types of valves (see Tables 1 and 8). Also observe that if predictions of future failures for valves in different safety systems were made using estimated population parameters from the previous model, the predictions would have been proportional to the number of valves in the specific safety system. The added information of valve type, therefore, refines the inference in this aspect.

If it is believed that the valves could be usefully divided into more than two categories, then this structure can be easily extended by introducing additional (η, β)'s and a Multinomial-Dirichlet combination for C_i's and p.

In a general situation, ordering the β parameters may be found as too restrictive, although it helps avoiding identifiability problems. Also it may not be easy to identify such orders in a multivariate parameter setup. In that case, instead, the "profiles" of the individuals may be used to fix/characterise the groups. For example, for the valve data, two valves may be identified to represent the two groups, namely "good" and "bad". This can easily be achieved by using predetermined values for C_i's of these two valves. The remaining valves may be modelled using the above mixture model. But note that now the estimation will be influenced by the profiles of the chosen "good" and "bad" valves which serve as fixed representatives of their categories.

5. Conclusions

The heterogeneity of failure behaviour of safety related components, such as valves in our case study, may have important implications for reliability analysis of safety systems. If such heterogeneity is not identified and taken into account, the decisions made to maintain or to enhance safety can be non-optimal or even erroneous. This non-optimality is more serious if the safety related decisions are made on the basis of failure histories of the components.

This work demonstrates how even rather simplistic models could describe the heterogeneous behaviour successfully. We were able to make an assessment of the quality (i.e. whether the valve is "good" or "bad") of individual valves, and we were also able to estimate the proportions of "good" and "bad" valves in the population. Furthermore, we could analyse how additional information of the valve quality helps in predicting the future failure behaviour of a considered valve. These predictions can be made both for individual valves and for the valve populations as a whole. Predictions of this kind can be used in developing maintenance programs and in predicting the cost of future maintenance. Another use of the results, in addition to qualitative failure analyses, can be the identification of weaknesses in earlier maintenance programs.

References

1. E. Arjas and M. Bhattacharjee, Modelling heterogeneity/selection and causal influence inreliability, *IISA-JSM 2000-2001 Conference, New Delhi, India*, (2001).
2. T. Bedford, R. M. Cooke, J. Dorrepaal, Mathematical review of swedish Bayesian methodology for nuclear plant reliability data bases, *The Practice of Bayesian Analysis. Eds, French, S, Smith J.Q.*, (1997).
3. H. W. Block and T. H. Savits, Burn-in, *Statistical Science*, **12**, 1-13 (1997).
4. D. R. Cox, Regression models and life tables, *Journal of the Royal Statistical Society, Series B*, **34**, 187-220 (1972).
5. E. Hofer, J. Peschke, Bayesian modelling of failure rates and initiating event frequencies, *Proceedings of the European Conference on Safety and Reliability (ESREL), Garching, Germany 13-17-September*, (1999).
6. P. Hougaard, *Analysis of Multivariate Survival Data*, (Springer) (2000).
7. N. J. Lynn and N. D. Singpurwalla, Comment: "Burn-in" makes us feel good, *Statistical Science*, **12**, 13-19 (1997).
8. K. Pörn, On empirical Bayesian inference applied to Poisson probability models, *Linköping studies in science and technology. Dissertations 234*, (1990).

9. U. Pulkkinen and K. Simola, Bayesian Models and Ageing Indicators for Analysing Random Changes in Failure Occurrence, *Reliability Engineering & System Safety*, **68**, 255-268 (2000).

10. K. Simola and K. Laakso, Analysis of Failure and Maintenance Experiences of Motor Operated Valves in a Finnish Nuclear Power Plant, *VTT Research Notes 1322* (1992).

23

COMPETING RISK PERSPECTIVE ON RELIABILITY DATABASES

Cornel Bunea and Roger Cooke

Delft University of Technology, Department of Mathematics,
Mekelweg 4, NL-2628 CD Delft, The Netherlands
E-mail: c.bunea@its.tudelft.nl, r.m.cooke@its.tudelft.nl

Bo Henry Lindqvist

Norwegian University of Science and Technology,
Department of Mathematical Sciences, N-7491 Trondheim, Norway
E-mail: bo@math.ntnu.no

The chapter considers competing risk models suitable for analysis of modern reliability databases. Commonly used models are reviewed and a new simple and attractive model is developed. This model grew out of a study of real failure data from an ammonia plant. The use of graphical methods as an aid in model choice is advocated in the chapter.

1. Introduction

Since Daniel Bernoulli's attempt in the 18th century to separate the risk of dying due to smallpox from other causes, the competing risk theory has spread through various fields of science such as statistics, medicine and reliability analysis.

The competing risk approach in reliability is closely related to the development of modern reliability databases (RDB's) in the second half of the last century. RDB's consist of several data fields containing information on failure modes, failure causes, maintenance/repair actions, severity of failure, component characteristics and operating circumstances. For every specific field we can often distinguish ten or more competing risks, which compete to terminate a service sojourn of the component. The general theory of competing risk is advanced here as a proper mathematical language for modeling reliability data.

355

To check the performance of different probabilistic models, we discuss a dataset coming from two identical compressor units at an ammonia plant of Norsk Hydro, covering an observation period from 1968 to 1989 (Erlingsen[8]). The competing risk models available in the reliability literature are especially developed for the nuclear sector, where strict regulations are imposed. Consequently these models are often not appropriate for the various fields of the compressor unit data. Due to the fact that the compressor unit consists of several heterogeneous sub-components, we therefore introduce another competing risk model, called the "mixture of exponentials model", to interpret the competing risks between different failure modes.

The article is organized as follows. Section 2 first introduces some basic concepts and definitions, and then goes on to present specific models including the new model mentioned above. Section 3 is devoted to the analysis of the compressor unit data. The purpose of that section is to show how graphical plots can be used as a guide to choose a model. Some concluding remarks are given in Section 4.

2. Competing Risks

In the competing risks approach we model the data as a sequence of i.i.d. pairs (Z_i, D_i), $i = 1, 2, \ldots$. Each Z is the minimum of two or more variables, corresponding to the competing risks. In many cases we can reduce the problem to the analysis of two competing risks, described by two random variables X and Y such that $Z = \min(X, Y)$. Typically X will be the minimum of failure times corresponding to a set of failure modes which are of primary interest, while Y is a censoring time corresponding to termination of observation by other causes, for example preventive maintenance or non-critical failure. In addition to the time Z one observes the indicator variable $D = I(X < Y)$ which describes the cause of the termination of observation. For simplicity we assume that $P(X = Y) = 0$. Examples of the present setup are given in Section 3 in connection with the ammonia plant data.

It is well known (Tsiatis[11]) that from observation of (Z, D) we can identify only the subsurvival functions of X and Y,

$$S_X^*(t) = P(X > t, X < Y) = P(Z > t, D = 1)$$
$$S_Y^*(t) = P(Y > t, Y < X) = P(Z > t, D = 0),$$

but not in general the true survival functions of X and Y, $S_X(t)$ and $S_Y(t)$. Note that $S_X^*(t)$ depends on Y, though this fact is suppressed in the notation. Note also that $S_X^*(0) = P(X < Y) = P(D = 1)$ and $S_Y^*(0) = P(Y < X) = P(D = 0)$, so that $S_X^*(0) + S_Y^*(0) = 1$.

The conditional subsurvival functions are defined as the survival functions conditioned on the occurrence of the corresponding type of event. Assuming continuity of $S_X^*(t)$ and $S_Y^*(t)$ at zero, these functions are given by

$$CS_X^*(t) = P(X > t | X < Y) = P(Z > t | D = 1) = S_X^*(t)/S_X^*(0)$$
$$CS_Y^*(t) = P(Y > t | Y < X) = P(Z > t | D = 0) = S_Y^*(t)/S_Y^*(0).$$

Closely related to the notion of subsurvival functions is the probability of censoring beyond time t,

$$\Phi(t) = P(Y < X | Z > t) = P(D = 0 | Z > t) = \frac{S_Y^*(t)}{S_X^*(t) + S_Y^*(t)}.$$

This function seems to have some diagnostic value, aiding us to choose among competing risk models to fit the data. Note that $\Phi(0) = P(D = 0) = S_Y^*(0)$.

Suppose we have a finite set of data, $(Z_1, D_1), \ldots, (Z_n, D_n)$. The empirical subsurvival functions and the conditional subsurvival functions are defined as[6]

$$\widehat{S}_X^*(t) = \frac{\text{number of pairs } (Z_i, D_i) \text{ with } Z_i > t,\ D_i = 1}{n}$$

$$\widehat{S}_Y^*(t) = \frac{\text{number of pairs } (Z_i, D_i) \text{ with } Z_i > t,\ D_i = 0}{n}$$

$$\widehat{CS}_X^*(t) = \frac{\text{number of pairs } (Z_i, D_i) \text{ with } Z_i > t,\ D_i = 1}{\sum_{i=1}^{n} D_i}$$

$$\widehat{CS}_Y^*(t) = \frac{\text{number of pairs } (Z_i, D_i) \text{ with } Z_i > t,\ D_i = 0}{n - \sum_{i=1}^{n} D_i}$$

As already mentioned, without any additional assumptions on the joint distribution of X and Y, it is impossible to identify the marginal survival functions $S_X(t)$ and $S_Y(t)$. However, by making extra assumptions, one may restrict to a bunch of models in which the survival functions are identifiable. Such models will be considered in the following subsections.

Peterson[10] derived bounds on the survival function $S_X(t)$ by noting that

$$P(X \le t, X \le Y) \le P(X \le t) \le P(\min(X, Y) \le t),$$

which entails

$$S_X^*(t) + S_Y^*(0) \ge S_X(t) \ge S_X^*(t) + S_Y^*(t).$$

Note that the quantities on the left and right hand sides can always be estimated consistently from data. The failure rate $r_X(t)$ for X is

$$r_X(t) = -(dS_X(t)/dt)/S_X(t) = -d(\ln(S_X(t)))/dt.$$

Thus the time average failure rate as a function of t is

$$(1/t) \int_0^t r_X(u)du = -(1/t)\ln(S_X(t)).$$

and the Peterson bounds hence imply

$$-\frac{\ln(S_X^*(t) + S_Y^*(0))}{t} \leq \frac{\int_0^t r_X(u)du}{t} \leq -\frac{\ln(S_X^*(t) + S_Y^*(t))}{t}.$$

A similar relation holds for the failure rate of Y. These bounds, which we shall later call the Peterson bounds, can give us valuable indications on the monotonicity of the failure rates for X and Y. This is demonstrated in Section 3 where estimated Peterson bounds are graphed as functions of t.

2.1. *Independent Exponential Competing Risks*

A classical result on competing risks states that, assuming independence of X and Y, we can determine uniquely the survival functions of X and Y from the joint distribution of (Z, D). In this case the survival functions of X and Y are said to be identifiable from the censored data (Z, D). Hence, an independent model is always consistent with data, but an independent exponential model is of course not in general. One can derive a very sharp criterion for independence and exponentiality in terms of the subsurvival functions (Cooke[3]):

Theorem 1: Let X and Y be independent life variables. Then any two of the following conditions imply the others:

$$S_X(t) = \exp(-\lambda t)$$
$$S_Y(t) = \exp(-\gamma t)$$
$$S_X^*(t) = \frac{\lambda}{\lambda + \gamma} \exp(-(\lambda + \gamma)t)$$
$$S_Y^*(t) = \frac{\gamma}{\lambda + \gamma} \exp(-(\lambda + \gamma)t)$$

Thus if X and Y are independent exponential life variables with failure rates λ and γ, then the conditional subsurvival functions of X and Y are equal and correspond to exponential distributions with failure rate $\lambda + \gamma$. Moreover, the probability of censoring beyond time t is constant. Thus

$$CS_X^*(t) = CS_Y^*(t) = \exp(-(\lambda + \gamma)t)$$
$$\Phi(t) = \frac{\gamma}{\lambda + \gamma}.$$

2.2. *Random Signs Censoring*

Perhaps the simplest dependent competing risk model which leads to identifiable marginal distribution of X is random signs censoring[3]. Consider a component subject to right censoring at Y, and let X denote the time at which the component would expire if not censored. Suppose that the event that the life of the component be censored is independent of the age X at which the component would expire, but given that the component is censored, the time at which it is censored may depend on X. This might arise if a component emits a warning before expiring; if the warning is seen then the component is taken out, thus censoring its life, otherwise it fails. The random signs model assumes that the probability of seeing the warning is independent of the component's age. This situation is captured in the following definition:

Definition 2: Let X and Y be life variables with $Y = X - W\delta$, where $0 < W < X$ is a random variable and δ is a random variable taking values $\{1, -1\}$, with X and δ independent. The variable $Z \equiv [\min(X, Y), I(X < Y)]$ is called a random sign censoring of X by Y.

Note that in this case

$$S_X^*(t) = Pr\{X > t, \delta = -1\} = Pr\{X > t\}Pr\{\delta = -1\} =$$
$$= S_X(t)Pr\{Y > X\} = S_X(t)S_X^*(0).$$

Hence $S_X(t) = CS^*(t)$ and it follows in particular that the distribution of X is identifiable under random signs censoring.

Cooke[3] proved that a joint distribution of (X, Y) which satisfies the random signs requirement, exists if and only if $CX^*(t) > CY^*(t)$ for all $t > 0$. In this case the probability of censoring beyond time t, $\Phi(t)$, is maximum at the origin.

2.3. *Conditional Independence Model*

Another model from which we have identifiability of marginal distributions is the conditional independence model introduced by Hokstad, see e.g. Hokstad and Jensen[9] or Dorrepaal et al.[7]. This model considers the competing risk variables X and Y to be sharing a common quantity, V, and to be independent given V. More precisely, the assumption is that

$$X = V + W, \quad Y = V + U,$$

where V, U, W are mutually independent. Hokstad and Jensen[9] derived explicit expressions for the case when V, U, W are exponentially distributed:

Theorem 3: Let V, U, W be independent with $S_V(t) = e^{-\lambda_V t}, S_U(t) = e^{-\lambda_U t}, S_W(t) = e^{-\lambda_W t}$. Then

$$S_X^*(t) = \frac{\lambda_V \lambda_W e^{-(\lambda_U + \lambda_W)t}}{(\lambda_U + \lambda_W)(\lambda_V - \lambda_W - \lambda_U)} - \frac{\lambda_W e^{-\lambda_V t}}{\lambda_V - \lambda_W - \lambda_U}$$

$$S_Y^*(t) = \frac{\lambda_V \lambda_U e^{-(\lambda_U + \lambda_W)t}}{(\lambda_U + \lambda_W)(\lambda_V - \lambda_W - \lambda_U)} - \frac{\lambda_U e^{-\lambda_V t}}{\lambda_V - \lambda_W - \lambda_U}$$

$$CS_X^*(t) = CS_Y^*(t) = S_X^*(t) + S_Y^*(t)$$

$$\Phi(t) = \frac{\lambda_U}{\lambda_U + \lambda_W}$$

Moreover, if V has an arbitrary distribution such that $P(V \geq 0) = 1$, and V is independent of U and W, then still we have

$$CS_X^*(t) = CS_Y^*(t)$$

Thus, as in the case of independent exponential competing risks we have equal conditional subsurvival functions, and the probability of censoring beyond time t, $\Phi(t)$, is constant.

2.4. *Mixture of Exponentials Model*

Suppose that $S_X(t)$ is a mixture of two exponential distributions with parameters λ_1, λ_2 and mixing coefficient p, and that the censoring survival distribution $S_Y(t)$ is exponential with parameter λ_y:

$$S_X(t) = p \exp\{-\lambda_1 t\} + (1 - p) \exp\{-\lambda_2 t\}$$
$$S_Y(t) = \exp\{-\lambda_y t\}.$$

The properties of the corresponding competing risk model is given by the next theorem.

Theorem 4: Let X and Y be independent life variables with the above distributions. Then,

$$S_X^*(t) = p\frac{\lambda_1}{\lambda_y + \lambda_1}\exp\{-(\lambda_y + \lambda_1)t\} + (1-p)\frac{\lambda_2}{\lambda_y + \lambda_2}\exp\{-(\lambda_y + \lambda_2)t\}$$

$$S_Y^*(t) = p\frac{\lambda_y}{\lambda_y + \lambda_1}\exp\{-(\lambda_y + \lambda_1)t\} + (1-p)\frac{\lambda_y}{\lambda_y + \lambda_2}\exp\{-(\lambda_y + \lambda_2)t\}$$

$$CS_X^*(t) = \left(\exp\{-(\lambda_y + \lambda_1)t\} + \frac{1-p}{p}\frac{\lambda_2}{\lambda_1}\frac{\lambda_y + \lambda_1}{\lambda_y + \lambda_2}\exp\{-(\lambda_y + \lambda_2)t\}\right)$$
$$/\left(1 + \frac{1-p}{p}\frac{\lambda_2}{\lambda_1}\frac{\lambda_y + \lambda_1}{\lambda_y + \lambda_2}\right)$$

$$CS_Y^*(t) = \left(\exp\{-(\lambda_y + \lambda_1)t\} + \frac{1-p}{p}\frac{\lambda_y + \lambda_1}{\lambda_y + \lambda_2}\exp\{-(\lambda_y + \lambda_2)t\}\right)$$
$$/\left(1 + \frac{1-p}{p}\frac{\lambda_y + \lambda_1}{\lambda_y + \lambda_2}\right)$$

$$CS_X^*(t) \le CS_Y^*(t)$$

Moreover, $\Phi(t)$ is minimal at the origin, and is strictly increasing when $\lambda_1 \neq \lambda_2$.

Proof: Using the independence of X and Y one can show that

$$S_X^*(t) = -\int_t^\infty S_Y(u)dS_X(u), \quad S_Y^*(t) = -\int_t^\infty S_X(u)dS_Y(u).$$

The first four formulas of the theorem follow from this by straightforward integration.

To prove the inequality between the conditional subsurvival functions we can rewrite the conditional subsurvival functions in a condenced form as

$$CS_X^*(t) = \frac{A + \frac{\lambda_2}{\lambda_1}BC}{1 + \frac{\lambda_2}{\lambda_1}B}, \qquad CS_Y^*(t) = \frac{A + BC}{1 + B}.$$

Hence,

$$\frac{S_X^*(t)}{S_X^*(0)} \le \frac{S_Y^*(t)}{S_Y^*(0)}$$

is equivalent to

$$A(1 - \frac{\lambda_2}{\lambda_1}) \le C(1 - \frac{\lambda_2}{\lambda_1}),$$

or

$$\exp\{-\lambda_1 t\}(\lambda_1 - \lambda_2) \le \exp\{-\lambda_2 t\}(\lambda_1 - \lambda_2).$$

But this clearly holds for all λ_1, λ_2 and the announced inequality follows.

From $S_X^*(t)/S_X^*(0) \leq S_Y^*(t)/S_Y^*(0)$ it follows immediately that $\Phi(t)$ is minimal at the origin. That $\Phi(t)$ is strictly increasing is seen from

$$\Phi'(t) = C\frac{pS_1(t) \cdot (1-p)S_2(t)}{(pS_1(t) + (1-p)S_2(t))^2},$$

where $C = \lambda_y(\lambda_1 - \lambda_2)^2/[(\lambda_y + \lambda_1)(\lambda_y + \lambda_2)]$ and $S_1(t) = \exp\{-\lambda_1 t\}$, $S_2(t) = \exp\{-\lambda_2 t\}$. □

It is interesting to note that the survival function of Z becomes

$$P(Z > t) = P(X > t, Y > t) = P(X > t)P(Y > t) =$$
$$= p\exp\{-(\lambda_1 + \lambda_y)t\} + (1-p)\exp\{-(\lambda_2 + \lambda_y)t\}$$

This is also a mixture of two exponential distributions with parameters $\lambda_1 + \lambda_y$ and $\lambda_2 + \lambda_y$, and mixing coefficient p.

The expectation and variance of Z are:

$$E(Z) = \int_0^\infty z f_z(z)dz = \int_0^\infty P(Z > t)dt =$$
$$= p\frac{1}{\lambda_y + \lambda_1} + (1-p)\frac{1}{\lambda_y + \lambda_2},$$

respectively

$$\frac{1}{(\lambda_y + \lambda_2)^2} + 2p\frac{1}{\lambda_y + \lambda_1}\left(\frac{1}{\lambda_y + \lambda_1} - \frac{1}{\lambda_y + \lambda_2}\right) - p^2\left(\frac{1}{\lambda_y + \lambda_1} - \frac{1}{\lambda_y + \lambda_2}\right)^2$$

Using the above equations, $S_Y^*(0)$ and $\Phi'(t)$ we can obtain an estimation for the unknown parameters. For example $\lambda_y = S_Y^*(0)/E(Z)$ and the values of p, λ_1 and λ_2 can be obtained numerically. Note that the same solution for λ_y is obtained using the maximum likelihood method.

For an observation $(Z, D) = (t, 1)$ the likelihood contribution is

$$-dS_X^*(t)/dt = \exp\{-\lambda_y t\}[p\lambda_1 \exp\{-\lambda_1 t\} + (1-p)\lambda_2 \exp\{-\lambda_2 t\}],$$

while for an observation $(Z, D) = (t, 0)$ it is

$$-dS_Y^*(t)/dt = \lambda_y \exp\{-\lambda_y t\}[p\exp\{-\lambda_1 t\} + (1-p)\exp\{-\lambda_2 t\}].$$

Thus with data $(Z_1, D_1), \ldots, (Z_n, D_n)$ we get the likelihood function

$$\lambda_y^{n - \sum D_i} \exp\{-\lambda_y \sum Z_i\} \prod_i [p\lambda_1 \exp\{-\lambda_1 t\} + (1-p)\lambda_2 \exp\{-\lambda_2 t\}]^{D_i}$$
$$\times [p\exp\{-\lambda_1 t\} + (1-p)\exp\{-\lambda_2 t\}]^{1-D_i}$$

Maximizing this with respect to the parameters we get the explicit solution $\hat{\lambda}_y = (n - \sum D_i)/\sum Z_i$. To obtain the other parameter estimates we need to use numerical methods.

2.5. *Model selection*

The probability of censoring after time t, $\Phi(t)$, seems to be an important indicator for model selection, together with the conditional subsurvival functions $CS_X^*(t)$ and $CS_Y^*(t)$. In practice one estimates these functions from data and consider their graphs (see examples in Section 3). The following statements, which follow from the results of the previous subsections, may then guide in model selection.

- If the risks are exponential and independent, then the conditional subsurvival functions are equal and correspond to exponential distributions. Moreover, $\Phi(t)$ is constant.
- Under random signs censoring, $\Phi(0) > \Phi(t)$ and $CS_X^*(t) > CS_Y^*(t)$ for all $t > 0$.
- If the conditional independence model holds with exponential marginals, then the conditional subsurvival functions are equal and $\Phi(t)$ is constant
- If the mixture of exponentials model holds, then $\Phi(t)$ is strictly increasing and $CS_X^*(t) \leq CS_Y^*(t)$ for all $t > 0$.

3. Analysis of a Competing Risks Dataset

The dataset used to discuss different competing risk models comes from an ammonia plant of Norsk Hydro. Detailed failure data are given for two identical compressor units for the observation period 2-10-68 to 25-6-89. This yields 21 years of observation and more than 370 events. In accordance with modern reliability databases, the data contain the following compressor unit history:

- Time of component failure.
- Failure mode: leakage, no start, unwanted start, vibration, warming, overhaul, little gas stream, great gas stream, others.
- Degree of failure: critical, non-critical.
- Down time of component.
- Failure at compressor unit level: 1 - first unit failed, 2 - second unit failed, 3 - both units failed.
- System and sub-system failure.
- Action taken: immediate reparation, immediate replacement, adjustment, planned overhaul, modification, others.
- Revision periods: 18 revision periods with durations from 4 to 84 days.

                 C. Bunea, R. Cooke and B.H. Lindqvist

Date	Calendar time	Operation time	Revision time	Failure mode	Degree of failure	Failure time	Degree time	Sys. Time	Sys. Degree time	Down time (h)	Unit	System	Subsystem	Subfailure mode	Action taken
10/2/68	0	0	0			0	0	0	0	0					
10/3/68	1	1	1	UST	C	1	1	1	1	1.25	1	SMO	PUM	UST	AKR
10/6/68	4	4	4	UST	C	3	3	4	4	135	2	SPE	ANN	LIG	AKR
10/6/68	4.5	4.5	4.5	UST	C	0.5	0.5	0.5	0.5	0.08	1	SPE	ANN	LIG	AKR
11/7/68	36	36	36	VIB	N	32	36	35	36	0	3	SMO	PUM	VIB	AKR
12/8/68	67	67	67	LEK	N	31	31	63	67	0	2	SPE	TAN	LEK	AKU
1/2/69	92	92	92	UST	C	25	88	25	88	5.33	2	SPE	FOR	ANN	JUS
1/2/69	112	R	E	V	I	S	I	O	N	1272		3/16/69			
4/23/69	203	150	38.2	LEK	N	58	83	150	150	0	1	KOM	ROR	LEK	AKR
4/23/69	203.25	150.25	38.4	LEK	N	0.25	0.25	0.25	0.25	0	2	KOM	ROR	LEK	AKR
4/23/69	203.5	150.5	38.6	LEK	N	0.25	0.25	150	150	0	2	INS	TRY	LEK	AKR
4/23/69	203.75	150.75	38.8	LEK	N	0.25	0.25	58	83	0	2	SPE	VEN	LEK	AKR
7/18/69	289	236	124	LEK	N	86	86	86	86	0	2	SPE	VEN	LEK	JUS
8/3/69	305	252	140	UST	C	16	160	252	252	154	1	ELM	ANN	UST	AKR
8/18/69	320	267	155	OVH	N	15	31	31	31	0	3	SPE	FIL	OVH	OVH
8/20/69	322	269	157	LEK	N	2	2	233	233	0	3	SMO	SEP	LEK	AKR
8/28/69	330	277	165	UST	C	8	25	10	185	0.5	1	SPE	VEN	LEK	AKR
8/28/69	330.5	277.5	165	UST	C	0.5	0.5	127	277	1.25	1	INS	NIV	ISI	AKU
9/4/69	337	284	172	UST	C	7	7	15	283	2.5	3	SMO	PUM	UST	JUS
10/4/69	367	R	E	V	I	S	I	O	N	720		11/3/69			
12/1/69	425	342	28	LEK	N	58	73	65	75	0	2	SPE	ROR	LEK	AKR
12/29/69	453	370	56	VIB	N	28	28	28	28	0	2	SPE	PUM	VIB	AKU
1/2/70	457	374	60	LEK	C	4	90	224	427	15	2	KOM	ROR	LEK	AKR

Failure mode	
LEK	leakage
IST	no start
UST	unwanted start
LIG	little gas stream
STG	great gas stream
VIB	vibration
VAR	warming
OVH	overhaul
ANN	others
Degree of failure	
N	non-critical
C	critical
Action taken	
AKR	immediate reparation
AKU	immediate replacement
JUS	adjusment
OVH	planned overhaul
MOD	modification
ANN	others
UNIT	
1	compressor 1 failed
2	compressor 2 failed
3	both compressor failed
System	
SMO	

Figure 1. Compressor units data

The analysis performed on the data was both statistical and probabilistic. Nevertheless, the most time consuming operation was "cleaning data". Experience shows that 2/3 of the time is spent on extracting the most relevant information from the huge amount of information gathered in a reliability database. The main preparations and analyses of the data are presented in Bunea et al.[1]

A software tool was developed in the higher order programming languages Visual Basic and Excel. The operator can choose two exhaustive classes of competing risks (corresponding to the X and Y in Section 2) for the four fields of interest: failure mode, degree of failure, action taken, system. The analysis may be performed for the complete observation period or for a certain time window by specifying the limiting dates of interest.

Figures 2-5 illustrate the different field analyses performed on the data. The upper graphs show the empirical subsurvival functions, $\widehat{S}_X^*(t), \widehat{S}_Y^*(t)$, empirical conditional subsurvival functions, $\widehat{CS}_X^*(t), \widehat{CS}_Y^*(t)$, and the estimated probability of censoring beyond time t, $\widehat{\Phi}(t)$. The bottom graphs show the estimated Peterson bounds for the average failure rate, for both classes of risk, as given in Section 2. Cooke and Bedford[4] explained how to use such graphical displays to choose an appropriate competing risk model. In Bunea et al.[2] we proposed a statistical test to check whether an independent exponential model is appropriate for the data, against the alternative of the random signs model or the mixture of exponentials model. The one sided Kolmogorov-Smirnov (KS) test was used to test the null hypothesis that the two empirical conditional subsurvival functions are drawn from two identical populations, against the alternative that one conditional subsurvival function is larger than the other. If the null hypothesis is not rejected at the chosen significance level, then the empirical conditional subsurvival functions may come from two identical populations, and the independent exponential model or the conditional independence model may be appropriate.

3.1. *Failure Mode*

The risk engineer tries to avoid a functional failure of the system. Discussions with maintenance personnel indicate that the most dangerous failure modes among the listed failure mechanism are "leakage", "no start", "unwanted start" and "vibration". We let this be the risk class corresponding to X in the notation of Section 2. These risks are censored by the other failure modes, which together correspond to Y. The total number of failures is 359 out of which 166 are censoring events and 193 are events that we want to prevent. The large number of unwanted events can be explained by a poor maintenance policy.

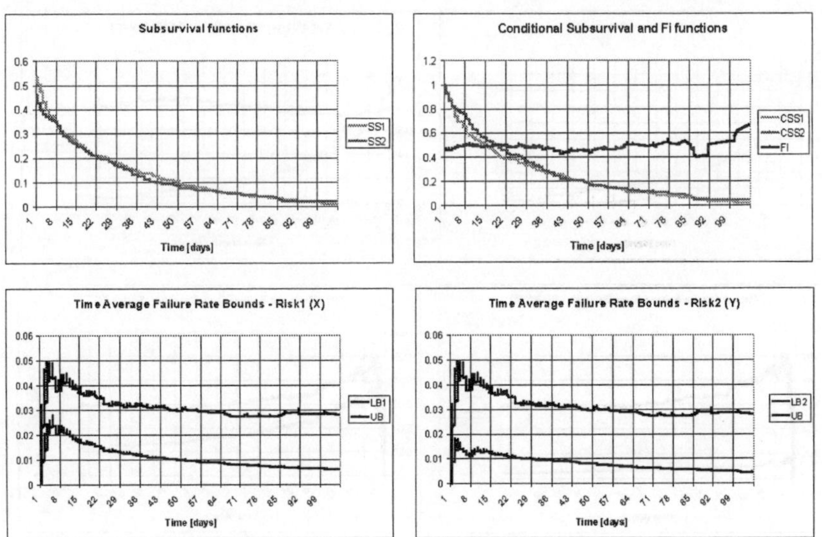

Figure 2. Analysis of "failure effect" field: Risk1 (X) - "leakage", "no start", "unwanted start", "vibration"; Risk2 (Y) - "warming", "overhaul", "little gas stream", "great gas stream", "others"

Figure 2 shows a slightly increasing probability of censoring after time t, which is not consistent with previously used models. The bottom graphs indicate that a constant time average failure rate is not reasonable for the distribution of the first competing risk class. Hence, the independent exponential model is not appropriate in this case. In fact, choosing the significance level $\alpha = 0.05$, gives the critical value of the KS statistic $d_\alpha = 1.2230$. The empirical KS statistic calculated from data is 1.24783, and hence the null hypothesis is rejected at the level $\alpha = 0.05$. The graphs seem to indicate that the mixture of exponentials model may be appropriate in this case.

3.2. *Action Taken*

The maintenance engineer is mainly interested in this field. His goal is to avoid the most expensive maintenance operation, immediate replacement, which corresponds to X in the model. He will prefer every other maintenance operation, together corresponding to Y, to this one. 85 replacements and 274 other maintenance actions are found. The empirical conditional subsurvival functions are crossing once and the estimated probability of censoring after time t has an inflexion point (Figure 3).

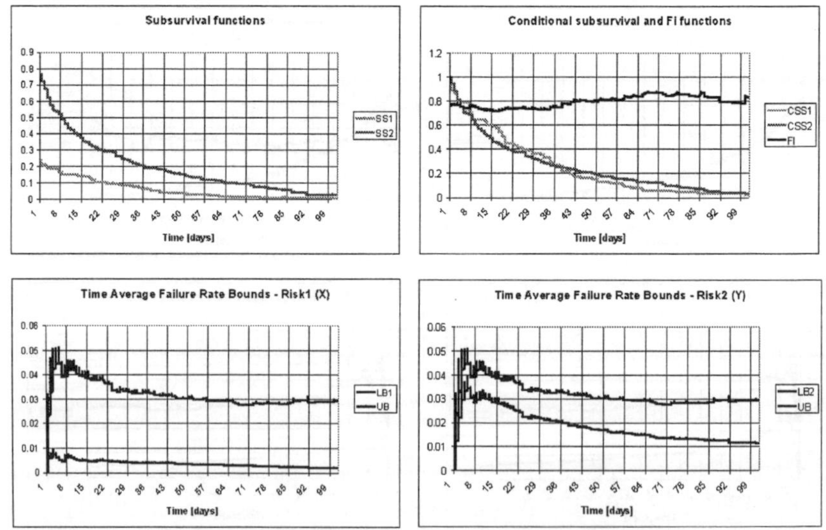

Figure 3. Analysis of "action taken" field: Risk1 (X) - "immediate replacement"; Risk2 (Y) - "immediate reparation", "adjustment", "planned overhaul", "modification", "others".

Thus, the graphical interpretation of data might suggest that the independent exponential model does not hold. However, the Kolmogorov-Smirnov test does not reject the hypothesis that the empirical conditional subsurvival functions are coming from the same population. Thus, the independent exponential model might be applied. Given the uncertainty involved in the graphical visualization and the fact that no model is available in the literature for this case, we regard these risks as independent.

3.3. *Degree of Failure*

The risk engineer and the maintenance engineer are both interested in this field. Obviously, they want to prevent a "critical" failure. From the risk point of view this is the event with major consequences on the state of the system. On the other hand a critical failure is usually associated with a corrective maintenance and a non-critical with a preventive maintenance. To keep the maintenance costs lower a corrective maintenance should be avoided. 92 critical failures (X) and 267 non-critical failures (Y) are recorded in the database. The empirical conditional subsurvival functions are more or less equal and the estimated probability of censoring after time t is roughly constant (Figure 4).

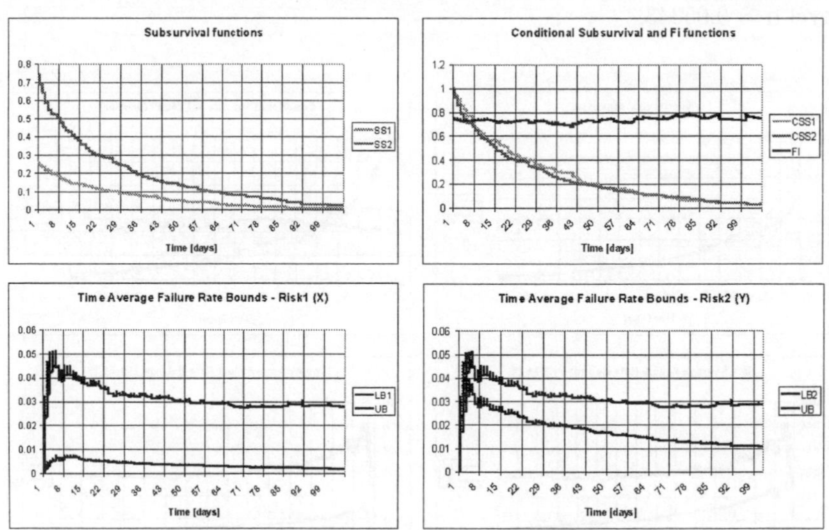

Figure 4. Analysis of "degree of failure" field: Risk1 (X) - "critical failure"; Risk2 (Y) - "non-critical failure"

The exponential independence model seems to be appropriate to fit the data, but the estimated Peterson bounds seem to be inconsistent with an exponential distribution for the critical failure. The KS-test does not reject the hypothesis that the empirical conditional subsurvival functions are equal. Recalling that the conditional independence model provides equal conditional subsurvival functions, this model seems most compliant with the graphical and statistical analyses.

3.4. *System*

A third type of engineer is interested in this field: the designer. He is concerned with avoiding the failure of an expensive component. Since the electrical components are more expensive than mechanical components, their failure should be considered as the X, while the mechanical components correspond to Y. 208 failures of electrical components and 151 failures of mechanical components are detected. The empirical conditional subsurvival of the censoring variable dominates the one of the unwanted event and the probability of censoring after time t is increasing (Figure 5). This is the opposite case of the random signs model, which models the behaviour of a very competent maintenance team. For the X-related risk, the exponential distribution is not compatible with the Peterson bounds in Figure 5. The KS-test also rejects the exponential independent model at the significance level $\alpha = 0.05343$.

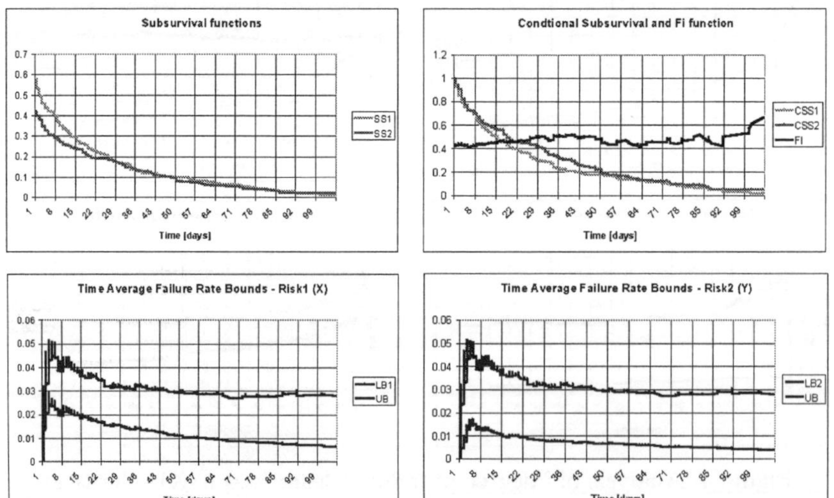

Figure 5. Analysis of "system" field: Risk1 (X) - electrical components; Risk2 (Y) - mechanical components

Based on this we used the mixture of exponentials model of Section 2.4. The following estimates were obtained by the moment method: failure rate of the censoring variable, $\lambda_y = 0.0015$; failure rates of the mixture, $\lambda_1 = 0.03757$, $\lambda_2 = 0.00936$, mixing coefficient, $p = 0.59$. Estimated parametric (continuous line) and nonparametric (dotted line) curves are presented in Figure 6. The figure indicates very good model fit for these data.

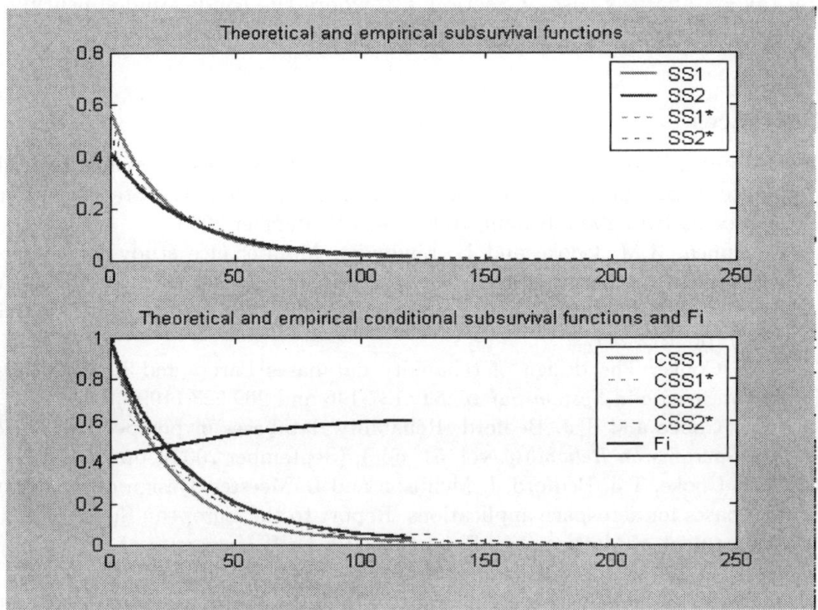

Figure 6. Model validation

The good fit of the mixture of exponentials model can be explained by heterogeneity. The compressor unit is vast and the component histories constituting the data come from pumps, valves, electro-motors, etc., which are expected to have very different failure characteristics. The Peterson bounds suggest a mixture of a few components with very high average failure rates and other components with low average failure rates. A similar behavior of data has been found in the pressure relief data from Swedish nuclear facilities (Cooke et al.[5]).

4. Conclusions

A new simple and attractive model has been developed for the case when the conditional subsurvival functions of the censoring variable dominates the conditional subsurvival functions of the other risks. The model agrees with empirical findings for the "Failure Mode" and "System" fields, where we observe a decreasing time average failure rate for X and a roughly constant one for Y. Such simple data analyses should be performed routinely by engineers. A suitable model may then be fit after this primary analysis of the data.

For future work, it remains to find a model appropriate for the "action taken" analysis case of Section 3.2, where the conditional subsurvival functions are crossing (see Figure 3).

References

1. C. Bunea, R.M. Cooke and B. Lindqvist, Analysis tools for competing risk failure data, European Network for Business and Industrial Statistics Conference, ENBIS 2002, Rimini, Italy, 23-24 September 2002.
2. C. Bunea, R.M. Cooke and B. Lindqvist, Maintenance study for components under competing risks, in *Safety and Reliability, 1st volume*, (European Safety and Reliability Conference, ESREL 2002, Lyon, France, 18-21 March 2002), pp. 212-217.
3. R.M. Cooke, The design of reliability databases Part I and II, *Reliability Engineering and System Safety*, **51**, 137-146 and 209-223 (1996).
4. R.M. Cooke and T.J. Bedford, Reliability databases in perspective, *IEEE Transactions on Reliability*, vol. 51, no 3, (September 2002), pp. 294-310.
5. R.M. Cooke, T.J. Bedford, I. Meilijson and L. Meester, Design of reliability databases for aerospace applications, Report to the European Space agency, Department of Mathematics Report 93-110, Delft University of Technology, 1993.
6. J. Dorrepaal, Analysis tools for reliability databases, Report TU Delft and RISØ, 1996.
7. J. Dorrepaal, P. Hokstad, R.M. Cooke and J.L. Paulsen, The effect of preventive maintenance on component reliability, in *Advances in Safety and Reliability*, Ed. C.G. Soares (Proceedings of the ESREL '97 conference, 1997), pp. 1775-1781.
8. S.E. Erlingsen, Using reliability data for optimizing maintenance, unpublished master's thesis, Norwegian Institute of Technology, 1989.
9. P. Hokstad and R. Jensen, Predicting the failure rate for components that go through a degradation state, *Reliability Engineering and System Safety*, **53**, 389-396 (1998).
10. A.V. Peterson, Bounds for a joint distribution function with fixed subdistribution functions: Application to competing risks, *Proceedings of the National Academy of Sciences, USA*, **73**, 11-13 (1976).
11. A. Tsiatis, A nonidentifiability aspect of the problem of competing risks, *Proceedings of the National Academy of Sciences, USA*, **72**, 20-22 (1975).

24

STOCHASTIC MODELLING OF THE JOINT EFFECT OF PREVENTIVE AND CORRECTIVE MAINTENANCE ON REPAIRABLE SYSTEMS RELIABILITY

Laurent Doyen and Olivier Gaudoin

Institut National Polytechnique de Grenoble
Laboratoire LMC
BP 53 - 38 041 Grenoble Cedex 9 France
Laurent.Doyen@imag.fr, Olivier.Gaudoin@imag.fr

The aim of this chapter is to propose a general model of the joint effect of corrective and preventive maintenance on repairable systems. The model is defined by the failure intensity of a stochastic point process. It takes into account the possibility of dependent maintenance times with different effects. The modelling is based on a competing risks approach. The likelihood function is derived, so parameter estimation and assessment of maintenance efficiency are possible.

1. Introduction

All important industrial systems are subjected to corrective and preventive maintenance actions which are supposed to extend their functioning life. Corrective maintenance (CM), also called repair, is carried out after a failure and intends to put the system into a state in which it can perform its function again. Preventive maintenance (PM) is carried out when the system is operating and intends to slow down the wear process and reduce the frequency of occurrence of system failures. The assessment of the efficiency of these maintenance actions is of great practical interest, but it has seldomly been studied.

Several models of maintenance effect have been proposed (see for example a review in Pham and Wang[14]). Most of them, known as repair models, consider only the effect of CM. Others consider only the effect of PM. In fact, the same kinds of assumptions are done in both cases, so the same models can be considered for the effect of both kinds of maintenances. Usually, these models are defined by the conditional distributions of succes-

371

sive interfailure times. For comparison purpose and in view of a statistical analysis, we think, as in Lindqvist[12], that it is more convenient to define a model by the failure intensity of a stochastic point process. In this case, maintenance effect is characterized by a change on the failure intensity. Section 2 presents this approach for the case where only CM are considered. Some usual repair models are mentioned as particular cases of the general modelling. Section 3 presents some models taking into account both CM and deterministic planned PM. In section 4, we propose a general modelling mixing CM and random PM, based on a competing risk approach. Within this modelling, the distribution of the time to next failure and likelihood function are derived.

2. Corrective Maintenance Models

2.1. *Stochastic Modelling of the Failure Process*

Let $\{T_i\}_{i\geq 1}$ be the successive failure times of a repairable system not subjected to PM actions, starting from $T_0 = 0$. Let $X_i = T_i - T_{i-1}, i \geq 1$, be the successive interfailure times and N_t be the number of failures observed up to time t. The repair times are assumed to be negligible or not taken into account, so failure times are equal to CM times. When the occurence of a failure at any time is assumed to depend only on the own past of the failure process, this process is a self excited point process, and it is characterized by its failure intensity, defined as:

$$\forall t \geq 0, \quad \lambda_t^N = \lim_{dt \to 0} \frac{1}{dt} P(N_{t+dt} - N_t = 1 | \mathcal{F}_t(N))$$

where $\mathcal{F}_t(N) = \sigma(\{N_s\}_{0 \leq s \leq t})$ is the internal history of the failure process at time t. But external variables can also influence failures. If $E = \{E_s\}_{s \geq 0}$ denotes the process of external variables, the failure intensity relative to E is defined as:

$$\forall t \geq 0, \quad \lambda_t^N(E) = \lim_{dt \to 0} \frac{1}{dt} P(N_{t+dt} - N_t = 1 | \mathcal{F}_t(N, E))$$

where $\mathcal{F}_t(N, E) = \sigma(\{N_s, E_s\}_{0 \leq s \leq t})$ is here the joint history of the failure process and the external process. In the case of a self excited point process, the failure intensity completely characterizes the failure process. Otherwise, informations about the joint distribution of processes N and E are needed.

2.2. *Basic Models*

The basic assumptions on maintenance efficiency are known as minimal repair or As Bad As Old (ABAO) and perfect repair or As Good As New

(AGAN). In the ABAO case, each maintenance restores the system to the state it was before failure. The corresponding random processes are the Non Homogeneous Poisson Processes (NHPP). These processes are such that the failure intensity is only a function of time:

$$\lambda_t^N = \lambda(t)$$

In the AGAN case, each maintenance perfectly repairs the system and restores it as if it were new. The corresponding random processes are the Renewal Processes (RP). These processes are such that the failure intensity is defined as:

$$\lambda_t^N = \lambda(t - T_{N_t})$$

Obviously, reality is between these two extreme cases: standard maintenance reduces failure intensity but does not restore the system to as good as new. This is sometimes known as better-than-minimal repair.

2.3. *The Dorado-Hollander-Sethuraman Model*

Many basic repair models can be considered as particular cases of the recent model proposed by Dorado, Hollander and Sethuraman (DHS).[7] This model assumes that there exists two sequences $\{A_i\}_{i \geq 1}$ and $\{\Theta_i\}_{i \geq 1}$ called respectively the effective ages and life supplements, such that $A_1 = 0, \Theta_1 = 1$ and the conditional distributions of the interfailure times are given by:

$$P(X_i > x | A_i, \Theta_i, X_1, ..., X_{i-1}) = \frac{\overline{F}(\Theta_i \, x + A_i)}{\overline{F}(A_i)}$$

where $\overline{F}$ is the survival function of the first failure time X_1. This means that after the ith CM, the system behaves as a new one having survived until A_i, and its wear-out speed is weighted by a factor Θ_i. Denoting by λ the failure rate of X_1, the failure intensity of the DHS model is :

$$\lambda_t^N(A, \Theta) = \Theta_{N_t+1} \lambda(A_{N_t+1} + \Theta_{N_t+1}[t - T_{N_t}])$$

By giving particular values to A_i and Θ_i for $i \geq 2$, several CM models appear to be members of the DHS family.

When $\Theta_i = 1$ for all i, we obtain the virtual age models proposed by Kijima[10]. NHPP and RP belong to this family of models for respectively $A_i = T_{i-1}$ and $A_i = 0$.

Kijima defined two classes of models, where maintenance effect is expressed by iid random variables D_j over $[0, 1]$. Type I models are such that

$A_i = \sum_{j=1}^{i-1} D_j X_j$. When the D_j's are deterministic and constant equal to $1 - \rho$, the failure intensity is simply :

$$\lambda_t^N = \lambda(t - \rho T_{N_t})$$

The model is the Kijima Morimura and Suzuki[11] model, which happens to be the same as the Malik[13] model known as the proportional age reduction (PAR) model. The parameter ρ is known as the improvement factor. Type II models are such that $A_i = \sum_{j=1}^{i-1} \prod_{k=j}^{i-1} D_k X_j$. When the distribution of the D_j's is Bernoulli, the model is the Brown and Proschan[4] model. When the D_j's are deterministic and constant equal to $1 - \rho$, the model happens to be the same as the Brown, Mahoney and Sivazlian[3] model.

Kijima models with $D_j = 1 - \rho$ can be understood as particular cases of the Arithmetic Reduction of Age models with memory m (ARA_m), proposed by Doyen and Gaudoin[8], defined by the intensity:

$$\lambda_t^N = \lambda(t - \rho \sum_{j=0}^{\min(m-1,N_t-1)} (1 - \rho)^j T_{N_t-j})$$

There are also some DHS models for which $\Theta_i \neq 1$. For example, the model proposed by Wang and Pham[15] can be understood as a DHS model with $\Theta_i = 1/\alpha^{i-1}$ and $A_i = 0$. The failure process is a quasi-renewal process: the interfailure times are independent but not identically distributed.

2.4. *Other Non-DHS Models*

A few other CM models are not included in the DHS family. Other quasi-renewal processes can be defined by the intensity :

$$\lambda_t^N = h(N_t, t - T_{N_t})$$

which is not necessarily of the DHS form.

An alternative to the Arithmetic Reduction of Age models is the Arithmetic Reduction of Intensity models (ARI_m), proposed in Doyen and Gaudoin[8], defined by the intensity :

$$\lambda_t^N = \lambda(t) - \rho \sum_{j=0}^{\min(m-1,N_t-1)} (1 - \rho)^j \lambda(T_{N_t-j})$$

The Chan and Shaw[6] model appears to be a ARI_∞ model.

Finally, the Trend-Renewal process (Lindqvist[12]) is defined by a failure intensity of the form :

$$\lambda_t^N = \mu(t)\, z \left(\int_{T_{N_t}}^t \mu(s)ds \right)$$

3. Mixing Corrective and Deterministic Preventive Maintenance

In practice, for important systems of high security, failures are very rare, so there are much more PM than CM actions. Then it is necessary to include preventive maintenance into the above modelling.

The basic assumptions are that PM are planned at deterministic times $\{\tau_i\}_{i\geq 1}$, and that their efficiency obeys to one of the models presented in section 2. In addition, CM are supposed to be ABAO, then the failure process is simply an NHPP.

For example, the failure intensity of a model with ARA_m assumption on PM is:

$$\lambda_t^N = \lambda(t - \rho \sum_{i=0}^{\nu_t} (1 - \rho)^i \tau_{\nu_t - i})$$

where ν_t is the number of observed PM at time t.

Another interesting particular case of these assumptions is the model proposed by Canfield[5], assuming periodic PM of periodicity Δt and with translated ARA efficiency characterized by $\tau < \Delta t$. The model is defined by the failure intensity :

$$\lambda_t^N = \lambda(t - \tau \lfloor \frac{t}{\Delta t} \rfloor) + \sum_{i=1}^{\lfloor \frac{t}{\Delta t} \rfloor} \lambda(i(\Delta t - \tau) + \tau) - \lambda(i(\Delta t - \tau))$$

This model has a continuous failure intensity.

Other models mixing CM and deterministic PM actions have been proposed (Pham and Wang[14]).

4. Mixing CM and Random PM

In practice, PM times are not necessarily planned: for example they can be determined according to the state of the failure process or to wear-out controls. Then, models mixing CM and random PM have been proposed. Even if PM times are not deterministic, they are generally assumed to be independent of CM times. In fact, it is desirable to obtain a certain kind of dependence between PM and CM times : PM will be optimal if PM actions are carried out "just before" failures. In this section, we propose a modelling mixing CM and random PM, where PM and CM times are not necessarily independent and PM and CM effects are not necessarily the same.

4.1. *A Virtual Age Model for PM and CM*

Let $\{T_i\}_{i \geq 1}$ and $\{\tau_i\}_{i \geq 1}$ be the CM and PM times. Let $\{N_t\}_{t \geq 0}$ and $\{M_t\}_{t \geq 0}$ be the cumulative number of CM and PM. Let $\{X_i\}_{i \geq 1}$ and $\{\chi_i\}_{i \geq 1}$ be the inter-CM and inter-PM times.

We propose a generalized virtual age model including PM and CM effects, defined by the failure intensity:

$$\lambda_t^N(A, M) = \lambda(A_{M_t, N_t} + t - max(\tau_{M_t}, T_{N_t}))$$

In this model, the virtual age at time t is equal to the time elapsed since the last maintenance: $t - max(\tau_{M_t}, T_{N_t})$, plus the effective age: A_{M_t, N_t}. The effective age represents the virtual age just after maintenance and is a function of the previous observed PM and CM times.

Most of basic assumptions on the effect of PM and CM can be expressed by a particular form of effective age. For example, ABAO PM and CM are obtained when the effective age is equal to the last maintenance time, $A_{M_t, N_t} = max(\tau_{M_t}, T_{N_t})$ and so the failure intensity is simply that of a NHPP:

$$\lambda_t^N(M) = \lambda(t)$$

With an effective age equal to zero, PM and CM are AGAN but the failure process is not a renewal process:

$$\lambda_t^N(M) = \lambda(t - max(\tau_{M_t}, T_{N_t}))$$

We can also have different effects for PM and CM. For example, to have ABAO PM (resp. CM) and AGAN CM (resp. PM) the effective age have to be equal to the time elapsed since the last PM (resp. CM), and the failure intensity is then equal to:

$$\lambda_t^N(M) = \lambda(t - T_{N_t}) \quad \left(\text{resp. } \lambda_t^N(M) = \lambda(t - \tau_{N_t})\right)$$

By analogy with repair models presented in section 2, we can build models mixing CM and PM effect on the failure process. For example, we can choose a proportional age reduction model for PM and CM with the same improvement factor ρ:

$$\lambda_t^N(M) = \lambda(t - \rho \, max(\tau_{M_t}, T_{N_t}))$$

We can also have different improvement factors ρ_P for PM and ρ_C for CM. The failure intensity is then equal to:

$$\lambda_t^N(M) = \lambda(t - \rho_P \sum_{j=0}^{N_t-1} (\tau_{M_{T_{j+1}}} - T_j) - \rho_C \sum_{j=1}^{N_t} (T_j - \tau_{M_{T_j}}) - \delta_{M_t,N_t} \rho_P (\tau_{M_t} - T_{N_t}))$$

where $\delta_{M_t,N_t} = \mathbb{1}_{\{\tau_{M_t} > T_{N_t}\}}$ indicates if the last maintenance is a PM or a CM.

In this approach, in which only the failure intensity relative to the CM process is given, the PM-CM process is not completely defined: more information on PM times and dependence between PM and CM times is needed.

4.2. *Stochastic Modelling of the PM-CM Process*

4.2.1. *A Competing Risk Approach*

A convenient way for modelling the PM-CM process is the competing risk approach, developed for example in Bedford and Mesina[2] and more recently in Bedford and Cooke[1]. When the system is restored after maintenance, the time to next maintenance is either a CM time or a PM time. Let $\{W_i\}_{i \geq 1}$ denote the times between successive maintenances. Let Y_i be the time to next PM in case that no CM occur before and Z_i be the time to next CM in case that no PM occur before. Then, the observations are :

$$W_i = \min(Y_i, Z_i) \qquad \text{and} \qquad U_i = \mathbb{1}_{\{Y_i \leq Z_i\}}$$

where U_i indicates if the ith maintenance is a CM or a PM.

In the classical competing risk problem, each maintenance is supposed to be AGAN and the marginal distributions of Y and Z are studied. The main result is that observations do not allow to estimate these distributions without making additional non-testable assumptions. The approach developed in this chapter is very different: we are not interested in the distribution of Y and Z, but in the distribution of the observed CM and PM processes. Then, we can make assumptions on maintenance effect much larger than AGAN.

For complete proofs of the lemmas and theorems developed in this section, see Doyen and Gaudoin [9].

4.2.2. *Notations*

Let us recall or introduce all the notations needed in the following.

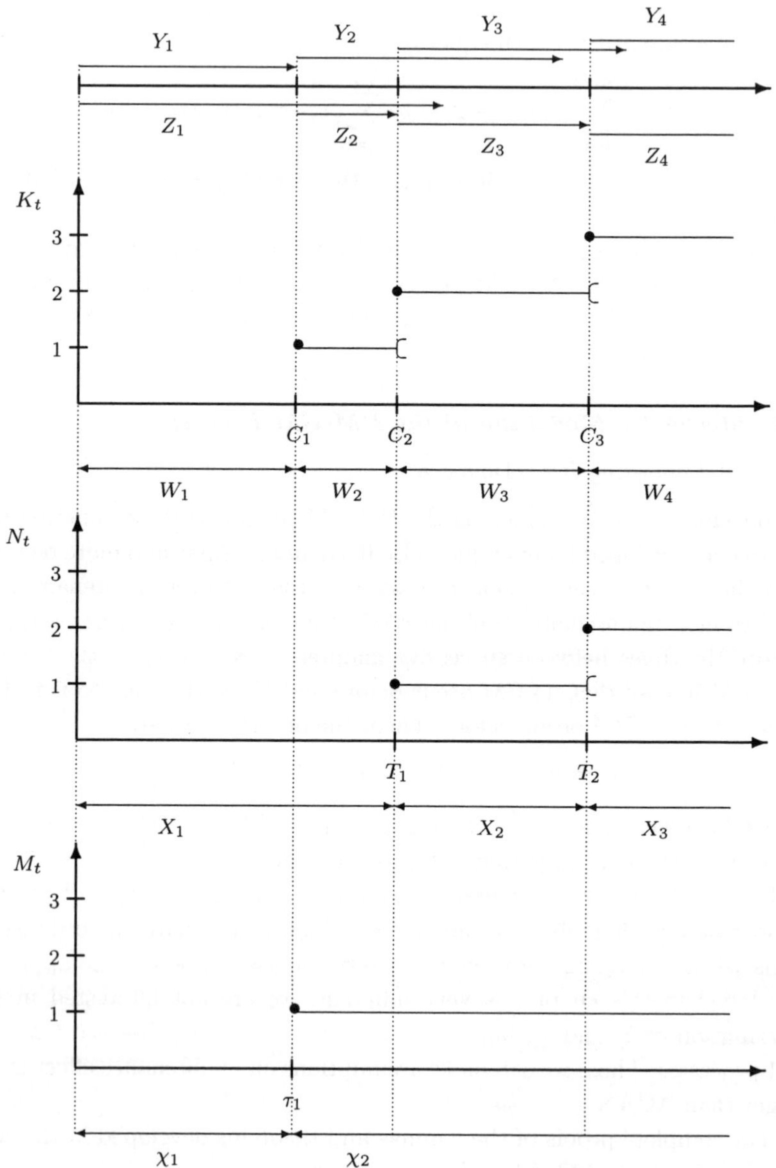

Fig. 1. Example of a PM-CM process defined by its Y and Z processes with the corresponding PM and CM times

For the global maintenance process (PM and CM):

- $\{U_i\}_{i\geq 1}$ the indicators of types of maintenance, $U_i = \mathbb{1}_{\{Y_i \leq Z_i\}}$
- $\{W_i\}_{i\geq 1}$ the times between maintenances
- $\{C_i\}_{i\geq 1}$ the maintenance times, $C_i = \sum_{j=1}^{i} W_j$
- $\{K_t\}_{t\geq 0}$ the counting maintenance process,
  $K_t = \max\{k \in \mathbb{N} | C_k \leq t\}$

For the failure or CM process:

- $\{N_t\}_{t\geq 0}$ the counting CM process, $N_t = \sum_{k=1}^{K_t}(1 - U_k)$
- $\{T_i\}_{i\geq 1}$ the CM times, $T_i = \min\{t \in \mathbb{R}^+ | N_t = i\}$
- $\{X_i\}_{i\geq 1}$ the times between CM, $X_i = T_{i-1} - T_i$

Finally for the PM process:

- $\{M_t\}_{t\geq 0}$ the counting PM process, $M_t = \sum_{k=1}^{K_t} U_k$
- $\{\tau_i\}_{i\geq 1}$ the PM times, $\tau_i = \min\{t \in \mathbb{R}^+ | M_t = i\}$
- $\{\chi_i\}_{i\geq 1}$ the times between PM, $\chi_i = \tau_{i-1} - \tau_i$

Figure 1 presents an example of a PM-CM process with its corresponding notations.

Remark: The indicator of maintenance type U_i assimilates the case where $Y_i = Z_i$ to a PM time, that is to say, if there is a failure at a preventive maintenance time, then the failure is not detected and only PM is observed.

4.2.3. *Characterization of the PM-CM Process*

In order to characterize the PM-CM process, we define three intensities, relative to the whole maintenance history.

Definition 1:

- The CM intensity:

$$\lambda_t^N(K,U) = \lim_{dt \to 0} \frac{1}{dt} P(N_{t+dt} - N_t = 1 | K_t, W_1, U_1..., W_{K_t}, U_{K_t})$$

- The PM intensity:

$$\lambda_t^M(K,U) = \lim_{dt \to 0} \frac{1}{dt} P(M_{t+dt} - M_t = 1 | K_t, W_1, U_1..., W_{K_t}, U_{K_t})$$

- The (global) maintenance intensity:

$$\lambda_t^K(U) = \lim_{dt \to 0} \frac{1}{dt} P(K_{t+dt} - K_t = 1 | K_t, W_1, U_1..., W_{K_t}, U_{K_t})$$

The maintenance intensity can be easily deduced from PM and CM intensities thanks to the following property.

Lemma 2:

$$\lambda_t^K(U) = \lambda_t^N(K,U) + \lambda_t^M(K,U)$$

4.2.4. *A Generalized Sub-Survival Function*

The basis of the classical competing risk approach is the study of the sub-survival functions defined as:

$$S_Y^*(t) = P(Y > t, Y < Z) \text{ and } S_Z^*(t) = P(Z > t, Z < Y)$$

We generalize this notion to our context.

Definition 3: For a PM-CM process we define two sub-survival functions:

$$S_{Z_{k+1}}^*(t; w_1, ..., u_k) = P(Z_{k+1} > t, Z_{k+1} < Y_{k+1}|W_1 = w_1, ..., U_k = u_k)$$
$$= P(W_{k+1} > t, U_{k+1} = 0|W_1 = w_1, ..., U_k = u_k)$$
$$S_{Y_{k+1}}^*(t; w_1, ..., u_k) = P(Y_{k+1} > t, Y_{k+1} \leq Z_{k+1}|W_1 = w_1, ..., U_k = u_k)$$
$$= P(W_{k+1} > t, U_{k+1} = 1|W_1 = w_1, ..., U_k = u_k)$$

These functions represent the probability that the next maintenance will not take place within the next t unit of times and that it will be a corrective (resp. preventive) maintenance, when the whole past of the maintenance process is known. They completely define the joint distribution of the observations, that is to say the processes $\{W_i\}_{i \geq 1}$ and $\{U_i\}_{i \geq 1}$. These two sub-survival functions can also be expressed with the two PM-CM intensities.

Lemma 4:

$$S_{Z_{k+1}}^*(t; w_1, ..., u_k) = \int_t^{+\infty} \lambda_{c_k+v}^N(k, w_1, ..., u_k) e^{-\int_0^v \lambda_{c_k+s}^K(k,w_1,...,u_k)ds} dv$$

$$S_{Y_{k+1}}^*(t; w_1, ..., u_k) = \int_t^{+\infty} \lambda_{c_k+v}^M(k, w_1, ..., u_k) e^{-\int_0^v \lambda_{c_k+s}^K(k,w_1,...,u_k)ds} dv$$

$$\text{where } c_k = \sum_{j=1}^k w_j$$

Therefore, the two PM-CM intensities completely define the distribution of the observations.

For the demonstration of the model properties, we have to define the notion of sub-density function.

Definition 5: The sub-densities functions of a PM-CM process are:

$$f^*_{k+1,0}(t; w_1, ..., u_k) = -\frac{d}{dt} S^*_{Z_{k+1}}(t; w_1, ..., u_k)$$

$$f^*_{k+1,1}(t; w_1, ..., u_k) = -\frac{d}{dt} S^*_{Y_{k+1}}(t; w_1, ..., u_k)$$

We easily deduce from the previous lemma that the sub-densities can be expressed as functions of the PM-CM intensities.

Corollary 6:

$$f^*_{k+1,0}(t; ...) = \lambda^N_{c_k+t}(k, ...)e^{-\int_0^t \lambda^K_{c_k+s}(k,...)ds}$$

$$f^*_{k+1,1}(t; ...) = \lambda^M_{c_k+t}(k, ...)e^{-\int_0^t \lambda^K_{c_k+s}(k,...)ds}$$

4.2.5. *Distribution of the Time to the Next Failure*

PM and CM intensities can be used to determine the distribution of the time to the next failure, when the whole maintenance history at the time of the last failure is given. This distribution is defined by the probability:

$$S_{X,k}(x; w_1, ..., 0) = P(X_{N_{c_k}+1} > x | W_1 = w_1, ..., U_k = 0)$$

Lemma 7:

$$S_{X,k}(x; w_1, ..., 0) =$$

$$\sum_{j=0}^{+\infty} \int \cdots \int_{\sum_{i=1}^j y_i < x} S_{W_{k+j+1}}\left(x - \sum_{i=1}^j y_i, w_1, ..., 0, y_{k+1}, 1, ..., y_{k+j}, 1\right)$$

$$\prod_{i=1}^j f^*_{k+i,1}(y_{k+1}, w_1, ..., 0, y_{k+1}, 1, ..., y_{k+i-1}, 1) \; dy_{k+j}...dy_{k+1}$$

where $S_{W_{k+1}}(x; w_1, ..., u_k) = S^*_{Y_{k+1}}(x; w_1, ..., u_k) + S^*_{Z_{k+1}}(x; w_1, ..., u_k)$

4.2.6. *Likelihood Function*

In a parametric approach, we can estimate the parameters of the PM and CM intensities with the likelihood function. This function has a simple expression depending on PM and CM intensities.

Theorem 8: The likelihood function corresponding to the observation of a PM-CM process over $[0, t]$ is:

$$\mathcal{L}(\theta; k, w_1, ..., u_k) =$$

$$\left[\prod_{i=1}^{k} \lambda_{c_i}^{u_i}(i - 1, w_1, ..., u_{i-1}) \right] e^{\displaystyle -\sum_{j=1}^{k+1} \int_{c_{j-1}}^{c_j} \lambda_s^K (j - 1, w_1, ..., u_{j-1})ds}$$

$$\text{where } c_0 = 0, \ c_{k+1} = t, \ \lambda_t^u(...) = \begin{cases} \lambda_t^N(...) & \text{if } u = 0 \\ \lambda_t^M(...) & \text{if } u = 1 \end{cases}$$

4.2.7. *Example 1:PM AGAN and CM AGAN*

As a particular case of our model, AGAN PM and CM represent the classical competing risk approach. In this approach, used by Bedford and Mesina[2], the model is completely defined by the joint distribution of Y and Z:

$$S(y, z) = P(Y_1 > y, Z_1 > z)$$

But it is well known that the observations do not allow to estimate this function.

Theorem 9: For the classical competing risk model, we have:

$$\lambda_t^N(K, U) = \frac{-s^Z(t - C_{K_t})}{S(t - C_{K_t})}; \qquad \lambda_t^M(K, U) = \frac{-s^Y(t - C_{K_t})}{S(t - C_{K_t})}$$

where $S(t) = S(t, t); s^Y(t) = \left[\dfrac{\partial}{\partial y} S(y, z) \right](t, t)$ and

$$s^Z(t) = \left[\frac{\partial}{\partial z} S(y, z) \right](t, t)$$

Then, the sub-densities, the distribution of time to the next failure and the likelihood function can be derived.

Corollary 10: *For the classical competing risk model:*

- $f_0^*(z) = -s^Z(z); \qquad f_1^*(z) = -s^Y(z)$

- $S_X(x) = \left[S \otimes \sum_{k=0}^{+\infty} (-1)^k \, s^{Y \otimes^k} \right](x)$
- *in the case of time truncated data*

$$\mathcal{L}(\theta; k, w_1, u_1, ..., w_k, u_k) = S(t - c_k) \prod_{j=1}^{k} (s^Y(w_j))^{u_j} \, (s^Z(w_j))^{1-u_j}$$

Both PM and CM intensities depend on the joint distribution function of Y and Z only "around the identity line". This illustrates the well known property saying that different competing risk models can lead to the same distribution of the observation.

4.2.8. *Example 2: ABAO PM and ABAO CM*

Any kind of assumptions on maintenance effect can be made. For example, a basic case is when PM and CM are ABAO.

Theorem 11: When PM and CM are ABAO, the PM-CM intensities are:

$$\lambda_t^N(K,U) = \frac{-s^Z(t)}{S(t)}; \qquad \lambda_t^M(K,U) = \frac{-s^Y(t)}{S(t)}$$

Then both observed PM and CM processes are NHPP that only depend on the value of the joint distribution function of Y and Z around the identity line. The following quantities are easily obtained.

Corollary 12:

-

$$f_0^*(z;c) = \frac{-s^Z(z+c)}{S(c)}; \qquad f_1^*(z;c) = \frac{-s^Y(z+c)}{S(c)}$$

-

$$S_X(x;c) = \frac{S(x+c)}{S(c)} \, exp\left[\int_{c_k}^{c_k+x} \frac{s^Y(u)}{S(u)} du \right]$$

- *in the case of time truncated data*

$$\mathcal{L}(\theta; k, w_1, u_1, ..., w_k, u_k) = S(t) \prod_{j=1}^{k} \frac{-(s^Y(c_j))^{u_j}(s^Z(c_j))^{1-u_j}}{S(c_{j-1})}$$

384   *L. Doyen and O. Gaudoin*

4.2.9. *Example 3: PAR PM and PAR CM*

Finally, the model is applied to the case where PM and CM effect are of the Proportional Age Reduction (PAR) type.

Theorem 13: When PM and CM are PAR, the PM-CM intensities are:

$$\lambda_t^N(K,U) = \frac{-s^Z(t - \rho\, C_{K_t})}{S(t - \rho\, C_{K_t})}; \qquad \lambda_t^M(K,U) = \frac{-s^Y(t - \rho\, C_{K_t})}{S(t - \rho\, C_{K_t})}$$

As particular cases, both previous models corresponds respectively to $\rho = 1$ and $\rho = 0$. All the following quantities can be derived.

Lemma 14:

- $$f_0^*(z;c) = \frac{-s^Z(z + (1 - \rho)c)}{S((1 - \rho)c)}; f_1^*(z;c) = \frac{-s^Y(z + (1 - \rho)c)}{S((1 - \rho)c)}$$

- $$S_X(x;c) = \frac{1}{S((1 - \rho)t)} S(c + x - \rho c) +$$
$$\sum_{j=1}^{+\infty} \int \cdots \int_{c \le y_1 \le \dots \le y_j \le c+x} S(c + x - \rho y_j) \prod_{i=1}^{j} \frac{f_1^*(y_i - \rho y_{i-1})}{S((1 - \rho)y_i)} dy_j ... dy_1$$

  where $y_0 = c$
- *for time truncated data*
$$\mathcal{L}(\theta; k, w_1, u_1, ..., w_k, u_k) =$$
$$S(t - \rho\, c_k) \prod_{j=1}^{k} \frac{-(s^Y(c_j - \rho\, c_{j-1}))^{u_j} (s^Z(c_j - \rho\, c_{j-1}))^{1 - u_j}}{S((1-\rho)\, c_{j-1})}$$

Therefore, maintenance efficiency can be estimated in that case.

When PM and CM efficiencies are assumed to be of the PAR type with different improvement factors ρ_P and ρ_C, similar results can be derived but the complexity increases notably.

5. Conclusion

This study proposes a new framework for the modelling of possibly linked random PM and CM. Thanks to this modelling, it is possible to establish links between different approaches such as imperfect repair modelling and competing risk problem, and then develop new results.

The likelihood function of the model has been derived. We now have to assess the quality of the corresponding estimators and develop goodness-of-fit tests for the models. The final goal is to apply this modelling to real data and estimate maintenance efficiency on complex repairable systems.

References

1. T. Bedford and R. Cooke, Reliability databases in perspective, *IEEE Transactions on Reliability*, **51**, 294-310 (2002).
2. T. Bedford and C. Mesina, The impact of modeling assumptions on maintenance optimization, in *Second International Conference on Mathematical Methods in Reliability*, (MMR, Bordeaux, 2000), pp. 171-174.
3. J. F. Brown, J. F. Mahoney and B. D. Sivazlian, Hysteresis repair in discounted replacement problems, *IIE Transactions* , **15**, 156-165 (1983).
4. M. Brown and F. Proschan, Imperfect repair, *Journal of Applied Probability* , **20**, 851-859 (1983).
5. R. V. Canfield, Cost optimisation of periodic preventive maintenance, *IEEE Transactions on Reliability* , **35**, 78-81 (1986).
6. J. K. Chan and L. Shaw, Modeling repairable systems with failure rates that depend on age and maintenance, *IEEE Transactions on Reliability* , **42**, 566-571 (1993).
7. C. Dorado, M. Hollander and J. Sethuraman, Nonparametric estimation for a general repair model, *The Annals of Statistics* , **25**, 1140-1160 (1997).
8. L. Doyen and O. Gaudoin, Modelling and assessment of maintenance efficiency for repairable systems, in *13th European Safety and Reliability International Conference*, (ESREL, Lyon, 2002), pp. 145-148.
9. L. Doyen and O. Gaudoin, Stochastic modelling of the joint effect of corrective and preventive maintenance, *LMC research report* (2003).
10. M. Kijima, Some results for repairable systems with general repair, *Journal of Applied Probability* , **26**, 89-102 (1989).
11. M. Kijima, H. Morimura and Y. Suzuki, Periodical replacement problem without assuming minimal repair, *European Journal of Operational Research* , **37**, 194-203 (1988).
12. B. Lindqvist, Statistical modelling and analysis of repairable systems, in *Statistical and Probabilistic Models in Reliability*, Eds. D. C. Ionescu and N. Limnios, (Birkhaüser, Boston, 1999) pp. 3-25.
13. M. A. K. Malik, Reliable preventive maintenance policy, *AIIE Transactions* , **11**, 221-228 (1979).
14. H. Pham and H. Wang, Imperfect maintenance, *European Journal of Operational Research* , **94**, 452-438 (1996).
15. H. Wang and H. Pham, A quasi-renewal process and its application in imperfect maintenance, *International Journal of System Science* , **27**, 1055-1062 (1996).

25

MODELING THE INFLUENCE OF MAINTENANCE ACTIONS

Waltraud Kahle

Department of Water Management
University of Applied Sciences Magdeburg-Stendal
P.O. Box 3680, D-39011 Magdeburg, Germany
E-mail: waltraud.kahle@wasserwirtschaft.hs-magdeburg.de

Charles E. Love

Faculty of Business Administration, Simon Fraser University
Burnaby, British Columbia, Canada V5A 1S6
E-mail: ernie.love@sfu.ca

An operating system (machine) is observed to undergo failures. On failure, one of three actions was taken: failures were minimally repaired, given a minor repair or given a major repair. Furthermore, periodically the machine was stopped for either minor maintenance action or major maintenance action. In addition to the kind of maintenance action, the length of duration for each repair action is known. Either on failure or maintenance stoppage, both types of repairs are assumed to impact the intensity following a virtual age process of the general form proposed by Kijima. There are several possibilities for assumptions of the impact of repair: it can be assumed that a minor or major repair impact the virtual age of the item to an unknown fixed part. It is also possible to assume that the impact of repair depends on the repair time. The issue in this research is to identify not only the virtual aging process associated with repairs but also the form of the failure intensity associated with the system. A series of models appropriate for such an operating/maintenance environment are developed and estimated in order to identify the most appropriate statistical structure. Field data from an industrial setting are used to fit the models.

1. Introduction

In this research, we are concerned with the statistical modeling of repairable systems. Our particular interest is the operation of electrical generating systems. As a repairable system, we assume the failure intensity at a point in time depends on the history of repairs. In the environment under investigation, it was observed that maintenance decisions were regularly carried out. We assume that such actions impacted the failure intensity. Specifically we assume that maintenance actions served to adjust the virtual age of the system in a Kijima Type manner [6] [7]. Kijima proposed that the state of the machine just after repair can be described by its so-called virtual age which is smaller (younger) than the real age. In his framework, the failure rate depends on the virtual age of the system. Furthermore a repair, following a failure, can be a minimal repair or a maintenance action.

Kijima proposed two repair effect models. In his first model he assumed that repairs served only to remove damage created in the last sojourn (a Kijima Type 1 virtual age process). In his second model he assumed that the repair action could remove all damage accumulated up to that point in time (a Kijima Type 2 virtual age process). That is, such repairs reset the virtual age of the unit to somewhere between that of a completely restored unit (good-as-new repair) and a minimally repaired unit, inclusively. The concept of minimal repair upon failure is well understood in the literature (Brown [1]) and the resultant non-homogeneous Poisson process has been applied extensively to describe the operation of repairable systems. Incomplete repair processes provide a more general framework for the modeling of failure-repair process (Guo, Love [5] [10]).

Consistent with the data being used to test this model, two levels of maintenance action were observed. Periodically the unit was stopped for minor repairs. Less frequently the unit was stopped for more extensive, major repairs. Unlike the more typical treatments we do not assume in this research that major repairs served to reset the failure intensity of the system (a good-as-new repair). Such an assumption appears to be an excessive constraint on the calibration of such systems. Hence we treat the impact of major repairs as also producing a reduction in virtual age lying between renewal and minimal. We will further assume that major repairs have more impact than minor repairs. Despite regular shutdowns for maintenance however, the system was observed to occasionally fail. Repair activity for such failures took one of three forms. With some failures, no repair work was reported and as such they are assumed to have been minimally repaired.

Following other failures however repair work was undertaken and was reported as either a minor or a major repair. The purpose of this chapter is to provide a model to estimate the impacts of these various activities.

2. Modeling the System

Consider the impact of repairs. The simplest idea is to assume that each type of repair has its own impact.

A system (machine) starts working with an initial prescribed failure rate $\lambda_1(t) = \lambda(t)$. Let t_1 denote the random time of the first sojourn. At this time t_1 the item will be repaired with the degree $\xi_{11}\xi_{21}$. The two degrees are used to distinguish between three possible types of repair: When the system is minimally repaired then both degrees are equal to one. With a minor repair we set $\xi_{21} = 1$, $\xi_{11} = \xi_1$, where ξ_1 is the (unknown) impact of a minor repair. With a major repair we have the composite impact $\xi_{21} = \xi_2$, $\xi_{11} = \xi_1$. Notice here we assume that major repairs are modeled as a composite effect $(\xi_1\xi_2)$ not simply ξ_2.

The virtual age of the system at the time t_1, following the repair, is $v_1 = \xi_{11}\xi_{21}t_1$, implying the age of the system is reduced by maintenance actions. The distribution of the time until the next sojourn then has failure intensity $\lambda_2(t) = \lambda(t - t_1 + v_1)$. Assume now that t_k is the time of the kth $(k \geq 1)$ sojourn and that $\xi_{1k}\xi_{2k}$ is the degree of repair at that time. We assume that $0 \leq \xi_{1k} \leq 1$, $0 \leq \xi_{2k} \leq 1$, $\xi_{1k} \in \{1, \xi_1\}$, $\xi_{2k} \in \{1, \xi_2\}$ and $\xi_{1k}\xi_{kn} \in \{1, \xi_1, \xi_1\xi_2\}$ for $k \geq 1$.

After repair the failure intensity during the $(k+1)^{th}$ sojourn is determined by

$$\lambda_{k+1}(t) = \lambda(t - t_k + v_k) \quad , t_k \leq t < t_{k+1}, k \geq 0,$$

where the virtual age v_k is for Kijima's Type II imperfect repair model

$$v_k = \xi_{1k}\xi_{2k}(v_{k-1} + (t_k - t_{k-1})),$$

that is, the repair resets the intensity of failure proportional to the virtual age.

In Kijima's Type I imperfect repair model, he suggested that upon failure, the repair undertaken could serve to reset the intensity only as far back as the virtual age at the start of the last failure. That is:

$$v_k = t_{k-1} + \xi_{1k}\xi_{2k}(t_k - t_{k-1}).$$

The process defined by $v(t, \xi_1, \xi_2) = t - t_k + v_k$, $t_k \leq t < t_{k+1}$, $k \geq 0$ is called the *virtual age process* (Last, Szekli [8]).

For estimation purpose it is necessary to differentiate between sojourn numbers associated with failure times and those associated with maintenance (censor) times. Let δ_k be an indicator with

$$\delta_k = \begin{cases} 1 & \text{if the sojourn number } k \text{ is related to a failure time,} \\ 0 & \text{if the sojourn number } k \text{ is related to a censored observation.} \end{cases}$$

In this chapter we assume that the baseline failure intensity of the system follows a Weibull distribution

$$\lambda(x) = \frac{\beta}{\alpha}\left(\frac{x}{\alpha}\right)^{\beta-1}, \ \beta > 0, \ \alpha > 0 \,.$$

Our purpose is to estimate 4 parameters; α, β, ξ_1 and ξ_2.

3. Parameter Estimation

The log likelihood function for observation of point processes is of the form (Liptser, Shiryayev [9], Gasmi, Kahle [3])

$$\ln L(t;\theta) = \sum_{k=1}^{N(t)} \ln\left(\lambda(t_k - t_{k-1} + v_{k-1})\right)^{\delta_k} + \int_{t_0}^{t} (1 - \lambda(v(x)))dx.$$

The first term contains all failures and the second term contains the information about working periods without failures. To simplify the notation let us denote by $t_{N(t)+1} = t$ the end of the observations. Define $\mathcal{I}$ as $\int_{t_0}^{t} \lambda(v(x))dx$. To calculate $\mathcal{I}$ we can use the fact that $v(x)$ is linear between two consecutive failure times. Thus we get the following log likelihood function:

$$\ln L(t;\theta) = \tilde{N}(t)(\ln\beta - \beta\ln\alpha) + (\beta-1)\sum_{k=1}^{N(t)} \delta_k \cdot \ln(t_k - t_{k-1} + v_{k-1})$$

$$-\frac{1}{\alpha^\beta}\sum_{k=1}^{N(t)+1}\left((t_k - t_{k-1} + v_{k-1})^\beta - (v_{k-1})^\beta\right) + t, \tag{1}$$

where $\tilde{N}(t)$ denotes the number of failures until t, $\tilde{N}(t) = \sum_{k=1}^{N(t)} \delta_k$. Using the standard maximum likelihood approach to maximize equation (1) some analytical simplifications of the structure are possible. First we can see that it is possible to explicitly determine the parameter estimate $\hat{\alpha}$:

$$\hat{\alpha} = \left(\frac{S_2}{\tilde{N}(t)}\right)^{1/\beta}, \quad S_2 = \sum_{k=1}^{N(t)+1}\left\{(t_k - t_{k-1} + v_{k-1})^\beta - (v_{k-1})^\beta\right\}$$

and the remaining log likelihood function is

$$\ln L(t, \theta) \sim \tilde{N}(t) \ln \beta - \tilde{N}(t) \ln S_2 + (\beta - 1) \sum_{k=1}^{N(t)} \delta_k \ln(t_k - t_{k-1} + v_{k-1}) \,,$$

which depends on the 3 unknown parameters β, ξ_1, and ξ_2. The parameters ξ_1 and ξ_2 implicitly define the virtual ages v_1, v_2, $\ldots$, $v_{N(t)}$. Now for any set of observations it is possible to maximize the log likelihood function numerically (specifically in our case, the use of the GAUSS package [4]).

4. Analysis of a Hydro-electric Turbine Model

In this section, we provide some numerical results using data from a selected hydro-electric turbine unit within the British Columbia Hydro-Electric Power Generation System. The data collected over the period January 1977 to December 1999 contains 496 sojourns with 160 failures. As well, two types of repairs are recorded by maintenance personnel, major repairs and minor repairs. The classification of repairs into these two categories is made at the time of the repair. Within this period, 50 major repairs and 96 minor repairs were conducted. All 50 major repairs occurred from a censor decision (i.e., a decision to shut the system down). Furthermore, of the 96 minor repairs, 1 of them was undertaken immediately following a failure. The remaining 95 are censored minor repairs. In addition to sojourn and censor times of these stoppages, the data also included the times to repair the system. These times ranged from a smallest of 1 minute to a largest of 66,624 minutes (or approximately 46 days).

The following models were utilized in an attempt to find the best model to fit this data.

(a) All repairs (major and minor) impact the virtual age using Kijima Type I imperfect form.
(b) All repairs requiring less than 50,000 minutes impact virtual age using Kijima Type I form. However, repairs above this cut-off point are assumed to renew the system.
(c) All repairs requiring less than 20,000 minutes impact virtual age using Kijima Type I form. However, repairs above this cut-off point are assumed to renew the system.
(d) All repairs are assumed to renew the system.
(e) All repairs (major and minor) impact the virtual age using Kijima Type II imperfect form.

(f) All repairs requiring less than 50,000 minutes impact virtual age using Kijima Type II form. However, repairs above this cut-off point are assumed to renew the system.

(g) All repairs requiring less than 20,000 minutes impact virtual age using Kijima Type II form. However, repairs above this cut-off point are assumed to renew the system.

The parameter estimates are given in Table 1.

Table 1.   Parameter Estimates from 7 models

Case	α	β	ξ_1	ξ_2
(a)	121,592.68	1.1365	0.17	0.000
(b)	69,544.28	0.99976	0.4991	0.4828
(c)	68,057.52	0.9982	0.4621	0.475
(d)	69,036.60	0.99685	1.0	0.5
(e)	105,800.22	1.2561	0.7873	0.5817
(f)	97,204.20	1.2139	0.7351	0.5321
(g)	68,977.81	0.998161	1.0	1.0

Several observations are noteworthy from Table 1. First, the Kijima Type I impact when all repairs are assumed to be imperfect [Case (a)] is to set minor repairs to have an almost good-as-new (or renewal) impact (0.17) and the major repairs are estimated to have a good-as-new (renewal) impact (0.0). Contrast this with Case (e) for Type II repair effects. Here, minor impacts (0.7873) and major impacts (0.5817) are much less severe. That is, they are closer to bad-as-old than good-as-new. Secondly, notice when a cut-off time of $20,000$ minutes is selected, all repairs (requiring less than $20,000$ minutes to complete) following a Kijima Type II form [Case g)] generate a minimal or bad-as-old impact on the failure intensity. This makes sense since it seems limiting the rise in the failure intensity is assured by virtue of those repairs taking longer than $20,000$ minutes to complete. Such repairs serve to reset the failure intensity to 0, (a renewal). For this same cut-off time of $20,000$ minutes, with Type I repairs [Case (c)], again the impacts move upwards towards minimal effects (0.4621, for minor and 0.475 for major repairs).

However, more importantly, notice the shape factor of the baseline Weibull. In all cases when a cut-off time is selected which is not ∞, a value for β is approximately 1. That is, the intensity is more or less exponential (i.e., random), once one takes into account those repairs taking a long time. If the intensity is not perfectly reset by repairs [Cases (a) or (e)], then the best estimate is that the system is experiencing rising intensity ($\beta > 1$). The maximum log likelihood is nearly the same for all models, it differs from -2,581.11 to -2,581.75.

5. Simulations

In the last section we have seen that our assumption about constant degrees of repair (that means we have a degree of repair ξ_1 for *each* minor repair and a degree of repair $\xi_1 \cdot \xi_2$ for *each* major repair) is too restrictive for modeling real failure-repair processes. In real situations, an engineer utilizes maintenance policies that depend on the state of the system. It is conceivable that a good maintenance policy makes the observed failure rate exponential. This result agrees with the results obtained in the last section. Of course, if the observed failure rate is exponential, then it implies there is no difference between major, minor and minimal repairs. In this case, the estimate of the degree parameters ξ_1 and ξ_2 leads to an ill conditioned problem.

We have tried to fit other models, too. As an example, Dorado, Hollander and Sethuraman [2] considered the following general repair model. There is a family of distribution functions

$$\bar{F}_a^\theta(x) = \bar{F}(\theta x + a)/\bar{F}(a)$$

and two sequences

$$\{A_j\}_{j \geq 1} \quad \text{and} \quad \{\Theta_j\}_{j \geq 1}.$$

The sequence $\{A_j\}_{j \geq 1}$ characterizes the effective age of the item and the sequence $\{\Theta_j\}_{j \geq 1}$ is called the sequence of life supplements. The failure intensity of this general repair process is given by

$$\lambda(t|\mathcal{F}_{t-}) = \theta(t) \cdot \lambda^o(a(t) + \theta(t)(t - T_{N(t-)}))$$

There are several possibilities to fit this model to our real data. We have considered 3 cases:

- Each minor repair gives a constant life supplement and each major repair leads to a younger effective age,
- We get two different life supplements for both, major and minor repair,
- Each minor repair leads to a younger effective age and each major repair gives an additional constant life supplement.

In all these cases the estimate of the shape parameter of the baseline Weibull, β, was approximately 1.

On the other hand, our model works very well if the impact of repair is the same for all minor and major maintenance actions. This fact is illustrated by the two following examples.

For this, samples of size $n = 500$ were simulated with true parameters $\alpha = 1$, $\beta = 2$, $\xi_1 = \xi_2 = 0.5$ and two different distances between maintenance actions. In the first simulated sample the distance between two maintenance actions is .8. We assume that a major repair with degree $\xi_1 \cdot \xi_2 = 0.25$ following Kijima Type II model is made at these maintenance time points. After a failure, the item is repaired with degree $\xi_1 = 0.5$, that is a minor repair. In the simulated sample we got 219 failures, that is nearly the same number of failures as in our set of real data.

In figure 1 the likelihood function both as a contour and a surface plot is shown for the two variables ξ_1 and ξ_2 and for fixed $\beta = 2$. Figure 2 shows the same likelihood function for the two variables ξ_1 and β and for fixed $\xi_2 = 0.5$.

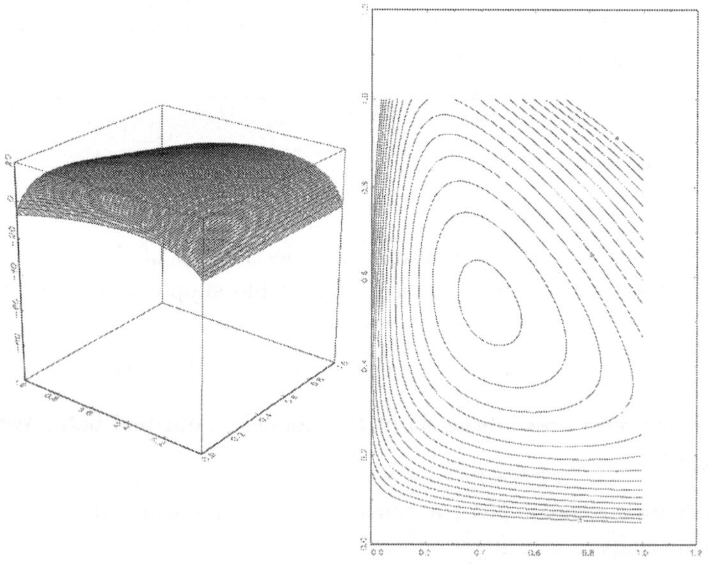

Fig. 1. Likelihood function for simulated data from example 1: $\xi_1 - \xi_2$

It can be seen in both figures that the likelihood function has a unique

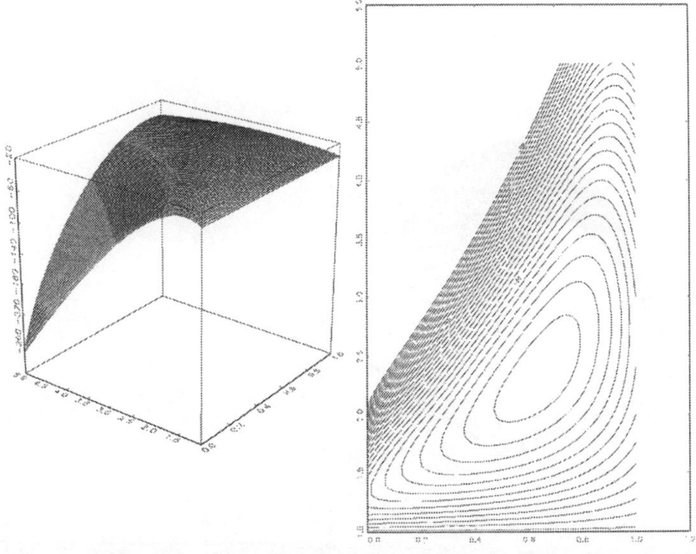

Fig. 2.   Likelihood function for simulated data from example 1: $\xi_1 - \beta$

maximum which is located close to the true parameter values. Developing similar plots for the various cases (a)-(g) in table would no doubt yield similar observations, although in the interests of space we do not present such results.

In the second example a sample of size $n = 500$ was simulated again with true parameters $\alpha = 1$, $\beta = 2$, $\xi_1 = \xi_2 = 0.5$. In this sample we assumed a distance between two major repairs of .3. For such a distance the sample contains a smaller number of failures (66) and the other 434 times are censor times where the system was stopped for a maintenance action.

In figure 3 again the likelihood function both as a contour and a surface plot is shown for the two variables ξ_1 and β and for fixed ξ_2. Here we can see that a very small number of failures can lead to bad estimates of the degree of repair. Nevertheless for the shape parameter of Weibull intensity we get a relatively good estimate.

If we look at the plot of the likelihood function for our real data set (Figure 4 shows a contour and a surface plot of the likelihood function for the two variables ξ_1 and β and for fixed $\xi_2 = 0.5$) then we get a very

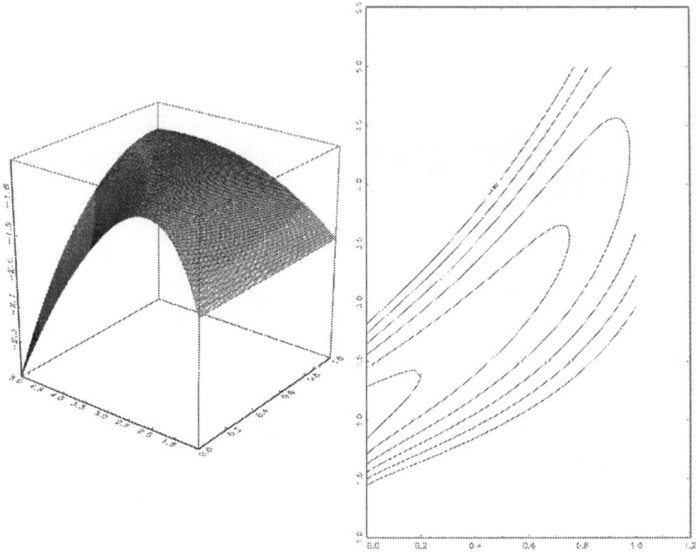

Fig. 3.   Likelihood function for simulated data from example 2: $\xi_1 - \beta$

different picture. The likelihood is decreasing in β.

One of the reasons for this could be that for real systems, maintenance actions taken are dependent on the state of the system.

6. Estimating the Degree of Repair From Times to Repair the System

In section 4 it was mentioned that the data file contains information on the times to repair the system ranging from a smallest of 1 minute to a largest of 66,624 minutes (or approximately 46 days). The range of these times is very large and the information from repair times might be not very reliable. Nevertheless, our first trials show that these times contain very useful information about the degree of repair. The idea is to estimate the degree of repair from these times to repair the system. Figure 5 shows a histogram of all times to repair the system on a logarithmic scale.

If we assume now that each maintenance action has its own degree of repair which is assumed to be

$$\xi_k = 1 - \Phi(\log(r_k) - 2.4) \, ,$$

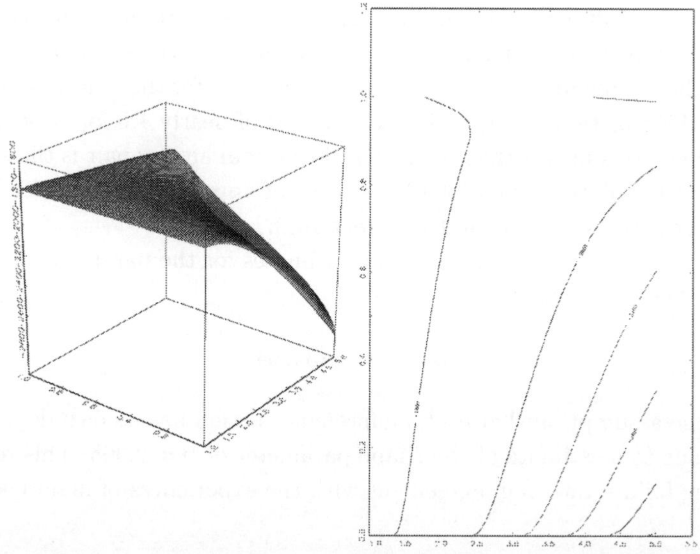

Fig. 4.   Likelihood function for the real data set: $\xi_1 - \beta$

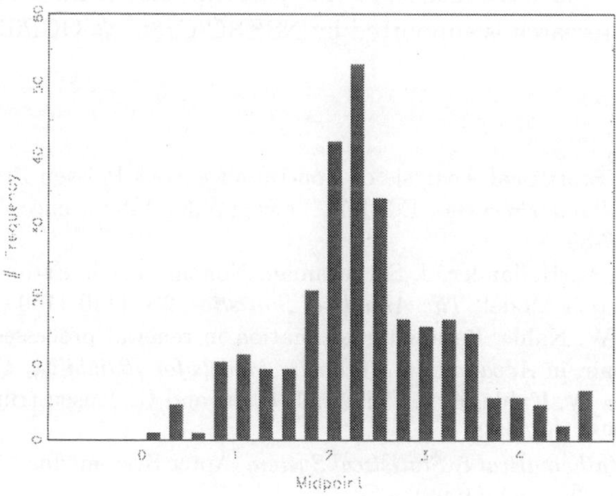

Fig. 5.   Repair times at a logarithmic scale

where ξ_k is the degree of repair after the kth failure or shut down, r_k is the repair time after the kth sojourn and Φ is the distribution function of the standard normal distribution. The constant 2.4 is the estimated mean value of the log repair times. The estimated variance for these data is about 1. From this function we get a degree of repair of nearly 1.0 for very small repair times (this means that the age of the system after repair is the same as before the failure or shutdown) and a ξ_k of approximately 0.0 for long repair times (the system is perfectly repaired).

For this model we get the following estimates for the parameters of the baseline Weibull intensity:

$$\hat{\beta} = 2.305 \quad \hat{\alpha} = 134,645 \,,$$

that is, the assumption that each maintenance action has its own degree of repair leads to a estimate of the shape parameter of $\hat{\beta} = 2.305$. This really increasing failure rate is in agreement with the experiences of maintenence engineers.

Acknowledgments

The authors would like to express their appreciation to the reliability staff of B.C. Hydro for their introduction to this problem and for making the data available. This research is supported by NSERC Grant # OCP0197319.

References

1. M. Brown, Statistical Analysis of Non-Homogeneous Poisson Processes, in *Stochastic Point Processes*, Ed. P.A. Lewis (Wiley-Interscience, New York, 1972) pp 67-89.
2. C. Dorado, M. Hollander, J. Sethuraman, Nonparametric Estimation for a General Repair Model, *The Annals of Statistics*, **25**, 1140-1160 (1997).
3. S. Gasmi, W. Kahle, Parameter estimation in renewal processes with imperfect repair, in *Advances in Stochastic Models for Reliability, Quality and Safety*, Eds. W. Kahle, E. v. Collani, J. Franz and U. Jensen (Birkhhauser, Boston, 1998) pp 53-65.
4. GAUSS, *Mathematical & Statistical System*, Aptec Systems Inc., Maple Valley, WA, 98038, USA (1996).
5. R. Guo, C.E. Love, Simulating non-homogeneous Poisson processes with proportional intensities, *Naval Research Logistics Quarterly*, **41**, 507-522 (1994).
6. M. Kijima, Some results for repairable systems with general repair, *Journal of Applied Probability*, **26**, 89-102 (1989).

7. M. Kijima, H. Morimura, Y. Suzuki, Periodical replacement problem without assuming minimal repair, *European Journal of Operational Research*, **37** 194-203 (1988).
8. G. Last, R. Szekli, Stochastic comparison of repairable systems by coupling, *Journal of Applied Probability*, **35**, 348-70 (1995).
9. R.S. Liptser, A.N. Shiryayev, *Statistics of Random Processes, vol II* (Springer, New York, 1978).
10. C.E. Love, R. Guo, Simulation strategies to identify the failure parameters of repairable systems under the influence of general repair, *Quality and Reliability Engineering International*, **10**, 37-47 (1994).

26

A CLASS OF TESTS FOR RENEWAL PROCESS VERSUS MONOTONIC AND NONMONOTONIC TREND IN REPAIRABLE SYSTEMS DATA

Jan Terje Kvaløy

Stavanger University College
P.O. Box 8002
N-4068 Stavanger, Norway
E-mail: jan.t.kvaloy@tn.his.no

Bo Henry Lindqvist

Department of Mathematical Sciences
Norwegian University of Science and Technology
N-7491 Trondheim, Norway
E-mail:bo@math.ntnu.no

A class of statistical tests for trend in repairable systems data based on the general null hypothesis of a renewal process is proposed. This class does in particular include a test which is attractive for general use by having good power properties against both monotonic and nonmonotonic trends. Both the single system and the several systems cases are considered.

1. Introduction

In recurrent event data it is often of interest to detect possible systematic changes in the pattern of events. An example is a repairable system for which it is important to detect systematic changes in the pattern of failures. Such changes can for instance be caused by various aging effects or reliability growth. We say that there is a trend in the pattern of failures if the inter-failure times tend to alter in some systematic way, which means that the inter failure times are not identically distributed. By using statistical trend tests it is possible to decide whether such an alteration is statistically significant or not.

In this paper we focus on the repairable system example, but the methods presented are relevant for any kind of recurrent event data which can be modeled by the same statistical models as used for repairable systems in this paper. A trend in the pattern of failures of a repairable system can be either monotonic, corresponding to an improving or deteriorating system, or non-monotonic, corresponding to for instance cyclic variations or a so called bathtub trend characterizing a system going through the three phases "burn-in frailty", "useful life" and "wearout".

Common statistical models for systems without trend are the homogeneous Poisson process (HPP) and the renewal process (RP). The HPP, assuming independent identically exponentially distributed interarrival times, is a special case of the far more general RP, assuming independent identically distributed interarrival times. For both models there exist a number of tests with power against monotonic trends but few tests with power against nonmonotonic trends. A test with power both against monotonic and nonmonotonic trends based on an HPP model was studied by Kvaløy & Lindqvist.[5] The problem, however, with this test and many other tests based on the null hypothesis of an HPP is that even if the null hypothesis is rejected, it does not necessarily indicate a trend, just that the process is not an HPP. An RP may for instance cause rejection with high probability (Lindqvist, Kjønstad and Meland[8]).

In this paper we present a class of tests for trend based on the null hypothesis of an RP. This class includes the well known Lewis-Robinson test, and an Anderson-Darling type test which turns out to be a test with attractive properties for general use by having good power properties against both monotonic and nonmonotonic trends.

2. Notation and Models

2.1. *Notation and Terminology*

We first present the basic ideas for the situation with one system which is observed from time $t = 0$. Generalizations to several systems are discussed in Section 4. The successive failure times are denoted $T_1, T_2, \ldots$ and the corresponding interfailure or interarrival times are denoted $X_1, X_2, \ldots$ where $X_i = T_i - T_{i-1}$, $i = 1, 2, \ldots$. Another way to represent the same information is by the counting process representation $N(t) =$ number of failures in $(0, t]$.

A system is said to be failure censored if it is observed until a given number n of failures has occurred, and time censored if it is observed until a given time τ. In the present paper we consider the failure censored case.

The obtained asymptotic results may, however, in a straightforward manner be generalized to the case when τ tends to infinity.

We say that a system inhibits no trend if the marginal distributions of all interarrival times are identical, otherwise there is a trend. If the expected length of the interarrival times is monotonically increasing or decreasing with time, corresponding to an improving or a deteriorating system, there is a monotone trend (a decreasing or an increasing trend), otherwise the trend is nonmonotone.

2.2. *Renewal Process*

The stochastic process $T_1, T_2, \ldots$ is an RP if the interarrival times $X_1, X_2, \ldots$ are independent and identically distributed. The conditional intensity of an RP given the history $\mathcal{F}_{t-}$ up to, but not including time t, can be written

$$\gamma(t|\mathcal{F}_{t-}) = z(t - T_{N(t-)})$$

where $z(t)$ is the hazard rate of the distribution of the interarrival times and $T_{N(t-)}$ is the last failure time strictly before time t.

2.3. *Trend-Renewal Process*

The trend-renewal process (TRP) is presented by Lindqvist, Elvebakk and Heggland[7], and is a generalization of the RP to a process with trend. The RP, HPP and nonhomogeneous Poisson process (NHPP) are all special cases of this process.

Let $\lambda(t)$ be a nonnegative function, called the trend function, defined for $t \geq 0$ and let $\Lambda(t) = \int_0^t \lambda(u)du$. Then the process $T_1, T_2, \ldots$ is a TRP if $\Lambda(T_1), \Lambda(T_2), \ldots$ is an RP.

The conditional intensity of a TRP may be written[7]

$$\gamma(t|\mathcal{F}_{t-}) = z(\Lambda(t) - \Lambda(T_{N(t-)}))\lambda(t)$$

where $z(t)$ is the hazard rate of the distribution of the underlying RP. In other words the intensity depends on both the age of the system and the time since last failure.

The NHPP follows as the special case when the RP part of the TRP is an HPP with intensity 1. In this case the hazard rate $z(t) \equiv 1$ and thus $\gamma(t|\mathcal{F}_{t-}) = \lambda(t)$. The special case of an RP follows when $\lambda(t)$ is constant, and the HPP follows when both $\lambda(t)$ and $z(t)$ are constant.

3. The Class of Tests for Trend

The derivation of the class of tests considered is based on Donsker's theorem (e.g Billingsley[3]). Let $\xi_1, \ldots, \xi_n$ be n independent identically distributed random variables with $\mathbf{E}(\xi_i) = 0$ and $\mathbf{Var}(\xi_i) = \sigma^2$, let $[u]$ denote the integer part of u, let $Z_j = \xi_1 + \cdots + \xi_j$ and define

$$Q_n(t) = \frac{Z_{[nt]}}{\sigma\sqrt{n}} + \frac{nt - [nt]}{\sigma\sqrt{n}}\xi_{[nt]+1}$$

for $0 \le t \le 1$. Then according to Donsker's theorem $Q_n \xrightarrow{d} W$, where $\xrightarrow{d}$ denotes convergence in distribution and W is a Brownian motion. Now consider a failure censored RP with interarrival times $X_1, \ldots, X_n$. Define $\xi_i = X_i - \mu$ where $\mu = \mathbf{E}(X_i)$ and let $Z_j = \xi_1 + \cdots + \xi_j = \sum_{i=1}^{j} X_i - j\mu = T_j - j\mu$. It follows from Donsker's theorem that $V_{n,\mu,\sigma} \xrightarrow{d} W$ as $n \to \infty$ where

$$V_{n,\mu,\sigma}(t) = \frac{T_{[nt]} - [nt]\mu}{\sigma\sqrt{n}} + \frac{nt - [nt]}{\sigma\sqrt{n}}(X_{[nt]+1} - \mu)$$

and $\sigma^2 = \mathbf{Var}(X_i)$. Further defining $W^0(t) = W(t) - tW(1)$, so that W^0 is a Brownian bridge, we will have that $V_{n,\sigma}^0 \xrightarrow{d} W^0$, where

$$V_{n,\sigma}^0(t) = V_{n,\mu,\sigma}(t) - tV_{n,\mu,\sigma}(1) = \frac{\sqrt{n}\bar{X}}{\sigma}\left(\frac{T_{[nt]}}{T_n} + (nt - [nt])\frac{X_{[nt]+1}}{T_n} - t\right) \quad (1)$$

where $\bar{X} = T_n/n$. Notice that μ cancels in (1). Replacing σ by a consistent estimator $\hat{\sigma}$, we conclude that $V_{n,\hat{\sigma}}^0 \xrightarrow{d} W^0$.

It follows from the above that under the null hypothesis of no trend, $V_{n,\hat{\sigma}}^0$ will be approximately a Brownian bridge. On the other hand, if there is a trend in the data, $V_{n,\hat{\sigma}}^0$ is likely to deviate from the Brownian bridge. Thus tests for trend can be based on measures of deviation from a Brownian bridge of $V_{n,\hat{\sigma}}^0$. Some suggestions for such measures are the Kolmogorov-Smirnov type statistic $\sup_t V_{n,\hat{\sigma}}^0(t)$, the Cramér-von Mises type statistic $\int_0^1 V_{n,\hat{\sigma}}^0(t)^2 dt$, the Anderson-Darling type statistic $\int_0^1 V_{n,\hat{\sigma}}^0(t)^2/(t(1-t))dt$ or simply $\int_0^1 V_{n,\hat{\sigma}}^0(t)dt$. Based on previous experience with these kinds of statistics we will focus on the two last ones.

3.1. *The Generalized Anderson-Darling Test for Trend*

With the Anderson-Darling type measure we have that

$$GAD \equiv \int_0^1 \frac{V_{n,\hat{\sigma}}^0(t)^2}{t(1-t)}dt \xrightarrow{d} \int_0^1 \frac{W^0(t)^2}{t(1-t)}dt$$

which has the usual limit distribution of the Anderson-Darling statistic.[1,2] It is easily realized that a test which rejects the null hypothesis of renewal process for large values of GAD has sensitivity against both monotonic and nonmonotonic trends. Straightforward but tedious calculations show that

$$GAD = \frac{n\bar{X}^2}{\hat{\sigma}^2} \sum_{i=1}^{n} \left[q_i^2 \ln(\frac{i}{i-1}) + (q_i + r_i)^2 \ln(\frac{n-i+1}{n-i}) - \frac{r_i^2}{n} \right]$$

where $q_i = (T_i - iX_i)/T_n$ and $r_i = nX_i/T_n - 1$. We call this test the generalized Anderson-Darling test for trend to distinguish it from the Anderson-Darling test for trend based on an HPP null hypothesis studied by Kvaløy & Lindqvist.[5] Any consistent estimator of σ^2 can be used. Rather than the usual estimator, $S^2 = \sum_{i=1}^{n}(X_i - \bar{X})^2/(n-1)$, we propose using the estimator

$$\hat{\sigma}^2 = \frac{1}{2(n-1)} \sum_{i=1}^{n-1} (X_{i+1} - X_i)^2. \tag{2}$$

This estimator is consistent under the null hypothesis of independent and identically distributed interarrival times, but tends to be smaller than S^2 in cases with positive dependence between subsequent interarrival times. Thus using this variance estimator yields a test statistic which is likely to have larger power against alternatives like monotonic and bathtub shaped trends than the test using the estimator S^2. This has been confirmed by simulations, see Section 5

3.2. *The Lewis-Robinson Test*

We now consider the statistic $\int_0^1 V_{n,\hat{\sigma}}^0(t)dt$ where $\int_0^1 V_{n,\hat{\sigma}}^0(t)dt \xrightarrow{d} \int_0^1 W^0(t)dt$ which is normally distributed with expectation 0 and variance $1/12$. This statistic will primarily have power against deviations from a Brownian bridge caused by monotonic trends. By scaling the test statistic to be asymptotically standard normally distributed, straightforward calculations yield the test statistic

$$LR = \sqrt{12} \int_0^1 V_{n,\hat{\sigma}}^0(t)dt = \frac{\bar{X}}{\hat{\sigma}} \frac{\sum_{i=1}^{n-1} T_i - \frac{n-1}{2}T_n}{T_n\sqrt{\frac{n}{12}}}.$$

This is simply a variant of the classical Lewis-Robinson test.[6] The original derivation by Lewis & Robinson[6] differs from the derivation presented here by taking a different starting point and using other arguments to prove the asymptotic normality. Lewis & Robinson[6] also consider more general settings.

4. Several Systems

We now consider the case when we have data from more than one system and want to do a simultaneous analysis of trend in the data. Let T_{ij} denote the jth failure time in the ith system and let X_{ij} denote the corresponding interarrival time. Let m be the number of systems and let n_i be the number of failures observed in the ith system. Under the null hypothesis of no trend we assume that the failures in the different systems occur as independent RPs. Again we study failure truncated systems.

When we consider several systems we need to take into account the possibility of heterogeneities between the systems. For instance, differences in working environment or manufacturing process may lead to differences in the distribution of failure times for different systems. However, performing a common trend analysis might still be relevant. It is easy to generalize the test approach presented in Section 3 to this situation. By adding the Brownian bridge type processes for the various systems and scaling by the square root of the number of processes we obtain a new pooled process which converges to a Brownian bridge. More precisely, under the null hypothesis of m independent (nonidentical) RPs we will have that $V^0_{n_1,\ldots,n_m,\sigma_1,\ldots,\sigma_m} \overset{d}{\to} W^0$ as $n_1,\ldots,n_m \to \infty$, where

$$V^0_{n_1,\ldots,n_m,\sigma_1,\ldots,\sigma_m}(t)$$
$$= \frac{1}{\sqrt{m}} \sum_{i=1}^{m} \left[\frac{\sqrt{n_i}\bar{X}_i}{\sigma_i} \left(\frac{T_{i[n_it]}}{T_{in_i}} + (n_it - [n_it]) \frac{X_{i[n_it]+1}}{T_{in_i}} - t \right) \right] \quad (3)$$

Here $\bar{X}_i = T_{in_i}/n_i$. Replacing $\sigma_1,\ldots,\sigma_m$ by consistent estimators, we obtain tests for trend by applying measures of deviation from Brownian bridge to $V^0_{n_1,\ldots,n_m,\sigma_1,\ldots,\sigma_m}$. Note that this approach allows for completely different distributions of the interarrival times from the different systems. Making the same calculations as in Section 3 leads to the natural generalization of the Lewis-Robinson test statistic, namely

$$LR = \frac{1}{\sqrt{m}} \sum_{i=1}^{m} \frac{\bar{X}_i}{\hat{\sigma}_i} \frac{\sum_{j=1}^{n_i-1} T_{ij} - \frac{n_i-1}{2} T_{in_i}}{T_{in_i}\sqrt{\frac{n_i}{12}}}$$

For the generalized Anderson-Darling test for trend the computations become a bit more complicated, although straightforward. A simple special case is when the same number of failures, n, is observed in all processes. In this case

$$GAD = \frac{n}{m} \sum_{j=1}^{n} \left[q_j^2 \ln(\frac{j}{j-1}) + (q_j + r_j)^2 \ln(\frac{n-j+1}{n-j}) - \frac{r_j^2}{n} \right]$$

where we have $q_j = \sum_{i=1}^{m}(\bar{X}_i/\hat{\sigma}_i)(T_{ij}/T_{in} - jX_{ij}/T_{in})$ and $r_j = \sum_{i=1}^{m}(\bar{X}_i/\hat{\sigma}_i)(nX_{ij}/T_{in} - 1)$.

In some cases the considered systems can be assumed to have identical (and independent) failure distributions. This is a reasonable assumption when the systems are both manufactured and operate under equal conditions. In such cases a common estimator for σ based on all the observed interarrival times, could be applied. Simulations have shown, however, that there is generally little or no gain in power by using such a common estimator. Thus we shall in the following always use the tests based on (3), which allow for heterogeneities between systems.

5. Simulation Study

We have performed a simulation study to compare the properties of trend tests based on GAD to tests based on LR, and also to the well known Mann test[10]. The Mann test is a rank test based on counting the number of reverse arrangements, M, among the interarrival times $X_1, \ldots, X_n$. We have a reverse arrangement if $X_i < X_j$ for $i < j$, and M is thus

$$M = \sum_{i=1}^{n-1} \sum_{j=i+1}^{n} I(X_i < X_j)$$

where $I(A) = 1$ or 0 according to whether the event A occurs or not. The test statistic M is approximately normally distributed under the null hypothesis, with expectation $n(n-1)/4$ and variance $(2n^3 + 3n^2 - 5n)/72$ for $n \geq 10$. For $n < 10$ there exist tables of the distribution of M. If the normal approximation is used, then the test is easily extended to the case of several systems by adding together the test statistics for each system.

In our study we estimate rejection probabilities by simulating 100 000 data sets for each choice of model and parameter values, and recording the relative number of rejections of each test. Let $\hat{p}$ denote an estimated rejection probability. Then the standard deviation of $\hat{p}$ is $\sqrt{\hat{p}(1-\hat{p})/100000} \leq 0.0016$. All simulations are done in C++. The nominal significance level has been set to 5%, and all simulations are done for failure censored processes. The alternative variance estimator (2) is used for the GAD and LR statistics.

5.1. *Level Properties*

First the level properties of the tests are studied by generating datasets from Weibull RPs with shape parameters respectively 0.75 and 1.5, corresponding respectively to a process which is overdispersed and a process which is underdispersed relative to an NHPP. In Figure 1 the simulated level of the tests for single systems with from 3 to 80 failures and for 5 systems with from 3 to 50 failures in each are reported. The plots in Figure 1

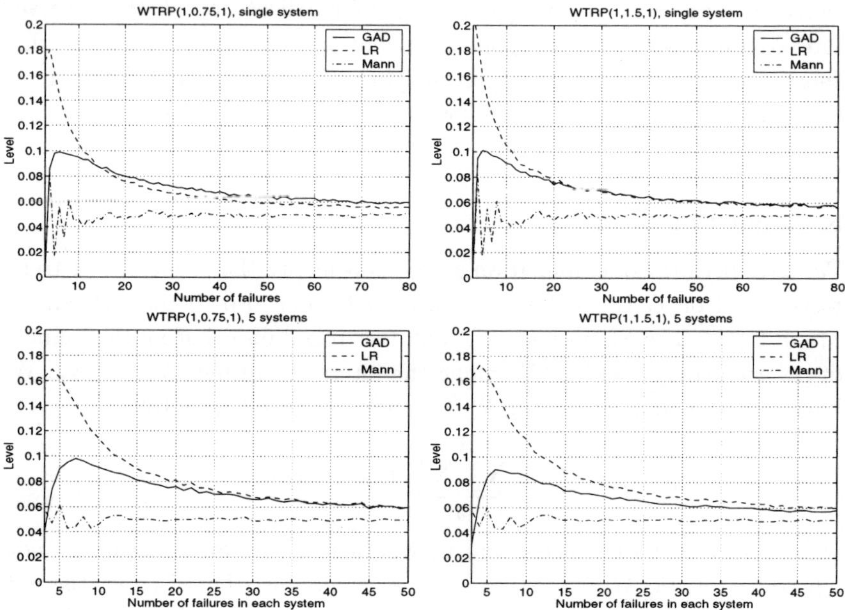

Fig. 1.   Simulations of Weibull RPs with shape parameters respectively 0.75 (overdispersed TRP) and 1.5 (underdispersed TRP). In the first row a single system with from 3 to 80 failures and in the second row 5 system with from 3 to 50 failures in each.

show that the generalized Anderson-Darling test and the Lewis-Robinson test both have poor level properties here. One reason for this is the use of the alternative variance estimator (2) which has larger variance than the ordinary estimator S^2. If S^2 is used instead, then the level properties of the tests are far better. However, instead of using the less powerful tests based on S^2 as variance estimator, we have adjusted the test statistics based on the alternative variance estimator by multiplying the test statistics by level correcting factors found by empirical explorations. More precisely we have multiplied the GAD statistic by the factor $1 - 4/n + 10/n^2$ and the LR statistic by the factor $1 - 2/n + 2/n^2$. The resulting level properties are reported in Figure 2. We see that the generalized Anderson-Darling test and the Lewis-Robinson test now have very good level properties. The level

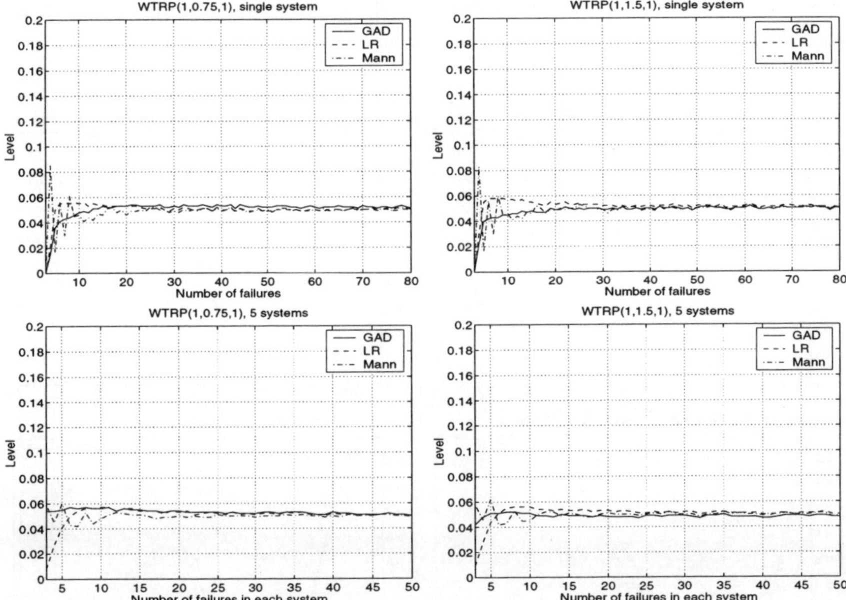

Fig. 2. Simulations of Weibull RPs with shape parameters respectively 0.75 (overdispersed TRP) and 1.5 (underdispersed TRP). In the first row a single system with from 3 to 80 failures and in the second row 5 system with from 3 to 50 failures in each. Level-adjusted tests.

corrected versions of the tests are therefore used in the rest of the simulation study. For the Mann test the normal approximation is used for all sample sizes. The level properties of this test for samples smaller than 10 are better, however, if critical values from tables are used.

5.2. *Power Properties*

5.2.1. *Monotonic Trend*

Datasets with a monotonic trend are generated by simulating data from TRPs with the underlying RP distribution being Weibull and the trend function $\lambda(t)$, governing the trend of the process, being on the power law form $\lambda(t) = bt^{b-1}$. The conditional intensity of the process can then be written

$$\gamma(t|\mathcal{F}_{t-}) = \alpha\beta(t^b - T^b_{N(t-)})^{\beta-1}bt^{b-1}. \tag{4}$$

Rejection probabilities are computed by simulation for single systems with 25 failures and for 5 systems with 5 failures in each, as functions of b. Here

$b < 1$ corresponds to a decreasing trend, $b = 1$ corresponds to no trend and $b > 1$ corresponds to an increasing trend. This is repeated for two

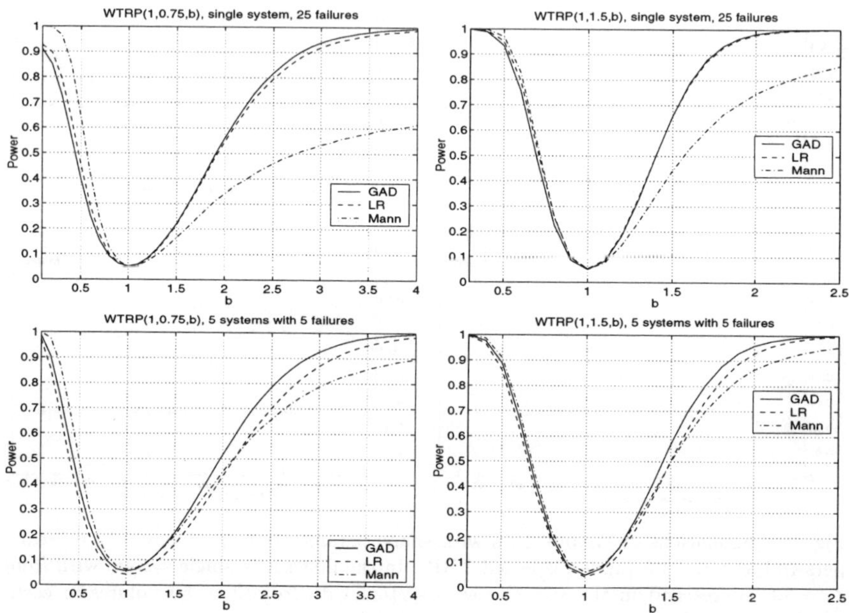

Fig. 3.   Simulations of TRPs with underlying Weibull RPs with shape parameters respectively 0.75 (overdispersed TRP) and 1.5 (underdispersed TRP), trend function $\lambda(t) = bt^{b-1}$. In the first row a single system with 25 failures and in the second row 5 system with 5 failures in each.

different values of the shape parameter β of the underlying Weibull distribution, $\beta = 0.75$ and $\beta = 1.5$. The scale parameter α is set to 1. The results are displayed in Figure 3. We see in this figure that the generalized Anderson-Darling test and the Lewis-Robinson test have fairly similar power properties, with the generalized Anderson-Darling test being a bit more powerful than the Lewis-Robinson test against increasing trend and the opposite against decreasing trend. The Mann test is clearly less powerful than the other tests against increasing trend in the single systems case, but is the most powerful test against decreasing trend in the overdispersed cases. In the several systems case the Mann test is a bit unconservative. The difference in power by observing 5 failures in 5 systems instead of 25 failures in one system is not large, but except for the Mann test against increasing trend, the tests have a slightly larger power in the single systems case.

5.2.2. *Nonmonotonic Trend*

Datasets with a bathtub trend are generated by simulating data from TRPs with trend function $\lambda(t)$ on the form displayed in Figure 4. Here d represents

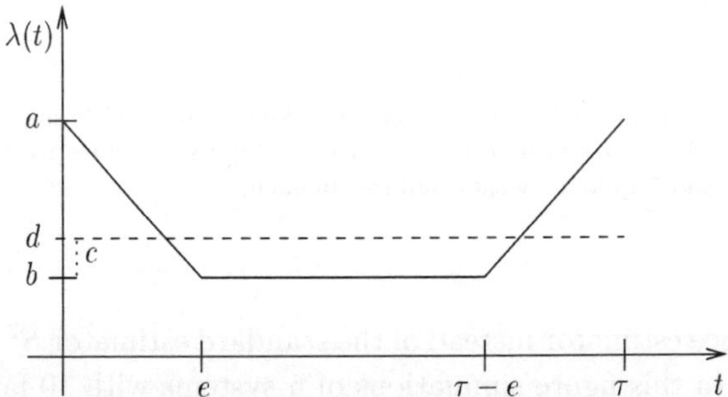

Fig. 4.   Bathtub-shaped trend function.

the average of the function $\lambda(t)$ over the interval $[0, \tau]$. The degree of bathtub shape can be expressed by the parameter c, with $c = 0$ corresponding to a horizontal line (no trend).

The rejection probability as a function of c is simulated for a single system with 25 failures and for 5 systems with 5 failures in each, with τ in each case set to a value such that the expected number of failures in $[0, \tau]$ is approximately 25 in the single system case and approximately 5 in the several systems case. The parameter e is adjusted according to c to make the expected number of failures in each phase (decreasing, no, increasing trend) equal. The shape parameter of the underlying Weibull distribution is set to $\beta = 1.5$ (underdispersed). The results are displayed in Figure 5.

We see in Figure 5 that the generalized Anderson-Darling test for trend has clearly better power properties than the other tests in the single system case, while none of the tests have much power in the several systems case. Intuitively, the difficulty to detect a bathtub trend with only 5 failures in each system is not surprising. Similar patterns are seen if we consider overdispersed TRPs, with the difference that in the latter case none of the tests are particularly powerful for small sample sizes, even in the single system case.[9] This is due to the larger variance of the underlying RP.

Finally, simulations which illustrate the gain in power by using the alter-

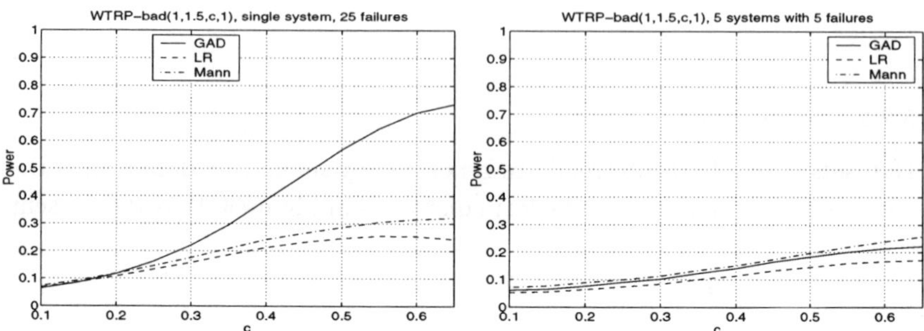

Fig. 5. Simulations of TRPs with underlying Weibull RPs with shape parameter 1.5 (underdispersed TRP), trend function displayed in Figure 4 and respectively one system with 25 failure and 5 systems with 5 failures in each.

native variance estimator instead of the standard estimator S^2 are displayed in Figure 6. In this figure simulations of 5 systems with 10 failures in each

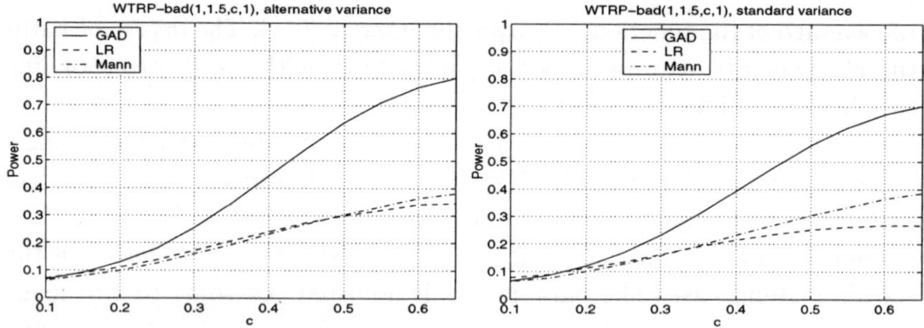

Fig. 6. Simulations of TRPs with underlying Weibull RPs with shape parameter 1.5 (underdispersed TRP), trend function displayed in Figure 4, 5 systems with 10 failures in each and respectively the alternative variance estimator and the standard variance estimator used for the generalized Anderson-Darling test and the Lewis-Robinson test.

and the same bathtub trend as used above are reported. In the left plot the alternative variance estimator is used for the generalized Anderson-Darling test and the Lewis-Robinson test, while the standard variance estimator is used in the right plot. We see that there is a clear gain in power by using the alternative variance estimator for both tests.

5.3. *Discussion*

In the presented simulations the generalized Anderson-Darling test for trend is the best test against increasing and bathtub shaped trend, while the other tests are slightly better against decreasing trend. It is not surprising that the generalized Anderson-Darling test is better than the other tests against bathtub shaped trend since this test, contrary to the other tests, is constructed to have power against both monotonic and nonmonotonic alternatives. The fact that the generalized Anderson-Darling test also performs equally well as the other tests against monotonic alternatives makes this test an attractive choice as a trend test for general use.

Instead of introducing the level correcting factors an alternative is to use bootstrapping to estimate the null distribution of the test statistics.[4,9] In some cases bootstrapping may also improve the power properties of the tests.[9]

Note that for the Lewis-Robinson test and the Mann test, an alternative generalization of the tests to several systems could be used. Instead of combining standardized test statistics for each system as is done here, the tests could instead be computed by first summarizing the unstandardized test statistics from each system and then standardizing this sum to an approximately standard normally distributed statistic by subtracting the mean and dividing by the standard deviation of this sum. For the Lewis-Robinson test this alternative statistic becomes

$$LR = \frac{\sum_{i=1}^{m} \sum_{j=1}^{n_i-1} T_{ij} - \sum_{i=1}^{m} \frac{n_i-1}{2} T_{in_i}}{\sqrt{\sum_{i=1}^{m} \frac{n_i^3}{12} \hat{\sigma}_i^2}}.$$

For situations with different numbers of failures in the systems the values of the test statistics calculated from the two approaches will be somewhat different due to the different weightings of observations from different systems implied by the two approaches. A problem with standardizing after summarizing for the Lewis-Robinson test is that in cases with strong heterogeneities across the systems the data from the systems with largest variance will be given most weight. Except from this a limited simulation study has indicated that the two approaches in most cases give quite similar results. It could however be of interest to investigate this further.

6. Conclusion

A class of tests for trend in repairable systems, based on the general null hypothesis of an RP, has been presented. In particular, an Anderson-Darling type test in this class turns out to be an attractive test for general use by having good power properties against both monotonic and nonmonotonic trends.

References

1. T. W. Anderson and D. A. Darling, Asymptotic theory of certain goodness of fit criteria based on stochastic processes, *Annals of Mathematical Statistics*, **23**, 193-212 (1952).
2. T. W. Anderson and D. A. Darling, A test of goodness of fit, *Journal of the American Statistical Association*, **49**, 765-769 (1954).
3. P. Billingsley, *Convergence of Probability Measures* (John Wiley, New York, 1968).
4. G. Elvebakk, Analysis of Repairable Systems Data: Statistical Inference for a Class of Models Involving Renewals, Heterogeneity and Time Trends, PhD thesis, Norwegian University of Science and Technology, 1999.
5. J. T. Kvaløy and B. H. Lindqvist, TTT-based tests for trend in repairable systems data, *Reliability Engineering and System Safety*, **60**, 13-28 (1998).
6. P. A. Lewis and D. W. Robinson, Testing for a monotone trend in a modulated renewal process, in *Reliability and Biometry*, Eds. F. Proschan and R.J Serfling (SIAM, Philadelphia, 1974) pp. 163-182.
7. B. H. Lindqvist, G. Elvebakk and K. Heggland, The trend-renewal process for statistical analysis of repairable systems, *Technometrics* **45**, 31-44, (2003).
8. B. H. Lindqvist, G. A. Kjønstad and N. Meland, Testing for trend in repairable systems data, *Proceedings of ESREL'94*, La Baule, France, May 30 - June 3, 1994.
9. H. Malmedal, Trend testing, Master Thesis, Norwegian University of Science and Technololgy, 2001.
10. H. B. Mann, Nonparametric tests against trend, *Econometrica*, **13**, 245-259 (1945).

27

A MAINTENANCE MODEL FOR COMPONENTS EXPOSED TO SEVERAL FAILURE MECHANISMS AND IMPERFECT REPAIR

Helge Langseth and Bo Henry Lindqvist

Department of Mathematical Sciences
Norwegian University of Science and Technology
N-7491 Trondheim, Norway
E-mail: helgel@math.ntnu.no, bo@math.ntnu.no

We investigate the mathematical modelling of maintenance and repair of components that can fail due to a variety of failure mechanisms. Our motivation is to build a model, which can be used to unveil aspects of the quality of the maintenance performed. The model we propose is motivated by imperfect repair models, but extended to model preventive maintenance as one of several "competing risks". This helps us to avoid problems of identifiability previously reported in connection with imperfect repair models.

1. Introduction

In this chapter we employ a model for components which fail due to one of a series of "competing" failure mechanisms, each acting independently on the system. The components under consideration are repaired upon failure, but are also preventively maintained. The preventive maintenance (PM) is performed periodically with some fixed period τ, but PM can also be performed out of schedule due to casual observation of an evolving failure. The maintenance need not be perfect; we use a modified version of the imperfect repair model by Brown and Proschan[7] to allow a flexible yet simple maintenance model. Our motivation for this model is to estimate quantities which describe the "goodness" of the maintenance crew; their ability to prevent failures by performing thorough maintenance at the correct time. The data required to estimate the parameters in the model we propose are the intermediate failure times, the "winning" failure mechanism associated with each failure (i.e. the failure mechanism leading to the failure), as well as the maintenance activity. This data is found in most modern reliability data banks.

The rest of this chapter is outlined as follows: We start in Section 2 with the problem definition by introducing the type of data and parameters we consider. Next, the required theoretical background is sketched in Section 3, followed by a complete description of the proposed model in Section 4. We make some concluding remarks in Section 5.

2. Problem Definition, Typical Data and Model Parameters

Consider a mechanical component which may fail at random times, and which after failure is immediately repaired and put back into service. In practice there can be several root causes for the failure, e.g. vibration, corrosion, etc. We call these causes failure mechanisms and denote them by $M_1, \ldots, M_k$. It is assumed that each failure can be classified as the consequence of exactly one failure mechanism.

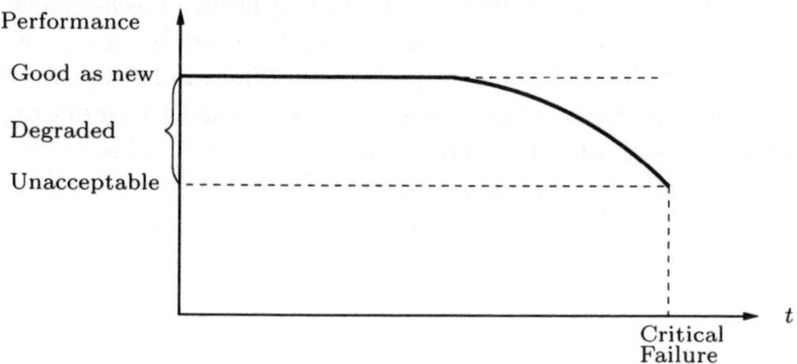

Fig. 1.   Component with degrading performance.

The component is assumed to undergo preventive maintenance (PM), usually at fixed time periods $\tau > 0$. In addition, the maintenance crew may perform unscheduled preventive maintenance of a component if required. The rationale for unscheduled PM is illustrated in Fig. 1: We assume that the component is continuously deteriorating when used, so that the performance gradually degrades until it falls outside a preset acceptable margin. As soon as the performance is unacceptable, we say that the component experiences a critical failure. Before the component fails it may exhibit inferior but admissible performance. This is a "signal" to the maintenance crew that a critical failure is approaching, and that the inferior component may be repaired. When the maintenance crew intervenes and repairs a component before it fails critically, we call it a degraded failure, and the repair action is called (an unscheduled) preventive maintenance. On the other hand, the repair activity performed after a critical failure is called a corrective maintenance.

The history of the component may in practice be logged as shown in Table 1. The events experienced by the component can be categorized as either (*i*) Critical failures, (*ii*) Degraded failures, or (*iii*) External events (component taken out of service, periodic PM, or other kind of censoring).

Table 1. Example of data describing the history of a fictitious component.

Time	Event	Failure mech.	Severity
0	Put into service	—	—
314	Failure	Vibration	Critical
8.760	(Periodic) PM	External	—
17.520	(Periodic) PM	External	—
18.314	Failure	Corrosion	Degraded
20.123	Taken out of service	External	—

The data for a single component can now formally be given as an ordered sequence of points

$$(Y_i, K_i, J_i); \ i = 1, 2, \ldots, n, \tag{1}$$

where each point represents an event. Here

$$Y_i = \text{inter-event time, i.e. time since previous event}$$
$$\text{(time since start of service if } i = 1)$$

$$K_i = \begin{cases} m \text{ if failure mechanism } M_m \ (m = 1, \ldots, k) \\ 0 \text{ if external event} \end{cases}$$

$$J_i = \begin{cases} 0 \text{ if critical failure} \\ 1 \text{ if degraded failure} \\ 2 \text{ if external event.} \end{cases} \tag{2}$$

The data in Table 1 can thus be coded as (with $M_1 = $ Vibration, $M_2 = $ Corrosion),

$$(314, 1, 0), \ (8446, 0, 2), \ (8760, 0, 2), \ (794, 2, 1), \ (1809, 0, 2).$$

A complete set of data will typically involve events from several similar components. The data can then be represented as

$$(Y_{ij}, K_{ij}, J_{ij}); \ i = 1, 2, \ldots, n_j; \ j = 1, \ldots, r, \tag{3}$$

where j is the index which labels the r component.

In practice there may also be observed covariates with such data. The models considered in this chapter will, however, not include this possibility even though they could easily be modified to do so.

Our aim is to present a model for data of type (1) (or (3)). The basic ingredients in such a model are the hazard rates $\omega_m(t)$ at time t for each failure mechanism M_m, for a component which is new at time $t = 0$. We assume that $\omega_m(t)$ is a continuous and integrable function on $[0, \infty)$. In practice it will be important to estimate $\omega_m(\cdot)$ since this information may, e.g., be used to plan future maintenance strategies.

The most frequently used models for repairable systems assume either perfect repair (renewal process models) or minimal repair (nonhomogeneous Poisson-process models). Often none of these may be appropriate, and we shall here adopt the idea of the imperfect repair model presented by Brown and Proschan[4]. This will introduce two parameters per failure mechanism:

$$p_m = \text{probability of perfect repair for a preventive maintenance of } M_m$$
$$\pi_m = \text{probability of perfect repair for a corrective maintenance of } M_m.$$

These quantities are of interest since they can be used as indications of the quality of maintenance. The parameters may in practice be compared between plants and companies, and thereby unveil maintenance improvement potential.

Finally, our model will take into account the relation between preventive and corrective maintenance. It is assumed that the component gives some kind of "signal", which will alert the maintenance crew to perform a preventive maintenance before a critical failure occurs. Thus it is not reasonable to model the (potential) times for preventive and corrective maintenance as stochastically independent. We shall therefore adopt the random signs censoring of Cooke[5]. This will eventually introduce a single new parameter q_m for each failure mechanism, with interpretation as the probability that a critical failure is avoided by a preceding unscheduled preventive maintenance.

In the cases where there is a single failure mechanism, we shall drop the index m on the parameters above.

3. Basic Ingredients of the Model

In this section we describe and discuss the two main building blocks of our final model. In Section 3.1 we consider the concept of imperfect repair, as defined by Brown and Proschan[4]. Then in Section 3.2 we introduce our basic model for the relation between preventive and corrective maintenance. Throughout the section we assume that there is a single failure mechanism ($k = 1$).

3.1. *Imperfect Repair*

Our point of departure is the imperfect repair model of Brown and Proschan[4], which we shall denote BP in the following. Consider a single sequence of failures, occurring at successive times $T_1, T_2, \ldots$ As in the previous section we let the Y_i be times between events, see Fig. 2. Furthermore, $N(t)$ is the number of events in $(0, t]$, and $N(t^-)$ is the number of events in $(0, t)$.

For the explanation of imperfect repair models it is convenient to use the conditional intensity

$$\lambda(t \mid \mathcal{F}_{t-}) = \lim_{\Delta t \downarrow 0} \frac{P(\text{event in } [t, t + \Delta t) \mid \mathcal{F}_{t-})}{\Delta t},$$

where $\mathcal{F}_{t-}$ is the history of the counting process[2] up to time t. This notation enables us to review some standard repair models. Let $\omega(t)$ be the hazard rate of a component of "age" t. Then perfect repair is modelled by $\lambda(t \mid \mathcal{F}_{t-}) = \omega(t - T_{N(t-)})$ which means that the age of the component at time t equals $t - T_{N(t-)}$, the time elapsed since the last event. Minimal repair is modelled by $\lambda(t \mid \mathcal{F}_{t-}) = \omega(t)$, which means that the age at any time t equals the calendar time t. Imperfect repair can be modelled by $\lambda(t \mid \mathcal{F}_{t-}) = \omega(\Xi_{N(t-)} + t - T_{N(t-)})$ where $0 \leq \Xi_i \leq T_i$ is some measure of the effective age of the component immediately after the ith event, more precisely, immediately after the corresponding repair. In the BP model, Ξ_i is defined indirectly by letting a failed component be given perfect repair with probability p, and minimal repair with probability $1 - p$.

For simplicity of notation we follow Kijima[8] and introduce random variables D_i to denote the outcome of the repair immediately after the ith event. If we put $D_i = 0$ for a perfect repair and $D_i = 1$ for a minimal one, it follows that

$$\Xi_i = \sum_{j=1}^{i} \left(\prod_{k=j}^{i} D_k \right) Y_j.$$

The BP model with parameter p corresponds to assuming that the D_i are *i.i.d.* and independent of $Y_1, Y_2, \ldots$, with $P(D_i = 0) = p$, $P(D_i = 1) = 1 - p$, $i = 1, \ldots, n$.

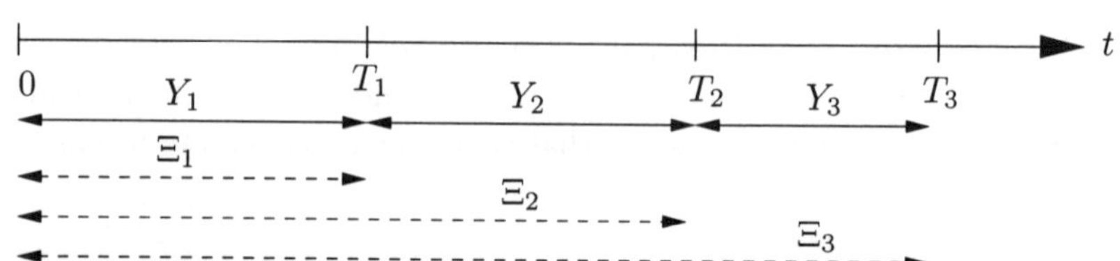

Fig. 2. In imperfect repair models there are three time dimensions to measure the age of a component: Age versus calendar time T_i, age versus inter-event times Y_i, and effective age Ξ_i. The values of Ξ_i, $i > 1$, depend upon both inter-event times and maintenance history. This is indicated by dotted lines for the Ξ_i.

BP type models have been considered by several authors, including Block *et al.*[3] who extended the model to allow the parameter p to be time varying, Kijima[8] who studied two general repair models for which BP is a special case, Hollander *et al.*[7] who studied statistical inference in the model, Dorado *et al.*[6] who proposed a more general model with BP as a special case, and most notably for the present work, Whitaker and Samaniego[10] whose results we discuss in further detail below.

Whitaker and Samaniego[10] found non-parametric maximum likelihood estimators for (p, F) in the BP model, where F is the distribution function corresponding to the hazard $\omega(\cdot)$. They noted that p is in general not identifiable if only the inter-event times Y_i are observed. The problem is related to the memoryless property of the exponential distribution, and is hardly a surprise. To ensure identifiability, Whitaker and Samaniego made strong assumptions about data availability, namely that the type of repair (minimal or perfect) is reported for each repair action (i.e., the variables D_j are actually observed). In real applications, however, exact information on the type of repair is rarely available. As we shall see in Section 4.2, identifiability of p is still possible in the model by appropriately modelling the maintenance actions.

In order to illustrate estimation in the BP model based on the Y_i alone, we consider the failure times of Plane 7914 from the air conditioner data of Proschan[9] given in Table 2. These data were also used by Whitaker and Samaniego[10]. The joint density of the observations $Y_1, \ldots, Y_n$ can be calculated as a product of conditional densities,

$$f(y_1, \ldots, y_n) = f(y_1)f(y_2|y_1) \cdots f(y_n|y_1, \ldots, y_{n-1}).$$

For computation of the ith factor we condition on the unobserved

Table 2. Proschan's air conditioner
data; inter-event times of plane 7914.

50	44	102	72	22	39	3	15
197	188	79	88	46	5	5	36
22	139	210	97	30	23	13	14

$D_1, \ldots, D_{i-1}$, getting

$$f(y_i \mid y_1, \ldots, y_{i-1}) = \sum_{d_1,\ldots,d_{i-1}} f(y_i \mid y_1, \ldots, y_{i-1}, d_1, \ldots, d_{i-1})$$

$$\times f(d_1, \ldots, d_{i-1} \mid y_1, \ldots, y_{i-1})$$

$$= \sum_{j=1}^{i} f(y_i \mid y_1, \ldots, y_{i-1}, d_{j-1} = 0, d_j = \cdots = d_{i-1} = 1)$$

$$\times P(D_{j-1} = 0, D_j = \cdots = D_{i-1} = 1)$$

$$= \sum_{j=1}^{i} \omega \left(\sum_{k=j}^{i} y_k \right) e^{-\left[\Omega\left(\sum_{k=j}^{i} y_k\right) - \Omega\left(\sum_{k=j}^{i-1} y_k\right) \right]} (1 - p)^{i-j} p^{\delta(j>1)},$$

where $\Omega(x) = \int_0^x \omega(t)dt$ is the cumulative hazard function and $\delta(j > 1)$
is 1 if $j > 1$ and 0 otherwise. The idea is to partition the set of vectors
$(d_1, \ldots, d_{i-1})$ according to the number of 1s immediately preceding the ith
event.

Let the cumulative hazard be given by $\Omega(x) = \mu x^\alpha$ for unknown μ and
α. The profile log likelihoods of the single parameter p and the pair (α, p)
are shown in Fig. 3a) and Fig. 3b) respectively. The maximum likelihood
estimates are $\hat{\alpha} = 1.09$, $\hat{\mu} = \exp(-4.81)$, and $\hat{p} = 0.01$. However, the data
contain very little information about p; this is illustrated in Fig. 3a). It is
seen that both $p = 0$, corresponding to an NHPP, and $p = 1$, corresponding to a Weibull renewal process are "equally" possible models here. The
problem is closely connected to the problem of unidentifiability of p, noting
that the maximum likelihood estimate of α is close to 1. Indeed, the exponential model with $\alpha = 1$ fixed gives the maximum log likelihood -123.86
while the maximum value in the full model (including μ, α and p) is only
marginally larger, -123.78.

3.2. *Modelling Preventive versus Corrective Maintenance*

Recall from Section 2 that PM interventions are basically periodic with
some fixed period τ, but that unscheduled preventive maintenance may
still be performed within a PM period, reported as degraded failures. Thus

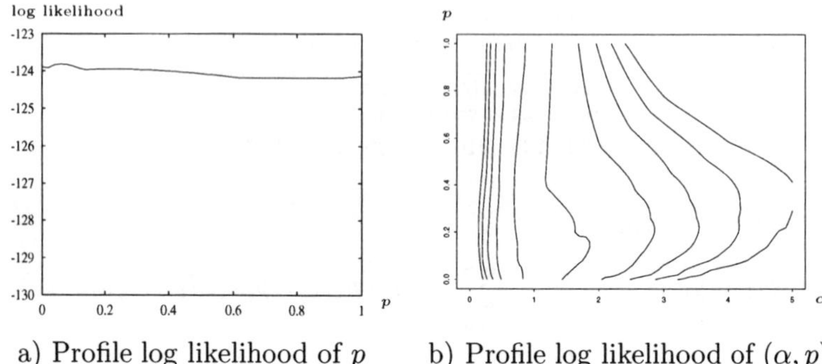

a) Profile log likelihood of p b) Profile log likelihood of (α, p)

Fig. 3.   Profile log likelihoods for the data in Table 2. Fig. 3a) shows the profile likelihood of p, Fig. 3b) shows the (α, p)-profile likelihood.

degraded failures may censor critical failures, and the two types of failure may be highly correlated.

A number of possible ways to model interaction between degraded and critical failures are discussed by Cooke[5]. We adopt one of these, called random signs censoring. In the notation introduced in Section 2 we consider here the case when we observe pairs (Y_i, J_i) where the Y_i are inter-event times whereas the J_i are indicators of failure type (critical or degraded). For a typical pair (Y, J) we let Y be the minimum of the potential critical failure time X and the potential degraded failure time Z, while $J = I(Z < X)$ is the indicator of the event $\{Z < X\}$ (assuming that $P(Z = X) = 0$ and that there are no external events). Thus we have a competing risk problem. However, while X and Z would traditionally be treated as independent, random signs censoring makes them dependent in a special way.

The basic assumption of random signs censoring is that the event of successful preventive maintenance, $\{Z < X\}$, is stochastically independent of the potential critical failure time X. In other words, the conditional probability $q(x) = P(Z < X | X = x)$ does not depend on the value of x.

Let X have hazard rate function $\omega(x)$ and cumulative hazard $\Omega(x)$. In addition to the assumption of random signs censoring, we will assume that conditionally, given $Z < X$ and $X = x$, the distribution of the intervention time Z satisfies

$$P(Z \leq z \mid X = x, Z < X) = \frac{\Omega(z)}{\Omega(x)}, \ 0 \leq z \leq x. \tag{4}$$

To see why (4) is reasonable, consider Fig. 4. When "Nature" has chosen in favour of the crew and has selected the time to critical failure,

$X = x$, which the crew will have to beat, she first draws a value u uniformly from $[0, \Omega(x)]$. Then the time for preventive maintenance is chosen as $Z = \Omega^{-1}(u)$, where $\Omega^{-1}(\cdot)$ is the inverse function of $\Omega(\cdot)$. Following this procedure makes the conditional density of Z proportional to the intensity of the underlying failure process. This seems like a coarse but somewhat reasonable description of the behaviour of a competent maintenance crew.

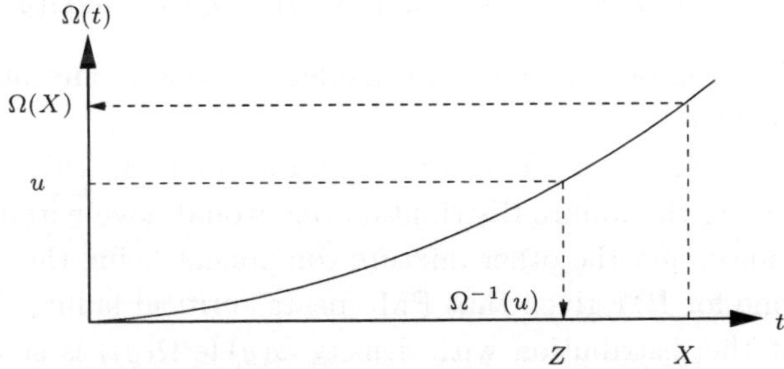

Fig. 4.   Time to PM conditioned on $\{Z < X, X = x\}$.

Our joint model for (X, Z) is thus defined from the following:

(i) X has hazard rate $\omega(\cdot)$.
(ii) $\{Z < X\}$ and X are stochastically independent.
(iii) Z given $Z < X$ and $X = x$ has distribution function (4).

These requirements determine the distribution of the observed pair (Y, J) as follows. First, by (ii) we get

$$P(y \leq Y \leq y + dy, J = 0) = P(y \leq X \leq y + dy, X < Z)$$
$$= (1 - q)\, \omega(y)\, \exp(-\Omega(y))\, dy$$

where we introduce the parameter $q = P(Z < X)$. Next,

$$P(y \leq Y \leq y + dy, J = 1)$$
$$= P(y \leq Z \leq y + dy, Z < X)$$
$$= \int_y^\infty P(y \leq Z \leq y + dy | X = x, Z < X)$$
$$\times P(Z < X | X = x)\, \omega(x) \exp(-\Omega(x))\, dx$$
$$= q\, \omega(y)\, dy \int_y^\infty \omega(x)\, \exp(-\Omega(x)) / \Omega(x)\, dx$$
$$= q\, \omega(y)\, \mathrm{Ie}(\Omega(y))\, dy,$$

where $\text{Ie}(t) = \int_t^{\infty} \exp(-u)/u\,du$ is known as the exponential integral[1].

It is now straightforward to establish the density and distribution function of Y,

$$f_Y(y) = (1 - q)\,\omega(y)\,\exp\left(-\Omega(y)\right) + q\,\omega(y)\,\text{Ie}(\Omega(y)) \tag{5}$$

and

$$F_Y(y) = P(Y \leq y) = 1 - \exp(-\Omega(y)) + q\,\Omega(y)\,\text{Ie}(\Omega(y)). \tag{6}$$

Note that the proposed maintenance model introduces only one new parameter, namely q.

The distribution (5) for Y is a mixture distribution, with one component representing the failure distribution one would have without preventive maintenance, and the other mixture component being the conditional density of time for PM given that PM "beats" critical failure. It is worth noticing that the distribution with density $\omega(y)\,\text{Ie}(\Omega(y))$ is stochastically smaller than the distribution with density $\omega(y)\,\exp(-\Omega(y))$; this is a general consequence of random signs censoring.

4. General Model

Recall that the events in our most general setting are either critical failures, degraded failures or external events; consider Fig. 2. We shall assume that corrective maintenance is always performed following a critical failure, while preventive maintenance is performed both after degraded failures and external events. Moreover, in the case of several failure mechanisms, any failure is treated as an external event for all failure mechanisms except the one failing.

4.1. *Single Failure Mechanism*

In this case the data for one component are (Y_i, J_i); $i = 1, \ldots, n$ with J_i now defined as in (2) with three possible values. Suppose for a moment that all repairs, both corrective and preventive, are perfect. Then we shall assume that the (Y_i, J_i) are *i.i.d.* observations of (Y, J) where $Y = \min(X, Z, U)$, (X, Z) is distributed as in Section 3.2, and U is the (potential) time of an external event. The U is assumed to be stochastically independent of (X, Z) and to have a distribution which does not depend on the parameters of our model. It follows that we can disregard the terms corresponding to U in the likelihood calculation. The likelihood contribution from an observation

(Y, J) will therefore be as follows (see section 3.2):

$$f(y, 0) = (1 - q)\,\omega(y)\,\exp\left(-\Omega(y)\right)$$
$$f(y, 1) = q\,\omega(y)\,\mathrm{Ie}(\Omega(y)) \tag{7}$$
$$f(y, 2) = \exp\left(-\Omega(y)\right) - q\,\Omega(y)\,\mathrm{Ie}(\Omega(y)).$$

The last expression follows from (6) and corresponds to the case where all we know is that $\max(X, Z) > y$.

To the model given above we now add imperfect repair. Recall that in the BP model there is a probability p of perfect repair ($D_i = 0$) after each event. We shall here distinguish between preventive maintenance and corrective maintenance by letting D_i equal 0 with probability p if the ith event is a preventive maintenance or an external event, and with probability π if the ith event is a critical failure. Moreover, we shall assume that for all i we have $D_1, \ldots, D_i$ conditionally independent given $y_1, \ldots, y_i, j_1, \ldots, j_i$.

From this we are able to write down the likelihood of the data as a product of the following conditional distributions. The derivation is a straightforward extension of the one in Section 3.1.

$$f\left((y_i, j_i) \mid (y_1, j_1), \ldots, (y_{i-1}, j_{i-1})\right)$$
$$= \sum_{d_1, \ldots, d_{i-1}} f((y_i, j_i) \mid (y_1, j_1), \ldots, (y_{i-1}, j_{i-1}), d_1, \ldots, d_{i-1})$$
$$\times f(d_1, \ldots, d_{i-1} \mid (y_1, j_1), \ldots, (y_{i-1}, j_{i-1}))$$
$$= \sum_{j=1}^{i} f\left((y_i, j_i) \,\middle|\, \xi_{i-1} = \sum_{k=j}^{i-1} y_k\right)$$
$$\times P\left(D_{j-1} = 0, D_j = \cdots = D_{i-1} = 1 \mid j_1, \ldots, j_{i-1}\right).$$

Here $P(D_{j-1} = 0, D_j = \cdots = D_{i-1} = 1 \mid j_1, \ldots, j_{i-1})$ is a simple function of p and π. Thus, what remains to be defined are the conditional densities $f((y_i, j_i) \mid \xi_{i-1})$, i.e. the conditional densities of (Y_i, J_i) given that the age of the component immediately after the $(i-1)$th event is ξ_{i-1}. We shall define these to equal the conditional densities given no event in $(0, \xi_{i-1})$, of the distribution given in (7). Thus we have

$$f\left((y_i, 0) \mid \xi_{i-1}\right) = \frac{(1 - q)\,\omega(\xi_{i-1} + y_i)\,\exp(-(\Omega(\xi_{i-1} + y_i)))}{\exp(-\Omega(\xi_{i-1})) - q\,\Omega(\xi_{i-1})\,\mathrm{Ie}(\Omega(\xi_{i-1}))}$$

$$f\left((y_i, 1) \mid \xi_{i-1}\right) = \frac{q\,\omega(\xi_{i-1} + y_i)\,\mathrm{Ie}(\Omega(\xi_{i-1} + y_i))}{\exp(-\Omega(\xi_{i-1})) - q\,\Omega(\xi_{i-1})\,\mathrm{Ie}(\Omega(\xi_{i-1}))}$$

$$f\left((y_i, 2) \mid \xi_{i-1}\right) = \frac{\exp(-\Omega(\xi_{i-1} + y_i)) - q\,\Omega(\xi_{i-1} + y_i)\,\mathrm{Ie}(\Omega(\xi_{i-1} + y_i))}{\exp(-\Omega(\xi_{i-1})) - q\,\Omega(\xi_{i-1})\,\mathrm{Ie}(\Omega(\xi_{i-1}))}.$$

If we have data from several independent components, the complete likelihood is given as the product of the individual likelihoods.

4.2. *Identifiability of Parameters*

The present discussion of identifiability is inspired by the corresponding discussion by Whitaker and Samaniego[10], who considered the simple BP model.

Refer again to the model of the previous subsection. We assume here that, conditional on $(Y_1, J_1), (Y_2, J_2), \ldots, (Y_{i-1}, J_{i-1})$, the (potential) time to the next external event is a random variable U with continuous distribution G and support on all of $(0, \tau]$ where τ as before is the regular maintenance interval. Moreover, the distribution G does not depend on the parameters of the model, and it is kept fixed in the following.

We also assume that $\omega(x) > 0$ for all $x > 0$ and that $0 < q < 1$. The parameters of the model are ω, q, p, π. These, together with G, determine a distribution of $(Y_1, J_1), \ldots, (Y_n, J_n)$ which we call $\mathcal{F}_{(\omega, q, p, \pi)}$. Here n is kept fixed.

The question of identifiability can be put as follows: Suppose

$$\mathcal{F}_{(\omega, q, p, \pi)} = \mathcal{F}_{(\omega^*, q^*, p^*, \pi^*)}, \tag{8}$$

which means that the two parameterizations lead to the same distribution of the observations $(Y_1, J_1), \ldots, (Y_n, J_n)$. Can we from this conclude that $\omega = \omega^*$, $q = q^*$, $p = p^*$, $\pi = \pi^*$?

First note that (8) implies that the distribution of (Y_1, J_1) is the same under the two parameterizations; $Y_1 = \min(X, Z, U)$. It is clear that each of the following two types of probabilities are the same under the two parameterizations,

$$P(x \leq X \leq x + dx, Z > x, U > x)$$
$$P(z \leq Z \leq z + dz, X > z, U > z).$$

By independence of (X, Z) and U, and since $P(U > x) > 0$ if and only if $x < \tau$, we conclude that each of the following two types of probabilities are equal under the two parameterizations,

$$P(x \leq X \leq x + dx, Z > x); \; x < \tau$$
$$P(z \leq Z \leq z + dz, X > z); \; z < \tau.$$

These probabilities can be written respectively

$$(1 - q)\, \omega(x)\, e^{-\Omega(x)}\, dx; \; x < \tau$$
$$q\, \omega(z)\, \mathrm{Ie}(\Omega(z))\, dz; \; z < \tau.$$

Thus, by integrating from 0 to x we conclude that (8) implies for $x \leq \tau$

$$(1-q)\left(1 - e^{-\Omega(x)}\right) = (1-q^*)\left(1 - e^{-\Omega^*(x)}\right) \tag{9}$$

$$q\left(1 - e^{-\Omega(x)} + \Omega(x)\mathrm{Ie}(\Omega(x))\right) = q^*\left(1 - e^{-\Omega^*(x)} + \Omega^*(x)\mathrm{Ie}(\Omega^*(x))\right). \tag{10}$$

We shall now see that this implies that $q = q^*$ and $\Omega(x) = \Omega^*(x)$ for all $x \leq \tau$. Suppose, for contradiction, that there is an $x_0 \leq \tau$ such that $\Omega(x_0) < \Omega^*(x_0)$. Then since both $1 - \exp(-t)$ and $1 - \exp(-t) + t\,\mathrm{Ie}(t)$ are strictly increasing in t, it follows from respectively (9) and (10) that $1 - q > 1 - q^*$ and $q > q^*$. But this is a contradiction. In the same manner we get a contradiction if $\Omega(x_0) > \Omega^*(x_0)$. Thus $\Omega(x) = \Omega^*(x)$ for all $x \leq \tau$ (so $\omega(x) = \omega^*(x)$ for all $x \leq \tau$) and hence also $q = q^*$.

We shall see below that in fact we have $\Omega(x) = \Omega^*(x)$ on the interval $(0, n\tau)$, but first we shall consider the identifiability of p and π. For this end we consider the joint distribution of $(Y_1, J_1), (Y_2, J_2)$. In the same way as already demonstrated we can disregard U in the discussion, by independence, but we need to restrict y_1, y_2 so that $y_1 + y_2 \leq \tau$. First, look at

$$P\Big(y_1 \leq Y_1 \leq y_1 + dy_1, J_1 = 0, y_2 \leq Y_2 \leq y_2 + dy_2, J_2 = 0\Big) \tag{11}$$

$$= (1-q)\,\omega(y_1)\,e^{-\Omega(y_1)}\left[\pi(1-q)\omega(y_2)e^{-\Omega(y_2)}\right.$$

$$\left. + (1-\pi)(1-q)\,\frac{\omega(y_1+y_2)\exp(-\Omega(y_1+y_2))}{\exp(-\Omega(y_1)) - q\,\Omega(y_1)\,\mathrm{Ie}(\Omega(y_1))}\right]dy_1\,dy_2.$$

This is a linear function of π with coefficient of π proportional to

$$\omega(y_2)\exp(-\Omega(y_2)) - \frac{\omega(y_1+y_2)\exp(-\Omega(y_1+y_2))}{\exp(-\Omega(y_1)) - q\,\Omega(y_1)\,\mathrm{Ie}(\Omega(y_1))}. \tag{12}$$

Using the assumption that $0 < q < 1$ we thus conclude that $\pi = \pi^*$ unless (12) equals 0 for all y_1 and y_2 with $y_1 + y_2 \leq \tau$. Making the similar computation, putting $J_2 = 1$ instead of $J_2 = 0$ in (11), we can similarly conclude that $\pi = \pi^*$ unless

$$\omega(y_2)\mathrm{Ie}(\Omega(y_2)) - \frac{\omega(y_1+y_2)\,\mathrm{Ie}(\Omega(y_1+y_2))}{\exp(-\Omega(y_1)) - q\,\Omega(y_1)\,\mathrm{Ie}(\Omega(y_1))} \tag{13}$$

equals 0 for all y_1 and y_2 with $y_1 + y_2 \leq \tau$. Now, if both (12) and (13) were 0 for all y_1 and y_2 with $y_1 + y_2 \leq \tau$, then we would necessarily have

$$\frac{\exp(-\Omega(y_2))}{\mathrm{Ie}(\Omega(y_2))} = \frac{\exp(-\Omega(y_1+y_2))}{\mathrm{Ie}(\Omega(y_1+y_2))} \tag{14}$$

for all y_1 and y_2 with $y_1 + y_2 \leq \tau$. Since we have assumed that $\omega(\cdot)$ is strictly positive, (14) would imply that $\exp(-t)/\mathrm{Ie}(t)$ is constant for t in some interval (a, b). This is of course impossible by the definition of $\mathrm{Ie}(\cdot)$, and it follows that not both of (12) and (13) can be identically zero. Hence π is identifiable.

Identifiability of p is concluded in the same way by putting $J_1 = 1$ instead of $J_1 = 0$ in (11).

So far we have concluded equality of the parameters q, p, π under the two parameterizations, while we have concluded that $\Omega(x) = \Omega^*(x)$ for all $x \leq \tau$. But then, putting $y_1 = \tau$ in (11), while letting y_2 run from 0 to τ, it follows that $\Omega(x) = \Omega^*(x)$ also for all $\tau < x \leq 2\tau$. By continuing we can eventually conclude that $\Omega(x) = \Omega^*(x)$ for all $0 < x \leq n\tau$.

If $\tau = \infty$, then of course the whole function $\omega(\cdot)$ is identifiable. However, even if $\tau < \infty$ we may have identifiability of all of $\omega(\cdot)$. For example, suppose $\Omega(x) = \mu x^\alpha$ with μ, α positive parameters. Then the parameters are identifiable since (9) in this case implies that $\mu x^\alpha = \mu^* x^{\alpha^*}$ for all $x \leq \tau$. This clearly implies the pairwise equality of the parameters.

4.3. *Several Failure Mechanisms*

We now look at how to extend the model of Section 4.2 to $k > 1$ failure mechanisms and data given as in (1) or (3).

Our basic assumption is that the different failure mechanisms $M_1, \ldots, M_k$ act independently on the component. More precisely we let the complete likelihood for the data be given as the product of the likelihoods for each failure mechanism. Note that the set of events is the same for all failure mechanisms, and that failure due to one failure mechanism is treated as an external event for the other failure mechanisms.

The above assumption implies a kind of independence of the maintenance for each failure mechanism. Essentially we assume that the pairs (X, Z) are independent across failure mechanisms. This is appropriate if there are different maintenance crews connected to each failure mechanisms, or could otherwise mean that the "signals" of degradation emitted from the component are independent across failure mechanisms.

Another way of interpreting our assumption is that, conditional on

$$(y_1, k_1, j_1), \ldots, (y_{i-1}, k_{i-1}, j_{i-1})$$

the next vector (Y_i, K_i, J_i) corresponds to a competing risk situation involving m independent risks, one for each failure mechanism, and each with properties as for the model given in Section 4.1.

The parameters (ω, q, p, π) may (and will) in general depend on the failure mechanism. As regards identifiability of parameters, this will follow from the results for single failure mechanisms of Section 4.2 by the assumed independence of failure mechanisms.

If we have data from several independent components of the same kind, given as in (3), then the complete likelihood is given as the product of the likelihoods for each component.

5. Concluding Remarks

In this chapter we have proposed a simple but flexible model for maintained components which are subject to a variety of failure mechanisms. The proposed model has the standard models of perfect and minimal repair as special cases. Moreover, some of the parameters we estimate (namely p_m, π_m and q_m) can be used to examine the sufficiency of these smaller models. "Small" values of $\hat{q}_m$ accompanied by "extreme" values of all $\hat{p}_m$ and $\hat{\pi}_m$ (either "close" to one or zero) indicate that reduced models are detailed enough to capture the main effects in the data. Making specific model assumptions regarding the preventive maintenance we are able to prove identifiability of all parameters.

Our motivation has been to build a model that could be used to estimate the effect of maintenance, where "effect" has been connected to the model parameters q_m, p_m and π_m. For a given level of maintenance activity, we can use q_m as an indication of the crew's efficiency; their ability to perform maintenance at the correct times to try to stop evolving failures. The p_m and π_m indicate the crew's thoroughness; their ability to actually stop the failure development. The proposed model indirectly estimates the naked failure rate.

We make modest demands regarding data availability: Only the inter-failure times and the failure mechanisms leading to the failure accompanied by the preventive maintenance program are required. This information is available in most modern reliability data banks.

Acknowledgements

We would like to thank Tim J. Bedford for an interesting conversation about the model for PM versus critical failures and Roger M. Cooke for helpful comments to an early version of this chapter. The first author was supported by a grant from the Research Council of Norway.

References

1. M. Abramowitz and I. A. Stegun, *Handbook of Mathematical Functions* (Dover Publishers, New York, 1965).
2. P. Andersen, Ø. Borgan, R. Gill, and N. Keiding, *Statistical models based on counting processes* (Springer, New York, 1992).
3. H. Block, W. Borges, and T. Savits, Age dependent minimal repair, *Journal of Applied Probability*, **22**, 370–385 (1985).
4. M. Brown and F. Proschan, Imperfect repair, *Journal of Applied Probability*, **20**, 851–859 (1983).
5. R. M. Cooke, The design of reliability data bases, Part I and Part II, *Reliability Engineering and System Safety*, **52**, 137–146 and 209–223 (1996).
6. C. Dorado, M. Hollander, and J. Sethuraman, Nonparametric estimation for a general repair model, *Annals of Statistics*, **25**, 1140–1160 (1997).
7. M. Hollander, B. Presnell, and J. Sethuraman, Nonparametric methods for imperfect repair models, *Annals of Statistics*, **20**, 879–896 (1992).
8. M. Kijima, Some results for repairable systems with general repair, *Journal of Applied Probability*, **26**, 89–102 (1989).
9. F. Proschan, Theoretical explanation of observed decreasing failure rate, *Technometrics*, **5**, 375–383 (1963).
10. L. R. Whitaker and F. J. Samaniego, Estimating the reliability of systems subject to imperfect repair, *Journal of American Statistical Association*, **84**, 301–309 (1989).

Part VIII

STATISTICAL INFERENCE IN SURVIVAL ANALYSIS

28

EMPIRICAL PLUG-IN CURVE AND SURFACE ESTIMATES

Jiancheng Jiang

Department of Probability & Statistics,
Peking University, Beijing 100871
E-mail: Jiang@math.pku.edu.cn

Kjell A. Doksum

Department of Statistics, University of Wisconsin-Madison,
Madison, WI 53706, USA
E-mail: doksum@stat.wisc.edu

We obtain estimates of a surface $\theta : \mathcal{R}^d \to \mathcal{R}$ by first finding the coefficient vector $\beta(\mathbf{x})$ that minimizes a distance between $\theta(\mathbf{z})$ and an approximating function $\theta(\mathbf{z} - \mathbf{x}; \beta)$ for $\mathbf{z}$ in a neighborhood of a given point $\mathbf{x} \in \mathcal{R}^d$, next expressing this $\beta(\mathbf{x})$ as a functional $\beta(\mathbf{x}; G)$ of a surface $G : \mathcal{R}^q \to \mathcal{R}$ that admits an empirical estimate $\hat{G}(\cdot)$, and then using the empirical plug-in approach with $\hat{\beta}(\mathbf{x}) = \beta(\mathbf{x}; \hat{G})$ and $\hat{\theta}(\mathbf{x}) = \theta(\mathbf{0}, \hat{\beta}(\mathbf{x}))$. Examples of G's are the multivariate distribution function and the integrated hazard function. With this approach the random part of the estimation error can be expressed in terms of empirical processes of the form $\int v(\cdot, \mathbf{x}) d[\hat{G}(\cdot) - G(\cdot)]$, which facilitates asymptotic analysis. This approach gives simple derivations of old and new curve and surface estimates as well as their properties. It is particularly useful for providing simple and highly efficient estimates in boundary regions. We apply the approach to density estimation, univariate and multivariate; to hazard estimation; and to nonparametric regression.

1. Introduction

Let $\theta(\cdot) : \mathcal{R}^d \to \mathcal{R}$ be a nonparametric surface and let $\Theta = \{\theta(\cdot; \beta) : \mathcal{R}^d \to \mathcal{R}, \ \beta \in \mathcal{R}^q\}$ be a class of approximating functions. The class Θ may or may not be a sieve. It is assumed that Θ contains all constants. Because $\theta(\cdot)$ is unknown, estimators are often constructed by minimizing a distance

433

between $\theta(\cdot; \beta)$ and some data based function or value that is a proxy for $\theta(\cdot)$. In the case of regression where $\theta(\mathbf{x}) = E(Y|\mathbf{X} = \mathbf{x})$, it is natural to use Y_i as a proxy of $\theta(\mathbf{X}_i)$. However, in other cases, it is not clear what data based proxy for $\theta(\cdot)$ one should use. For instance, for the cases where $\theta(\cdot)$ are a density, a hazard function, and a hazard regression function, Jones[7] (c.f. Section 2.7.2 of Fan and Gijbels[4]), Nielsen and Tanggaard[11] and Jiang and Doksum,[6] and Nielsen,[10] respectively, used data based Dirac functions to construct proxies for $\theta(\cdot)$. However, Dirac functions do not exist in the ordinary sense, they only exist in the space of Schwartz distributions.

Here we consider a simple approach that only uses ordinary functions and gives the same solutions as the Dirac function approach when it applies. This approach also gives the same solutions as the one based on proxies that are ordinary functions. One version of the approach is simply to find the $\beta = \beta(\mathbf{x})$ that locally, near x, minimizes a distance based on $\theta(\mathbf{z}) - \theta(\mathbf{z} - \mathbf{x}; \beta)$. This β depends on the unknown $\theta(\cdot)$, however, in many interesting cases it is possible to express this dependence through dG, where G is a function that admits an empirical estimate $\hat{G}$ such that $\hat{G}$ and integrals with respect to $d\hat{G}$ exist in the ordinary sense. Examples of $\hat{G}$ are the empirical distribution, the Kaplan Meier estimate and the Nelson-Aalen estimate. Let $\hat{\beta}(\mathbf{x})$ be the estimate of $\beta(\mathbf{x})$ obtained by plugging in the empirical estimate $\hat{G}$ for G, then $\hat{\theta}(\mathbf{x}) = \theta(\mathbf{0}; \hat{\beta}(\mathbf{x}))$ is the empirical plug-in estimate of $\theta(\cdot)$ evaluated at $\mathbf{x}$. Properties of $\hat{\theta}(\cdot)$ can then be developed using well established properties of empirical estimates $\hat{G}$. In the sections that follow we will apply this approach to old and new function estimation problems.

Here are some further details on how to carry out the approach: First note that we are faced with the usual overfitting problem because for each $\mathbf{x}$ there is a β such that $\theta(\mathbf{x}; \beta) = \theta(\mathbf{x})$. One way to avoid overfitting would be to introduce a roughness penalty. For instance, we could minimize the penalized distance

$$\int [\theta(\mathbf{x}) - \theta(\mathbf{x}; \beta)]^2 w(\mathbf{x}) d\mathbf{x} + \lambda \int ||\nabla^m \theta(\mathbf{x}; \beta)||^2 d\mathbf{x},$$

where $||\nabla^m \theta(\mathbf{x}; \beta)||$ is the Euclidean norm of the mth order partial derivatives with respect to $\mathbf{x}$ of $\theta(\mathbf{x}; \beta)$, and $w(\mathbf{x})$ is a weight function. However, in this chapter we will use kernel smoothing which insists on closeness of $\theta(\mathbf{z} - \mathbf{x}; \beta)$ to $\theta(\mathbf{z})$ for $\mathbf{z}$ in a neighborhood of $\mathbf{x}$. For a given $\mathbf{x}$, we find the member $\theta(\cdot - \mathbf{x}, \beta(\mathbf{x}))$ of Θ that best approximates $\theta(\cdot)$ in a neighborhood of $\mathbf{x}$ by setting

$$\beta(\mathbf{x}) = \arg\min D_{\mathbf{x}}(\theta(\cdot), \theta(\cdot - \mathbf{x}; \beta)),$$

where

$$D_{\mathbf{x}}(\theta(\cdot), \theta(\cdot - \mathbf{x}; \beta)) = \int [\theta(\mathbf{z}) - \theta(\mathbf{z} - \mathbf{x}; \beta)]^2 K_{\mathbf{H}}(\mathbf{z} - \mathbf{x}) w(\mathbf{z}) d\mathbf{z}.$$

Here $K_{\mathbf{H}}(\mathbf{t}) = K(\mathbf{H}^{-1}\mathbf{t})/|\mathbf{H}|$, $K(\cdot)$ is a d-variate probability density (the kernel) with unique mode $\mathbf{0}$ and $\int \mathbf{u}K(\mathbf{u})d\mathbf{u} = \mathbf{0}$, $\mathbf{H}$ is a nonsingular $d \times d$ matrix (the bandwidth matrix), and $w(\cdot)$ is a nonnegative weight function which is continuous and nonzero at $\mathbf{x}$. The best local approximation to $\theta(\mathbf{x})$ in Θ is $\theta(\mathbf{0}, \hat{\beta}(\mathbf{x}))$.

A minimizer $\beta(\mathbf{x})$ of $D_{\mathbf{x}}$ exists if

$$\lim_{\|\beta\| \to \infty} \theta(\mathbf{x}; \beta) = \infty,$$

where $\| \cdot \|$ is the Euclidean norm. If further $\theta(\mathbf{x}; \beta)$ is differentiable in β, then $\beta = \beta(\mathbf{x})$ must satisfy the equations (for $j = 1, \cdots, q$)

$$\int \theta(\mathbf{z}) K_{\mathbf{H}}(\mathbf{z} - \mathbf{x}) w(\mathbf{z}) \nabla_j \theta(\mathbf{z} - \mathbf{x}; \beta) d\mathbf{z}$$

$$= \int \theta(\mathbf{z} - \mathbf{x}; \beta) \nabla_j \theta(\mathbf{z} - \mathbf{x}; \beta) K_{\mathbf{H}}(\mathbf{z} - \mathbf{x}) w(\mathbf{z}) d\mathbf{z}, \tag{1}$$

where ∇_j denotes the partial derivative with respect to β_j. It turns out that in many interesting cases we can choose $w(\cdot)$ so that the solution $\beta(\mathbf{x})$ to these equations can be expressed explicitly in terms of integrals of the form

$$\int v(\mathbf{z}) K_{\mathbf{H}}(\mathbf{z} - \mathbf{x}) dG(\mathbf{z}),$$

where $v(\mathbf{z})$ is a weight function and G is a function that admits a nonparametric empirical estimate $\hat{G}$. Thus by plugging in $d\hat{G}$ for dG, we obtain an empirical plug-in estimate $\hat{\beta}(\mathbf{x})$ of $\beta(\mathbf{x})$ and $\hat{\theta}(\mathbf{x}) = \theta(\mathbf{0}; \hat{\beta}(\mathbf{x}))$ is the empirical plug-in estimate of $\theta(\mathbf{x})$. In particular, if $\theta(\mathbf{x}; \beta)$ is a polynomial of order p, then $\partial^k \theta(\mathbf{z}; \beta)/\partial \mathbf{z}^k|_{\mathbf{z}=\mathbf{0}, \beta=\hat{\beta}(\mathbf{x})}$ is the empirical plug-in estimate of $\partial^k \theta(\mathbf{x})/\partial \mathbf{x}^k$.

Whenever there is a natural data based proxy $\hat{\theta}(\mathbf{x})$ for $\theta(\cdot)$, e.g. $\theta(\cdot)$ admits a nonparametric empirical estimate $\hat{\theta}(\mathbf{x})$, then we can choose $\tilde{\beta} = \tilde{\beta}(\mathbf{x})$ to minimize

$$\rho_{\mathbf{x}}(\hat{\theta}(\cdot), \theta(\cdot - \mathbf{x}, \beta)) = \int [\hat{\theta}(\mathbf{z}) - \theta(\mathbf{z} - \mathbf{x}; \beta)]^2 K_{\mathbf{H}}(\mathbf{z} - \mathbf{x}) w(\mathbf{z}) d\mathbf{z}. \tag{2}$$

This $\tilde{\beta}(\mathbf{x})$ will typically coincide with $\hat{\beta}(\mathbf{x})$. Thus the empirical plug-in approach is general in the sense that it is applicable when a data based proxy for $\theta(\cdot)$ is available as well as when no such proxy exists in the ordinary sense.

With the empirical plug-in approach, it is convenient to analyze the asymptotic properties of the resulting estimates $\hat{\beta}(\mathbf{x})$, because the random part of the estimation error can be expressed in terms of empirical processes of the form $\int v(\mathbf{z}) K_{\mathbf{H}}(\mathbf{z} - \mathbf{x}) d[\hat{G}(\cdot) - G(\cdot)]$, and the bias of the estimates can be easily obtained from the difference between $\int v(\mathbf{z}) K_{\mathbf{H}}(\mathbf{z} - \mathbf{x}) dG(\cdot)$ and the estimand $\theta(\cdot)$ and its derivatives. We will show that the estimates are boundary adaptive and highly efficient.

2. Plug-in Density Estimation

2.1. *The One Dimensional Case*

We assume that $X_1, \cdots, X_n$ are i.i.d. with distribution function F and density $f = F'$, $X_i \in \mathcal{R}$. Let

$$f(z - x; \beta) = \beta_0 + \sum_{j=1}^{p} \beta_j (z - x)^j$$

be a local polynomial approximation to $f(x)$ for z mear x. Now $q = p + 1$, $K_b(\cdot) = b^{-1} K(\cdot/b)$, where K is a unimodal density with $\int u K(u) du = 0$, and

$$D_x(f(\cdot), f(\cdot - x; \beta)) = \int [f(z) - f(z - x; \beta)]^2 K_b(z - x) w(z) dz.$$

The minimizer $\beta(x)$ of D_x satisfies the equations (for $\ell = 0, \cdots, p$)

$$\int (z - x)^{\ell} K_b(z - x) w(z) dF(z) = \sum_{j=0}^{p} \beta_j \int (z - x)^{j+\ell} K_b(z - x) w(z) dz.$$

That is, $\beta(x)$ can be expressed as a functional of $F(x)$

$$\mathbf{B}\beta(x) = \mathbf{C}^{-1} \mathbf{a}(F), \tag{3}$$

where $\mathbf{B} = \operatorname{diag}(1, b, \cdots, b^p)$, $\mathbf{C} = (c_{j\ell})$, $\mathbf{a}(F) = (a_0(F), \cdots, a_p(F))^T$,

$$c_{j\ell} = \int \left(\frac{z - x}{b}\right)^{j+\ell} K_b(z - x) w(z) dz,$$

$$a_{\ell}(F) = \int \left(\frac{z - x}{b}\right)^{\ell} K_b(z - x) w(z) dF(z), \quad j, \ell = 0, \cdots, p.$$

Replacing dF with $d\hat{F}$ where $\hat{F}$ is the empirical distribution gives us

$$\hat{\beta}(x) = \mathbf{B}^{-1} \mathbf{C}^{-1} \mathbf{a}(\hat{F}), \tag{4}$$

which estimates the density function $f(x)$ and its derivatives $\mathbf{f}(x) = (f(x), \cdots, f^{(p)}(x)/p!)^T$. The empirical plug-in estimate of $f^{(k)}(x)$ (for $k = 0, \cdots, p$) is $\hat{f}^{(k)}(x) \equiv k! \partial^k f(u; \beta)/\partial u^k |_{u=0, \beta=\hat{\beta}(x)} = k! \hat{\beta}_k$.

Assume that the supports of $f(x)$ and $K(u)$ are $[0, 1]$ and $[-1, 1]$, respectively. Let $c = \min(1, \max(0, x/b))$ and $d = \max(0, \min(1, (1 - x)/b))$ for $x \in [0, 1]$. Let $s_\ell = \int_{-c}^{d} K(u) u^\ell du$, $v_\ell = \int_{-c}^{d} u^\ell K^2(u) du$, $\mathbf{c}_p = (s_{p+1}, \cdots, s_{2p+1})^T$, $\mathbf{c}_{p+1} = (s_{p+2}, \cdots, s_{2p+2})^T$, $\mathbf{S} = (s_{i+j})$ and $\mathbf{V}^* = (v_{i+j})$ $(0 \leq i \leq p; 0 \leq j \leq p)$. Then we have the following theorems which describe the asymptotic properties of the empirical plug-in estimates for the density and its derivatives.

Theorem 2.1: Under conditions (A1)-(A4) in Appendix I, for $x \in [0, 1]$

$$\sqrt{nb} \ \{\mathbf{B}[\hat{\beta}(x) - \mathbf{f}(x)] - \frac{f^{(p+1)}(x) b^{p+1}}{(p+1)!} \mathbf{S}^{-1} \mathbf{c}_p$$

$$- \frac{f^{(p+2)}(x) b^{p+2}}{(p+2)!} \mathbf{S}^{-1} \mathbf{c}_{p+1} + o(b^{p+2})\} \xrightarrow{\mathcal{L}} \mathcal{N}\left(0, \mathbf{S}^{-1} \mathbf{V}^* \mathbf{S}^{-1} f(x)\right).$$

Remark 2.1: From the proof of Theorem 2.1, one sees that, when estimating a density function which is a polynomial of order p on an interval, the finite sample bias of the plug-in estimators on the interval is zero. This contrasts with the kernel density estimates based on higher order kernels,[8] for which the respective zero bias only holds true asymptotically.

Note that when x is an interior point (i.e. $x \in (b, 1 - b)$) and all odd-order moments of $K(\cdot)$ vanish, that is, $\int_{-1}^{1} u^\ell K(u) du = 0$ for ℓ odd, some of the entries in $\mathbf{S}^{-1} \mathbf{c}_p$ are zero and Theorem 2.1 is not very informative. We now take a closer look at this issue. Let $\mathbf{e}_k = (0, \cdots, 1, \cdots, 0)^T$, where $\mathbf{e}_k$ has one in the $(k+1)-$th component and zeros in the others. Then from Theorem 2.1

Theorem 2.2: Suppose $K(\cdot)$ is symmetric. Under conditions (A1)-(A4) in Appendix I, for an interior point $x \in (b, 1 - b)$

(i) For $p - k$ odd

$$\sqrt{nb^{2k+1}} \left\{ \frac{1}{k!} [\hat{f}^{(k)}(x) - f^{(k)}(x)] - \frac{f^{(p+1)}(x) b^{p+1-k}}{(p+1)!} \mathbf{e}_k^T \mathbf{S}^{-1} \mathbf{c}_p (1 + o(1)) \right\}$$

$$\xrightarrow{\mathcal{L}} \mathcal{N}\left(0, \mathbf{e}_k^T \mathbf{S}^{-1} \mathbf{V}^* \mathbf{S}^{-1} \mathbf{e}_k f(x)\right).$$

(ii) For $p - k$ even

$$\sqrt{nb^{2k+1}} \left\{ \frac{1}{k!}[\hat{f}^{(k)}(x) - f^{(k)}(x)] - \frac{f^{(p+2)}(x)b^{p+2-k}}{(p+2)!} \mathbf{e}_k^T \mathbf{S}^{-1} \mathbf{c}_{p+1}(1 + o(1)) \right\}$$

$$\xrightarrow{\mathcal{L}} \mathcal{N}\left(0, \mathbf{e}_k^T \mathbf{S}^{-1} \mathbf{V}^* \mathbf{S}^{-1} \mathbf{e}_k f(x)\right).$$

Remark 2.2: From Theorem 2.2, the asymptotic mean squared error ($AMSE_k$) for estimating $f^{(k)}(x)/k!$ ($k = 0, \cdots, p$) is

$$b^{2(p+1-k)}[\mathbf{e}_k^T \mathbf{S}^{-1} \mathbf{c}_p \frac{f^{(p+1)}(x)}{(p+1)!}]^2 + \frac{1}{nb^{2k+1}} \mathbf{e}_k^T \mathbf{S}^{-1} \mathbf{V}^* \mathbf{S}^{-1} \mathbf{e}_k f(x),$$

for $p - k$ odd; and

$$b^{2(p+2-k)}[\mathbf{e}_k^T \mathbf{S}^{-1} \mathbf{c}_{p+1} \frac{f^{(p+2)}(x)}{(p+2)!}]^2 + \frac{1}{nb^{2k+1}} \mathbf{e}_k^T \mathbf{S}^{-1} \mathbf{V}^* \mathbf{S}^{-1} \mathbf{e}_k f(x),$$

for $p - k$ even. Therefore, the optimal local bandwidth $b_{k,opt}(x)$ for estimating the kth derivative of $f(x)$ at x, in the sense of minimizing $AMSE_k(b, x)$, is

$$n^{-\frac{1}{2p+3}} \left(\frac{[(p+1)!]^2 \mathbf{e}_k^T \mathbf{S}^{-1} \mathbf{V}^* \mathbf{S}^{-1} \mathbf{e}_k f(x)}{2(p+1-k)[f^{(p+1)}(x)]^2 (\mathbf{e}_k^T \mathbf{S}^{-1} \mathbf{c}_p)^2} \right)^{\frac{1}{2p+3}}, \tag{5}$$

if $p - k$ is odd; and

$$n^{-\frac{1}{2p+5}} \left(\frac{[(p+2)!]^2 \mathbf{e}_k^T \mathbf{S}^{-1} \mathbf{V}^* \mathbf{S}^{-1} \mathbf{e}_k f(x)}{2(p+2-k)[f^{(p+2)}(x)]^2 (\mathbf{e}_k^T \mathbf{S}^{-1} \mathbf{c}_{p+1})^2} \right)^{\frac{1}{2p+5}}, \tag{6}$$

if $p - k$ is even.

In parallel to Theorem 2.2, consider the estimation of $f^{(k)}(x)$ at the left boundary point $x \in [0, b)$. Since $\mathbf{e}_k^T \mathbf{S}^{-1} \mathbf{c}_p$ does not vanish, we have the following result from Theorem 2.1:

Theorem 2.3: Suppose $K(\cdot)$ is symmetric. Under conditions (A1)-(A4) in Appendix I, for $x \in [0, b)$

$$\sqrt{nb^{2k+1}} \left\{ \frac{1}{k!}[\hat{f}^{(k)}(x) - f^{(k)}(x)] - \frac{1}{(p+1)!} f^{(p+1)}(0)b^{p+1-k} \mathbf{e}_k^T \mathbf{S}^{-1} \mathbf{c}_p \right.$$

$$\left. + o(b^{p+1-k}) \right\} \xrightarrow{\mathcal{L}} \mathcal{N}\left(0, \mathbf{e}_k^T \mathbf{S}^{-1} \mathbf{V}^* \mathbf{S}^{-1} \mathbf{e}_k f(0)\right).$$

Remark 2.3: From Theorems 2.2 and 2.3, the plug-in estimate of $f^{(k)}(\cdot)$ (for $p - k$ odd) is boundary adaptive in the sense that it automatically achieves the same convergence rate $O(n^{-\frac{2(p-k+1)}{2p+3}})$ on boundary points as in interior regions if a symmetric kernel and the optimal bandwidths $b_{k,opt} = O(n^{-\frac{1}{2p+3}})$ in (5) are employed. For $p - k$ even, the estimator is also consistent (even if $p = 0$ is employed), but the bias at the boundary is of order lower than at interior points if the optimal bandwidths $b_{k,opt}$ in (6) is used. This is similar to the case of the well-known local polynomial regression.

2.2. *The Multivariate Case*

Let $\mathbf{X}_1, \cdots, \mathbf{X}_n$ be i.i.d. as $\mathbf{X} \sim F$, where $\mathbf{X} \in \mathcal{R}^d$. The empirical distribution function is $\hat{F}(\mathbf{x}) = n^{-1} \sum_{i=1}^{n} 1[\mathbf{X}_i \leq \mathbf{x}]$, where $\mathbf{X} \leq \mathbf{x}$ is componentwise inequality, $\mathbf{x} \in \mathcal{R}^d$. We let $K(\mathbf{u})$ denote a d-variate kernel which is a d-variate density function with the marginals of K are all equal, say to $K_1(u)$, and $\int u_i u_j K(\mathbf{u}) d\mathbf{u} = 1[i = j]\mu_2(K)$ where $\mu_2(K) = \int u^2 K_1(u) du$ and u_i is the ith component of $\mathbf{u}$. Let $\mathbf{H} = \mathrm{diag}(h_1, \cdots, h_d)$ with $h_j > 0$ and set $K_\mathbf{H}(\mathbf{u}) = |\mathbf{H}|^{-1} K(\mathbf{H}^{-1}\mathbf{u})$. Assume that the support $\Omega = \{\mathbf{x} : f(\mathbf{x}) > 0\}$ of f is compact. Let $\alpha(\mathbf{x})$ and $\beta(\mathbf{x}) = (\beta_1(\mathbf{x}), \cdots, \beta_d(\mathbf{x}))^T$ be the minimizers of

$$D_\mathbf{x}(f(\cdot), f(\cdot - \mathbf{x}; \alpha, \beta)) \equiv \int_\Omega \{f(\mathbf{t}) - [\alpha + \beta^T(\mathbf{t} - \mathbf{x})]\}^2 K_\mathbf{H}(\mathbf{t} - \mathbf{x}) dt \quad (7)$$

and define the boundary correcting kernel estimate as $\hat{f}(\mathbf{x}) = \hat{\alpha}(\mathbf{x})$ and $\widehat{\nabla f}(\mathbf{x}) = \hat{\beta}(\mathbf{x})$, where $(\hat{\alpha}(\mathbf{x}), \hat{\beta}^T(\mathbf{x}))$ are the empirical plug-in estimates and $\nabla f(\mathbf{x})$ denotes the column vector of first-order partial derivatives of $f(\mathbf{x})$. That is, $(\hat{\alpha}(\mathbf{x}), \hat{\beta}^T(\mathbf{x}))$ is the solution to the following linear equation

$$\mathbf{S}_{\mathbf{x},\mathbf{H}} \cdot \begin{bmatrix} \alpha \\ \beta \end{bmatrix} = n^{-1} \sum_{i=1}^{n} \begin{bmatrix} 1 \\ \mathbf{X}_i - \mathbf{x} \end{bmatrix} K_\mathbf{H}(\mathbf{X}_i - \mathbf{x}), \quad (8)$$

where

$$\mathbf{S}_{\mathbf{x},\mathbf{H}} = \begin{bmatrix} \int_\Omega K_\mathbf{H}(\mathbf{t} - \mathbf{x}) dt & \int_\Omega (\mathbf{t} - \mathbf{x})^T K_\mathbf{H}(\mathbf{t} - \mathbf{x}) dt \\ \int_\Omega (\mathbf{t} - \mathbf{x}) K_\mathbf{H}(\mathbf{t} - \mathbf{x}) dt & \int_\Omega (\mathbf{t} - \mathbf{x})(\mathbf{t} - \mathbf{x})^T K_\mathbf{H}(\mathbf{t} - \mathbf{x}) dt \end{bmatrix}.$$

It is worthwhile to investigate the boundary behavior of the empirical plug-in estimates particular at the multivariate case. We will use the convenient mathematical formulation of the edge effect problem given by Ruppert and Wand.[12] Let $\varepsilon_{\mathbf{x},\mathbf{H}} = \{\mathbf{z} : \mathbf{H}^{-1}(\mathbf{x} - \mathbf{z}) \in \mathrm{supp}(K)\}$ be the support of $K_\mathbf{H}(\mathbf{x} - \cdot)$. We call $\mathbf{x}$ interior if $\varepsilon_{\mathbf{x},\mathbf{H}} \in \mathrm{supp}(f)$. Otherwise, $\mathbf{x}$ is called a boundary point. Let $\mathcal{D}_{\mathbf{x},\mathbf{H}} = \{\mathbf{z} : (\mathbf{x} + \mathbf{H}\mathbf{z}) \in \text{supp}(f) \cap \text{supp}(K)\}$. Then $\mathcal{D}_{\mathbf{x},\mathbf{H}} = \text{supp}(K)$ if and only if $\mathbf{x}$ is an interior point. When $\max|h_j| \to 0$ as $n \to \infty$, a fixed point $\mathbf{x}$ in the interior of $\text{supp}(f)$ is an interior point for large n. Denote by $\mathbf{x}_\partial$ a point on the boundary of $\text{supp}(f)$. Assume that there is a convex set $\mathcal{C}$ with nonnull interior and containing $\mathbf{x}_\partial$ such that

$$\inf_{\mathbf{x} \in \mathcal{C}} f(\mathbf{x}) > 0. \tag{9}$$

Simple algebra gives that

$$\mathbf{S}_{\mathbf{x},\mathbf{H}} = \mathbf{A}\mathbf{N}\mathbf{A},$$

where

$$\mathbf{A} = \begin{bmatrix} 1 & 0 \\ 0 & \mathbf{H} \end{bmatrix} \quad \text{and} \quad \mathbf{N} = \int_{\mathcal{D}_{\mathbf{x},\mathbf{H}}} \begin{bmatrix} 1 \\ \mathbf{u} \end{bmatrix} [1 \ \ \mathbf{u}^T] K(\mathbf{u}) d\mathbf{u}.$$

The solution to the linear equation (8) is then

$$(\hat{\alpha}(\mathbf{x}), \hat{\beta}^T(\mathbf{x}))^T = \mathbf{A}^{-1}\mathbf{N}_{\mathbf{x}}^{-1}\mathbf{A}^{-1}\mathbf{S}_n(\mathbf{x}), \tag{10}$$

where

$$\mathbf{A} = \begin{bmatrix} 1 & 0 \\ 0 & \mathbf{H} \end{bmatrix},$$

$$\mathbf{N}_{\mathbf{x}} = \int_{\mathcal{D}_{\mathbf{x},\mathbf{H}}} \begin{bmatrix} 1 \\ \mathbf{u} \end{bmatrix} [1 \ \ \mathbf{u}^T] K(\mathbf{u}) d\mathbf{u},$$

and

$$\mathbf{S}_n(\mathbf{x}) = n^{-1} \sum_{i=1}^{n} \begin{bmatrix} 1 \\ \mathbf{X}_i - \mathbf{x} \end{bmatrix} K_{\mathbf{H}}(\mathbf{X}_i - \mathbf{x}).$$

Note that for an interior point $\mathbf{x}$, $\mathcal{D}_{\mathbf{x},\mathbf{H}} = \text{supp}(K)$ and if $K(\cdot)$ is a d-variate density with $\int K(\mathbf{u})d\mathbf{u}=1$, $\int \mathbf{u}K(\mathbf{u})d\mathbf{u}=0$, then

$$\mathbf{N}_{\mathbf{x}} = \begin{bmatrix} 1 & 0 \\ 0 & \int \mathbf{u}\mathbf{u}^T K(\mathbf{u})d\mathbf{u} \end{bmatrix}$$

and the solution to (8) is

$$\hat{f}(\mathbf{x}) = n^{-1} \sum_{i=1}^{n} K_{\mathbf{H}}(\mathbf{X}_i - \mathbf{x}),$$

which is the usual multivariate kernel density estimate. However, for a boundary point, $\mathbf{x} = \mathbf{x}_\partial + \mathbf{H}\mathbf{c}$ say (for a fixed $\mathbf{c}$ in $\text{supp}(K)$), the plug-in estimate is different but automatically boundary-corrected, as shown in the following theorem.

Let $\mathcal{H}_f(\mathbf{x})$ be the $d \times d$ Hessian matrix of $f(\mathbf{x})$.

Theorem 2.4: Suppose that (9) holds. Then under conditions (B1)-(B3) in Appendix I, for any $\mathbf{x} \in \Omega$

$$\sqrt{n|\mathbf{H}|} \left\{ \begin{bmatrix} \hat{\alpha}(\mathbf{x}) - f(\mathbf{x}) \\ \mathbf{H}[\hat{\beta}(\mathbf{x}) - \nabla f(\mathbf{x})] \end{bmatrix} - \frac{1}{2}\mathbf{N}_{\mathbf{x}}^{-1}\mathbf{C}_{\mathbf{x}}(1 + o(1)) \right\}$$

$$\xrightarrow{\mathcal{L}} \mathcal{N}\left(0, \mathbf{N}_{\mathbf{x}}^{-1}\mathbf{T}_{\mathbf{x}}\mathbf{N}_{\mathbf{x}}^{-1} f(\mathbf{x})\right),$$

where

$$\mathbf{C}_{\mathbf{x}} = \int_{\mathcal{D}_{\mathbf{x},\mathbf{H}}} \begin{bmatrix} 1 \\ \mathbf{u} \end{bmatrix} K(\mathbf{u})\mathbf{u}^T \mathbf{H}\mathcal{H}_f(\mathbf{x})\mathbf{H}u\,d\mathbf{u},$$

$$\mathbf{T}_{\mathbf{x}} = \int_{\mathcal{D}_{\mathbf{x},\mathbf{H}}} \begin{bmatrix} 1 \\ \mathbf{u} \end{bmatrix} [1 \ \ \mathbf{u}^T]K^2(\mathbf{u})d\mathbf{u}.$$

Corollary 2.1: *For a symmetric density kernel $K(\cdot)$ and an interior point* $\mathbf{x}$,

$$\sqrt{n|\mathbf{H}|} \left[\hat{f}(\mathbf{x}) - f(\mathbf{x}) - \frac{1}{2}\mu_2(K)tr\{\mathbf{H}\mathcal{H}_f(\mathbf{x})\mathbf{H}\}(1 + o(1)) \right]$$

$$\xrightarrow{\mathcal{L}} \mathcal{N}\left(0, f(\mathbf{x})\int K^2(\mathbf{u})d\mathbf{u}\right).$$

From the theorem, we can see that the empirical plug-in estimate of the multivariate density is automatically boundary corrected. The optimal bandwidth can be developed in the same way as in the one dimensional case. Since the well known curse of dimensionality, care should be taken for direct use of the method in high dimensional case. In any case, the empirical plug-in estimate provides a manageable method for estimating densities in lower dimensional case.

3. Hazard Estimation with Censored Data

Let $T_1, \cdots, T_n$ be i.i.d. lifetimes distributed as T where $T \geq 0$, T has density f, distribution function F, hazard function $\lambda(x) = \Lambda'(x) = f(x)/[1 - F(x)]$ and integrated hazard function $\Lambda(x) = -\log[1 - F(x)]$. We are interested in estimating $\lambda(\cdot)$ based on $X_i = \min(T_i, C_i)$ and δ_i, where $C_1, \cdots, C_n$ are i.i.d. censoring times with distribution function G, $\delta_i = 1[T_i \geq C_i] = 1[T_i = C_i]$, $i = 1, \cdots, n$, and the C's are independent of the X's.

Let L denote the distribution function of X_i, then $\bar{L} = \bar{F}\bar{G}$, where for any distribution function H, $\bar{H} = 1 - H$ is the corresponding survival function. We consider the problem of estimating $\lambda^{(k)}(x)$ on the interval $[0, T]$ (for $k = 0, 1, \cdots$), where $T^* < \inf\{x : L(x) = 1\}$.

For a given kernel $K(\cdot)$, let $c = \min(1, \max(0, x/b))$, $s_{\ell,c} = \int_{-c}^{1} K(u)u^{\ell}du$, $v_{\ell,c} = \int_{-c}^{1} u^{\ell}K^2(u)du$, $\mathbf{S}_c = (s_{i+j,c})$ and $\mathbf{V}_c^* = (v_{i+j,c})$, $\mathbf{c}_{p,c} = (s_{p+1,c}, \cdots, s_{2p+1,c})^T$, $\mathbf{c}_{p+1,c} = (s_{p+2,c}, \cdots, s_{2p+2,c})^T$, $(0 \leq i \leq p;\ 0 \leq j \leq p;\ 0 \leq \ell \leq 2p + 2)$.

3.1. *The "Classic" Approach*

Consider finding $\beta(x) = \beta(x, \Lambda)$ that minimizes

$$\min_{\beta} \int_{u \geq 0} K_b(u - x)[\lambda(u) - \sum_{j=0}^{p} \beta_j(u - x)^j]^2 du. \qquad (11)$$

By Taylor expansion, the solution of the optimization problem, denoted by $\beta(x) \equiv (\beta_0, \cdots, \beta_p)^T$, approximates $\mathbf{a}(x) \equiv (\lambda(x), \cdots, \lambda^{(p)}(x)/p!)^T$. Our basic empirical estimating equations can be written as

$$\int_{u \geq 0} K_b(u - x)(u - x)^{\ell}d\Lambda(u) = \sum_{j=0}^{p} \beta_j \int_{u \geq 0} (u - x)^{j+\ell}K_b(u - x)du.$$

We will use the Nelson-Aalen's estimator of $\Lambda(x)$:

$$\hat{\Lambda}(x) = \int_0^x (1 - L_n(u))^{-1}dL_{1n}(u) = \sum_{i:X_{(i)} \leq x} \delta_{(i)}/(n - i + 1),$$

where $X_{(i)}$ are the order statistics of X_i, and $\delta_{(i)}$ is the concomitant of $X_{(i)}$, i.e. $\delta_{(i)}$ and $X_{(i)}$ are from the same observation. It follows that the empirical plug-in estimate can be written as

$$\hat{\beta}(x) = \beta(x, \hat{\Lambda}) = \mathbf{H}^{-1}\mathbf{S}_c^{-1} \cdot \mathbf{T}(x, \hat{\Lambda}), \qquad (12)$$

where $\mathbf{T}(x, \hat{\Lambda}) = (T_0(x, \hat{\Lambda}), \cdots, T_p(x, \hat{\Lambda}))^T$, and for $\ell = 0, \cdots, p$

$$T_\ell(x, \hat{\Lambda}) = \int_{u \geq 0} K_b(u - x)(\frac{u - x}{b})^{\ell}d\hat{\Lambda}(u)$$

$$= \sum_{i=1}^{n} K_b(X_{(i)} - x)\left(\frac{X_{(i)} - x}{b}\right)^{\ell} \frac{\delta_{(i)}}{n - i + 1}.$$

The plug-in estimator $\hat{\beta}(x)$ is exactly the same as the locally polynomial estimator of Jiang and Doksum.[6] Its asymptotic distribution can now be derived (see Jiang and Doksum[6]) by using well known properties of the process $\sqrt{n}[\hat{\Lambda}(x) - \Lambda(x)]$. In particular,

Theorem 3.1: If $K(\cdot)$ is symmetric, and if $p - k$ is odd, then under conditions (C1) – (C4) in Appendix I,

(i) If x is an interior point, i.e. $x \in (b, T^*]$,

$$\sqrt{nb^{2k+1}} \left\{ \frac{1}{k!}[\hat{\lambda}^{(k)}(x) - \lambda^{(k)}(x)] - \frac{\lambda^{(p+1)}(x)b^{p+1-k}}{(p+1)!} \mathbf{e}_k^T \mathbf{S}^{-1} \mathbf{c}_p(1 + o(1)) \right\}$$

$$\xrightarrow{\mathcal{L}} \mathcal{N}\left(0, \mathbf{e}_k^T \mathbf{S}^{-1} \mathbf{V}^* \mathbf{S}^{-1} \mathbf{e}_k \lambda(x)/\bar{L}(x)\right),$$

where $\mathbf{S}$, $\mathbf{V}^*$ and $\mathbf{c}_p$ are the same as $\mathbf{S}_c$, $\mathbf{V}_c^*$ and $\mathbf{c}_{p,c}$ but with $c = 1$

(ii) If x is a boundary point, $x = cb$ ($0 \le c < 1$), then

$$\sqrt{nb^{2k+1}} \left\{ \frac{1}{k!}[\hat{\lambda}^{(k)}(x) - \lambda^{(k)}(x)] - \frac{\lambda^{(p+1)}(0)b^{p+1-k}}{(p+1)!} \mathbf{e}_k^T \mathbf{S}_c^{-1} \mathbf{c}_{p,c}(1 + o(1)) \right\}$$

$$\xrightarrow{\mathcal{L}} \mathcal{N}\left(0, \mathbf{e}_k^T \mathbf{S}_c^{-1} \mathbf{V}_c^* \mathbf{S}_c^{-1} \mathbf{e}_k \lambda(0)/\bar{L}(0)\right).$$

3.2. *The Counting Process Approach*

The counting process approach, due to Aalen,[1,2] allows for more general frameworks in hazard estimation. For n independently observed cases, let $N_i(t)$ be 1 if the ith case failed in the interval $(0, t]$ and 0 otherwise; and let $Y_i(t)$ be 1 if the ith case is still at risk at time t and 0 otherwise. In our previous notation,

$$N_i(t) = 1(X_i \le t, \delta_i = 1), \quad Y_i(t) = 1(X_i \ge t),$$

where $(N_i(t), Y_i(t))$; $i = 1, \cdots, n$, are i.i.d.. See page 137 of Andersen, Borgan, Gill and Keiding.[3] In this section we compare our approach to that of Nielsen[10] and Nielsen and Tanggaard,[11] who proposed the estimate

$$\hat{\lambda}(t) = \arg\min_{\Theta} \sum_{i=1}^{n} \int_0^T [\Delta N_i(s) - \beta_0 - \beta_1(t-s)]^2 K_b(t-s)W(s)Y_i(s)ds, \quad (13)$$

where $\Delta N_i(s)$ is the Dirac "derivative" of $N_i(s)$ defined by the property

$$\int v(s)\Delta N_i(s)ds = \int v(s)dN_i(s)$$

for every integrable function $v(\cdot)$. Our empirical plug-in approach would be to replace $N_i(s)$ by its expected value $L_1(s) = E[N_i(s)]$, replace $\Delta N_i(s)$ with the subdistribution density $\ell_1(s) = L_1'(s)$ for the uncensored observations, replace $\beta_0 + \beta_1(t-s)$ by $\sum_{j=0}^{p} \beta_j(s-t)^j$, solve the new (13) for β in terms of $dL_1(\cdot)$, and to replace $dL_1(t)$ with $dN^{(n)}(t) = n^{-1} \sum_{i=1}^{n} dN_i(t)$ to get $\hat{\ell}_1(t) = \hat{\beta}_0(t)$. However,

$$L_1(t) = E[N_i(t)] = \int_0^t \bar{G}(v)f(v)dv,$$

$$\ell_1(t) = \bar{G}(t)f(t) \neq \lambda(t) = f(t)/\bar{F}(t).$$

Thus our approach would give an estimate of $\ell_1(t) = \bar{G}(t)f(t) \neq \lambda(t)$. To remedy this, note that $\bar{L}(t) = P(X_i \geq t) = \bar{F}(t)/\bar{G}(t)$ and $\lambda(t) = \ell(t)/\bar{L}(t)$, if we set $\hat{\lambda}_1(t) = \hat{\ell}_1(t)/\hat{\bar{L}}(t)$, where $\hat{\bar{L}}(t) = n^{-1} \sum_{i=1}^{n} 1(X_i \geq t)$, then $\hat{\lambda}_1(t)$ provides an estimate of $\lambda(t)$.

Another empirical plug-in approach would be to replace (13) with

$$\hat{\ell}_1(t) = \arg\min \int [\ell_1(s) - \sum_{j=0}^{p} \beta_j(s-t)^j]^2 K_b(s-t)v(s)ds, \qquad (14)$$

where $v(s)$ is a weight function which can be a random process, e.g. $v(s) = W(s)Y^{(n)}(s)$ where $Y^{(n)}(s) = \sum_{i=1}^{n} Y_i(s)$ and $W(s)$ is as in (13). Note that (14) can be solved explicitly for $\beta = \beta(t, dL_1)$ and the empirical plug-in estimate of $\lambda(t)$ is

$$\hat{\lambda}_2(t) = \hat{\beta}_0(t, dN^{(n)}(t))/\hat{\bar{L}}(t).$$

If $v(s) = W(s)Y^{(n)}(s)$, then some algebra gives $\hat{\lambda}_1(t) = \hat{\lambda}_2(t)$.

As in Nielsen and Tanggaard,[11] we assume that the weight function $W(s)$ and the exposure have local limits $\gamma(s)$ and $w(s)$ which are continuous at time t, i.e.

$$\sup_{s \in \Delta_{t,b}} |Y^{(n)}(s)/n - \gamma(s)| \xrightarrow{P} 0, \qquad (15)$$

$$\sup_{s \in \Delta_{t,b}} |W(s) - w(s)| \xrightarrow{P} 0, \qquad (16)$$

where $\Delta_{t,b} = [t - 10b, t + 10b]$, and $\gamma(s)$ and $w(s)$ are continuous at time t.

Theorem 3.2: Under conditions (C1)-(C4) in Appendix I and (15)-(16), for p odd, $i = 1, 2$ and $t \in [0, T^*]$

$$\sqrt{nb}[\hat{\lambda}_i(t) - \lambda(t) - \frac{\ell_1^{(p+1)}(t)b^{p+1}}{(p+1)!\bar{L}(t)} \mathbf{e}_1^T \mathbf{S}_c^{-1} \mathbf{c}_{p,c}(1 + o(1))] \xrightarrow{\mathcal{L}} \mathcal{N}(0, \sigma^2(t)),$$

where $\sigma^2(t) = \mathbf{e}_1^T \mathbf{S}_c^{-1} \mathbf{V}_c^* \mathbf{S}_c^{-1} \mathbf{e}_1 \lambda(t)/\bar{L}(t)$.

The plug-in estimate has the same asymptotic variance as that in Theorem 3.1, but their asymptotic bias are different although they are of the same order. Note that the above plug-in estimates are not smooth since $\hat{\bar{L}}(t)$ is empirical estimate which is not continuous. To get a smooth $\hat{\lambda}(t)$, one can employ a smooth technique for estimating $\bar{L}(t)$ such as the one in Jiang, et al.[5] Since the resulted $\hat{\bar{L}}(t)$ is $\sqrt{n}$-consistent, the asymptotic normality remains the same as before.

4. Plug-in Estimation of the Regression Function

Given i.i.d observations $\{X_i, Y_i\}_{i=1}^n$ of (X, Y), where (X, Y) have density $f(\cdot, \cdot)$ and distribution function $F(\cdot, \cdot)$, consider estimation of the regression function $m(x) = E[Y|X = x]$. In the following, we will illustrate that our empirical plug-in approach reduces to the kernel smoother, the local linear smoother, and others.

4.1. *Local Polynomial Regression As a Special Case of Empirical Plug-in Estimation*

Let $m(z - x; \beta) = \beta_0 + \sum_{j=1}^p \beta_j (z - x)^j$. Then a familiar approach is to minimize

$$\sum_{i=1}^n [\hat{m}(X_i) - m(X_i - x; \beta)]^2 K_h(X_i - x). \tag{17}$$

We can choose $\hat{m}(X_i) = Y_i$ (i.e. Fan and Gijbels,[4] page 58), which is the same as the empirical plug-in estimate obtained by computing $E_{\hat{P}}(Y|X_i) = Y_i$, where $\hat{P}$ is the empirical probability which assigns probability $1/n$ to each (X_i, Y_i), $i = 1 \cdots, n$. When $\hat{m}(X_i) = Y_i$, the solution to (17) is (e.g. Fan and Gijbels,[4] page 59),

$$\hat{\beta}(x) = (\mathbf{X}^T \mathbf{W}_K \mathbf{X})^{-1} \mathbf{X}^T \mathbf{W}_K \mathbf{Y}, \tag{18}$$

where $\mathbf{X}$ is a matrix with (i, j)th element $(X_i - x)^{j-1}$ for $1 \le i \le n$ and $1 \le j \le p + 1$, $\mathbf{Y} = (Y_1, \cdots, Y_n)^T$, and $\mathbf{W}_K = \text{diag}(K_h(X_1 - x), \cdots, K_h(X_n - x))$. The empirical plug-in approach yields the same $\hat{\beta}(x)$ as a special case. To see this set $w(z) = f_1(z)v(z)$, where $f_1(z)$ is the marginal density of X, and $v(z) \ge 0$ is a weight function which is continuous and positive at x.

Then the left hand side of (1) becomes

$$\int \left[\int y \frac{f(z,y)}{f_1(z)} dy \right] K_h(z-x) f_1(z) v(z) \nabla_j m(z-x;\beta) dz$$

$$= \int y K_h(z-x) v(z) \nabla_j m(z-x;\beta) dF(z,y),$$

where $\nabla_j m(z-x;\beta)$ denotes the partial derivative of $m(z-x;\beta)$ with respect to β_j. Let F_1 denote the marginal distribution function of X, then the right hand side of (1) becomes

$$\int m(z-x;\beta) \nabla_j m(z-x;\beta) K_h(z-x) v(z) dF_1(z).$$

If we choose $v(z) = $ constant > 0, replace F and F_1 with their empirical estimates, we find that (18) solves (1).

4.2. *A Different Plug-in Estimate of Regression*

We define $G(x) = \int y f(x,y) dy$; then $m(x) = G(x)/f_1(x)$. In Section 2, we defined the plug-in estimate $\hat{f}_1(\cdot)$. Similarly, we can define a plug-in estimate of $G(\cdot)$. Assume that the support of $f_1(x)$ is compact, $[0,1]$ say. Let us consider the following optimization problem: for $p = 0,1,2,\cdots$,

$$\min_\beta \int_0^1 [G(u) - \sum_{j=0}^p \beta_j (u-x)^j]^2 K_h(u-x) du. \tag{19}$$

By Taylor expansion, the solution of the optimization problem, denoted by $\beta(x) \equiv (\beta_0, \cdots, \beta_p)^T$, approximates $(G(x), \cdots, G^{(p)}(x)/p!)^T$. The left hand side of (1) becomes

$$\int K_h(u-x)(u-x)^\ell y dF(x,y)$$

and the plug-in estimate $\hat{\beta}(x)$ is the solution to the linear equations: for $\ell = 0, \cdots, p$,

$$n^{-1} \sum_{i=1}^n K_h(X_i - x)(X_i - x)^\ell Y_i = \sum_{j=0}^p \beta_j \int_0^1 (u-x)^{j+\ell} K_h(u-x) du. \tag{20}$$

Let $c = \min(1, \max(0, x/h))$ and $d = \max(0, \min(1, (1-x)/h))$ for $x \in [0,1]$. Then by a change of variable (20) can be rewritten as

$$n^{-1} \sum_{i=1}^n K_h(X_i - x)(X_i - x)^\ell Y_i = \sum_{j=0}^p \beta_j \int_{-c}^d (th)^{j+\ell} K(t) dt. \tag{21}$$

It follows that the plug-in estimator $\hat{\beta}(x)$ admits the following form

$$\mathbf{H}\hat{\beta}(x) = \mathbf{S}(x)^{-1} \cdot \mathbf{S}_n(x), \tag{22}$$

where $\mathbf{H} = \text{diag}(1, h, \cdots, h^p)$, $\mathbf{S}(x) = (s_{i+j})$ is a $(p+1) \times (p+1)$ matrix with $s_{i+j} = \int_{-c}^{d} K(u)u^{i+j} du$ for $i, j = 0, \cdot, \ell$, $\mathbf{S}_n(x) = (S_{n0}(x), \cdots, S_{np}(x))^T$, and

$$S_{n\ell}(x) = \sum_{i=1}^{n} K_h(X_i - x)(\frac{X_i - x}{h})^{\ell} Y_i, \quad \text{for } \ell = 0, \cdots, p. \tag{23}$$

Similarly, we have the following expression for $\hat{\mathbf{f}}_1(x) \equiv (\hat{f}_1, \cdots, \hat{f}_1^{(p)}/p!)^T$ (see Section 2.1):

$$\mathbf{H}\hat{\mathbf{f}}_1(x) = \mathbf{S}(x)^{-1} \cdot \mathbf{T}_n(x), \tag{24}$$

where

$$T_{n\ell}(x) = \sum_{i=1}^{n} K_h(X_i - x)(\frac{X_i - x}{h})^{\ell}, \quad \text{for } \ell = 0, \cdots, p. \tag{25}$$

Then

$$\hat{m}(x) = \hat{G}(x)/\hat{f}_1(x) = \frac{\mathbf{e}_1^T \mathbf{S}(x)^{-1} \cdot \mathbf{S}_n(x)}{\mathbf{e}_1^T \mathbf{S}(x)^{-1} \cdot \mathbf{T}_n(x)},$$

where $\mathbf{e}_1^T = (1, 0, \cdots, 0)$. In particular, when $p = 0$ is employed, the estimator is

$$\hat{m}(x) = S_{n0}(x)/T_{n0}(x) = \frac{\sum_{i=1}^{n} K_h(X_i - x)Y_i}{\sum_{i=1}^{n} K_h(X_i - x)},$$

which is just the Nadaraya-Watson (N-W) estimate (Nadaraya[9]; Watson[13]). For $p = 1$, we have

$$\hat{m}(x) = \frac{s_2 S_{n0}(x) - s_1 S_{n1}(x)}{s_2 T_{n0}(x) - s_1 T_{n1}(x)}. \tag{26}$$

If a symmetric kernel $K(\cdot)$ is employed, then for an interior point $x = 0$ (i.e. $x \in [h, 1 - h]$), $s_1 = 0$ and $\hat{m}(x) = S_{n0}(x)/T_{n0}(x)$ is also the N-W estimator; for a boundary point x (i.e. $x \in [0, h) \cup (1 - h, 1]$), $s_1 \neq 0$ and the estimator is different from the N-W estimator. Therefore, for $p = 1$ the plug-in estimator $\hat{m}(x)$ can be regarded as boundary corrected N-W estimator, which shares some advantages with the N-W estimator, such as a simple closed formula for analysis and computation.

In the following, we focus only on the case with $p = 1$, which will show that the corresponding estimator $\hat{m}(x)$ keeps its convergence rate for boundary points, i.e. the estimator automatically corrects the boundary effect.

4.2.1. *Asymptotic Properties*

The following theorem describes the asymptotic behavior and the boundary adaptation of the estimator $\hat{m}(x)$ in (26).

Theorem 4.1: Suppose conditions (D1)-(D5) in Appendix I holds. Then for $x \in [0,1]$, $\sqrt{nh}[\hat{m}(x) - m(x) - B(x)]$ is asymptotically normal with mean zero and variance

$$D(x) = \frac{\sigma^2 \int_c^d (s_2 - u s_1)^2 K^2(u) du}{f_1(x)(s_0 s_2 - s_1^2)^2},$$

where

$$B(x) = \frac{h^2}{2} \frac{s_2^2 - s_1 s_3}{s_0 s_2 - s_1^2} \left[m''(x) + \frac{2 f_1'(x) m'(x)}{f_1(x)} \right] + o_p(h^2).$$

Remark 4.1: A consequence of this theorem is that the estimator $\hat{m}(x)$ keeps its convergence rate for the boundary points. That is, the estimator is automatically boundary adaptive.

From Theorem 4.1, the asymptotic mean squared error (*AMSE*) for estimating $m(x)$ is

$$
\begin{aligned}
AMSE(h,x) = {} & \frac{h^4 (s_2^2 - s_1 s_3)^2}{4(s_0 s_2 - s_1^2)^2} [m''(x) + \frac{2 m'(x) f_1'(x)}{f_1(x)}]^2 \\
& + \frac{\sigma^2 \int_c^d (s_2 - u s_1)^2 K^2(u) du}{nh f_1(x)(s_0 s_2 - s_1^2)^2}.
\end{aligned}
\tag{27}
$$

Suppose $x = ch$, $0 < c < 1$. Then for h small, $d = 1$, and (27) coincides with the *AMSE* expression for the locally linear estimator (18) except for the $2m'(x) f_1'(x)/f_1(x)$ term (see Fan and Gijbels,[4] p.17). Thus (26) is for some models more efficient than the locally linear estimate at boundary and interior points. The optimal local bandwidth for estimating $m(x)$ at x, in the sense of minimizing $AMSE(h, x)$, is

$$
h_{opt}(x) = n^{-\frac{1}{5}} \left(\frac{\sigma^2 \int_c^d (s_2 - u s_1)^2 K^2(u) du}{f_1(x)(s_2^2 - s_1 s_3)^2 [m''(x) + \frac{2 m'(x) f_1'(x)}{f_1(x)}]^2} \right)^{\frac{1}{5}}.
$$

Acknowledgments

This work was supported in part by the Chinese NSF grants 10001004 and 39930160, and by NSF grant DMS-9971301. We would like to thank Peter Bickel for valuable remarks on the early drafts of this chapter.

Appendix I: Conditions

(A1) The density function $f(x)$ has a continuous $(p+2)th$ derivative at the point x in its compact support $[0,1]$, say.

(A2) The kernel function K is a continuous function with bounded support $[-1,1]$, say.

(A3) The weight function $w(\cdot)$ is nonzero and has a continuous derivative at the point x.

(A4) $h \to 0$ and $nh \to +\infty$ as $n \to +\infty$.

(B1) The density function $f(\mathbf{x})$ has a continuous second derivative at the point $\mathbf{x}$ in its compact support Ω.

(B2) The kernel function K is a continuous function with bounded support.

(B3) $h \to 0$ and $nh \to +\infty$ as $n \to +\infty$

(C1) The hazard function $\lambda(x)$ has a continuous $(p+2)th$ derivative at the point x.

(C2) The bandwidth b tends to zero such that $nb \to +\infty$ as $n \to +\infty$. Let $\mathbf{B} = \mathrm{diag}(1, b, \cdots, b^p)$.

(C3) $L(x)$ is continuous at the point x.

(C4) The kernel function K is a continuous function of bounded variation and with bounded support $[-1,1]$.

(D1) The kernel function $K(x)$ is continuous with bounded support $[-1,1]$, say.

(D2) The density function $f(x)$ is bounded away from 0 and has a second derivative and compact support $[0,1]$, say.

(D3) The second derivative of regression function $m(x)$ is bounded and continuous.

(D4) The bandwidth sequence $h_n \to 0$ and satisfies $nh_n^3 \to \infty$ as $n \to \infty$.

(D5) $\mathrm{Var}(\varepsilon) = \sigma^2 < \infty$, where $\varepsilon = Y - m(X)$.

Appendix II: Proof of the theorems

Proof of Theorem 2.1. By (3) and (4), we have

$$\mathbf{B}[\hat{\beta}(x) - \beta(x)] = \mathbf{C}^{-1}[a(\hat{F}_n) - a(F)].$$

Note that because $supp(K) = [-1, 1]$, we need only consider $|z - x| \leq b$ and $|X_i - x| \leq b$. By a change of variable and the continuity of $w(x)$

$$c_{j\ell} = \int \left(\frac{z-x}{b}\right)^{j+\ell} K_b(z-x)w(z)dz = s_{j+\ell}w(x)(1 + o(1)),$$

$$a_\ell(F) = \int_{-c}^{d} t^\ell K(t)f(x+tb)w(x+tb)dt$$

$$= \int_{-c}^{d} [\sum_{k=0}^{p+2} \frac{1}{k!}f^{(k)}(x)(tb)^k + o(b^{p+2})t^{p+2}]t^\ell K(t)w(x+tb)dt$$

$$= (c_{0,\ell}, \cdots, c_{p,\ell})\mathbf{Bf} + \sum_{j=1}^{2} b^{p+j}s_{\ell+p+j}w(x)f^{(p+j)}(x)/(p+j)! + o(b^{p+2}).$$

Then $\mathbf{C} = \mathbf{S}w(x)(1 + o(1))$, and

$$a(F) = \mathbf{CBf} + \sum_{j=1}^{2} b^{p+j}\mathbf{c}_{p+j-1}w(x)f^{(p+j)}(x)/(p+j)! + o(b^{p+2}),$$

so that

$$\mathbf{B}\beta(x) = \mathbf{C}^{-1}a(F)$$
$$= \mathbf{Bf} + b^{p+1}\mathbf{S}^{-1}\mathbf{c}_p w(x)f^{(p+1)}(x)/(p+1)!$$
$$+ b^{p+2}\mathbf{S}^{-1}\mathbf{c}_{p+1}w(x)f^{(p+2)}(x)/(p+2)! + o(b^{p+2}). \quad (A.1)$$

Note that

$$a_\ell(\hat{F}) = n^{-1} \sum_{i=1}^{n} \left(\frac{X_i - x}{b}\right)^\ell K_b(X_i - x)w(X_i)$$

is an i.i.d.'s sum. Applying the central limit theorem and the Cramér-Wold device, we obtain

$$\sqrt{nb}[a(\hat{F}) - a(F)] \to \mathcal{N}(0, \mathbf{V}^*w^2(x)f(x)).$$

Then $\sqrt{nb}\mathbf{C}^{-1}[a(\hat{F}) - a(F)] \to \mathcal{N}(0, \mathbf{S}^{-1}\mathbf{V}^*\mathbf{S}^{-1}f(x))$, which combined with (A.1) completes the proof of the theorem. ◇

Proof of Theorem 2.4. Using the first order Taylor expansion and the same argument as for Theorem 2.1, we obtain the bias term

$$E\begin{bmatrix} \hat{\alpha}(\mathbf{x}) - f(\mathbf{x}) \\ \mathbf{H}[\hat{\beta}(\mathbf{x}) - \nabla f(\mathbf{x})] \end{bmatrix} = \frac{1}{2}\mathbf{N}_\mathbf{x}^{-1}\mathbf{C}_\mathbf{x}(1 + o(1)). \quad (A.2)$$

It is straightforward to show that

$$\mathrm{Cov}(\mathbf{A}^{-1}\mathbf{S}_n(\mathbf{x}), \mathbf{A}^{-1}\mathbf{S}_n(\mathbf{x})) = n^{-1}|\mathbf{H}|^{-1}f(\mathbf{x})\mathbf{T}_{\mathbf{x}}(1 + o(1)).$$

Note that $\mathbf{A}^{-1}\mathbf{S}_n(\mathbf{x})$ is an i.i.d.'s sum, Then using the central limit theorem and the Cramér-Wold device, one gets

$$\sqrt{n|\mathbf{H}|}\begin{bmatrix} \hat{\alpha}(\mathbf{x}) - f(\mathbf{x}) \\ \mathbf{H}[\hat{\beta}(\mathbf{x}) - \nabla f(\mathbf{x})] \end{bmatrix} \to \mathcal{N}(0, f(\mathbf{x})\mathbf{T}_{\mathbf{x}}).$$

This together with (A.2) yields the result of the theorem. $\diamond$

Proof of Corollary 2.1. For an interior point $\mathbf{x}$, simple algebra gives $\mathbf{e}_1^T\mathbf{N}_{\mathbf{x}}^{-1}\mathbf{C}_{\mathbf{x}} = \int K(\mathbf{u})\mathbf{u}^T\mathbf{H}\mathcal{H}_f\mathbf{H}\mathbf{u}d\mathbf{u} = \mu_2(K)tr(\mathbf{H}\mathcal{H}_f\mathbf{H})$, and $\mathbf{e}_1^T\mathbf{N}_{\mathbf{x}}^{-1}\mathbf{T}_{\mathbf{x}}\mathbf{N}_{\mathbf{x}}^{-1}\mathbf{e}_1 = \int K^2(\mathbf{u})d\mathbf{u}$. The corollary now follows from Theorem 2.4. $\diamond$

Proof of Theorem 3.2. Note that (14) is for estimating the density function $\ell(t)$ for the uncensored observations, by Theorem 2.1 we have

$$\sqrt{nb}\{[\hat{\ell}_1(t) - \ell_1(t)] - \frac{\ell_1^{(p+1)}(t)b^{p+1}}{(p+1)!}\mathbf{S}_c^{-1}\mathbf{c}_{p,c} + o(b^{p+1})\}$$

$$\xrightarrow{\mathcal{L}} \mathcal{N}\left(0, \mathbf{e}_1^T\mathbf{S}^{-1}\mathbf{V}^*\mathbf{S}^{-1}\mathbf{e}_1\ell_1(t)\right).$$

This combined with $\hat{\bar{L}}(t) = \bar{L}(t) + O_p(1/\sqrt{n})$ yields the result of the theorem. $\diamond$

Proof of Theorem 4.1. Note that

$$S_{n\ell}(x) = n^{-1}\sum_{i=1}^n (X_i - x)^\ell h^{-\ell}K_h(X_i - x)\, Y_i$$

$$= n^{-1}\sum_{i=1}^n (X_i - x)^\ell h^{-\ell}K_h(X_i - x)\, [m(X_i) + \varepsilon_i]$$

$$\equiv B_{n\ell} + V_{n\ell}.$$

Then by (26)

$$\hat{m}(x) - m(x) = \frac{s_2(B_{n0} - m(x)T_{n0}) - s_1(B_{n1} - m(x)T_{n1})}{s_2T_{n0} - s_1T_{n1}} + \frac{s_2V_{n0} - s_1V_{n1}}{s_2T_{n0} - s_1T_{n1}}$$

$$\equiv B_n + V_n.$$

We will show that the first term above contributes the bias of the estimator $\hat{m}(x)$ and the second term the variance:

Since $K(\cdot)$ vanishes outside of $[-1, 1]$, we need only consider $|X_i - x| \le h$. Note that for $\ell = 0, 1$

$$B_{n\ell} - m(x)T_{n\ell} = n^{-1}\sum_{i=1}^{n}(X_i - x)^{\ell}h^{-\ell}K_h(X_i - x)\,[m(X_i) - m(x)]. \quad (A.3)$$

By Taylor expansion, one gets

$$m(X_i) - m(x) = \sum_{k=1}^{2} h^k m^{(k)}(x)/k!\,[(X_i - x)/h]^k + o(h^2),$$

uniformly for $|X_i - x| \le h$. This combined with (A.3) yields

$$B_{n\ell} - m(x)T_{n\ell} = hm'(x)n^{-1}\sum_{i=1}^{n}[(X_i - x)/h]^{\ell+1}K_h(X_i - x)$$

$$+\frac{h^2 m''(x)}{2}n^{-1}\sum_{i=1}^{n}[(X_i - x)/h]^{\ell+2}K_h(X_i - x) + o(h^2).$$

By directly computing the mean and variance we have

$$B_{n\ell} - m(x)T_{n\ell} = hs_{\ell+1}m'(x)f(x) + h^2 s_{\ell+2}[m'(x)f'(x) + \frac{1}{2}m''(x)f(x)]$$

$$+o(h^2) + O_p(h/\sqrt{nh}). \quad (A.4)$$

Note that

$$s_2 T_{n0} - s_1 T_{n1} = (s_2 s_0 - s_1^2)f(x) + \frac{h^2}{2}(s_2^2 - s_1 s_3)f''(x)$$

$$+o(h^2) + O_p(h/\sqrt{nh}). \quad (A.5)$$

It follows from (A.4) and (A.5) that

$$B_n = \frac{h^2}{2}\frac{s_2^2 - s_1 s_3}{s_0 s_2 - s_1^2}\frac{m''(x)f(x) + 2m'(x)f'(x)}{f(x)} + o(h^2).$$

Note that $E(V_{n\ell}) = 0$, and by a change of variable

$$\mathrm{Var}(V_{n\ell}) = \frac{\sigma^2}{nh^2}E[(X_i - x)^{2\ell}h^{-2\ell}K^2(\frac{X_i - x}{h})]$$

$$= \frac{\sigma^2}{nh^2}\int_0^1 (t - x)^{2\ell}h^{-2\ell}K^2(\frac{t - x}{h})f(t)dt$$

$$= \frac{\sigma^2}{nh}f(x)\int_{-c}^{d} u^{2\ell}K^2(u)du(1 + o(1)), \quad (A.6)$$

$$\text{Cov}(V_{n0}, V_{n1}) = \frac{\sigma^2}{nh^2} \int_0^1 \frac{t-x}{h} K^2(\frac{t-x}{h}) f(t) dt$$

$$= \frac{\sigma^2}{nh} f(x) \int_{-c}^d u K^2(u) du (1 + o(1)). \qquad (A.7)$$

It follows from (A.6) and (A.7) that

$$\text{Var}(s_2 V_{n0} - s_1 V_{n1}) = \frac{\sigma^2}{nh} f(x) \int_{-c}^d (s_2 - u s_1)^2 K^2(u) du (1 + o(1)). \qquad (A.8)$$

Since $s_2 V_{n0} - s_1 V_{n1}$ is a sum of iid random variables of mean zero, it is asymptotically normal with mean zero and variance as in (A.8). This together with (A.5) gives the variance of V_n, and the result of the theorem follows. $\diamond$

References

1. O. O. Aalen, Statistical Inference for a Family of Counting Processes, Ph.D. thesis, University of California, Berkeley, (1975).

2. O. O. Aalen, Nonparametric inference for a family of counting processes, *Annals of Statistics*, **6**, 701–726 (1978).

3. P. K. Andersen, Ø. Borgan, R. D. Gill and N. Keiding, *Statistical Models Based on Counting Processes* (Springer, New York, 1993).

4. J. Fan and I. Gijbels, *Local Polynomial Modelling and Its Applications* (Chapman and Hall, London, 1996).

5. J. Jiang, B. Cheng and X. Wu, On estimation of survival function under random censoring model, *Science in China*, **45**, 503–511 (2002).

6. J. Jiang and K. Doksum, On Local Polynomial Estimation of Hazard Rates and Their Derivatives under Random Censoring, *Constance van Eeden Volume*, IMS (to appear) (2002).

7. M. C. Jones, Simple boundary correction for kernel density estimation, *Statistics and Computing*, **3**, 135 – 146 (1993).

8. H.-G. Müller, Smooth optimum kernel estimators of densities, regression curves and modes, *Annals of Statistics*, **12**, 766–774 (1984).

9. E. A. Nadaraya, On estimating regression, *Theory Prob. Appl.*, **10**, 186 –190 (1964).

10. J. P. Nielsen, Marker dependent kernel hazard estimation from local linear estimation, *Scandinavian Actuarial Journal*, **2**, 113–124 (1998).

11. J. P. Nielsen and C. Tanggaard, Boundary and bias correction in kernel hazard estimation, *Scandinavian Journal of Statistics*, **28**, 675–698 (2001).

12. D. Ruppert and M. P. Wand, Multivariate locally weighted least squares regression, *Annals of Statistics*, **22**, 1346–1370 (1994).

13. G. S. Watson, Smooth regression analysis, *Sankhyá*, Series A, **26**, 359–372 (1964).

29

SOME EXACT AND APPROXIMATE CONFIDENCE REGIONS FOR THE RATIO OF PERCENTILES FROM TWO DIFFERENT DISTRIBUTIONS

Richard A. Johnson and Li-Fei Huang

Department of Statistics,
University of Wisconsin-Madison
1210 West Dayton, Madison, WI 53706, U.S.A.
E-mail: rich@stat.wisc.edu, lfhuang@stat.wisc.edu

Engineers commonly use the ratio of means to compare the strengths of two materials or even the deterioration of the strength of one material undergoing aging or various stresses. However, wood engineers also use the ratio of the fifth or other lower percentiles when comparing lumber of two different grades, moisture content, or sorted under two different grading systems. Motivated by this application, we develop some exact, and some approximate, confidence regions for the ratio of percentiles from two different populations. This work generalizes earlier work on inferences concerning the ratio of two means.

Our exact sampling results are limited to a normal linear model case, a related log-normal case, and negative exponential distributions. Other general distributions are treated by large sample methods.

We also consider the exact nonparametric Bayesian approach to obtain regions for the ratio of percentiles.

1. Introduction

Engineers often express a comparison of two materials in terms of the ratio of their mean strengths. In the wood industry, it is common practice to compare lumber of two different dimensions, grades or species. For example, they could consider the ratio of mean bending strengths for lumber of two different dimensions. Because lumber standards in the United States are given in terms of population fifth percentiles, the ratio is often expressed in terms of the fifth percentiles rather than means. For example, Aplin et. al.(1986) give point estimates of the ratio of dry to green values of the fifth percentiles of the modulus of rupture.

Estimates of a ratio can also be used to establish allowable properties for lumber in specified grade-size-species combinations that were never tested. It is our purpose here to address procedures for constructing confidence intervals for the ratios of percentiles from two different distributions for life-length or a strength property.

Most of the existing literature on ratios deals with the ratio of means and, in particular, ratios of means of normal distributions. Fieller(1954) treats bivariate normal distributions. McDonald(1981) obtained confidence regions for the ratio of means in the normal theory setting of fitting a straight line. Hwang(1995) used a resampling approach to construct confidence regions for this same ratio. Huang and Johnson(2002) study the two sample normal theory case. Evans, Johnson, Green and Verrill (2002) obtain large sample confidence intervals for the ratio of Weibull percentiles.

2. Some exact results

Throughout this paper, we let $X_1, \ldots, X_{n_1}$ be a random sample of size n_1 from the cumulative distribution function(cdf) $F_1(\cdot)$, $Y_1, \ldots, Y_{n_2}$ be a random sample of size n_2 from $F_2(\cdot)$, and let the two random samples be independent. Although it is common to compare the same percentiles, initially we allow for different percentiles, or p_1 and p_2, to be specified.

$$\xi_{1p_1} = \inf\{x : F_1(x) \geq p_1\} \qquad \text{and} \qquad \xi_{2p_2} = \inf\{y : F_2(y) \geq p_2\}$$

The ratio of population percentiles is $\theta = \xi_{1p_1}/\xi_{2p_2}$.

We first review a newly developed confidence interval procedure for θ when the two distributions are normal.

2.1. *Normal distributions with equal variances*

Suppose that F_i is a normal distribution with mean μ_i and variance σ_i^2, for $i = 1, 2$. Then, $\xi_{ip_i} = \mu_i + \sigma_i \Phi^{-1}(p_i)$ is the lower $100p_i$-th percentile $i = 1, 2$ where $\Phi(\cdot)$ is the standard normal cdf. We also let z_{p_i} be the upper $100p_i$-th percentile of the standard normal distribution, so $\xi_{ip_i} = \mu_i - z_{p_i}\sigma_i$ for $i = 1, 2$.

Under the further restriction of equal variances, Huang and Johnson(2002) obtain confidence regions for θ by establishing that the random quantity

$$T(\theta) = \left(\sqrt{\frac{1}{n_1} + \frac{\theta^2}{n_2}} \right)^{-1} \frac{(\,\overline{X} - \theta\overline{Y}\,)}{\sqrt{\dfrac{\sum_{i=1}^{n_1}(\,X_i - \overline{X}\,)^2 + \sum_{i=1}^{n_2}(\,Y_i - \overline{Y}\,)^2}{n_1 + n_2 - 2}}}$$

follows the non-central t distribution with $n_1 + n_2 - 2$ degrees of freedom and non-centrality parameter

$$\delta(\theta) = \left(\sqrt{\frac{1}{n_1} + \frac{\theta^2}{n_2}} \right)^{-1} (z_{p_1} - \theta z_{p_2})$$

A $100(1 - \alpha)\%$ confidence region for θ consists of all θ_0 such that the equal-tailed level α test of $H_0 : \theta = \theta_0$ versus $H_1 : \theta \neq \theta_0$ does not reject H_0. The confidence region can be a finite interval, the complement of an interval, or the real line.

2.2. *Log normal distributions*

Suppose the two populations are lognormal. Taking logarithms, we convert the problem to one about the difference of percentiles. An exact procedure can be obtained when the two distributions have equal variances on the log scale. We present the sampling distribution in a slightly more general form where the ratio of standard deviations $\mathrm{std}(\ln(X_1)) \,/\, \mathrm{std}(\ln(Y_1)) = \sigma_1/\sigma_2 = k$ is known.

Theorem 1: The random quantity

$$T(\theta, k) = \left(\sqrt{\frac{k^2}{n_1} + \frac{1}{n_2}} \right)^{-1}$$

$$\frac{\overline{\ln(X)} - \overline{\ln(Y)} - \ln(\theta)}{\sqrt{\dfrac{k^{-2}\sum_{i=1}^{n_1}(\,\ln(X_i - \overline{\ln(X)}\,)^2 + \sum_{i=1}^{n_2}(\,\ln(Y_i - \overline{\ln(Y)}\,)^2}{n_1 + n_2 - 2}}}$$

follows the non-central t distribution with $n_1 + n_2 - 2$ degrees of freedom and non-centrality parameter

$$\delta(\theta, k) = \left(\sqrt{\frac{k^2}{n_1} + \frac{1}{n_2}} \right)^{-1} (k z_{p_1} - z_{p_2})$$

Here the non-centrality parameter does not depend on θ. Moreover, the distribution becomes a central t if $p_1 = p_2$ and $k = 1$.

The $100(1-\alpha)\%$ confidence region for $\ln(\theta)$ obtained from the two-sided tests of $H_0 : \theta = \theta_0$, can be transformed, by the exponential function, to a confidence region for θ.

2.3. *Exponential and Weibull Distributions*

Let $F_i(\cdot)$ have a negative exponential distribution with scale parameter α_i. Then $\theta = \alpha_1[-\ln(1-p_1)] \, / \, \alpha_2[-\ln(1-p_2)]$ and the maximum likelihood estimator is $\hat{\theta} = \overline{X}\,[-\ln(1-p_1)] \, / \, \overline{Y}\,[-\ln(1-p_2)]$. A $100(1-\alpha)\%$ confidence interval for θ is given by

$$\left(\frac{\hat{\theta}\ln(1-p_1)}{\ln(1-p_2)]\, F_{2n_1,2n_2}(\alpha/2)} \, , \quad \frac{\hat{\theta}\ln(1-p_1)}{\ln(1-p_2)\, F_{2n_1,2n_2}(1-\alpha/2)} \right)$$

where $F_{2n_1,2n_2}(\alpha/2)$ is the upper $100\alpha/2$ percentile of the $F-$distribution with n_1 and n_2 degrees of freedom.

Under two-parameter Weibull distributions,

$$\overline{F}_i(x) = \exp[-(x \, / \, \alpha_i)^{\beta_i}] \quad i = 1, 2 \tag{1}$$

taking logarithms converts the problem to a difference of percentiles. Lawless (1982) gives exact distributions for estimators of percentiles for a single sample. Applying this result separately for each sample, we could then use numerical integration or simulation to obtain the sampling distribution of $\hat{\theta}$. However, in the context of the single sample problem, Lawless(1982) suggests that maximum likelihood estimation gives approximately the same answers except possibly for small samples. We consider a maximum likelihood ratio approach below.

3. Large Sample Approach-Weibull Distributions

When the underlying populations are each a two parameter Weibull distribution (1), we have the four-variate parameter $\phi = (\alpha_1, \beta_1, \alpha_2, \beta_2)$ and the likelihood is

$$L_{n_1, n_2}(\phi) = \prod_{i=1}^{n_1} \beta_1 (\frac{x_i}{\alpha_1})^{\beta_1} \exp[(\frac{x_i}{\alpha_1})^{\beta_1}] \ \prod_{i=1}^{n_2} \beta_2 (\frac{y_i}{\alpha_2})^{\beta_2} \exp[(\frac{y_i}{\alpha_2})^{\beta_2}]$$

A direct application of the delta method, and an application of Slutsky's result, leads to the large sample confidence interval

$$\left(\hat{\theta} - z_{\alpha/2} \sqrt{\frac{\hat{D}\hat{I}^{-1}\hat{D}}{n_1 + n_2}} , \quad \hat{\theta} + z_{\alpha/2} \sqrt{\frac{\hat{D}\hat{I}^{-1}\hat{D}}{n_1 + n_2}} \right)$$

where the empirical Fisher information is given by

$$\hat{I} = \begin{bmatrix} \dfrac{n_1}{n_1 + n_2} \hat{I}_{11} & 0 \\ 0 & \dfrac{n_2}{n_1 + n_2} \hat{I}_{22} \end{bmatrix}$$

with

$$\hat{I}_{11} = \frac{1}{n_1} \left(\frac{\partial^2 \ln L_{n_1, n_2}}{\partial \boldsymbol{\theta}_1 \partial \boldsymbol{\theta}_1'} \right) \bigg|_{\hat{\boldsymbol{\theta}}_1} \qquad \hat{I}_{22} = \frac{1}{n_2} \left(\frac{\partial^2 \ln L_{n_1, n_2}}{\partial \boldsymbol{\theta}_2 \partial \boldsymbol{\theta}_2'} \right) \bigg|_{\hat{\boldsymbol{\theta}}_2}$$

and

$$\hat{D} = \left[\frac{\partial \theta}{\partial \alpha_1}, \frac{\partial \theta}{\partial \beta_1}, \frac{\partial \theta}{\partial \alpha_2}, \frac{\partial \theta}{\partial \beta_2} \right] \bigg|_{\hat{\phi}}$$

$$= \hat{\theta} \left[\frac{1}{\hat{\alpha}_1}, -\frac{\ln[-\ln(1 - p_1)]}{\hat{\beta}_1^2}, -\frac{1}{\hat{\alpha}_2}, -\frac{\ln[-\ln(1 - p_2)]}{\hat{\beta}_2^2} \right]$$

This approach always produces an interval. Simulations suggest that their are many cases of moderate sample sizes, say 50 to 150, when the coverage rate is not close enough to the nominal value.

As an alternative, we parallel the normal theory approach in Huang and Johnson(2002) and propose alternative random quantities which are asymptotically standard normal.

$$Z_{n_1, n_2}(\theta) = \frac{\hat{\xi}_{1p_1} - \theta \, \hat{\xi}_{2p_2}}{\sqrt{\dfrac{1}{n_1} \hat{D}_1 \hat{I}_{11}^{-1} \hat{D}_1' + \theta^2 \dfrac{1}{n_2} \hat{D}_2 \hat{I}_{22}^{-1} \hat{D}_2'}} \tag{2}$$

where

$$\hat{D}_i = \left[\frac{\partial \xi_{ip_i}}{\partial \alpha_i}, \frac{\partial \xi_{ip_i}}{\partial \beta_i} \right] \bigg|_{\hat{\alpha}_i, \hat{\beta}_i} =$$

$$\left[[-\ln(1 - p_i)]^{1/\hat{\beta}_i}, \quad \hat{\alpha}_i [-\ln(1 - p_i)]^{1/\hat{\beta}_i} \ln[-\ln(1 - p_i)] \left(-\frac{1}{\hat{\beta}_i^2} \right) \right]$$

The large sample $100(1 - \alpha)\,\%$ confidence region consists of all values for θ_0 for which the null hypothesis $H_0 : \theta = \theta_0$ is not rejected at level α. Inspecting the form of $Z_{n_1,n_2}(\theta)$, we conclude that this region is

(i) an interval (θ_L, θ_U) when

$$
z_{\alpha/2} \; < \; \min\left\{ \frac{\hat{\xi}_{1p_1}}{\sqrt{\dfrac{1}{n_1}\hat{D}_1\hat{I}_{11}^{-1}\hat{D}_1'}} \;,\; \frac{\hat{\xi}_{2p_2}}{\sqrt{\dfrac{1}{n_2}\hat{D}_2\hat{I}_{22}^{-1}\hat{D}_2'}} \right\}
$$

(ii) a right-sided infinite interval (θ_L, ∞) if

$$
\frac{\hat{\xi}_{2p_2}}{\sqrt{\dfrac{1}{n_2}\hat{D}_2\hat{I}_{22}^{-1}\hat{D}_2'}} \; < \; z_{\alpha/2} \; < \; \frac{\hat{\xi}_{1p_1}}{\sqrt{\dfrac{1}{n_1}\hat{D}_1\hat{I}_{11}^{-1}\hat{D}_1'}}
$$

(iii) an interval $(0, \theta_U)$ if

$$
\frac{\hat{\xi}_{1p_1}}{\sqrt{\dfrac{1}{n_1}\hat{D}_1\hat{I}_{11}^{-1}\hat{D}_1'}} \; < \; z_{\alpha/2} \; < \; \frac{\hat{\xi}_{2p_2}}{\sqrt{\dfrac{1}{n_2}\hat{D}_2\hat{I}_{22}^{-1}\hat{D}_2'}}
$$

(iv) The whole real line if

$$
\max\left\{ \frac{\hat{\xi}_{1p_1}}{\sqrt{\dfrac{1}{n_1}\hat{D}_1\hat{I}_{11}^{-1}\hat{D}_1'}} \;,\; \frac{\hat{\xi}_{2p_2}}{\sqrt{\dfrac{1}{n_2}\hat{D}_2\hat{I}_{22}^{-1}\hat{D}_2'}} \right\} \; < \; z_{\alpha/2}
$$

In summary, this large sample confidence region is a bounded interval unless 0 can be rejected as the value of the denominator in θ, by a test of level $\alpha/2$. That is, unless

$$
\frac{\hat{\xi}_{2p_2}}{\sqrt{\dfrac{1}{n_2}\hat{D}_2\hat{I}_{22}^{-1}\hat{D}_2'}} \; < \; z_{\alpha/2}
$$

Our simulations showed that this form has better coverage properties when the sample sizes are only moderate. See Table 2 for one case.

By invariance, the coverages are the same for data (X, Y) and $(c_1 X, c_2 Y)$ for any $c_1, c_2 > 0$.

Table 1. Coverage Rates for the Ratio of 5-th Percentiles. Weibull populations with $\alpha_1 = .2 = \alpha_2, \beta_1 = 1.5$, and $\beta_2 = 1.25$. $m = n = 50$. 5000 replicates. Estimated standard errors in parentheses

Method	Coverage rate
Delta method	92.12 % (.33 %)
Modified $Z_{n_1,n_2}(\theta)$	95.74 % (.29 %)

4. Large Sample Nonparametric Approach

When sample sizes are large, we can also take a nonparametric approach. Let $[\,w\,]$ denote the smallest integer that is greater than or equal to w. We estimate ξ_{1p_1} by the order statistic $\hat{\xi}_{1p_1} = X_{(r_1)}$ where $r_1 = [\,n_1\,p_1\,]$ and estimate ξ_{2p_2} by the order statistic $\hat{\xi}_{2p_2} = Y_{(r_2)}$ where $r_2 = [\,n_2\,p_2\,]$.

Standard calculations show that, if F_i has a continuous derivative f_i in a neighborhood of ξ_{ip_i}, for $i = 1, 2$, and $\lim_{n_1,n_2 \to \infty} n_1(n_1+n_2)^{-1} = \lambda, 0 < \lambda < 1$, then the random quantity

$$\sqrt{n_1 + n_2}\; (\; \hat{\xi}_{1p_1} - \theta\,\hat{\xi}_{2p_2}\; - \;(\; \xi_{1p_1} - \theta\,\xi_{2p_2}\;)\;)$$

converges in distribution to the normal with mean 0 and variance

$$\frac{p_1(1 - p_1)}{\lambda[f_1(\xi_{1p_1})]^2} + \theta^2 \frac{p_2(1 - p_2)}{(1 - \lambda)[f_2(\xi_{2p_2})]^2}$$

Huang and Johnson(2002) estimate the value of density function $f_i(\xi_{ip_i})$ by the kernel estimate $\hat{f}_i(\hat{\xi}_{ip_i})$. Following Silverman(1986), they use a normal kernel with $h_i = 0.9 A_i n_i^{-\frac{1}{5}}$, where $A_i = min(s_i, \dfrac{\hat{R}_i}{1.34})$, $\hat{R}_i = \hat{\xi}_{i,0.75} - \hat{\xi}_{i,0.25}$, and s_i is the sample standard deviation.

Huang and Johnson(2002) show that this procedure has good coverage rates for normal distributions.

When estimating a lower percentile of a Weibull distribution we encountered difficulties with obtaining the nominal coverage rates. The basic difficulty is that the kernel estimates substantially underestimate the value of the density at the percentile. Figure 1 shows the difficulty for a typical sample from one Weibull distribution.

We also tried the Epanechnikov kernel (see Wand and Jones(1995)) but, although it fixed the support problem, it still typically underestimated the density at the 5-th percentile(see Figure 2). As suggested by Schuster(1985), the coverage improved when we replaced the normal kernel by the mirror

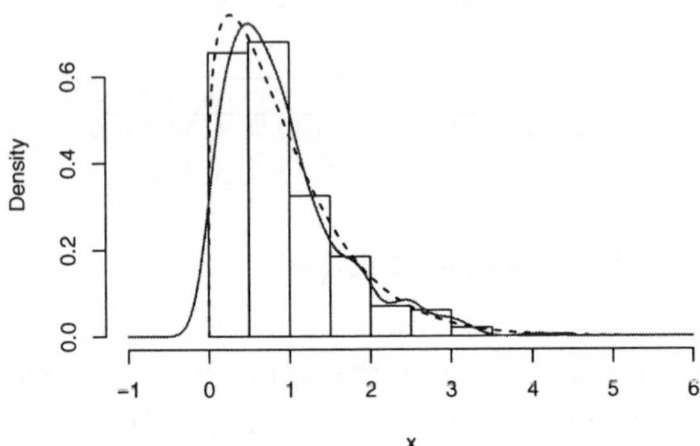

Fig. 1.   Histogram of 400 Weibull random values with shape parameter $= 1.25$ and scale $= 1$. Normal kernel density estimate.

image kernel

$$\frac{1}{\sqrt{2\pi}} \exp\{-\frac{1}{2}(\frac{\hat{\xi}_{2p_2} - y_j}{h_2})^2\} \ + \ \frac{1}{\sqrt{2\pi}} \exp\{-\frac{1}{2}(\frac{-\hat{\xi}_{2p_2} - y_j}{h_2})^2\}$$

See Figure 3 for a typical case.

A small simulation with these three estimators is summarized in Table 3. It confirms the substantial underestimation by all but the folded normal kernel. In the simulation, for the two normal kernels, we used the default bandwidth of the function `density` in the software package R (see Silverman(1986), page 48, eqn. 3.31). For the boundary Epanechnikov kernel, we used $2.36n^{-1/5} s$ (e.g. Bickel and Doksum(2003)).

Table 2.   Properties of Estimators of the population pdf at the percentile. Means and standard deviations for 1,000 trials of samples of size 400 from a Weibull distribution with shape 1.25 and scale 1. With $p = 0.05$, the true percentile $\xi_p = 0.0929$ and $f(\xi_p) = 0.6556$. The mean of estimated percentile is $\hat{\xi}_p = 0.0923$.

Kernel	mean(sd) of $\hat{f}(\xi_p)$	mean(sd) of $\hat{f}(\hat{\xi}_p)$
Normal	0.4610 (0.0368)	0.4604 (0.0321)
Folded Normal	0.6540 (0.0562)	0.6556 (0.0545)
Epanechnikov boundary	0.5091 (0.0722)	0.5072 (0.0625)

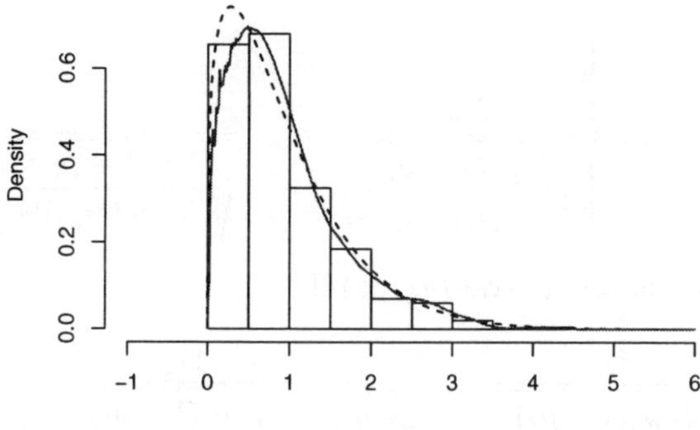

Fig. 2.   Histogram of 400 Weibull random values with shape parameter $= 1.25$ and scale $= 1$. Epanechnikov kernel density estimate.

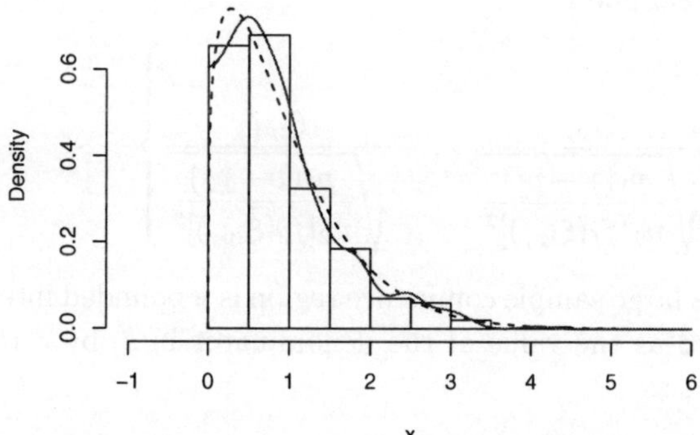

Fig. 3.   Histogram of 400 Weibull random values with shape parameter $= 1.25$ and scale $= 1$. Folded normal kernel density estimate.

The large sample test of $H_0 : \theta = \theta_0$, versus a two-sided alternative then leads to an approximate $(1 - \alpha)100\%$ confidence region for θ. For distributions whose support is contained in the positive real line, this region is

(i) an interval (θ_L, θ_U) when

$$z_{\alpha/2} < \min\left\{ \frac{\hat{\xi}_{1p_1}}{\sqrt{\dfrac{p_1(1-p_1)}{n_1[\hat{f}_1(\hat{\xi}_{1p_1})]^2}}} \, , \, \frac{\hat{\xi}_{2p_2}}{\sqrt{\dfrac{p_2(1-p_2)}{n_2[\hat{f}_2(\hat{\xi}_{2p_2})]^2}}} \right\}$$

(ii) a right-sided infinite interval (θ_L, ∞) if

$$\frac{\hat{\xi}_{2p_2}}{\sqrt{\dfrac{p_2(1-p_2)}{n_2[\hat{f}_2(\hat{\xi}_{2p_2})]^2}}} < z_{\alpha/2} < \frac{\hat{\xi}_{1p_1}}{\sqrt{\dfrac{p_1(1-p_1)}{n_1[\hat{f}_1(\hat{\xi}_{1p_1})]^2}}}$$

(iii) an interval $(0, \theta_U)$ if

$$\frac{\hat{\xi}_{1p_1}}{\sqrt{\dfrac{p_1(1-p_1)}{n_1[\hat{f}_1(\hat{\xi}_{1p_1})]^2}}} < z_{\alpha/2} < \frac{\hat{\xi}_{2p_2}}{\sqrt{\dfrac{p_2(1-p_2)}{n_2[\hat{f}_2(\hat{\xi}_{2p_2})]^2}}}$$

(iv) The whole real line if

$$\max\left\{ \frac{\hat{\xi}_{1p_1}}{\sqrt{\dfrac{p_1(1-p_1)}{n_1[\hat{f}_1(\hat{\xi}_{1p_1})]^2}}} \, , \, \frac{\hat{\xi}_{2p_2}}{\sqrt{\dfrac{p_2(1-p_2)}{n_2[\hat{f}_2(\hat{\xi}_{2p_2})]^2}}} \right\} < z_{\alpha/2}$$

In summary, this large sample confidence region is a bounded interval unless 0 can be rejected as the value of the denominator in θ, by a test of level $\alpha/2$. That is, unless

$$\frac{\hat{\xi}_{2p_2}}{\sqrt{\dfrac{p_2(1-p_2)}{n_2[\hat{f}_2(\hat{\xi}_{2p_2})]^2}}} < z_{\alpha/2}$$

Our simulations showed that this form has better coverage properties when the sample sizes are only moderate.

Based on 5000 replicates with $n_1 = n_2 = 100$ and the two Weibull distributions with $\overline{F}_i(x) = \exp(-(x/\beta_1)^{\beta_2})$ with $\beta_1 = 0.4$ and $\beta_1 = 1.25$, the estimate of the coverage probability was .9582 .

Remark 2: The use of an appropriate boundary kernel does not completely solve the practical problem which motivated this study. Weaker specimens of lumber break before they can be sold or tested. The three parameter Weibull, where the boundary of support is unknown, often provides better fits than the two parameter model.

5. A Bayesian Nonparametric Approach

Our Bayesian nonparametric approach begins by assigning a separate Dirichlet process prior to each of the two population distributions. Ferguson(1973) has established, that for each population, the posterior distribution of ξ_{ip_i} is given by

$$P_i\left[\xi_{ip_i} \le t\right] = \int_{p_i}^{t} \frac{\Gamma(M_i)}{\Gamma(u_i M_i)\,\Gamma((1-u_i)M_i)} z^{u_i M_i - 1}(1-z)^{(1-u_i)M_i - 1}\, dz \quad (3)$$

$i = 1, 2$ where $M_i = \alpha_i(R), u_i = \alpha_i(\,(-\infty, t])/\alpha_i(R)$ and R denotes the real line. Here $\alpha_i(-\infty, t] = (\ \#\ i\text{-th sample observations} \le t\ )\ + \alpha_{i0}(-\infty, t]$ where α_{i0} is the prior measure.

The posterior distribution of the ratio of percentiles $\theta = \xi_{1p_1}/\xi_{2p_2}$ can be evaluated by simulation.

By numerically inverting each posterior cdf, generate a value ξ_{1p_1} from $P_1[\xi_{1p_1} \le t]$ and independently generate a value ξ_{2p_2} from $P_2[\xi_{2p_2} \le t]$ and then take the ratio. Repeat a very large number of times and approximate the 95 % region with highest posterior density.

As an illustration, consider the ratio of 5th percentiles. As data, we generated a sample of size 100 from a $N(4, 1)$ population representing the first population. Let $x_{(1)} \le x_{(2)} \le \cdots \le x_{(100)}$ be the order statistics. We also independently generated a second sample of size 100 from the same normal population to represent the sample from the second population. Next, we select a value u from a uniform(0,1) distribution and then numerically solve $u = P_1\left[\xi_{1p_1} \le t\right]$ for t which is then an observed value from the posterior distribution of the 5th percentile from the first population.

Most of the time, the generated value is one of the order statistics near the sample 5th percentile. Table 1 shows the first 1000 values selected from the posterior distribution. Notice that 98.1 % are actual values in the data set.

A total of 10,000 values for the ratio θ were generated. The actual posterior distribution for the ratio of population 5th percentiles is a mixture of discrete and absolutely continuous parts. However, we will summarize the concentration of probability by a histogram. Because the data are far from

Table 3. Table of first 1,000 generated values from the first posterior distribution when the data are 100 $N(4,1)$ observations. Out of 1,000 generations, 981 are the data values.

Value of $\xi_{0.05}$	Freq	Property	Value of $\xi_{0.05}$	Freq	Property
1.13895	7	$x_{(1)}$	2.15480	1	interpolated
1.14322	1	interpolated	2.17881	193	$x_{(6)}$
1.50282	26	$x_{(2)}$	2.18925	1	interpolated
1.52061	1	interpolated	2.18941	1	interpolated
1.52337	1	interpolated	2.18973	1	interpolated
1.53013	1	interpolated	2.29040	151	$x_{(7)}$
1.53213	1	interpolated	2.29181	1	interpolated
1.53695	1	interpolated	2.30393	116	$x_{(8)}$
1.96986	71	$x_{(3)}$	2.33334	52	$x_{(9)}$
1.98814	152	$x_{(4)}$	2.33483	1	interpolated
2.00102	1	interpolated	2.34724	20	$x_{(10)}$
2.00158	1	interpolated	2.35767	1	interpolated
2.00182	1	interpolated	2.43491	14	$x_{(11)}$
2.00219	1	interpolated	2.46142	3	$x_{(12)}$
2.00276	1	interpolated	2.50924	1	$x_{(13)}$
2.00289	1	interpolated	2.59366	2	$x_{(14)}$
2.15238	173	$x_{(5)}$			

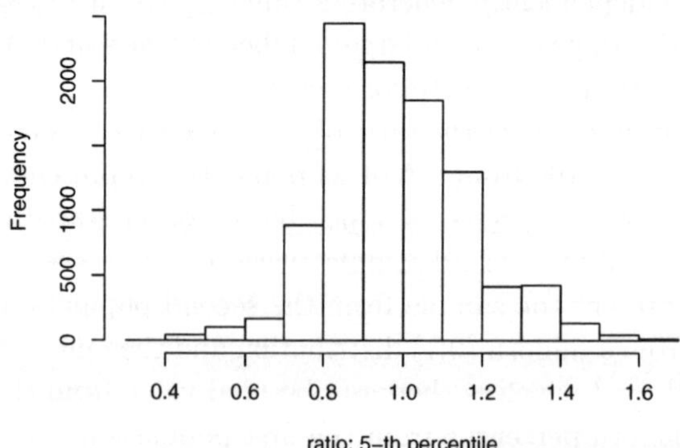

Fig. 4. Estimated posterior distribution of ratio ξ_1/ξ_2. Samples of size 100 from N(4, 1). 10,000 samples of θ.

0 and positive, the histogram in Figure 4 has a nice unimodal appearance. The interval, with 2.5 % of the 10,000 values in each tail, is (.653, 1.350). For comparison, the normal theory procedure in Section 2.1 gives the 95%

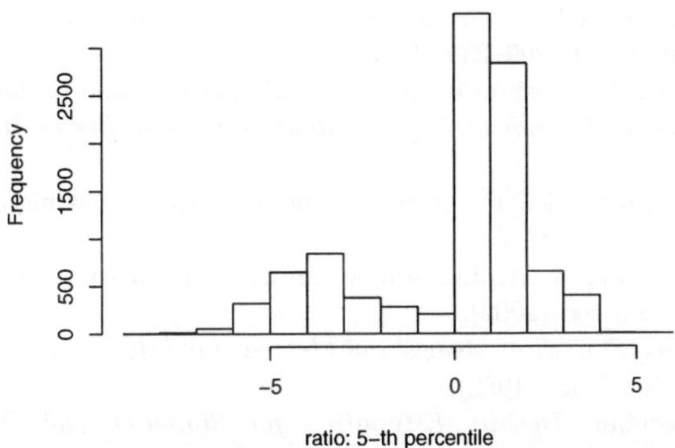

Fig. 5.   Estimated posterior distribution of ratio ξ_1/ξ_2. Samples of size 100 from N(2, 1). 10,000 samples of θ.

confidence interval (.837, 1.103).

This procedure was repeated but with the data generated from $N(2,1)$ populations. Here the histogram, see Figure 5, is bimodal suggesting the possibility of two disjoint finite intervals. For comparison, the normal theory procedure in Section 2.1, based on sampling theory, gives the 95% confidence region (.755, 27.567). However, the same normal theory 99% confidence region is the complement of the interval $(-3.383, .510)$. Here, any Bayesian interval, or union of two finite intervals, seems to be more sensible than either of these two normal theory regions.

Acknowledgments

We would like to thank Songyong Sim for preparing the graphs and performing some of the calculations.

References

1. Aplin, E. N. and Green, D. W. and Evans, J. W. and Barrett, J. D.(1986) The Influence of Moisture Content on the Flexural Properties of Douglas Fir Dimension Lumber. Madison, WI : U.S. Dept. of Agriculture, Forest Service, *Forest Products Laboratory*
2. P. Bickel and K. Doksum, *Mathematical Statistics*, Vol. II (Prentice Hall, New Jersey, 2003).

3. J. W. Evans, R. A. Johnson, D. W Green and S. Verrill, Applications of the Weibull Distribution in the Lumber Industry, in *Weibull Distributions*, Ed. N. Balakrishnan, to appear (2002).

4. T. S. Ferguson, A Bayesian analysis of some nonparametric problems, *Annals of Statistics* **1**, 209–230 (1973).

5. E. C. Fieller, Symposium on interval estimation: Some problems in interval estimation, *Journal of Royal Statistical Society, Series B* **16**, 175–185 (1954).

6. J. T. G. Hwang, Fieller's problems and resampling techniques, *Statistica Sinica 5*, 161–171 (1995).

7. L. F. Huang and R. A. Johnson, Confidence regions for the ratio of percentiles, Submitted (2002).

8. J. F. Lawless, *Statistical Models and Methods for Lifetime Data* (John Wiley and Sons, New York, 1982).

9. B. W.Silverman, *Density Estimation for Statistics and Data Analysis* (Chapman & Hall, London, 1986).

10. E. F. Schuster, Incorporating support constraints into nonparametric estimators of densities, *Communications in Statistics, Part A-Theory and Methods* **14**, 1123–1136 (1985)

11. M. P. Wand and M. C. Jones,. *Kernel Smoothing*, (Chapman and Hall, London, 1995)

30

EMPIRICAL ESTIMATORS OF RELIABILITY AND RELATED FUNCTIONS FOR SEMI-MARKOV SYSTEMS

Nikolaos Limnios

Division Mathématiques Appliquées,
Université de Technologie de Compiègne,
B.P. 529, 60205 Compiègne, France
E-mail: Nikolaos.Limnios@utc.fr

Brahim Ouhbi

Ecole Nationale Supérieure d'Arts et Métiers de Meknès, Morocco
E-mail: ouhbib@yahoo.co.uk

We consider a semi-Markov process with a finite state space, through one observation on a time interval $[0, T]$, for a fixed time T. The aim of this paper is to present empirical estimators of the semi-Markov kernel, reliability, availability, failure rate and rate of occurrence of failure (rocof) of a system described by the semi-Markov process. We also give asymptotic properties of the above estimators, as the uniform strong consistency and normality. This paper is an overview of our recent results.[19−16,11]

1. Introduction

Semi-Markov processes or equivalently Markov renewal processes[12,22,23] are widely used in many applied fields: reliability[12], survival analysis[2], queueing theory[8], etc.. Nonparametric statistical inference for semi-Markov kernels is considered in several papers. Moore and Pyke[13] studied empirical estimators for finite semi-Markov kernels; Lagakos, Sommer and Zelen[9] gave maximum likelihood estimators for nonergodic finite semi-Markov kernels; Gill[6] and Andersen et al.[2] studied Kaplan-Meier type estimators via point process theory; Greenwood and Wefelmeyer[7] studied efficiency of empirical estimators for linear functionals of semi-Markov kernels for general state spaces; Ouhbi and Limnios[19] studied empirical estimators of non-linear functionals of semi-Markov kernels including the Markov renewal matrix and reliability functions[21], the rate of occurrence of failure function[20] and the failure rate of a semi-Markov system[17]; Limnios[11] gave the invariance principle for the empirical estimator of semi-Markov kernels.

Here we give an overview of the results presented in the papers[19−16,11], specifically those concerning the empirical estimators. We present an empirical estimator of the semi-Markov kernel. This estimator yields also estimators for the Markov renewal function and the transition probability function. Then, the estimators for reliability metrics are derived using a general scheme, in which the state space E of the semi-Markov process is partitioned in two subsets, one containing the up states, say U, and the other containing the down states, say D. For instance, the reliability thus appears as the survival function of the hitting time to D.

The estimation is based on one observation (trajectory) of the semi-Markov process on a time interval $[0, T]$. Asymptotic properties are proven to hold when T goes to infinity. This model can easily be adapted in order to take into account several observations on $[0, T]$ and also for the lifetime function estimation.

In Section 2, we give some definitions and basic results concerning finite state space semi-Markov processes. In the following sections, we present the different estimators and their asymptotic properties: consistency and normality. Section 3 is devoted to the estimation of the semi-Markov kernel, the Markov renewal function and transition function. Section 4 is devoted to estimation of the reliability and availability functions. Section 5 is devoted to the failure rate estimation. Finally, Section 6 is devoted to the estimation of the rate of occurrence of failure (rocof) of the system.

2. Markov Renewal and Semi–Markov Processes

Let us here give basic results on finite semi-Markov processes following Limnios and Oprişan.[12] Consider a finite set, say $E = \{1, ..., s\}$, and an E-valued stochastic process $Z = (Z_t, t \in \mathbb{R}_+)$. Let $0 = S_0 \leq S_1 \leq ... \leq S_n \leq S_{n+1} \leq ...$ be the jump times of Z, and $J_0, J_1, J_2, ...$ the successive visited states of Z. Let us also define $X_n := S_n - S_{n-1}$, $n \geq 1$, the inter-jump times of Z.

Definition 1: The stochastic process $(J_n, S_n)_{n \in \mathbb{N}}$ is called a Markov renewal process (MRP), with state space E, if it verifies a.s., the following relation

$$\mathbb{P}(J_{n+1} = j, S_{n+1} - S_n \leq t \mid J_0, \ldots, J_n; S_1, \ldots, S_n)$$
$$= \mathbb{P}(J_{n+1} = j, S_{n+1} - S_n \leq t \mid J_n)$$

for all $j \in E$ and all $t \in \mathbb{R}_+$. In that case, Z is called a semi-Markov process (SMP).

Thus the process (J_n, S_n) is a Markov chain with state space $E \times \mathbb{R}_+$ and transition kernel $Q_{ij}(t) := \mathbb{P}(J_{n+1} = j, S_{n+1} - S_n \leq t \mid J_n = i)$, $i, j \in E$, $t \geq 0$, called the semi-Markov kernel. The process (J_n) is a Markov chain with state space E and transition probabilities $P(i,j) := Q_{ij}(\infty) = \mathbb{P}(J_{n+1} = j \mid J_n = i)$, which is called the embedded Markov chain (EMC) of Z. Then we can write:

$$Q_{ij}(t) := \mathbb{P}(J_{n+1} = j, X_{n+1} \leq t \mid J_n = i) = P(i,j)F_{ij}(t), \quad t \geq 0, i, j \in E,$$

where $F_{ij}(t) := \mathbb{P}(X_{n+1} \leq t \mid J_n = i, J_{n+1} = j)$ is the conditional distribution function of the inter-jump times. Let us also define the distribution function $H_i(t) := \sum_{j \in E} Q_{ij}(t)$ and its mean value m_i, which is the mean sojourn time of Z in state i. The function H_i is a distribution function, hence Q_{ij} is a subdistribution function, i.e., $H_i(\infty) = 1$, and $Q_{ij}(\infty) \leq 1$ with $Q_{ij}(0-) = H_i(0-) = 0$.

The semi-Markov process Z is connected to (J_n, S_n) through

$$Z_t = J_n, \quad \text{if} \quad S_n \leq t < S_{n+1}, \quad t \geq 0, \quad \text{and} \quad J_n = Z_{S_n}, \quad n \geq 0.$$

A Markov process with state space E and generating matrix $A = (a_{ij})_{i,j \in E}$ is a special semi-Markov process with semi-Markov kernel

$$Q_{ij}(t) = \frac{a_{ij}}{a_i}(1 - e^{-a_i t}),$$

where $a_i := -a_{ii}$, $i \in E$.

Let us also define the process $(N(t), t \in \mathbb{R}_+)$, by $N(t) := \sup\{n \geq 0 : S_n \leq t\}$, which counts the number of jumps of Z in the time interval $(0, t]$. Let $N_i(t)$ be the number of visits of Z to state $i \in E$ up to time t, and $N_{ij}(t)$ be the number of jumps from state i to state j up to time t, that is,

$$N_i(t) := \sum_{k=1}^{N(t)} \mathbf{1}_{\{J_k = i\}} = \sum_{k=1}^{\infty} \mathbf{1}_{\{J_k = i, S_k \leq t\}},$$

$$N_{ij}(t) := \sum_{k=1}^{N(t)} \mathbf{1}_{\{J_{k-1} = i, J_k = j\}} = \sum_{k=1}^{\infty} \mathbf{1}_{\{J_{k-1} = i, J_k = j, S_k \leq t\}}.$$

If we consider the (eventually delayed) renewal process $(S_n^i)_{n\geq 0}$ of successive times of visits to state i, then $N_i(t)$ is the counting process of jumps. We will denote by μ_{ii} the mean recurrence time of (S_n^i), i.e., $\mu_{ii} = \mathbb{E}[S_2^i - S_1^i]$.

Let $\phi(i,t)$, $i \in E, t \geq 0$, be a measurable real-valued function and define the convolution of ϕ by Q as follows

$$Q * \phi(i,t) := \sum_{k \in E} \int_0^t Q_{ik}(ds)\phi(k, t - s). \tag{1}$$

Consider now the n-fold convolution $Q_{ij}^{(n)}$ of Q by itself. For any $i, j \in E$, we have

$$Q_{ij}^{(n)}(t) = \begin{cases} \sum_{k \in E} \int_0^t Q_{ik}(ds)Q_{kj}^{(n-1)}(t-s) & n \geq 2 \\ Q_{ij}(t) & n = 1 \\ \delta_{ij}\mathbf{1}_{\{t \geq 0\}} & n = 0. \end{cases}$$

It is easy to see that

$$Q_{ij}^{(n)}(t) = \mathbb{P}_i(J_n = j, S_n \leq t). \tag{2}$$

Let us define the Markov renewal function $\psi_{ij}(t)$, $i, j \in E, t \geq 0$, by

$$\psi_{ij}(t) := \mathbb{E}_i[N_j(t)] = \sum_{n=1}^{\infty} Q_{ij}^{(n)}(t). \tag{3}$$

Another useful function is the semi-Markov transition function

$$P_{ij}(t) := \mathbb{P}(Z_t = j \mid Z_0 = i), \quad i, j \in E, t \geq 0,$$

which is the conditional marginal distribution of the process.

Definition 2: The semi-Markov process Z is said to be regular if

$$\mathbb{P}_i(N(t) < \infty) = 1,$$

for all $t \geq 0$ and all $i \in E$.

For regular semi-Markov processes, we have, for any $n \in \mathbb{N}$, $S_n < S_{n+1}$ and $S_n \to \infty$ (a.s.), as $n \to \infty$. The following theorem gives a criterion for regularity.

Theorem 3: If some real numbers, say $\alpha > 0$ and $\beta > 0$, exist, such that $H_i(\alpha) < 1 - \beta$, for all $i \in E$, then the semi-Markov process is regular.

In the following we are concerned with regular semi-Markov processes.

The Markov renewal equation is as basic in the theory of semi-Markov processes, as the renewal equation is basic in the theory of renewal process on the real-half line.

Let us write the Markov renewal function (3) in matrix form

$$\psi(t) = \sum_{n=1}^{\infty} Q^{(n)}(t). \tag{4}$$

From this, we can write also

$$\psi(t) = I(t) + Q \star \psi(t), \tag{5}$$

where $I(t) = I$ (the identity matrix) if $t \geq 0$ and $I(t) = 0$, if $t < 0$.

Equation (5) is a special kind of the so called Markov Renewal Equation (MRE). A general MRE is written as follows

$$\Theta(t) = L(t) + Q \star \Theta(t), \tag{6}$$

where $\Theta(t) = (\Theta_{ij}(t))_{i,j \in E}$, $L(t) = (L_{ij}(t))_{i,j \in E}$ are real measurable matrix-valued functions, with $\Theta_{ij}(t) = L_{ij}(t) = 0$ if $t < 0$. The function $L(t)$ is a given matrix-valued function and $\Theta(t)$ is the unknown matrix-valued function. We may also consider a vector version of Equation (6), i.e., consider the respective columns of the matrices Θ and L.

Let $\mathbf{B}$ be the space of all bounded on compact sets of $\mathbb{R}_+$ matrix-valued functions $\Theta(t)$, i.e., $\|\Theta(t)\| = \sup_{i,j} |\Theta_{i,j}(t)|$ is bounded on sets $[0, \xi]$ for all $\xi \in \mathbb{R}_+$.

Theorem 4: Equation (6) has a unique solution $\Theta = \psi \star L$ belonging to $\mathbf{B}$ if and only if $\psi \star L$ belongs to $\mathbf{B}$.

As an application of the above Markov renewal theorem, let us consider the following result.

By a renewal argument, it is easy to see that the transition function $P(t) = (P_{ij}(t))$ of the semi-Markov process Z satisfies the following MRE

$$P(t) = I(t) - H(t) + Q \star P(t),$$

whose unique solution is

$$P(t) = (I(t) - Q(t))^{(-1)} \star (I(t) - H(t)). \tag{7}$$

Here $H(t) = diag(H_i(t))$, and

$$(I(t) - Q(t))^{(-1)} = \sum_{n \geq 0} Q^{(n)}(t) = \psi(t).$$

The following Markov renewal theorem is a basic tool in the asymptotic theory of semi-Markov processes.

Here we use the concept of directly Riemann integrable functions (see, e.g., Limnios and Oprişan[12]). In the sequel, for any $\mathbb{R}^s$-valued function $h(u) = (h_i(u))_{i \in E}$, we set $\nu h(u) := \sum_{i \in E} \nu_i h_i(u)$, for all u, and $\hat{m} := \sum_{i \in E} \nu_i m_i$, where ν is the stationary distribution of the Markov chain (J_n). We have also $\hat{m} = \nu_i \mu_{ii}$ for all $i \in E$.

A finite state space MRP with irreducible EMC (J_n) is said to be irreducible. An irreducible MRP is said to be recurrent. If moreover $\hat{m} < \infty$, then it is positive recurrent. A state is said to be periodic or arithmetic if the distribution of the recurrence time in this state is arithmetic, i.e., is concentrated on a set $\{ic : i \in \mathbb{N}\}$, for some positive c.

Theorem 5: Let i be an aperiodic recurrent state. If h_i is a directly Riemann integrable function on $\mathbb{R}_+$, then, as $t \to \infty$,

$$\int_0^t \psi_{ji}(dy) h_i(t - y) \to \frac{1}{\mu_{ii}} \int_0^\infty h_i(u) du.$$

¿From the above theorem and Equation (7), we get, for any positive recurent MRP, and for any $i, j \in E$, as $t \to \infty$,

$$P_{ji}(t) \to \nu_i m_i / \hat{m} =: \pi_i. \tag{8}$$

3. Empirical Estimator of the Semi-Markov Kernel

Consider a finite set $E = \{1, ..., s\}$ and an E-valued semi-Markov process $(Z_t, t \geq 0)$ with semi-Markov kernel Q and $E \times \mathbb{R}_+$-valued embedded Markov renewal process $(J_n, S_n)_{n \in \mathbb{N}}$.

We need the following basic assumption for the results presented in this paper.

Assumption A:
- the process (Z_t) is positive recurrent; and
- there exists an $\varepsilon > 0$, such that $H_i(\varepsilon) < 1$ for all $i \in E$.

The problem here is to estimate the semi-Markov kernel of the above semi-Markov process by observing one sample path in a time interval $[0, T]$.

The observation of a sample path, in the time interval $[0, T]$, of a semi-Markov process may be given as

$$\mathcal{H}_T = \{Z_u, 0 \leq u \leq T\} = \{J_0, J_1, ..., J_{N(T)}, X_1, ..., X_{N(T)}, U_T\}, \tag{9}$$

where $U_T = T - S_{N(T)}$.

Let us define the empirical estimator $\hat{Q}_{ij}(t, T)$ of the semi-Markov kernel $Q_{ij}(t)$, related to $\mathcal{H}_T$, as follows

$$\hat{Q}_{ij}(t, T) := \frac{1}{N_i(T)} \sum_{k=1}^{N(T)} \mathbf{1}_{\{J_{k-1}=i, J_k=j, X_k \leq t\}}. \tag{10}$$

From this definition we obtain

$$\hat{Q}_{ij}(t, T) = \hat{p}_{ij}(T) \hat{F}_{ij}(t, T), \tag{11}$$

where

$$\hat{p}_{ij}(T) := \frac{N_{ij}(T)}{N_i(T)} \tag{12}$$

and

$$\hat{F}_{ij}(t, T) := \frac{1}{N_{ij}(T)} \sum_{k=1}^{N(T)} \mathbf{1}_{\{J_{k-1}=i, J_k=j, X_k \leq t\}}. \tag{13}$$

In the above formulas, when $N_i(T) = 0$, we have $\sum_{k=1}^{N(T)}(\cdot) = 0$ and then we set $0/0 = 0$. We set the same when $N_{ij}(t) = 0$.

Remark 6: In the case where the sojourn time distributions do not depend on the next visited state, that is $Q_{ij}(t) = P(i, j) H_i(t)$, this estimator is the Moore-Pyke estimator[13]. Actually, any MRP can be transformed in an MRP with sojourn time independent of the next visited state.[12] Nevertheless, this transformation is not always useful since it involves a three-dimensional process.

Let us give now the following asymptotic properties of the above empirical estimator (11): uniform strong consistency, invariance principle, and asymptotic normality.

Theorem 7: ([19]) The above empirical estimator (11) is uniformly strongly consistent, as $T \to \infty$, i.e.,

$$\max_{i,j} \sup_{0 \leq t < \infty} \left| \hat{Q}_{ij}(t, T) - Q_{ij}(t) \right| \xrightarrow{a.s.} 0.$$

In the following theorem the symbol $\Rightarrow$ denotes the weak convergence of random elements in the Skorohod space $D[0, \infty)$, and $W(t), t \geq 0$, denotes the standard Brownian motion.[4]

Let us consider the sequence of stochastic processes $(\hat{Q}_{ij}(x, nt), t \geq 0), n \geq 0$. Then we have the following invariance principle.

Theorem 8: (^{11}) For any fixed $i, j \in E$, and fixed $x \in (0, \infty)$, we have

$$n^{1/2}[\hat{Q}_{ij}(x, nt) - Q_{ij}(x)] \Rightarrow \sigma_{ij}(x)W(1/t), \quad n \to \infty,$$

provided that $\sigma_{ij}^2(x) > 0$, where $\sigma_{ij}^2(x) = \mu_{ii}Q_{ij}(x)[1 - Q_{ij}(x)]$.

For $t = 1$, using the continuous mapping theorem, we get as a particular case the following result.

Corollary 9: *The empirical estimator of the semi-Markov kernel (11) converges in distribution, as $T \to \infty$, to a normal centered r.v., i.e., for any fixed t,*

$$T^{1/2}[\hat{Q}_{ij}(t, T) - Q_{ij}(t)] \xrightarrow{d} N(0, \sigma_{ij}^2(t)),$$

where $\sigma_{ij}^2(t) = \mu_{ii}Q_{ij}(t)[1 - Q_{ij}(t)]$.

The estimator of the Markov renewal function of the semi-Markov process, obtained by substitution of the empirical kernel estimator,

$$\hat{\psi}_{ij}(t, T) := \sum_{n=0}^{\infty} \hat{Q}_{ij}^{(n)}(t, T), \tag{14}$$

verifies the following asymptotic properties.

Theorem 10: (^{19,16}) The estimator given by (14), of the Markov renewal function $\psi_{ij}(t)$, is uniformly strongly consistent, i.e., for any fixed $L \in \mathbb{R}_+$, as $T \to \infty$,

$$\max_{i,j} \sup_{0 \leq t < L} \left|\hat{\psi}_{ij}(t, T) - \psi_{ij}(t)\right| \xrightarrow{a.s.} 0.$$

Theorem 11: (^{19}) The estimator given by (14), of the Markov renewal function $\psi_{ij}(t)$, converges in distribution, for any fixed t, as $T \to \infty$, to a centered normal r.v.

$$T^{1/2}(\hat{\psi}_{ij}(t, T) - \psi_{ij}(t)) \xrightarrow{d} N(0, b_{ij}^2(t))$$

where $b_{ij}^2(t) = \sum_{r=1}^{s} \sum_{k=1}^{s} \mu_{rr}\{(\psi_{ir} * \psi_{kj})^2 * Q_{rk} - (\psi_{ir} * \psi_{kj} * Q_{rk})^2\}(t)$.

The estimator of the transition function of the semi-Markov process, obtained by substitution of the empirical kernel estimator,

$$\hat{P}(t, T) := \hat{\psi}_{ij} * (I - \hat{H})(t, T), \tag{15}$$

verifies the following asymptotic properties.

Theorem 12: ([19]) The estimator given by (15), of the transition function $P_{ij}(t)$, is uniformly strongly consistent, i.e., for any fixed $L \in \mathbb{R}_+$, as $T \to \infty$,

$$\max_{i,j} \sup_{0 \le t \le L} \left| \hat{P}_{ij}(t, T) - P_{ij}(t) \right| \xrightarrow{a.s.} 0.$$

Theorem 13: ([19]) The estimator given by (15), of the transition function $P_{ij}(t)$, converges in distribution, for any fixed $i, j \in E$, and fixed $t \ge 0$, as $T \to \infty$, to a centered normal r.v.

$$T^{1/2}(\hat{P}_{ij}(t, T) - P_{ij}(t)) \xrightarrow{d} N(0, c_{ij}^2(t))$$

where

$$c_{ij}^2(t) = \sum_{n,k=1}^{s} \left[(1 - H_j) * \psi_{in} * \psi_{kj} - \psi_{ij} 1_{\{n=j\}} \right]^2 * Q_{nk}(t)$$

$$- \left\{ \left[(1 - H_j) * \psi_{in} * \psi_{kj} - \psi_{ij} 1_{\{n=j\}} \right]^2 * Q_{nk}(t) \right\}^2.$$

4. Reliability Estimation

Let $Z_t, t \ge 0$, be a semi-Markov process with finite state space $E = \{1, 2, ..., s\}$, semi-Markov kernel $Q(t)$ and initial distribution α, describing the stochastic behavior of a repairable semi-Markov system. We assume that this process fulfills Assumption A, and that the Markov chain (J_n) is irreducible with stationary distribution ν. In reliability and survival analysis, the state space E is naturally partitioned into the set U of working (or 'up') states, and the set D of repair (or 'down') states, i.e., $E = U \cup D$ and $U \cap D = \emptyset$, and $U \ne \emptyset$, $D \ne \emptyset$. The transition from one state to another state means, physically speaking, the failure or the repair of one of the components of the system. The system is operational in U. No service is delivered if the system is in D. But, one repair will return the system from D to U. For more details, see Limnios and Oprişan.[12] In the sequel, we will consider the kernels and the functions defined in the previous sections $Q(t)$, $\psi(t)$, $P(t)$,... under their matrix form and we will denote their restrictions on the sets U, $U \times U$, $U \times D$,..., with index 0 for U and 1 for D. For example $Q_0(t)$ is the restriction of the matrix $Q(t)$ on $U \times U$, and α_0 is the restriction of the probability distribution α of the r.v. Z_0 on U.

The conditional and unconditional reliability and availability, $R_i(t), R(t)$, and $A_i(t), A(t)$, of a semi-Markov system are defined as follows:

$$R_i(t) := \mathbb{P}_i(Z_u \in U, \forall u \le t), \quad R(t) := \mathbb{P}(Z_u \in U, \forall u \le t), \quad t \ge 0,$$

with

$$R(t) = \sum_{i \in U} \alpha(i) R_i(t),$$

and

$$A_i(t) := \mathbb{P}_i(Z_t \in U), \quad A(t) := \mathbb{P}(Z_t \in U), \quad \text{and} \quad A(t) = \sum_{i \in E} \alpha(i) A_i(t).$$

The reliability and availability functions of a semi-Markov system verify Markov renewal equations, e.g., for the reliability

$$R_i(t) - \sum_{j \in U} \int_0^t Q_{ij}(ds) R_j(t - s) = 1 - H_i(t), \quad i \in U.$$

The solution of the above MRE is given by the following formula:

$$R(t) = \alpha_0 \psi_0 \star (I - H_0)(t)\mathbf{1},$$

where $\mathbf{1} = (1, ..., 1)'$.

By the same way, we get for the availability

$$A(t) = \alpha \psi \star (I - H)(t)\mathbf{e},$$

where $\mathbf{e} = (e_1, ..., e_s)'$ is an s-dimensional column-vector, with $e_i = 1$, if $i \in U$, and $e_i = 0$, if $i \in D$.

Note that when D is an absorbing set, which means that the considered system is non repairable, then we have $A(t) \equiv R(t)$. For this reason, the availability is not used in survival analysis.

We obtain estimators for the reliability and availability by substitution into the above formulas of the estimators of ψ and Q. For these estimators, we have the following properties.

Theorem 14: [21] The reliability and availability estimators obtained from the above empirical estimator are uniformly strongly consistent, i.e., for any fixed $L \in \mathbb{R}_+$,

$$\sup_{0 \leq t < L} \left| \hat{R}(t, T) - R(t) \right| \xrightarrow{a.s.} 0, \quad T \to \infty,$$

$$\sup_{0 \leq t < L} \left| \hat{A}(t, T) - A(t) \right| \xrightarrow{a.s.} 0, \quad T \to \infty.$$

Theorem 15: [21] For any fixed $t > 0$, we have

$$T^{1/2}(\hat{R}(t, T) - R(t)) \xrightarrow{d} N(0, \sigma_R^2(t)),$$

as $T \to \infty$, where

$$\sigma_R^2(t) = \sum_{i \in U} \sum_{j \in E} \mu_{ii} \{ (B_{ij} \mathbf{1}_{\{j \in U\}} - \sum_{r \in U} \alpha(r) \psi_{ri})^2 * Q_{ij}(t)$$

$$- [(B_{ij} \mathbf{1}_{\{j \in U\}} - \sum_{r \in U} \alpha(r) \psi_{ri}) * Q_{ij}(t)]^2 \}$$

and

$$B_{ij} = \sum_{n \in E} \sum_{k \in U} \alpha(i) \psi_{ni} * \psi_{jk} * (I - H_{kk}(t))).$$

Theorem 16: $(^{21})$ For any fixed $t > 0$, we have

$$T^{1/2}(\hat{A}(t, T) - A(t)) \xrightarrow{d} N(0, \sigma_A^2(t)),$$

as $T \to \infty$, where

$$\sigma_A^2(t) = \sum_{i \in U} \sum_{j \in E} \mu_{ii} \{ (B_{ij} \mathbf{1}_{\{j \in U\}} - \sum_{r \in U} \alpha(r) \psi_{ri})^2 * Q_{ij}(t)$$

$$- [(B_{ij} \mathbf{1}_{\{j \in U\}} - \sum_{r \in U} \alpha(r) \psi_{ri}) * Q_{ij}(t)]^2 \}$$

$$+ \sum_{i \in D} \sum_{j \in E} \mu_{ii} \{ B_{ij}^2 * Q_{ij}(t) - [B_{ij} * Q_{ij}(t)]^2 \}.$$

5. Failure Rate Estimation

An interesting introduction to the stochastic process approach of the failure rate is given by Aalen and Gjessing. [1]

The failure rate of a semi-Markov system is defined as follows

$$\lambda(t) := \lim_{h \downarrow 0} \frac{1}{h} \mathbb{P}(Z_{t+h} \in D \mid Z_u \in U, \forall u \le t). \tag{16}$$

From this definition we get:

$$\lambda(t) = \frac{\alpha_0 \psi_0 \star H_0'(t) \mathbf{1}}{\alpha_0 \psi_0 \star (I - H_0(t)) \mathbf{1}}, \tag{17}$$

where $H_0'(t)$ is the diagonal matrix of derivatives of $H_i(t)$, i.e., $H_0'(t) := diag(H_i'(t), i \in U)$.

By replacing Q, ψ, H by their estimators in Equation (17), we get the empirical estimator for failure rate λ, i.e.,

$$\hat{\lambda}(t, T) = \frac{\hat{\alpha}_0 \hat{\psi}_0 \star \hat{H}_0'(t, T) \mathbf{1}}{\hat{\alpha}_0 \hat{\psi}_0 \star (I - \hat{H}_0(t, T)) \mathbf{1}}, \tag{18}$$

where the derivative $\hat{H}'(t,T)$, at t, is estimated by

$$\frac{\hat{H}(t+\Delta,T) - \hat{H}(t,T)}{\Delta}.$$

Theorem 17: ([17]) If the semi-Markov kernel is continuously differentiable, then the estimator of the failure rate given by (18) is uniformly strongly consistent in the sense that for all $L \in \mathbb{R}_+$,

$$\sup_{0 \le t \le L} \left| \hat{\lambda}(t,T) - \lambda(t) \right| \xrightarrow{a.s.} 0, \quad \text{as} \quad T \to \infty.$$

Theorem 18: ([17]) If f_{ij} is twice continuously differentiable for any $i,j \in E$, then

$$T^{\frac{1-\delta}{2}}(\hat{\lambda}(t,T) - \lambda(t)) \xrightarrow{d} N(0,\sigma_\lambda^2), \quad \text{as} \quad T \to \infty,$$

where $0 < \delta < 1/2$, and

$$\sigma_\lambda^2 = \frac{1}{(R(t))^2} \sum_{j \in U} \mu_{jj} \left\{ [(\sum_{i \in U} \alpha(i)\psi'_{0;ij})^2 \star H_j](t) - (\sum_{i \in U} \alpha(i)[\psi'_{0;ij} \star H_j](t))^2 \right\}.$$

6. Rate of Occurrence of Failure

The rate of occurrence of failure[3] (or the failure rate of the process[14]) is an important reliability indicator.[3,10] Let $N_f(t), t \ge 0$, be the counting process of the transitions of the semi-Markov process from U to D, i.e., the number of failures of the system up to time t, and let $\psi_f(t) := \mathbb{E}[N_f(t)]$ be the expected number of failures that have occurred up to time t. The rate of occurrence of failure (rocof) is the intensity of the process $N_f(t), t \ge 0$, i.e., the derivative of the function $\psi_f(t)$ with respect to t. In the exponential case the rocof is equal to the failure rate.

The following theorem gives a closed-form formula of the rocof of semi-Markov systems.

Theorem 19: ([20]) If the semi-Markov process fulfills Assumption A, and the semi-Markov kernel is absolutely continuous with respect to the Lebesgue measure on $\mathbb{R}_+$, and the derivative of $\psi(t)$, i.e., $\psi'(t) := \sum_{n \ge 0}[Q^{(n)}]'(t)$ is finite for any fixed $t \ge 0$, then the rocof of the semi-Markov process at time t is given by:

$$ro(t) = \sum_{i \in U} \sum_{j \in D} \sum_{l=1}^{s} \int_0^t \alpha(l)\psi_{li}(du)q_{ij}(t-u), \tag{19}$$

where $q_{ij}(t) \equiv Q'_{ij}(t)$.

The asymptotic rocof is given by

$$r := \lim_{t \to \infty} ro(t) = \nu_0 P_{01} \mathbf{1}/\hat{m},$$

where P_{01} is the restriction of P on $U \times D$, and ν_0 the restriction of ν on U.

For example, for an alternating renewal process, with up state sequence X and down state sequence Y, such that $\mathbb{E}(X + Y) < \infty$, the asymptotic rocof is given by

$$r = \frac{1}{\mathbb{E}X + \mathbb{E}Y}.$$

Remark 20: The above formula constitute a correction to Corollary 4, in Ouhbi and Limnios.[20]

The natural estimator for rocof

$$\hat{ro}(t) = \frac{N_f(t + \Delta) - N_f(t)}{\Delta},$$

was given for a renewal process, by Cox and Lewis.[5] In the sequel, we will give an estimator of the rocof of semi-Markov systems.[20] This estimator is based on both the estimator of the Markov renewal matrix and the estimator of the derivative of the semi-Markov kernel.

Consider the empirical estimator function of the semi-Markov kernel $\hat{Q}_{ij}(x, T)$ and the empirical density kernel given by:

$$\hat{q}_{ij}(x, T) = \frac{\hat{Q}_{ij}(x + \Delta, T) - \hat{Q}_{ij}(x, T)}{\Delta},$$

where $\Delta = T^{-\alpha}$ and $0 < \alpha < 1$, and the estimator $\hat{\psi}_{ij}(x, T)$ of $\psi_{ij}(x)$, given by

$$\hat{\psi}_{ij}(x, T) = \sum_{n \geq 0} \hat{Q}_{ij}^{(n)}(x, T).$$

From Theorem 19, we propose the following estimator for the rocof of the semi-Markov system,

$$\hat{ro}(t, T) = \sum_{i \in U} \sum_{j \in D} \sum_{l=1}^{s} \alpha(l) \hat{\psi}_{li} * \hat{q}_{ij}(t, T). \tag{20}$$

Note that since we consider only one trajectory, there exists some $l \in E$ such that $\alpha(l) = 1$.

The following theorem gives the uniform strong consistency of the estimator of the rocof.

Theorem 21: ([20]) Under the assumptions of Theorem 19, Estimator (20) is uniformly strongly consistent in the sense that, for all $L \in \mathbb{R}_+$,

$$\sup_{t \in [0,L]} |\hat{ro}(t,T) - ro(t)| \xrightarrow{a.s.} 0, \quad T \to \infty.$$

The following theorem is the central limit theorem of the estimator of the rocof.

Theorem 22: ([20]) Under the assumptions of Theorem 19, and if moreover $q_{ij}(.)$ is twice continuously differentiable at t for all $i \in U$ and $j \in D$, then the estimator $\hat{ro}(t,T)$ is asymptotically normal with mean $ro(t)$ and variance

$$\sigma^2(t) := \sum_{i \in U} \sum_{j \in D} \mu_{ii} \frac{\sum_{l=1}^{s} \alpha(l) \cdot \psi_{li} * q_{ij}(t)}{T^{1-\delta}} + O(T^{-1}),$$

where μ_{ii} is the mean time between two successive visits to state i, and $0 < \delta < 1$.

Acknowledgment. We are grateful to an anonymous referee for his valuable comments.

References

1. O.O. Aalen, H.K. Gjessing Understanding the shape of the hazard rate: a process point of view, *Statistical Science*, **16**, 1-14 (2001).
2. P.K. Andersen, Ø. Borgan, R.D. Gill, N. Keiding, *Statistical Models Based on Counting Processes* (Springer, N.Y., 1993).
3. H. Ascher and H. Feingold, *Repairable Systems Reliability* (Marcel Dekker, New York, 1984).
4. P. Billingsley, *Convergence of Probability Measures*, (Wiley, New York, 1999).
5. D.R. Cox, P.A.W. Lewis, *The Statistical Analysis of Series of Events* (Methuen, London, 1966).
6. R.D. Gill, Nonparametric estimation based on censored observations of Markov renewal process, *Z. Wahrsch. verw. Gebiete*, **53**, 97–116 (1980).
7. P. E. Greenwood and W. Wefelmeyer, Empirical estimators for semi-Markov processes, *Math. Methods Statist.*, **5**, 299–315 (1996).
8. J. Janssen, N. Limnios (Eds.) *Semi-Markov Models and Applications*, (Kluwer Academic, Dordrecht, 1999).
9. S.W. Lagakos, C.J. Sommer and M. Zelen, Semi-Markov models for partially censored data, *Biometrika*, **65**, 311–317 (1978).

10. Y. Lam, The rate of occurrence of failures, *J. Appl. Prob.*, **34**, 234–247 (1997).

11. N. Limnios, A functional central limit theorem for the empirical estimator of a semi-Markov kernel, *J. Nonparametric Statist.*, (to appear) (2002).

12. N. Limnios and G. Oprişan *Semi-Markov Processes and Reliability* (Birkhäuser, Boston, 2001).

13. E.H. Moore and R. Pyke, Estimation of the transition distributions of a Markov renewal process, *Ann. Inst. Stat. Math.*, **20**, 411–424 (1968).

14. G. Oprişan, On the failure rate, in *Statistical and Probabilistic Methods in Reliability*, Eds. D.C. Ionescu and N. Limnios (Birkhauser, Boston, 1999).

15. B. Ouhbi and N. Limnios, Non-parametric estimation for semi-Markov kernels with application to reliability analysis, *Appl. Stoch. Models Data Anal.*, **12**, 209–220 (1996).

16. B. Ouhbi and N. Limnios, Comportement asymptotique de la matrice de renouvellement markovien, *C. R. Acad. Sci. Paris*, Série I, t. 325, 921–924 (1997).

17. B. Ouhbi and N. Limnios, Estimation of kernel, availability and reliability of semi-Markov systems, in *Statistical and Probabilistic Methods in Reliability*, Eds. D.C. Ionescu and N. Limnios (Birkhauser, Boston, 1998).

18. B. Ouhbi, N. Limnios. Failure rate estimation of semi-Markov systems, in *Semi-Markov Models and Applications*, J. Janssen and N. Limnios (Eds), pp 207–218 (Kluwer, Dordrecht, 1998).

19. B. Ouhbi and N. Limnios, Non-parametric estimation for semi-Markov processes based on their hazard rate, *Statist. Infer. Stoch. Processes*, **2**, 151–173 (1999).

20. B. Ouhbi and N. Limnios, The rate of occurrence of failures for semi-Markov processes and estimation, *Statist. Probab. Lett.*, **59**, 245–255 (2001).

21. B. Ouhbi and N. Limnios, Nonparametric reliability estimation of semi-Markov processes, *J. Statist. Plann. Infer.*, **109**, 155–165 (2003).

22. R. Pyke, Markov renewal processes: definitions and preliminary properties, *Ann. Math. Statist.*, **32**, 1231–1241 (1961).

23. R. Pyke and R. Schaufele, The existence and uniqueness of stationary measures for Markov renewal processes, *Ann. Math. Statist.*, **37**, 1439–1462 (1966).

Parallel Simulation of Petri-net by SIMD-like Vector Systems ... 481

19. ... Proceedings of Annals of Failures ... 1994, 55

16. N. Grumman, An area-conservative theorem on the rotational motion ... Math. Comput. 39, No. 2, 1984.
... J. C. ... Stochastic Processes and Including Problem Session, 2001.
... H. Moyer and R. Pyle, Estimation of the transition distributions of a Markov chain, J. Amer. Stat. Assoc., 50, 415-424, 1967.
... On the future of non-Statistical and Probabilistic Methods in ... Eds. U. Grenander and ..., London, Chapman-Hall, 1994.
... V. Larson, Non-parametric estimates for semi-Markov ...

... S. Csiszár and N. Barton, Semi-Markov ... maximum likelihood estimation ... CAL, Stat. Soc., ... 143, 1984.

... Littauer, Estimation of Kernel availability and reliability in ... Markov systems ... in Reliability, Eds. J. C. Laurent and B. ... Reidhaven, Boston, 1994.
16. B. Ouhbi, Non-parametric estimation of semi-Markov kernels ..., J. Nonparam. and ..., ... Fininnce, Eds. ..., ... Kluwer, 1998.

10. B. Ouhbi and N. Limnios, Non-parametric estimation for semi-Markov processes based on their hazard rate, Stat. Inference ..., 2, 151-173, 1999.
... Ouhbi and N. Limnios, Time and asymptotic estimation of failure rate for semi-Markov processes ... Stat. Probab. Lett., 50, 243-254, 2000.
... B. Ouhbi and N. Limnios, Asymptotic ... estimation ... Markov Processes, J. Statist. Planning and Inference, 118, 123-129, 2003.
... Ouhbi, Markov reproduction ... distributions and reliability ...

22. P. Pyke and H. Schaufele, Limit theorems and ... Markov renewal processes, Ann. Math. Stat., 35, 1746-1764, 1964.

31

CLASSES OF FIXED-ORDER AND ADAPTIVE SMOOTH GOODNESS-OF-FIT TESTS WITH DISCRETE RIGHT-CENSORED DATA

Edsel A. Peña

Department of Statistics
University of South Carolina
Columbia, SC 29208 USA
E-mail: pena@stat.sc.edu

Classes of hazard-odds based fixed-order and adaptive smooth goodness-of-fit tests for the composite hypothesis that an unknown discrete distribution belongs to a family of distributions using right-censored observations are presented. The proposed classes of tests generalize Neyman's [33] smooth class of tests. The class of fixed-order tests is the discrete analog of the hazard-based class of tests for continuous failure times studied in Peña[35]. The class of adaptive tests employs a modified Schwartz[40] Bayesian information criterion for choosing the order of the embedding class, with the modification on the criterion accounting for the incompleteness mechanism.

1. Introduction

Statistical goodness-of-fit (gof) testing has always been an active research area as evidenced by entering the phrase "goodness of fit" in the `MathSciNet` search engine. In its simplest form a random sample $T_1, T_2, \ldots, T_n$ from an unknown distribution function F is observed, and it is desired to determine if $F = F_0$, where F_0 is a specified distribution. The most well-known gof procedure is Pearson's[34] chi-square test which utilizes the statistic

$$\chi^2 = \sum_{j=1}^{K} \frac{(O_j - E_j)^2}{E_j}, \tag{1}$$

where K is the size of the partition of the support of F_0, O_j is the number of T_i's in the jth member of the partition, and E_j is the number of T_i's expected to be in the jth member of the partition when F_0 holds. The popularity of this test is partly due to its simplicity and the fact that it requires only critical values from the family of chi-square distributions. There are other tests for the simple gof problem, such as Kolmogorov-Smirnov (KS) type tests, Neyman's[33] smooth gof tests, Cramer-von Mises (CVM) type tests, and those by Khamaladze[20,21]. A review of some of these procedures could be found in Stephens[42]. Many of these tests have extensions to the composite null hypothesis setting, where the problem is to test whether $F \in C$, with C a specified (parametric) family of distributions, cf., Chernoff and Lehmann[6], Rao and Robson[37], D'Agostino and Stephens[9], and Greenwood and Nikulin[13]. Except for Pearson's[34] test, most of the above-mentioned procedures imposes the restriction that F is continuous, with this assumption typically made in order to facilitate the derivations of distributional results.

Though not as prevalent as the case with continuous distributions, gof tests for discrete distributions, or when data arose from grouping of continuous data, have also been considered. Kulperger and Singh[26] examined χ^2 gof tests for discrete distributions and considered the issue of random grouping. Cressie and Read[8] introduced the family of power divergence statistics for performing gof with multinomial data. Best and Rayner[4,5] proposed Neyman smooth gof tests for the null hypothesis that F is geometric and Poisson, respectively; while Eubank[10] proposed Neyman smooth-type tests for dealing with multinomial data. In Choulakian, Lockhart and Stephens[7] a test for the discrete uniform was presented; while in Spinelli and Stephens[41] CVM-type procedures were developed for testing a Poisson distribution. Kocherlakota and Kocherlakota[24], Rueda, Perez-Abreau and O'Reilly[38], Baringhaus and Henze[3], and Nakamura and Perez-Abreau[32] examined gof procedures for discrete data using the empirical probability generating function; in particular, tests for the Poisson distribution were developed. Empirical distribution-based methods were also considered for discrete models. Among papers adopting this approach were Henze[15] and Klar[23]. However, all of these papers dealing with goodness-of-fit for discrete models assume that $T_1, T_2, \ldots, T_n$ are completely observed.

In biomedical, engineering, reliability, and in other areas where the primary variable of interest is the time-to-occurrence of an event, hereon referred to as a failure time, it is typical that some of the failure times will be right-censored due to time constraints, limited resources, withdrawal from the study, loss to follow-up, etc. Numerous papers have appeared dealing with the modeling and analysis of failure times in the presence of incomplete observations. For continuous failure times, the problem of gof testing has been addressed in several papers with the aim of extending to censored data those procedures that were developed for complete data. Among these papers are those of Koziol and Green[25], Hyde[19], Hollander and Proschan[17], Nair[29,30,31], Gatsonis, Hsieh and Korwar[11], Habib and Thomas[14], Akritas[2], Hjort[16], Hollander and Peña[18], Li and Doss[28], and Kim[22]. An interesting goal in gof testing with censored data is to extend Pearson's test. The difficulty underlying such an extension is that the exact number of failures in a member of the partition is not observable. An attempt to extend Neyman's smooth gof procedure in the presence of right-censored data has also been made by Gray and Pierce[12]. Their approach parallels that of Neyman[33] where the density function is embedded in a wider class. A different extension of the smooth gof tests with continuous failure times, which adapts naturally to censored data and enables point process theory, was that in Peña[35] and Agustin and Peña[1], the latter dealing with reliability models for recurrent events.

Except for the test proposed in Hyde[19] which is a special case of the class of tests proposed in this chapter, the gof problem with right-censored discrete failure times does not seem to have been investigated extensively in the literature. The existing gof procedures for discrete and complete data mentioned earlier have not yet been extended for discrete and censored data, which is rather surprising since discrete failure times are ubiquitous in many studies. For instance, discrete failure times occur because of the intrinsic nature of the failure time process such as when failure is measured in terms of counts or the number of cycles, or due to an inherent limitation in the measurement process forcing subjects to be observed only at the end of specified intervals (e.g., weekly basis). Discrete failure times also manifest when the times are interval-censored as in biomedical studies, or when data is presented in a life-table format as is done in actuarial settings. Right-censoring occurs due to the withdrawal of subjects from the study, a fixed study period, or due to failure (death) from competing causes. In these situations, prior to performing higher-level statistical analysis such as estimation or hypothesis testing, it is desirable to know the parametric family of distributions or hazards to which F or Λ belongs since this will enable the use of more efficient inferential methods.

This chapter aims to provide a general class of gof tests for discrete failure times and in the presence of right-censoring for the composite null hypothesis. In Peña[36] a general approach for generating a class of tests for the simple null hypothesis case was presented, an approach which is a hazard-based extension of Neyman's[33] smooth goodness-of-fit tests. See

Rayner and Best[39] for an extensive discussion of the Neyman formulation of this class of smooth goodness-of-fit tests. The present chapter considers the parallel treatment of the composite null hypothesis case. The procedures presented in this chapter are discrete analogs of the intensity-based smooth goodness-of-fit tests developed in Peña[35] for continuous failure times. In this formulation, the sequence of odds associated with the hazard rates are embedded in a wider class, in contrast to the usual Neyman formulation where the sequence of probabilities are embedded, cf., Rayner and Best[39]. This intensity-based embedding facilitates the derivation of the smooth goodness-of-fit tests as score tests, thereby endowing the tests with certain local optimality properties. In contrast to the development of Pearson's test in which the vantage point is the time origin and the underlying question is: 'How many observations are expected to have values in a member of the partition of the support of F_0?' the current approach's vantage point is dynamic in that the relevant question is: 'Given that just before a certain time point there are a certain number of units at risk, how many are expected to fail at this time point?' Consequently, instead of dealing with global probabilities, the main focus are conditional probabilities, hazards, or intensities, which are the natural quantities when dealing with dynamic or time-evolving systems.

Due to space and time constraints, proofs of the propositions and theorems will not be presented in this chapter, but we focus instead on the proposed class of goodness-of-fit procedures. Results of simulation studies pertaining to the achieved levels and powers will be presented in the paper containing the proofs of the propositions and theorems. We mention that simulation studies performed for the tests associated with the simple null hypothesis case demonstrated the viability of the proposed class of tests and indicates that the proposed adaptive test using the modified Schwartz information criterion could be used as an omnibus test. Results of these simulation studies can be found in Peña[36].

2. Description of the Problem

Let $T_1, T_2, \ldots, T_n$ be independent and identically distributed (IID) random variables from an unknown discrete distribution F whose support is known to be $\mathcal{A} = \{a_1, a_2, \ldots\}$ with $a_i < a_{i+1}, i = 1, 2, \ldots$. The T_i's are not completely observed, but only the random vectors $(Z_1, \delta_1), (Z_2, \delta_2), \ldots, (Z_n, \delta_n)$ are observed with the interpretation that $\delta_i = 1$ implies $T_i = Z_i$, whereas $\delta_i = 0$ implies $T_i > Z_i$. Let $\lambda_j = \lambda_j(F), j = 1, 2, \ldots$ be the hazard of T at a_j, so $\lambda_j = \mathbf{P}(T = a_j | T \geq a_j) = \Delta F(a_j)/\bar{F}(a_j-)$, and let $\Lambda(t) = \sum_{j=1}^{\infty} \lambda_j I\{a_j \leq t\}, t \in \Re$, be the discrete hazard function associated with F. We assume in the sequel the independent censoring condition:

$$\mathbf{P}\{T = a_j | T \geq a_j\} = \lambda_j = \mathbf{P}\{T = a_j | Z \geq a_j\}, \quad j = 1, 2, \ldots. \tag{2}$$

The problem dealt with is to test the hypothesis that F belongs to a parametric class $\mathcal{F}_0$ of discrete distributions parameterized by a q-dimensional vector η taking values in Γ, an open set in $\Re^q$. Denote by $\mathcal{C}_0$ the class of hazard functions associated with $\mathcal{F}_0$ so $\mathcal{C}_0 = \{\Lambda_0(\cdot|\eta) : \eta \in \Gamma\}$, where the functional form of $\Lambda_0(\cdot|\eta)$ is known. The goodness-of-fit problem is to test the composite hypotheses

$$H_0 : \Lambda(\cdot) \in \mathcal{C}_0 \quad \text{versus} \quad H_1 : \Lambda(\cdot) \notin \mathcal{C}_0 \tag{3}$$

on the basis of the right-censored data $(Z_i, \delta_i), i = 1, 2, \ldots, n$. The simple null hypothesis case where interest is on testing

$$H_0 : \Lambda(\cdot) = \Lambda_0(\cdot) \quad \text{versus} \quad H_1 : \Lambda(\cdot) \neq \Lambda_0(\cdot) \tag{4}$$

with $\Lambda_0(\cdot)$ a fully specified discrete hazard function was dealt with in Peña[36]. The present chapter extends the results in Peña[36] to the composite case. Note that in (3), the parameter vector η is a nuisance parameter.

3. Hazard Embeddings and Likelihoods

Let $\lambda_j^0(\eta), j = 1, 2, \ldots$ be the hazards associated with $\Lambda_0(\cdot|\eta)$, so

$$\Lambda_0(t|\eta) = \sum_{\{j:a_j \leq t\}} \lambda_j^0(\eta).$$

Following Peña[36], for $\lambda_j < 1$ and $\lambda_j(\eta) < 1$, let the hazard odds be

$$\rho_j = \frac{\lambda_j}{1 - \lambda_j} \quad \text{and} \quad \rho_j^0(\eta) = \frac{\lambda_j^0(\eta)}{1 - \lambda_j^0(\eta)}.$$

For a fixed smoothing order $p \in \mathcal{Z}_+$, and for the $p \times 1$ vectors $\boldsymbol{\Psi}_j = \boldsymbol{\Psi}_j(\eta), j = 1, 2, \ldots, J$, we embed $\rho_j^0(\eta)$ into the hazard odds determined by

$$\rho_j(\theta, \eta) = \rho_j^0(\eta) \exp\{\theta^{\mathrm{t}} \boldsymbol{\Psi}_j(\eta)\}, \quad j = 1, 2, \ldots; \theta \in \Re^p. \tag{5}$$

This is equivalent to postulating that the logarithm of the hazard odds ratio is linear in $\boldsymbol{\Psi}_j(\eta)$, that is,

$$\log\left\{\frac{\rho_j(\theta, \eta)}{\rho_j^0(\eta)}\right\} = \theta^{\mathrm{t}} \boldsymbol{\Psi}_j(\eta), \quad j = 1, 2, \ldots.$$

490   *E. Peña*

Within this embedding, the partial likelihood of (θ, η) based on the observation period $(-\infty, a_J]$ for some fixed $J \in \mathcal{Z}_+$ is (cf., Peña[36])

$$L(\theta, \eta) = \prod_{j=1}^{J} \frac{\rho_j(\theta, \eta)^{O_j}}{[1 + \rho_j(\theta, \eta)]^{R_j}} \qquad (6)$$

where

$$O_j = \sum_{i=1}^{n} I\{Z_i = a_j, \delta_i = 1\} \quad \text{and} \quad R_j = \sum_{i=1}^{n} I\{Z_i \geq a_j\}.$$

Furthermore, within this hazard odds embedding, the composite goodness-of-fit problem simplifies to testing $H_0 : \theta = 0, \eta \in \Gamma$ versus $H_1 : \theta \neq 0, \eta \in \Gamma$, so η is a nuisance parameter. For our notation, we shall denote by $\nabla_v = \partial/\partial v$ the gradient operator with respect to a vector v. The test is to be anchored by the estimated score statistic

$$U_\theta(0, \hat{\eta}) = \nabla_\theta \log L(\theta, \eta)|_{\theta=0, \eta=\hat{\eta}},$$

where $\hat{\eta} = \hat{\eta}(\theta = 0)$ is the restricted partial likelihood maximum likelihood estimator (RPLMLE). This is the η that maximizes the restricted partial likelihood function

$$L_0(\eta) = L(0, \eta) = \prod_{j=1}^{J} \frac{\rho_j(0, \eta)^{O_j}}{[1 + \rho_j(0, \eta)]^{R_j}} = \prod_{j=1}^{J} [\lambda_j^0(\eta)]^{O_j} [1 - \lambda_j^0(\eta)]^{R_j - O_j} \quad (7)$$

with $\rho_j(0, \eta) = \rho_j^0(\eta) = \lambda_j^0(\eta)/[1 - \lambda_j^0(\eta)]$.

4. Restricted Partial Likelihood MLE

From (7), the logarithm of the partial likelihood function is

$$l_0(\eta) = \log L_0(\eta) = \sum_{j=1}^{J} \left\{ O_j \log \lambda_j^0(\eta) + (R_j - O_j) \log[1 - \lambda_j^0(\eta)] \right\}. \quad (8)$$

Consequently,

$$\nabla_\eta l_0(\eta) = \sum_{j=1}^{J} \mathbf{A}_j(\eta)[O_j - E_j^0(\eta)]$$

where, for $j = 1, 2, \ldots,$

$$E_j^0(\eta) = R_j \lambda_j^0(\eta) \quad \text{and} \quad \mathbf{A}_j(\eta) = \frac{\nabla_\eta \lambda_j^0(\eta)}{\lambda_j^0(\eta)[1 - \lambda_j^0(\eta)]}$$

are the $q \times 1$ 'standardized' gradients of $\lambda_j^0(\eta)$ with respect to η. We form the $J \times q$ matrix of standardized gradients

$$\mathbf{A}(\eta) = [\mathbf{A}_1(\eta), \mathbf{A}_2(\eta), \ldots, \mathbf{A}_J(\eta)]^t, \tag{9}$$

and define the $J \times 1$ vectors

$$\mathbf{O} = (O_1, O_2, \ldots, O_J)^t \quad \text{and} \quad \mathbf{E}^0(\eta) = \left(E_1^0(\eta), E_2^0(\eta), \ldots, E_J^0(\eta)\right)^t.$$

Then, in matrix form,

$$\nabla_\eta l_0(\eta) = \mathbf{A}(\eta)^t \left[\mathbf{O} - \mathbf{E}^0(\eta)\right].$$

The estimating equation for the RPLMLE $\hat{\hat{\eta}}$ is therefore

$$\mathbf{A}(\eta)^t \left[\mathbf{O} - \mathbf{E}^0(\eta)\right] = \mathbf{0}. \tag{10}$$

For example, suppose that $\mathcal{C}_0$ is the class of constant hazards, which corresponds to the class of geometric distributions. Then $\mathbf{A}(\eta) = \mathbf{1}_J/[\eta(1-\eta)]$ and $\mathbf{E}^0(\eta) = \mathbf{R}\eta$, so the estimating equation becomes

$$\{\eta(1-\eta)\}^{-1} \mathbf{1}_J^t (\mathbf{O} - \mathbf{R}\eta) = 0,$$

yielding the RPLMLE given by

$$\hat{\hat{\eta}} = \frac{\mathbf{1}_J^t \mathbf{O}}{\mathbf{1}_J^t \mathbf{R}} = \frac{\sum_{j=1}^J O_j}{\sum_{j=1}^J R_j}. \tag{11}$$

Clearly, in many situations, $\hat{\hat{\eta}}$ will need to be obtained iteratively or through numerical methods.

5. Asymptotics and Test Procedure

The logarithm of the partial likelihood function is given by

$$l(\theta, \eta) = \sum_{j=1}^J \left\{O_j \log \rho_j(\theta, \eta) - R_j \log[1 + \rho_j(\theta, \eta)]\right\}.$$

With $\mathbf{\Psi}(\eta) = [\mathbf{\Psi}_1(\eta), \mathbf{\Psi}_2(\eta), \ldots, \mathbf{\Psi}_J(\eta)]^t$, the score function for θ, evaluated at $\theta = \mathbf{0}$, is immediately obtained to be

$$\mathbf{U}_\theta(\theta = \mathbf{0}, \eta) = \nabla_\theta l(\theta, \eta)|_{\theta=\mathbf{0}} = \mathbf{\Psi}(\eta)^t[\mathbf{O} - \mathbf{E}^0(\eta)]. \tag{12}$$

Since η is unknown, this score function is estimated by

$$\hat{\mathbf{U}}_\theta = \mathbf{U}_\theta(\theta = \mathbf{0}, \hat{\hat{\eta}}) = \mathbf{\Psi}(\hat{\hat{\eta}})^t[\mathbf{O} - \mathbf{E}^0(\hat{\hat{\eta}})]. \tag{13}$$

To develop the test we need the asymptotic distribution of $\hat{\mathbf{U}}_\theta$. For this purpose, we first present the joint asymptotic distribution of the $(p+q) \times 1$ vector of scores

$$\mathbf{U}(\eta) = \begin{bmatrix} \boldsymbol{\Psi}(\eta)^{\mathrm{t}} \\ \mathbf{A}(\eta)^{\mathrm{t}} \end{bmatrix} [\mathbf{O} - \mathbf{E}^0(\eta)] \tag{14}$$

at $\eta = \eta_0$, the true value of η under H_0.

To achieve a more compact notation, for a vector $\mathbf{v}$, we denote by $\mathrm{Diag}(\mathbf{v})$ the diagonal matrix whose diagonal elements are those of $\mathbf{v}$. Let

$$\mathbf{D}(\eta) = \mathrm{Diag}\left(\lambda_j(\eta)[1 - \lambda_j(\eta)] : j = 1, 2, \ldots, J\right)$$

and $\lambda(\eta) = (\lambda_1(\eta), \lambda_2(\eta), \ldots, \lambda_J(\eta))^{\mathrm{t}}$. Then, the matrix of standardized gradients could be re-expressed via

$$\mathbf{A}(\eta) = \mathbf{D}(\eta)^{-1} \nabla_{\eta^{\mathrm{t}}} \lambda(\eta). \tag{15}$$

The asymptotic distribution of $\mathbf{U}(\eta)$ can be obtained by invoking Theorem 4 in Peña[36]. To describe this asymptotic distribution, we need to introduce more notation. Let

$$\mathbf{V}(\eta) = \mathrm{Diag}(\mathbf{R})\mathbf{D}(\eta)$$

and with $\mathbf{B}(\eta) = [\boldsymbol{\Psi}(\eta), \mathbf{A}(\eta)]$, define the $(p+q) \times (p+q)$ matrix

$$\boldsymbol{\Xi}(\eta) = \mathbf{B}(\eta)^{\mathrm{t}} \mathbf{V}(\eta) \mathbf{B}(\eta).$$

Furthermore, for $i = 1, 2, \ldots, n$ and $j = 1, 2, \ldots, J$, let

$$V_{ij} = I\{Z_i = a_j, \delta_i = 1\};$$
$$W_{ij} = I\{Z_i \geq a_j\};$$
$$U_{ij}(\eta) = V_{ij} - W_{ij} \lambda_j^0(\eta),$$

and

$$\mathcal{F}_j = \bigvee_{i=1}^{n} \sigma\{W_{i1}, V_{i1}, W_{i2}, V_{i2}, \ldots, W_{ij}, V_{ij}, W_{ij+1}\}.$$

From Theorem 4 in Peña[36] we obtain the following proposition.

Proposition 1: *Assume that H_0 holds and that the true value of η is η_0. Furthermore, suppose that p does not change with n and for $i = 1, 2, \ldots, n$ and $j = 1, 2, \ldots, J$, the following conditions hold:*

(i) *the jth row of $\mathbf{B}(\eta_0)$, which is $\mathbf{B}_j(\eta_0) = [\boldsymbol{\Psi}_j(\eta_0)^{\mathrm{t}}, \mathbf{A}_j(\eta_0)^{\mathrm{t}}]$, is $\mathcal{F}_{j-1}$-measurable and $E\{[\| \mathbf{B}_j(\eta_0) \| U_{ij}]^2\} < \infty;$*

(ii) there exists a $(p+q) \times (p+q)$ positive definite matrix $\mathbf{\Xi}^{(0)}(\eta_0)$ such that, as $n \to \infty$,

$$n^{-1}\mathbf{\Xi}(\eta_0) \xrightarrow{\text{pr}} \mathbf{\Xi}^{(0)}(\eta_0);$$

(iii) with $V_{jj}(\eta_0) = R_j\lambda_j(\eta_0)[1 - \lambda_j(\eta_0)]$, then as $n \to \infty$,

$$\max_{1 \le j \le J} trace\left\{[\mathbf{\Xi}(\eta_0)]^{-1}[\mathbf{B}_j(\eta_0)^{\text{t}}V_{jj}(\eta_0)\mathbf{B}_j(\eta_0)]\right\} \xrightarrow{\text{pr}} 0;$$

(iv) as $n \to \infty$, $\max_{1 \le j \le J} \| \mathbf{B}_j(\eta_0) \|^2 = O_p(1)$.

Then, as $n \to \infty$,

$$\frac{1}{\sqrt{n}}\mathbf{U}(\eta_0) = \frac{1}{\sqrt{n}}\mathbf{B}(\eta_0)^{\text{t}}[\mathbf{O} - \mathbf{E}^0(\eta_0)] \xrightarrow{\text{d}} N_{p+q}(\mathbf{0}, \mathbf{\Xi}^{(0)}(\eta_0)).$$

Marginalizing on the score function for θ, it follows from Proposition 1 that

$$\frac{1}{\sqrt{n}}\mathbf{\Psi}(\eta_0)^{\text{t}}[\mathbf{O} - \mathbf{E}^0(\eta_0)] \xrightarrow{\text{d}} N_p(\mathbf{0}, \mathbf{\Xi}_{11}^{(0)}(\eta_0)) \tag{16}$$

where $\mathbf{\Xi}_{11}^{(0)}(\eta_0)$ is the in-probability limit of $n^{-1}\mathbf{\Psi}(\eta_0)^{\text{t}}\mathbf{V}(\eta_0)\mathbf{\Psi}(\eta_0)$. Of course this result is not directly useful for constructing the test since η_0 is not known; however, it will become useful later when ascertaining the impact of the estimation of η_0 by $\hat{\eta}$. For later use, we also denote by $\mathbf{\Xi}_{12}^{(0)}(\eta_0) = \mathbf{\Xi}_{21}^{(0)}(\eta_0)^{\text{t}}$ the in-probability limit of $n^{-1}\mathbf{\Psi}(\eta_0)^{\text{t}}\mathbf{V}(\eta_0)\mathbf{A}(\eta_0)$ and by $\mathbf{\Xi}_{22}^{(0)}(\eta_0)$ the in-probability limit of $n^{-1}\mathbf{A}(\eta_0)^{\text{t}}\mathbf{V}(\eta_0)\mathbf{A}(\eta_0)$.

We are now ready to present the asymptotic result which will be useful for constructing the goodness-of-fit procedure.

Theorem 1: *Assume that the conditions of Proposition 1 hold, and in addition there exists a neighborhood Γ_0 of η_0 in Γ such that, as $n \to \infty$,*

(i) for each $j = 1, 2, \ldots, J$, $\lambda_j(\eta)$ is twice-differentiable with $\eta \mapsto \nabla_\eta\lambda_j(\eta)$ continuous at $\eta = \eta_0$; $\max_{1 \le j \le J} \| \nabla_\eta\lambda_j(\eta) \| = O_p(1)$, and for each $l, l' \in \{1, 2, \ldots, q\}$,

$$\max_{1 \le j \le J} \sup_{\eta \in \Gamma_0} \left| \frac{\partial^2}{\partial \eta_l \eta_{l'}}\lambda_j(\eta) \right| = o_p(n);$$

(ii) for each $i = 1, 2, \ldots, n$ and $j = 1, 2, \ldots, J$, $\mathbf{\Psi}_{ij}(\eta)$ is twice-differentiable with $\eta \mapsto \nabla_\eta\mathbf{\Psi}_{ij}(\eta)$ continuous at $\eta = \eta_0$;

$$\max_{1 \le i \le n} \max_{1 \le j \le J} \| \nabla_\eta\mathbf{\Psi}_{ij}(\eta) \| = O_p(1),$$

494 E. Peña

and for each $l, l' \in \{1, 2, \ldots, q\}$,

$$\max_{1 \le i \le n} \max_{1 \le j \le J} \sup_{\eta \in \Gamma_0} \left| \frac{\partial^2}{\partial \eta_l \eta_{l'}} \Psi_{ij}(\eta) \right| = o_p(n);$$

(iii) the limiting matrix $\mathbf{\Xi}_{22}^{(0)}(\eta_0)$ is nonsingular.

Then, under H_0 and as $n \to \infty$,

$$\frac{1}{\sqrt{n}} \mathbf{\Psi}(\hat{\hat{\eta}})^{\mathrm{t}} [\mathbf{O} - \mathbf{E}^0(\hat{\hat{\eta}})] \xrightarrow{\mathrm{d}} N_p\left(\mathbf{0}, \mathbf{\Xi}_{11.2}^{(0)}(\eta_0)\right),$$

where $\mathbf{\Xi}_{11.2}^{(0)}(\eta_0) = \mathbf{\Xi}_{11}^{(0)}(\eta_0) - \mathbf{\Xi}_{12}^{(0)}(\eta_0)\left\{\mathbf{\Xi}_{22}^{0)}(\eta_0)\right\}^{-1}\mathbf{\Xi}_{21}^{(0)}(\eta_0).$

Comparing this result with that in (16), we see the effect of estimating the unknown parameter η_0 by the RPLMLE $\hat{\hat{\eta}}$ is to decrease the covariance matrix by the term $\mathbf{\Xi}_{12}^{(0)}(\eta_0)\left\{\mathbf{\Xi}_{22}^{0)}(\eta_0)\right\}^{-1}\mathbf{\Xi}_{21}^{(0)}(\eta_0)$. Also, by recalling the definitions of the matrices $\mathbf{\Xi}_{ij}^{(0)}(\eta_0)$'s, it is immediate that the limiting covariance matrix $\mathbf{\Xi}_{11.2}^{(0)}(\eta_0)$ can be estimated consistently by

$$\hat{\mathbf{\Xi}}_{11.2}^{(0)} = \frac{1}{n}\left\{ \mathbf{\Psi}(\hat{\hat{\eta}})^{\mathrm{t}}\mathbf{V}(\hat{\hat{\eta}})\mathbf{\Psi}(\hat{\hat{\eta}}) - \right.$$
$$\left. [\mathbf{\Psi}(\hat{\hat{\eta}})^{\mathrm{t}}\mathbf{V}(\hat{\hat{\eta}})\mathbf{A}(\hat{\hat{\eta}})][\mathbf{A}(\hat{\hat{\eta}})^{\mathrm{t}}\mathbf{V}(\hat{\hat{\eta}})\mathbf{A}(\hat{\hat{\eta}})]^{-1}[\mathbf{A}(\hat{\hat{\eta}})^{\mathrm{t}}\mathbf{V}(\hat{\hat{\eta}})\mathbf{\Psi}(\hat{\hat{\eta}})]\right\}. \quad (17)$$

With $\mathbf{M}^-$ denoting a generalized inverse of a matrix $\mathbf{M}$, the test statistic for testing H_0 and a fixed smoothing order p is

$$\hat{S}_p^2 = \left\{ \frac{1}{\sqrt{n}}\mathbf{\Psi}(\hat{\hat{\eta}})^{\mathrm{t}}[\mathbf{O} - \mathbf{E}^0(\hat{\hat{\eta}})] \right\}^{\mathrm{t}} \left\{ \hat{\mathbf{\Xi}}_{11.2}^{(0)} \right\}^- \left\{ \frac{1}{\sqrt{n}}\mathbf{\Psi}(\hat{\hat{\eta}})^{\mathrm{t}}[\mathbf{O} - \mathbf{E}^0(\hat{\hat{\eta}})] \right\}. \quad (18)$$

Corollary 1: *Under the conditions of Theorem 1 and under H_0, as $n \to \infty$, $\hat{S}_p^2 \xrightarrow{\mathrm{d}} \chi_{p^*}^2$ with $p^* = rank(\mathbf{\Xi}_{11.2}^{(0)}(\eta_0))$. Therefore, an asymptotic α-level test of H_0 versus H_1 rejects H_0 whenever $\hat{S}_p^2 > \chi_{\hat{p}^*;\alpha}^2$ with $\hat{p}^* = rank(\hat{\mathbf{\Xi}}_{11.2}^{(0)})$, and where $\chi_{p;\alpha}^2$ is the $100(1-\alpha)$th percentile of a χ_p^2 distribution.*

To further simplify our notation, let

$$\mathbf{A}^*(\eta) = \mathbf{V}(\eta)^{\frac{1}{2}}\mathbf{A}(\eta) \quad \text{and} \quad \mathbf{\Psi}^*(\eta) = \mathbf{V}(\eta)^{\frac{1}{2}}\mathbf{\Psi}(\eta),$$

and for a full rank $J \times q$ (with $J > q$) matrix $\mathbf{X}$, let

$$P(\mathbf{X}) = \mathbf{X}(\mathbf{X}^{\mathrm{t}}\mathbf{X})^{-1}\mathbf{X}^{\mathrm{t}}$$

be the projection operator (matrix) on the linear subspace $\mathcal{L}(\mathbf{X})$ generated by $\mathbf{X}$ in $\Re^J$. Also, denote by

$$P^{\perp}(\mathbf{X}) = \mathbf{I} - P(\mathbf{X})$$

the projection operator on the orthocomplement of $\mathcal{L}(\mathbf{X})$. Using these notation, the estimator $\hat{\mathbf{\Xi}}_{11.2}^{(0)}$ can be reexpressed via

$$\hat{\mathbf{\Xi}}_{11.2}^{(0)} = \frac{1}{n}\mathbf{\Psi}^*(\hat{\hat{\eta}})^{\mathrm{t}} P^{\perp}(\mathbf{A}^*(\hat{\hat{\eta}}))\mathbf{\Psi}^*(\hat{\hat{\eta}}). \tag{19}$$

Let us also define the 'standardized' observed and *dynamic* expected frequencies via

$$\mathbf{O}^* = \mathbf{V}(\hat{\hat{\eta}})^{-\frac{1}{2}}\mathbf{O} = \left(\frac{O_j}{\sqrt{R_j\lambda_j^0(\hat{\hat{\eta}})[1 - \lambda_j^0(\hat{\hat{\eta}})]}} : j = 1, 2, \ldots, J\right)^{\mathrm{t}} ;$$

$$\mathbf{E}^*(\hat{\hat{\eta}}) = \mathbf{V}(\hat{\hat{\eta}})^{-\frac{1}{2}}\mathbf{E}^0(\hat{\hat{\eta}}) = \left(\frac{R_j\lambda_j^0(\hat{\hat{\eta}})}{\sqrt{R_j\lambda_j^0(\hat{\hat{\eta}})[1 - \lambda_j^0(\hat{\hat{\eta}})]}} : j = 1, 2, \ldots, J\right)^{\mathrm{t}}$$

with the convention that $0/0 = 0$.

Using these standardized quantities, and upon further simplification, the test statistic can be expressed as

$$\hat{S}_p^2 = [\mathbf{O}^* - \mathbf{E}^*(\hat{\hat{\eta}})]^{\mathrm{t}} \left[\mathbf{\Psi}^*(\hat{\hat{\eta}}) \left\{\mathbf{\Psi}^*(\hat{\hat{\eta}})^{\mathrm{t}} P^{\perp}(\mathbf{A}^*(\hat{\hat{\eta}}))\mathbf{\Psi}^*(\hat{\hat{\eta}})\right\}^{-} \mathbf{\Psi}^*(\hat{\hat{\eta}})^{\mathrm{t}}\right] [\mathbf{O}^* - \mathbf{E}^*(\hat{\hat{\eta}})]. \tag{20}$$

Under an orthogonality condition between $\mathbf{A}^*(\hat{\hat{\eta}})$ and $\mathbf{\Psi}^*(\hat{\hat{\eta}})$, we further obtain the more compact and norm-like nature of the statistic given in the following corollary. This corollary also implies that under the orthogonality condition, the estimation of η_0 by $\hat{\hat{\eta}}$ does not require any adjustments in the limiting covariance matrix relative to the case when η_0 is known, an 'adaptiveness' property.

Corollary 2: *If* $\mathbf{\Psi}^*(\hat{\hat{\eta}})$ *lies in* $\mathcal{L}(\mathbf{A}^*(\hat{\hat{\eta}}))^{\perp}$, *the orthocomplement of* $\mathcal{L}(\mathbf{A}^*(\hat{\hat{\eta}}))$, *then*

$$\hat{S}_p^2 = \| P(\mathbf{\Psi}^*(\hat{\hat{\eta}}))[\mathbf{O}^* - \mathbf{E}^*(\hat{\hat{\eta}})] \|^2 .$$

For purposes of studying the asymptotic local power properties of the test, Theorem 1 could be generalized to cover the behavior under local alternatives. This generalization is contained in the following theorem.

496   *E. Peña*

Theorem 2: *If the conditions of Theorem 1 hold, then under the sequence of local alternatives $H_1^{(n)} : \theta^{(n)} = n^{-\frac{1}{2}}\gamma(1+o(1))$ for $\gamma \in \Re^p$ and as $n \to \infty$,*

$$\frac{1}{\sqrt{n}}\boldsymbol{\Psi}(\hat{\eta})^{\mathrm{t}}[\mathbf{O} - \mathbf{E}^0(\hat{\eta})] \xrightarrow{\mathrm{d}} N_p\left(\boldsymbol{\Xi}_{11.2}^{(0)}(\eta_0)\gamma, \boldsymbol{\Xi}_{11.2}^{(0)}(\eta_0)\right).$$

As a consequence, the asymptotic local power of the test described above for the sequence of local alternatives specified in Theorem 2 is

$$\mathrm{ALP}(\gamma) = \mathbf{P}\left\{\chi_{p^*}^2(\delta^2(\gamma)) > \chi_{p^*;\alpha}^2\right\}, \tag{21}$$

where the noncentrality parameter is

$$\delta^2(\gamma) = \gamma^{\mathrm{t}}\boldsymbol{\Xi}_{11.2}^{(0)}(\eta_0)\gamma,$$

which could be consistently estimated by

$$\hat{\delta}^2 = \frac{1}{n}[\boldsymbol{\Psi}^*(\hat{\hat{\eta}})\gamma]^{\mathrm{t}}P^{\perp}(\mathbf{A}^*(\hat{\hat{\eta}}))[\boldsymbol{\Psi}^*(\hat{\hat{\eta}})\gamma].$$

Under the orthogonality condition of Corollary 2 this simplifies to

$$\hat{\delta}^2 = \frac{1}{n}\parallel \boldsymbol{\Psi}^*(\hat{\hat{\eta}})\gamma \parallel^2 \xrightarrow{\mathrm{pr}} \gamma^{\mathrm{t}}\boldsymbol{\Xi}_{11}(\eta_0)\gamma = \delta^2.$$

6. Some Choices of $\boldsymbol{\Psi}$

For a fixed smoothing order p, three particular choices of the $J \times p$ matrix $\boldsymbol{\Psi}(\eta)$ are provided below. The first specification is given by

$$\boldsymbol{\Psi}_1 = \left(\left(\frac{\mathbf{R}}{n}\right)^0, \left(\frac{\mathbf{R}}{n}\right)^1, \ldots, \left(\frac{\mathbf{R}}{n}\right)^{p-1}\right), \tag{22}$$

where

$$(\mathbf{R}/n)^k = ((R_1/n)^k, (R_2/n)^k, \ldots, (R_J/n)^k)^{\mathrm{t}}.$$

Note that this choice does not depend functionally on η, but its distribution depends on η. This choice has proven effective in goodness-of-fit testing for the simple null hypothesis setting for this discrete failure time setting[36], and as such we expect that this will also perform satisfactorily in this composite null hypothesis setting.

The second specification, which depends functionally on η, is

$$\boldsymbol{\Psi}_2(\eta) = \left([\lambda_0(\eta)]^0, [\lambda_0(\eta)]^1, \ldots, [\lambda_0(\eta)]^{p-1}\right), \tag{23}$$

where

$$[\lambda_0(\eta)]^k = ([\lambda_1^0(\eta)]^k, [\lambda_2^0(\eta)]^k, \ldots, [\lambda_J^0(\eta)]^k)^{\mathrm{t}}.$$

The analogous choice for the continuous failure time situation was quite effective in generating tests with commendable powers (cf., Peña[35]).

The third specification produces a test statistic which generalizes Pearson's statistic. Let $C_1, C_2, \ldots, C_p$ be a (disjoint) partition of $\mathcal{J} = \{1, 2, \ldots, J\}$. Define

$$\mathbf{\Psi}_3 = \left(\mathbf{1}_{C_1}, \mathbf{1}_{C_2}, \ldots, \mathbf{1}_{C_p}\right)^{\mathrm{t}} \tag{24}$$

where for $C \subseteq \mathcal{J}$, $\mathbf{1}_C$ is a $J \times 1$ vector whose jth element is $I\{j \in C\}$. Furthermore, define

$$O_{\bullet}(C) = \sum_{j \in C} O_j \quad \text{and} \quad \hat{E}_{\bullet}^0(C) = \sum_{j \in C} E_j^0(\hat{\eta}).$$

Also, with

$$\hat{V}_{\bullet}^*(C) = \mathbf{1}_C^{\mathrm{t}} \mathbf{V}(\hat{\hat{\eta}})^{1/2} P^{\perp}(\mathbf{A}^*(\hat{\hat{\eta}}) \mathbf{V}(\hat{\hat{\eta}})^{1/2} \mathbf{1}_C$$

the resulting test statistic for the specification (24) is given by

$$\hat{S}_p^2 = \sum_{i=1}^{p} \frac{\left[O_{\bullet}(C_i) - \hat{E}_{\bullet}^0(C_i)\right]^2}{\hat{V}_{\bullet}^*(C_i)}, \tag{25}$$

which is a Pearson-type test statistic.

However, these choices do not satisfy the orthogonality condition in Corollary 2, so the correction term for the covariance matrix will be required. It is possible to start with these choices to arrive at a $\mathbf{\Psi}'$ that satisfies the orthogonality condition using a Gram-Schmidt type of orthogonalization. But, as pointed out in Peña[35] in the continuous failure time setting, the benefits of such a programme may not outweigh the effort and difficulty in performing the orthogonalization.

7. Adaptive Choice of Smoothing Order

The testing procedure described in the preceding section requires that the smoothing order p be fixed. This introduces an arbitrariness in the procedure, and without a good prior knowledge of the class of hazards that holds if the class under the null hypothesis does not hold, there is a great potential of choosing a p that is far from optimal. Of course, repeated testing with different smoothing orders is unwise since it will inflate the Type I error rates. It is therefore imperative and important to have a data-driven or adaptive approach for determining the smoothing order p.

We propose a procedure that uses a modified Schwartz information criterion (Schwartz[40]) to decide on the smoothing order p. We mention that for the classical Neyman's smooth goodness-of-fit test, Ledwina[27] proposed the use of the Schwartz information criterion for adaptively determining the smoothing order. Let $L_p(\theta_p, \eta)$ denote the partial likelihood of (θ_p, η) when the smoothing order p is given in (6), and let $l_p(\theta_p, \eta) = \log L_p(\theta_p, \eta)$ be the associated log-partial likelihood function. Denote by $(\hat{\theta}_p, \hat{\eta})$ the partial likelihood maximum likelihood estimator (PLMLE), so that

$$L_p(\hat{\theta}_p, \hat{\eta}) = \sup_{\theta_p \in \Re^p; \ \eta \in \Gamma} L_p(\theta_p, \eta).$$

Clearly, as in the computation of the RPLMLE $\hat{\hat{\eta}}$, numerical techniques will be needed to compute the PLMLE. Let $\mathbf{U}_p(\theta_p, \eta)$ and $\mathbf{I}_p(\theta_p, \eta)$ be the score function vector and observed Fisher information matrix associated with $L_p(\theta_p, \eta)$, respectively. Thus,

$$\mathbf{U}_p(\theta_p, \eta) = \begin{bmatrix} \nabla_\theta l_p(\theta_p, \eta) \\ \nabla_\eta l_p(\theta_p, \eta) \end{bmatrix};$$

and

$$\mathbf{I}_p(\theta_p, \eta) = - \begin{bmatrix} \frac{\partial^2}{\partial \theta_p \partial \theta_p^t} l_p(\theta_p, \eta) & \frac{\partial^2}{\partial \theta_p \partial \eta^t} l_p(\theta_p, \eta) \\ \frac{\partial^2}{\partial \theta_p^t \partial \eta} l_p(\theta_p, \eta) & \frac{\partial^2}{\partial \eta \partial \eta^t} l_p(\theta_p, \eta) \end{bmatrix}.$$

A possible approach to iteratively computing the PLMLE $(\hat{\theta}_p, \hat{\eta})$ is via the Newton-Raphson updating given by

$$\begin{bmatrix} \hat{\theta}_p \\ \hat{\eta} \end{bmatrix} \leftarrow \begin{bmatrix} \hat{\theta}_p \\ \hat{\eta} \end{bmatrix} + [\mathbf{I}_p(\hat{\theta}_p, \hat{\eta})]^{-1} \mathbf{U}_p(\hat{\theta}_p, \hat{\eta}). \tag{26}$$

Denote by $\hat{\lambda}_{\max}$ the largest eigenvalue of $\mathbf{I}_p(\hat{\theta}_p, \hat{\eta})$. The modified Schwartz information criterion is defined to be

$$\mathrm{MSIC}(p) = l_p(\hat{\theta}_p, \hat{\eta}) - \frac{p}{2} \left[\log(n) + \log(\hat{\lambda}_{\max}) \right]. \tag{27}$$

The first two terms in the right-hand side of (27) is the usual Schwartz information criterion for complete data. The last term in (27) represents the correction arising from the right-censoring. The justification for this modification will be provided in more a general framework involving incomplete data in a forthcoming paper.

The order selection procedure and the associated goodness-of-fit test proceeds as follows: First, a value of $P_{\max}$, which represents the upper bound of the smoothing order is specified. We propose to set the value of $P_{\max}$ to 10, though it could be changed to some other value. Second, the smoothing order to be used in the test statistic, denoted by p^*, is the value of p that maximizes $\mathrm{MSIC}(p)$ for $p = 1, 2, \ldots, P_{\max}$, that is,

$$p^* = \arg\max_{1 \leq p \leq P_{\max}} \mathrm{MSIC}(p). \tag{28}$$

Of course, note that this p^* is also a function of $P_{\max}$, although we suppress writing this explicitly. Finally, the asymptotic adaptive α-level test of H_0 versus H_1 rejects H_0 in favor of H_1 whenever $\hat{S}_{p^*}^2 > \chi_{1;\alpha}^2$. The fact that the critical value is that associated with a one degree-of-freedom chi-square distribution follows from the following asymptotic result, whose proof will be presented in a forthcoming paper.

Theorem 3: *If the conditions of Theorem 1 hold, then under H_0 and as $n \to \infty$, $p^* \xrightarrow{\mathrm{pr}} 1$ and $\hat{S}_{p^*}^2 \xrightarrow{\mathrm{d}} \chi_1^2$.*

For practical purposes, instead of using the asymptotic critical value of $\chi_{1;\alpha}^2$, for small to moderate sample sizes, we recommend the use of the test which rejects H_0 in favor of H_1 whenever $\hat{S}_{p^*}^2 > \chi_{p^*;\alpha}^2$. Another possibility, as yet unexplored, is to approximate the appropriate critical value using a bootstrap procedure.

Acknowledgements

The author wishes to acknowledge the US National Science Foundation DMS Grant 0102870 which partially supported this research. Furthermore, he wishes to thank the Organizing Committee of MMR2002, Professors Bo Lindqvist and Kjell Doksum, for inviting him to present a talk in the Conference and to contribute this chapter in the Conference Proceedings.

References

1. Z. Agustin and E. Peña, Goodness-of-Fit of the Distribution of Time-to-First-Occurrence in Recurrent Event Models, *Lifetime Data Analysis*, **7**, 287-304 (2001).
2. M. Akritas, Pearson-type goodness-of-fit tests: the univariate case, *Journal of the American Statistical Association*, **83**, 222-230 (1988).
3. L. Baringhaus and N. Henze, A goodness of fit test for the Poisson distribution based on the empirical generating function, *Statistics and Probability Letters*, **13**, 269-274 (1992).
4. D. Best and J. Rayner, Goodness of fit for the geometric distribution, *Biometrical Journal*, **31**, 307-311 (1989).
5. D. Best and J. Rayner, Goodness of fit for the Poisson distribution, *Statistics and Probability Letters*, **44**, 259-265 (1999).

6. H. Chernoff and E. Lehmann, The use of maximum likelihood estimates in χ^2 tests for goodness of fit, *Annals of Mathematical Statistics*, **25**, 579–586 (1954).

7. V. Choulakian, R. Lockhart, and M. Stephens, Cramer-von Mises statistics for discrete distributions, *Canadian Journal of Statistics*, **22**, 125–137 (1994).

8. N. Cressie and T. Read, Multinomial goodness-of-fit tests, *Journal of the Royal Statistical Society, B*, **46**, 440–464 (1984).

9. R. D'Agostino and M. Stephens, *Goodness-of-Fit Techniques* (Marcel Dekker, Inc., New York, 1986).

10. R. Eubank, Testing goodness of fit with multinomial data, *Journal of the American Statistical Association*, **92**, 1084–1093 (1997).

11. C. Gatsonis, H. Hsieh and R. Korwar, Simple nonparametric tests for a known standard survival based on censored data, *Communications in Statistics — Theory & Methods A*, **14**, 2137–2162 (1985).

12. R. Gray and D. Pierce, Goodness-of-fit for censored survival data, *The Annals of Statistics*, **13**, 552–563 (1985).

13. P. Greenwood and M. Nikulin, *A Guide to Chi-Squared Testing* (John Wiley & Sons, Inc., New York, 1996).

14. M. Habib and D. Thomas, Chi-squared goodness-of-fit tests for randomly censored data, *The Annals of Statistics*, **14**, 759–765 (1986).

15. N. Henze, Empirical-distribution-function goodness-of-fit tests for discrete models, *Canadian Journal of Statistics*, **24**, 81–93 (1996).

16. N. Hjort, Goodness-of-fit tests in models for life history data based on cumulative hazard rates, *The Annals of Statistics*, **18**, 1221–1258 (1990).

17. M. Hollander and F. Proschan, Testing to determine the underlying distribution using randomly censored data, *Biometrics*, **35**, 193–401 (1979).

18. M. Hollander and E. Peña, A chi-squared goodness-of-fit test for randomly censored data, *Journal of the American Statistical Association*, **87**, 458–463 (1992).

19. J. Hyde, Testing survival under right censoring and left truncation, *Biometrika*, **64**, 225–230 (1977).

20. E. Khamaladze, A martingale approach in the theory of goodness-of-fit tests, *Teor. Veroyatnost. i Primenen*, **26**, 246–265 (1981) (English translation in *Theory Probab. Appl.*, **26**, 240–257).

21. E. Khamaladze, Goodness of fit problem and scanning innovation martingales, *The Annals of Statistics*, **21**, 798–829 (1993).

22. J. Kim, Chi-square goodness-of-fit tests for randomly censored data, *The Annals of Statistics*, **21**, 1621–1639 (1993).

23. B. Klar, Goodness-of-fit tests for discrete models based on the integrated distribution function, *Metrika*, **49**, 53–69 (1999).

24. S. Kocherlakota and K. Kocherlakota, Goodness of fit tests for discrete distributions, *Communications in Statistics — Theory & Methods A*, **15**, 815–829 (1986).

25. J. Koziol and S. Green, A Cramer-von Mises statistic for randomly censored data, *Biometrika*, **63**, 139-156 (1976).

26. R. Kulperger and A. Singh, On random grouping in goodness of fit tests

of discrete distributions, *Journal of Statistical Planning and Inference*, **7**, 109–115 (1982).

27. T. Ledwina, Data driven version of Neyman's smooth test of fit, *Journal of the American Statistical Association*, **89**, 1000–1005 (1994).

28. G. Li and H. Doss, Generalized Pearson-Fisher chi-square goodness-of-fit tests, with applications to models with life history data, *The Annals of Statistics*, **21**, 772–797 (1993).

29. V. Nair, Plots and tests for goodness of fit with randomly censored data, *Biometrika*, **68**, 99–103 (1982).

30. V. Nair, Goodness of fit tests for multiply right censored data, in *Nonparametric Statistical Inference, Vol. I, II* (Budapest, 1980), 653–666, Colloq. Math. Soc. János Bolyai, **32**(North-Holland, Amsterdam-New York, 1982).

31. V. Nair, Confidence Bands for Survival Functions with Censored Data: A Comparative Study, *Technometrics*, **26**, 265–275 (1984).

32. M. Nakamura and V. Perez-Abreau, Use of an empirical probability generating function for testing a Poisson model, *Canadian Journal of Statistics*, **21**, 149–156 (1993).

33. J. Neyman, "Smooth" test for goodness of fit, *Skand. Aktuarietidskr.*, **20**, 150–199 (1937).

34. K. Pearson, On the criterion that a given system of deviations from the probable in the case of a correlated system of variables is such that it can reasonably be supposed to have arisen from random sampling, *Philos. Mag.*, 5th ser., **50**, 157–175 (1900).

35. E. Peña, Smooth Goodness-of-Fit Tests for Composite Hypothesis in Hazard-Based Models, *The Annals of Statistics*, **26**, 1935–1971 (1998).

36. E. Peña, Intensity-Based Approach to Goodness-of-Fit Testing with Discrete Right-Censored Data. Technical report, Department of Statistics, University of South Carolina (2002).

37. K. Rao and D. Robson, A chi-square statistic for goodness-of-fit tests within the exponential family, *Communications in Statistics*, **3**, 1139–1153 (1974).

38. R. Rueda, V. Perez-Abreu and F. O'Reilly, Goodness of fit for the Poisson distribution based on the probability generating function, *Communications in Statistics — Theory & Methods A*, **20**, 3093–3110 (1991).

39. J. Rayner and D. Best, *Smooth Tests of Goodness of Fit* (Oxford University Press, New York, 1989).

40. G. Schwartz, Estimating the dimension of a model, *The Annals of Statistics*, **6**, 461–464 (1978).

41. J. Spinelli and M. Stephens, Cramer-von Mises tests of fit for the Poisson distribution, *Canadian Journal of Statistics*, **25**, 257–268 (1997).

42. M. Stephens, Introduction to Kolmogorov (1933) On the empirical determination of a distribution, in *Breakthroughs in Statistics 2: Methodology and Distribution*, Eds. S. Kotz and N. Johnson (Springer: New York, 1992) pp.93–105.

Part IX

SOFTWARE RELIABILITY METHODS

32

THEORETICAL AND PRACTICAL CHALLENGES IN SOFTWARE RELIABILITY AND TESTING

Philip J. Boland and Santos Faundez Sekirkin

Department of Statistics, National University of Ireland - Dublin
Belfield - Dublin 4, Ireland
E-mail: philip.j.boland@ucd.ie ; Santos.FaundezSekirkin@ucd.ie

Harshinder Singh

Department of Statistics, West Virginia University
Morgantown, WV, 26506-6330, USA
E-mail: hsingh@stat.wvu.edu

In an effort to address the continuing demand for high quality software, an enormous multitude of software reliability growth models have been proposed in recent years. In spite of the diversity and elegance of many of these, there is still a need for models which can be more readily applied in practice. One aspect which should perhaps be taken more account of is the complex and challenging nature of the testing of software. A brief overview of software reliability modelling and testing is given, together with an indication of the challenges that lie ahead if advances in theory are to address the practical needs of software developers.

1. Introduction

Software reliability is often defined (Musa and Okumoto[31]) as *the probability of failure free operation of a computer programme in a specified environment for a specified period of time.* Software reliability is an ever increasing concern in our rapidly developing information technology world. Faults in software are not physical in nature, and usually result as a consequence of errors in programming logic. Such faults or errors can lead to results which are inconsistent with performance specifications and expectations, sometimes with catastrophic results.[33] Knowledge about the number and technical aspects of the faults often influences important economic decisions about the development and release of the software.

It is more than 30 years since Jelinski and Moranda[20] introduced in 1972 a simple model for the discovery and removal of faults in computer software. Since then, more than 100 other models and modifications have been proposed in the literature, and software reliability continues to be a very active area for research amongst engineers, mathematicians and statisticians. A recent search in the Science Citation Index came up with 463 references on the single topic of software reliability since 1990, and there are many recent interesting books on software reliability.[38,27,34,41,30,29,32,12,22]

The importance of research in software reliability and testing is not surprising due to the continuing growth in the software industry, and the higher expectations and demands users have in new and updated versions of software. Software reliability will continue to be a consumer driven requirement, and the price of failure can be great. According to the Software Engineering Institute (Carnegie Mellon University in Pittsburgh, PA, operates the SEI, which is sponsored by the US Department of Defence), even experienced programmers inject about one defect into every 10 lines of code. In a recent internal memo to employees[19], Bill Gates coined the term *Trustworthy Computing*. Microsoft has recently been plagued by glitches, and he wrote that "as an industry leader, we can and must do better ... there must be a company wide emphasis on developing high-quality code – even if it comes at the expense of new features." Watts Humphrey of the SEI claims that most security problems in systems are caused by known defects in code, and that if they are left unrepaired it provides opportunities for hackers to break into them. According to Humphrey[19], if Microsoft is serious about addressing its quality problems, "it needs to change an engineering culture that relies too heavily on catching problems during testing, rather than preventing them in development." The German government announced in the spring of 2002 that it was abandoning the Microsoft Windows operating system in favor of the Linux system (which is the world's 2nd largest operating system). Linux is free software, but the main reason given was to boost the security of their federal government computer systems by avoiding a software monoculture.

A laudable standard which is aspired to by many modern manufacturers of commercial software is what is referred to as the **Five 9's** - software which works 99.999% of the time. One classification scheme breaks commercial software development into four general areas: desktop computing, client/server computing, web-deployed applications, and .net enterprise (e.g. banking on line). High standards of reliability may be harder to achieve in some of these areas than others. For example, a looming challenge for software developers are the high expectations in reliability for the forthcoming revolution in mobile phone software.

One somewhat worrying aspect of the research development in software reliability is the huge number of different mathematical models which have been proposed (although many are derivations of a few more basic models), and the lack of real data sets to base them on. It is perhaps understandable that data on faults and software development is usually sensitive and confidential because of the severe competition in the software industry, but this also seems to be the case for very specifically designed software like that used in space and telecommunications technology, and the military. Surely more progress in developing useful models would be made if software developers released more sample data sets? Perhaps another reason why more use is not made of the many models developed by software reliability researchers, is that they don't incorporate enough of the complexity involved in the actual testing and debugging of a piece of software. Finding a software fault is one thing and resolving or fixing it is another, and often there is a considerable time delay between the two.

In section 2, a very brief and subjective selection of some of the more influential and/or innovative models which have been proposed for software reliability growth are briefly mentioned. Given the vast literature in this area, a good survey in such a short discussion is impossible, but there are several recent and extensive reviews in the literature.[38,30,10,37,42] Our belief is that practical models for software reliability growth should take account of both the nature of software testing and how faults are resolved. In section 3, various features of software testing are discussed, including the debate over random versus partition testing. Ireland has experienced an economic boom in recent years, and a significant part of this has been due to the developments in the software industry. Some practical issues resulting from contact with the local software industry suggest new approaches are needed. In the testing of software, considerable data is usually collected for any faults which are found. This can perhaps provide useful information in the modelling and development of software.

2. Software Reliability Growth Models

Modelling the failure process in a piece of software is a very challenging exercise (partially because of the diverse and interlocking nature of faults that may exist, and also because of the methods that may be used to detect or discover them). Many of the models developed in recent years by mathematicians, statisticians and engineers have been somewhat basic and often assume that faults in software are found through testing over time - usually by one individual. They also usually assume that once a failure is detected, the underlying fault characteristic is corrected quickly and (normally) perfectly. As a result, over time, one usually sees reliability growth in the software as faults are eliminated. Good and accessible data sets on software failures are (as previously indicated) scarce, but one of the more classic data sets on which many software reliability growth models have been tested is the NTDS (Navel Tactical Data System) data on inter-failure times of a large military software system.[20,38] Table 1 gives the 34 times T_i in days between software failures over 4 phases of the development process (production, testing phase I, user experience and further testing) in one module of this system. As we shall see, failure data for modern commercial software is usually of a much more complex form.

Table 1.   Inter-failure times for NTDS military software.

Failure (i)	T_i	S_i	**Failure (i)**	T_i	S_i
Production Phase			Production Phase (cont'd)		
1	9	9	20	1	105
2	12	21	21	11	16
3	11	32	22	33	149
4	4	36	23	7	156
5	7	43	24	91	247
6	2	45	25	2	249
7	5	50	26	1	250
8	8	58	Test Phase I		
9	5	63	27	87	337
10	7	70	28	47	384
11	1	71	29	12	396
12	6	77	30	9	405
13	1	78	31	135	540
14	9	87	User Experience		
15	4	91	32	258	798
16	1	92	Further Testing		
17	3	95	33	16	814
18	3	98	34	35	849
19	6	104			

Using for example the classification scheme for models described by Singpurwalla and Wilson[38], most software reliability models fall into one of two broad categories. Type I models are those which are based on modelling the times between successive failures of the software, and these are further broken down into those which use the failure rate as a modelling tool (Type I-1) and those which model an inter-failure time as a function of previous inter-failure times (Type I-2). Type II models on the other hand are those where a counting (usually a non-homogeneous Poisson) process approach is used to model the number $M(t)$ of observed failures in the software by time t. In the following treatment, some of the more influential software reliability growth models which have been introduced in recent years are mentioned in order to give some indication of the diversity of approaches which have been proposed. This is of course a somewhat subjective selection, and there are many other interesting models which have been proposed.

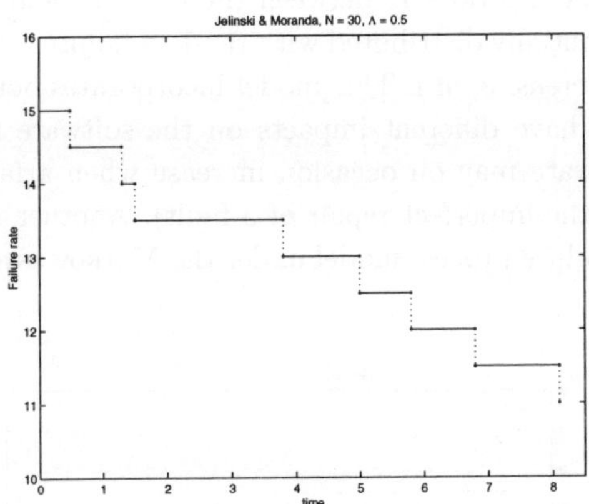

Fig. 1.   Failure rate function in the Jelinski Moranda model.

The Jelinski Moranda[20] model is a basic model of type I-1, where one assumes that there are a finite (but unknown) number N of bugs in the software, and that the failure rate of the software at any point in time is proportional to the number of bugs remaining in the software at that time. One also assumes that when a bug is found as a result of a failure, it is immediately and perfectly repaired. Essentially then the inter-failure time T_i between the $(i-1)^{st}$ and the ith observed failures is an exponentially distributed random variable with parameter (failure rate) $(N-i+1)\Lambda$ for some Λ. The $T_i's$ are also assumed to be independent. Figure 1 gives a possible realization of the failure rate in the Jelinski Moranda model when

$N = 30$ and $\Lambda = 0.5$. This model has proved seminal in the development of the general theory, partially because of its simplicity and mathematical tractability. The assumption in the Jelinski Moranda model that faults are perfectly repaired when encountered is in fact rarely satisfied in the testing of commercial (and perhaps any) software. Normally any valid software fault must pass a regression test before it is categorized as resolved, and in practice regression testing uncovers a significant number of imperfect fixes, and in some cases new faults.

Some modifications of the Jelinski Moranda model have been developed to incorporate imperfect repair[26,39,25,3]. Another criticism of the popular Jelinski Moranda model is the implicit assumption that all of the bugs remaining in the software at any point in time contribute an equal amount to the software failure rate. Littlewood and Verall[26] introduced a model where the inter-failure time T_i between the $(i-1)^{st}$ and the ith observed failures is exponentially distributed with random failure rate Λ_i, where Λ_i is stochastically decreasing in i. This model incorporates both the possibility that faults may have different impacts on the software failure rate, and that the failure rate may on occasion increase when a failure occurs (for example due to the imperfect repair of a fault). Another related approach is the software failure process model under the Markov assumption referred to in Xie[42].

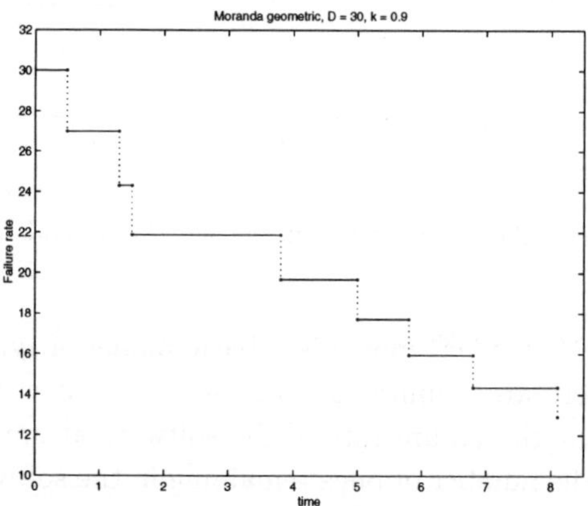

Fig. 2.   Failure rate function in the Moranda geometric de-eutrophication model.

Moranda[28] introduced his geometric de-eutrophication model (which is also of type I-1) to address the likely possibility that individual software faults make varying contributions to the failure rate of the software as a whole. Basic to his geometric model are the assumptions that the times between discovery of faults are independent and that the failure rate between the $(i-1)^{st}$ and the ith observed failures is of the form Dk^{i-1} for parameters D and k (and hence decrease geometrically after each failure). This model also tries to incorporate the situation where faults discovered early in testing reduce the failure rate more than those occurring later. Figure 2 gives a possible realization of the failure rate in the Moranda geometric de-eutrophication model for the situation when $D = 30$ and $k = 0.9$. The geometric model has considerable intuitive appeal, and the claim has been made[13] that it is conceptually well founded, but that it has not been studied as it should have been. If $M(t)$ represents the number of failures detected by time t in this model, then $\{M(t)\}_t$ is a pure birth stochastic process (but not a non-homogeneous Poisson process)[5] with mean function $\mu(t) = E(M(t))$ given by

$$\mu(t) = Dt + \sum_{j=1}^{\infty} (-1)^j \frac{(Dt)^{j+1}}{(j+1)!} \prod_{i=1}^{j} (1 - k^i). \tag{1}$$

Shick and Wolverton[35] introduced another interesting model of Type I-1 whereby the the failure rate function between failures is an increasing linear function of the time elapsed since the last failure (starting at 0). The motivation here is that the software is very reliable immediately following the successful detection and repair of a fault, but as time elapses confidence in it recedes. If S_i represents the time of the ith failure and T_i represents the time between failures $i - 1$ and i (that is $T_i = S_i - S_{i-1}$), then the conditional failure rate function for T_i given S_{i-1} takes the form

$$r_{T_i|S_{i-1}}(t - S_{i-1} \mid N, \Lambda) = \Lambda(N - i - 1)(t - S_{i-1}) \tag{2}$$

for $t > S_{i-1}$. Figure 3 gives a possible realization of such a failure rate function for the parameters $N = 10$ and $\Lambda = 1$. Note that the Schick Wolverton model assumes that the failure rate function (unlike any of the other models so far mentioned) actually increases between failures. Given that software, unlike hardware, does not age, this aspect of the model is considered unrealistic by some.

The random coefficient autoregressive approach to modelling software reliability was introduced by Singpurwalla and Soyer.[36] These Type I-2 models use a time series approach to the times T_i between successive failures, and assume that a relationship of the form $\ln T_i = \theta_i \ln T_{i-1} + \epsilon_i$ may

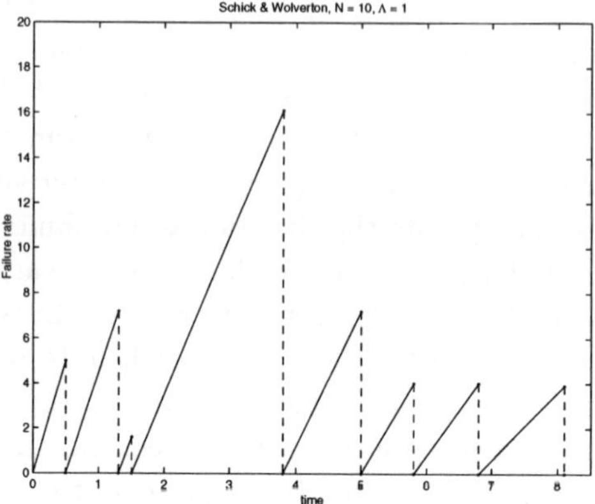

Fig. 3.   Failure rate function in the Schick Wolverton model.

be valid (where values of $\theta_i > 1$ are indicative of reliability growth). Another interesting model introduced by Mutairi, Chen and Singpurwalla[1] is the shot noise process approach, which is also a Type I-2 model. In contrast to the Shick Wolverton model, the failure rate function in this model takes a random jump (shot) upwards with the discovery of a fault, which then decreases deterministically until the next fault is discovered. Jumps in the failure rate are greater when failures are frequently found. Figure 4 gives a typical failure rate function for this shot noise process model.

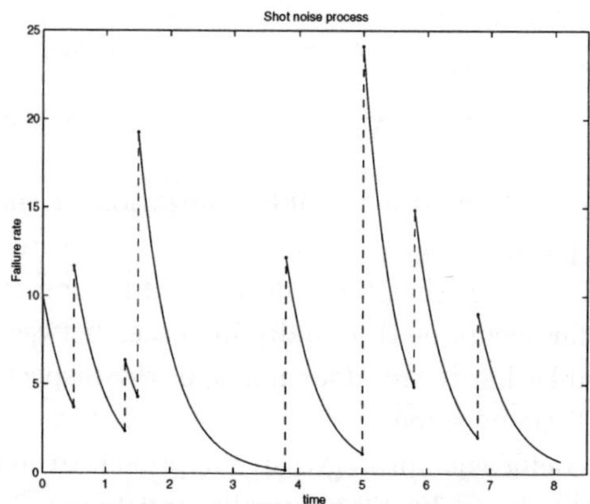

Fig. 4.   Shot Noise Failure rate Process.

The time dependent error detection model of Goel and Okumoto[14] is a classical model of type II, where one assumes that the number of faults $M(t)$ found by time t is a non-homogeneous Poisson process with mean value function given by $\mu(t) = E(M(t)) = a(1 - e^{-bt})$ (or equivalently the intensity function $\lambda(t)$ is of the form $\lambda(t) = abe^{-bt}$ for parameters a (mean of expected number of faults eventually found) and b). Another basic and popular Type II model is the logarithmic Poisson execution time model of Musa and Okumoto,[31] where one assumes that the intensity function of the fault counting process is related to the mean value function by $\lambda(t) = \lambda_0 e^{-\theta \mu(t)}$ (and in particular that log of the intensity function is a linear function of the expected number of faults discovered). This model incorporates the possibility of an infinite number of faults, and has an intensity function which decays more slowly than that of the Goel Okumoto model. Figure 5 plots the shape of intensity functions for the two models, and clearly indicates the slower decay of the intensity function for the Musa Okumoto model.

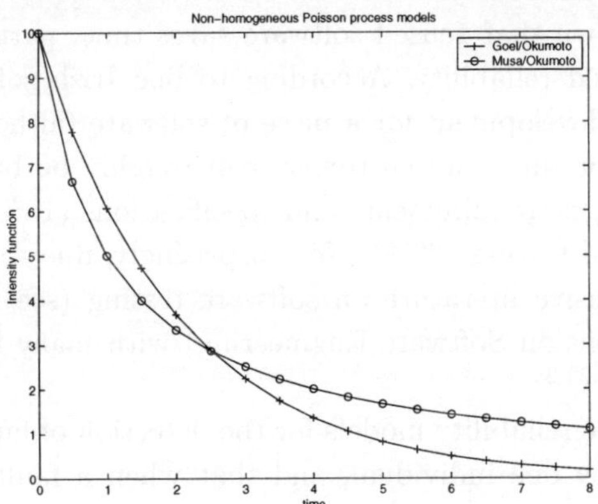

Fig. 5. Goel-Okumoto and Musa-Okumto non-homogeneous Poisson process models

As previously mentioned, the above discussion represents a brief and subjective selection of some of the more influential models which have been proposed and developed for use in software reliability. The complete list of proposed models is vast and continues to grow. This expansion has led to attempts to unify the theory somewhat, using for example Bayesian approaches[24] and the concept of self-exciting Poisson processes.[15,7]

In theory, software reliability models should be useful in many ways, including estimating development costs, determining optimal release times, and predicting quality and reliability of software. In spite of the fact that so many models have been proposed, relatively few of them are used extensively in practice. As Min Xie[42] has suggested, *software reliability is ready for a new phase of development where there is more emphasis on the potential for practical implementation.*

3. Software Testing

In the software life-cycle, testing is an important component which is strongly linked to software reliability. Perhaps software reliability models should take more account of testing methodologies? Some might say there is a gap between computer scientists and software developers on the one hand, and reliability analysts on the other. Computer scientists and software developers are trying to remove as many faults as possible from a piece of software, while reliability analysts are testing for the estimation or verification of its reliability.

It is well known that reused software saves time, particularly when it has an established reliability. According to one Irish software company, the life-cycle of development for a piece of software (although this clearly varies with the product and customer) can roughly be broken down into the following stages: requirements and specifications (18%), design (19%), coding (34%) and testing (29%). Not surprisingly, due to its importance, there is an extensive literature on software testing (see for example the IEEE Transactions on Software Engineering) with many interesting texts on the subject.[21,23,2]

Many software reliability models for the detection of faults assume that testing is done by one individual, and that when a fault is encountered it is properly fixed (perhaps with probability p) almost instantaneously. Furthermore, an assumption is often made that faults are usually discovered as a result of a random input procedure. In reality most software testing is done by teams of experts of varying ability and experience, and much of the testing procedure is very targeted in nature. The time devoted to software testing is a clearly a function of many variables including: tester quality, customer satisfaction, customer responsiveness, software quality, defect tolerance, tester availability, test-script quality, ability to fix, and the ability to find defects quickly and accurately.

As a consequence of studying the specifications for what a piece of software should do, a plan is normally developed as to how testing might proceed, often taking into account initially the potentially most problematic aspects of the software. In some cases, an experienced tester may be told to attempt to break the system. Although testing and developing are usually performed by separate groups, there is often considerable interaction between the two in improving the quality of the software. Often testers are not as experienced as developers, and a natural question is why would one be a tester if one has the opportunity of being more creative on the development side?

In order to assess the quality of a piece of software, test cases are selected from the input domain according to some testing strategy. Testers believe that directed testing, usually based on some form of partitioning, is more powerful (that is more likely to find faults) than random testing.[16,17,8,11,40] However there continues to be considerable debate in the testing literature concerning these methods of testing, and advances in the reasoning and efficacy of testing - particularly those which are both mathematically rigorous and applicable - are needed. In partition testing, the input domain is partitioned (usually according to some strategy) into k disjoint subdomains, and then n_i test cases are selected from the ith subdomain. Associated with each subdomain is a failure rate θ_i. In random testing, n test cases are selected at random from the entire input domain. A criterion frequently used for comparing test strategies is that of maximizing the probability of finding at least one fault. Hence partition testing is preferable to random testing if $P_p = 1 - \prod_1^k (1 - \theta_i)^{n_i} \geq 1 - (1 - \theta)^n = P_r$. Boland, Singh and Cukic[6] approached the problem of comparing partition and random testing using the tools of majorization and Schur functions. They showed that for equal sample sizes, P_p is a Schur convex function of the subdomain failure rates $(\theta_1, \ldots, \theta_k)$, and that partition testing is superior to random testing (using the criterion of maximizing the probability of finding at least one fault) if the arithmetic average $\overline{\theta}_a$ of the subdomain failure rates is larger than the overall failure rate of the program θ.

Although the probability of finding at least one fault is commonly used for comparing test strategies, it has been argued that this concept does not completely capture the procedure of software testing, and that other measures should be considered. When testing software, one is for example usually interested in finding and repairing as many (serious) faults as possible in a given time frame. One natural criterion to consider in comparing two testing methods is the expected number of faults detected in each case.[9] Another is to stochastically compare the number of faults found with two different methods. For example, if $X = \sum X_i$ (where X_i for $i = 1, \ldots, k$

is the number of faults found in the ith subdomain) and Y are the numbers of faults found in partition and random testing respectively, then partition testing is stochastically superior $(X \geq_{st} Y)$ to random testing if $P(X \geq r) \geq P(Y \geq r)$ for all $r = 1, \ldots$. Recent work of Boland, Singh and Cukic[4] on stochastic order comparisons of partition and random testing of software shows in particular that when equal sample sizes are used in each subdomain and the overall sample size is the same for both methods, then partition testing is stochastically superior to random testing if and only if the geometric average $\overline{\theta}_g$ of the subdomain failure rates is larger than the overall failure rate θ of the software. Results of this type can be helpful in identifying many situations where partition testing will be more effective than random testing, giving strength to the partition testing approach. There are, however, several important issues regarding partition and random testing which need to be more properly addressed by theoreticians and practitioners. For example, in most situations when one uses some method for *partitioning* software into subdomains, the resulting subdomains are overlapping (and therefore strictly speaking do not constitute a partition of the input domain). One should of course take account of the fact that faults are of different severities (and priorities) in comparing testing methods. Furthermore, one must distinguish between what is often referred to as the *debugging failure rate* (the failure rate which may be appropriate for one whose sole purpose is to find faults), and the *operational profile*[18] (the failure rate which is more related to that when a typical user is exercising the software).

In software testing, faults are usually logged into a database where an SPR (software problem report) is opened for each fault. Considerable information may be added to the report over time as progress on it develops, and the report is usually only closed when the fault is eventually resolved in some fashion. Faults are categorized in many ways, including by severity, priority, functional area, and who works on it. Information is also kept on the times when a report is opened, validated, pended, verified, regression tested and closed. For example, on the basis of over $36,000$ faults found over a three year period of testing software, one Irish company categorized 8% as severity 1 (crash), 58% as severity 2 (functional), and the remainder (34%) as severity 3 (cosmetic). These faults were found by expert user methods (40%), test scripts (42%), regression (from faulty fixing) (15%), and other (3%).

Is it not time to take into consideration more of this multitude of information which is being collected by software testers and developers in order to make more useful models for software reliability?

Acknowledgments

We wish to thank Paul McBride of Lionbridge Technologies, Ballina, County Mayo, Ireland, and George Clulow of the IBM Software Group in Dublin, Ireland working on Lotus software, for many interesting discussions on software testing. We wish to thank the Department of Statistics of Trinity College Dublin and particularly Simon Wilson for their assistance. Finally we appreciate the valuable comments made by an anonymous referee which have led to improvement in this chapter. This work was supported by Enterprise Ireland Research Grants SC/2000/123 and IC/2002/059.

References

1. D. Al-Mutairi, Y. Chen and N. Singpurwalla, An Adaptive Concatenated Failure Rate Model for Software Reliability, *Journal of the American Statistical Association*, **93**, 443, 1150-1163 (1998).
2. B. Beizer, *Black-Box Testing - Techniques for Functional Testing of Software and Systems*, (John Wiley and Sons, 1995).
3. P. J. Boland and N. Ní Chuív, Cost Implications of Imperfect Repair in Software Reliability, *International Journal of Reliability and Applications*, **2**, 3, 147-160 (2001).
4. P. J. Boland, H. Singh and B. Cukic, Stochastic Orders in Partition and Random Testing of Software, *Journal of Applied Probability*, **39**, 3, 555-565 (2002)
5. P. J. Boland and H. Singh, A Birth Process Approach to Moranda's Geometric Software Reliability Model, *IEEE Transactions in Reliability*, to appear (2003).
6. P. J. Boland, H. Singh and B. Cukic, Comparing Partition and Random Testing via Majorization and Schur Functions, *IEEE Transactions on Software Engineering* (to appear - 2003).
7. Y. Chen and N. D. Singpurwalla, Unification of Software Reliability Models Via Self-Exciting Point Processes, *Advances in Applied Probability*, **29**, 2, 337-352 (1997).
8. T. Y. Chen and Y.T. Yu, On the Relationship Between Partition and Random Testing, *IEEE Transactions on Software Engineering*, **20**, 12, 977-980 (1994).
9. T. Y. Chen and Y. T. Yu, On the expected number of failures detected by subdomain testing and random testing, *IEEE Transactions on Software Engineering*, **22**, 2, 109-119 (1996).
10. W. Farr, in Ref. 27, p. 71.
11. P. G. Frankl, R. G. Hamlet, B. Littlewood, and L. Strigini, Evaluating Testing Methods by Delivered Reliability, *IEEE Transactions on Software Engineering*, **24**, 8, 586-601 (1998).

12. M. A. Friedman and J. M. Voas, *Software Assessment: Reliability, Safety, Testability*, (Wiley Interscience, 1995).

13. O. Gaudoin and J. L. Soler, Statistical analysis of the geometric deeutrophication software reliability model, *IEEE Transactions in Reliability*, **41**, 4, 518-524 (1992).

14. A. L. Goel and K. Okumoto, Time-Dependent Error Detection Rate Model for Software Reliability and Other Performance Measures, *IEEE Transactions in Reliability* **28**, 206-211 (1979).

15. S. S. Gokhale, M. R. Lyu and K. S. Trivedi, Reliability Simulation of Component-Based Software Systems, in *Proceedings of the Ninth International Symposium on Software Engineering (ISSRE-98)*, IEEE Computer Society, Los Alamitos, CA, (1998), pp. 192-201.

16. W. J. Gutjahr, Partition Testing vs Random Testing, *IEEE Transactions on Software Engineering*, **25**, 661-674 (1999).

17. R. Hamlet amd R. Taylor, Partition Testing does not Inspire Confidence, *IEEE Transactions on Software Engineering*, **SE-16**, 1402-1411 (1990).

18. R. M. Hierons and M. P. Wiper, Estimation of Failure Rate using Random and Partition Testing, *Software Testing, Verification and Reliability*, **7**, 153-164 (1997).

19. G. V. Hulme, *InformationWeek.com* (URL:http://www.informationweek.com/story/IWK20020120S0003, 2002).

20. Z. Jelinski and P. Moranda, Software Reliability Research, in *Statistical Computer Performance Evaluation*, Ed. W. Freiberger (Academic, New York, 1972), pp. 465-484.

21. C. Kamer, H. Q. Nguyen and J. Falk, *Testing Computer Software - 2nd edition*, (John Wiley and Sons, 1999).

22. P. K. Kapur, R. B. Garg and S. Kumar, *Contributions to Hardware and Software Reliability* (World Scientific Pub. Co., 1999).

23. E. Kit and S. Finzi (Eds.), *Software Testing in the Real World: Improving the Process*, (Addison-Wesley Pub. Co., 1995).

24. N. Langberg and N. D. Singpurwalla, A Unification of Some Software Reliability Models, *SIAM Journal Scientific and Statistical Computing*, **6**, 781-790 (1985).

25. J. H. Lim and D. H. Park, Evaluation of average maintenance cost for imperfect-repair model, *IEEE Transactions in Reliability*, **48**, 2, 199-204 (1999).

26. B. Littlewood and J. L. Verall, A Bayesian Reliability Growth Model for Computer Software, *Applied Statistics* **22**, 332-346 (1973).

27. Michael R. Lyu (Ed.),*Handbook of Software Reliability Engineering*, (McGraw Hill Text, 1996).

28. P. B. Moranda, Prediction of Software Reliability and Its Applications, in *Proceedings on the* 1975 *Annual Reliability and Maintainability Symposium*, (1972), pp. 327-332.

29. J. D. Musa, *Software Reliability Engineering Testing* (McGraw-Hill Professional Publishing, 1998).

30. J. D. Musa, A. Iannino, and K. Okumoto, *Software Reliability, Measure-*

ment, Prediction, Application (Wiley, New York, 1987).

31. J. D. Musa and K. Okumoto, A Logarithmic Poisson Execution Time Model for Software Reliability Measurement, in *Proceedings of the 7th International Conference on Software Engineering* (Orlando, 1984), pp. 230-237.

32. D. Peled, *Software Reliability Methods* (Springer Verlag, 2001).

33. I. Peterson, *Fatal Defect: Chasing Killer Computer Bugs* (Vintage Books, 1996).

34. H. Pham, *Software Reliability* (Springer Verlag, 2000).

35. G. J. Shick and R. W. Wolverton, Assessment of Software Reliability, in *Proceedings in Operations Research*, Phsica-Verlag, Vienna, (1978), pp. 395-422.

36. N. D. Singpurwalla and R. Soyer, Assessing Software Reliability Growth Using a Random Coefficient Autoregressive Process and Its Ramifications, *IEEE Transactions Software Engineering*, **11**, 12, 1456-1464 (1985).

37. N. D. Singpurwalla and S. P. Wilson, Software Reliability Modeling, *International Statistical Review* **62**, 289-317 (1994).

38. N. D. Singpurwalla and S. Wilson, *Statistical Methods in Software Engineering* (Springer, New York, 1999).

39. H. Z. Wang and H. Pham, Optimal maintenance policies for several imperfect repair models, *International Journal of Systems Science*, **27**, 6, 543-549 (1996).

40. E. J. Weyuker and B. Jeng, Analyzing Partition Testing Strategies, *IEEE Transactions on Software Engineering*, **SE-17**, 703-711 (1991).

41. M. Xie, *Software Reliability Modelling*, (World Scientific Pub Co, 1991).

42. M. Xie, Software Reliability Models - Past, Present and Future, in *Recent Advances in Reliability Theory: Methodology, Practice and Inference*, Eds. N. Limnios and M. Nikulin (Birkhauser, 2000), pp. 325-340.

33

ON THE ASYMPTOTIC ANALYSIS OF LITTLEWOOD'S RELIABILITY MODEL FOR MODULAR SOFTWARE

James Ledoux

Centre de Mathmatiques INSA & IRMAR
INSA, 20 avenue des Buttes de Cosmes,
35043 Rennes Cedex,
France
E-mail: james.ledoux@insa-rennes.fr

We consider a Markovian model, proposed by Littlewood, to assess the reliability of a modular software. Specifically, we are interested in the asymptotic properties of the corresponding failure point process. We focus on its time-stationary version and on its behavior when reliability growth takes place. We prove the convergence in distribution of the failure point process to a Poisson process. Additionally, we provide a convergence rate using the distance in variation. This is heavily based on a similar result of Kabanov, Liptser and Shiryayev, for a doubly-stochastic Poisson process where the intensity is governed by a Markov process.

1. Introduction

Two approaches are used in the stochastic modeling of the failure process of a software. The prevalent approach adopts the so-called *black-box* view of the system, where only the interactions with the environment are considered. The *martingale methods*[3] for analyzing point processes, provide an unified framework for almost all published black-box models. To each model corresponds a specific *stochastic intensity* of the point process.[22,10] A second approach, called the *white-box* approach, incorporates information on the structure of the software in models. Littlewood's model[14] is the most popular model resulting from this approach. It has inspired most other works. Littlewood proposed a Markov-type model for the reliability assessment of modular software. For a software with a finite number of modules:

- the structure of the software is represented by an irreducible finite continuous time Markov chain $X = (X_t)_{t\geq 0}$, where X_t is the active module at time t.
- When module i is active, failure times are part of a homogeneous Poisson Process (HPP) with intensity $\mu(i)$.
- When control switches from module i to module j, a failure may happen with probability $\mu(i,j)$.
- When a failure appears, the execution is assumed to be resumed instantaneously. So, a failure does not affect the software.
- All failure processes are assumed to be independent, given a sequence of activated modules.

This could be considered to be a naive model. For instance, at any time t, the control of the execution is assumed to be in only one module. However, using a suitable definition of the state space of the Markov process X, the model may be adapted to a software system where the control is shared by several of its parts. In the Littlewood model, the failures do not affect the execution, the control structure is Markovian, ... A large number of theses assumptions are questionable, but some of them may be addressed. We refer the reader to [11] for a discussion on these issues and to [7] for a recent survey on the architecture-based software reliability.

An important issue in reliability theory, specifically for software systems, is what happens when the failure parameters tend to be smaller and smaller. Littlewood claimed:[14]

As all failure parameters $\mu(i),\mu(i,j)$ tend to zero, the failure process described above is asymptotically an HPP with intensity

$$\lambda = \sum_i \pi(i) \left(\sum_j Q(i,j)\mu(i,j) + \mu(i) \right) \tag{1}$$

where $Q = (Q(i,j))$, π are the generator and the stationary distribution of X, respectively. This statement is well-known in the community of software reliability and has widely supported the *hierarchical approach* for assessing the dependability of modular softwares.[7] However, to the best of our knowledge, no proof of this fact is reported in the applied probability literature.

The purpose of this chapter is to provide precise statements on the asymptotic properties of the failure point process associated with the Littlewood model. Firstly, we discuss its time-stationary version. Intuitively, this corresponds to place the origin of the observation of the process at an "arbitrary time" t (with large t). Secondly, we focus on the main contributions of this chapter, which are : first, a rigorous proof to the convergence results announced by Littlewood[14,15]; second, a rate of convergence. The derivation of our results will be based on the martingale approach for analyzing point processes.

The chapter is organized as follows. Section 2 gives the mathematical specification of the Littlewood model. Section 3 addresses the time-stationary version of the failure point process. Section 4 is devoted to its Poisson approximation. First, we give the *compensator* and the stochastic intensity of the point process. Then, we prove the convergence of the finite-dimensional distributions of the counting process to those of a homogeneous Poisson process with intensity λ in (1). In the final part, a convergence rate for the convergence of these finite-dimensional distributions is provided.

2. Mathematical Formulation of the Failure Point Process

Let us consider the successive times at which a failure occurs T_n, $n \geq 1$, and assume that $T_0 = 0$. We denote the number of failures in the interval $]0, t]$ by N_t

$$N_t = \begin{cases} \max\{n \geq 1 \ : \ T_n \leq t\} & \text{if } T_1 \leq t \\ 0 & \text{if } T_1 > t. \end{cases}$$

We denote the finite set of modules $\{1, \ldots, M\}$ by $\mathcal{M}$. X will be called the *environment* or *execution process*. Considering the bivariate process $Z = (N_t, X_t)_{t \geq 0}$ is very appealing from a mathematical and computational point of view. Indeed, it is easily seen from the assumptions on Littlewood's model, that Z is a jump Markov process over the state space $S = \mathbb{N} \times \mathcal{M}$. The infinitesimal generator of this Markov process is

$$G = \begin{pmatrix} D_0 & D_1 & 0 & \cdots \\ 0 & D_0 & D_1 & \ddots \\ \vdots & \ddots & \ddots & \ddots \end{pmatrix}$$

when the states in S are listed in lexicographic order. Matrices D_0 and D_1 are defined by

$$\text{if } i \neq j : D_0(i,j) = Q(i,j)(1 - \mu(i,j)) \qquad D_1(i,j) = Q(i,j)\mu(i,j),$$
$$D_0(i,i) = -\sum_{j \neq i} Q(i,j) - \mu(i) \quad D_1(i,i) = \mu(i).$$

Note that $Q = D_0 + D_1$ and $\lambda = \pi D_1 \mathbb{1}^t$ (see (1)), where $\mathbb{1}^t$ denotes the M-dimensional column vector having all components equal to 1. We point out that we deal with a failure point process that is a Markovian Arrival Process

(MAP) as defined by Neuts.[19,17] The Littlewood model has the (doubly-stochastic) Poisson process (driven) modulated by a Markov process as a special instance (setting parameters $\mu(\cdot,\cdot)$ to 0). We refer the reader to [12] for the computational issue of various reliability metrics using the bivariate process Z.

As the observation duration t tends to be large, we have the following asymptotic formula for the mean number of failures:[12]

$$E[N_t] = \lambda\, t + (\alpha - \pi)\big(\mathbb{I}^{\mathrm{t}}\pi - Q\big)^{-1} D_1 \mathbb{I}^{\mathrm{t}} + o(1),$$

where α is the probability distribution of the random variable X_0 and λ is defined in (1). We have as a result

$$\lim_{t \to +\infty} \frac{E[N_t]}{t} = \lambda = \pi D_1 \mathbb{I}^{\mathrm{t}}.$$

Thus, the mean value of the random variable N_t has a linear asymptote when t growths to infinity. As it can be expected from the last formula, the first moment of the counting process is $\lambda\, t$ when the environment process X is stationary (that is, the probability distribution of X_n's is π). In such a case, we are concerned with the time-stationary version of the point process. This the core of the next section.

3. Time-Stationary Version of the Failure Point Process

In this section, we discuss the time-stationary version of the failure point process. That is, we give conditions under which the counting processes $(N_{t+s} - N_t)_{s \geq 0}$ (the point process viewed from time t onwards) have the same distribution for all t. It is known that we get the time-stationary version of a MAP when its environment process X is stationary.[19,2] In fact, the question of time-stationarity for an MAP is related to the asymptotic properties of an associated Markov Renewal Process (MRP). Specifically, we need the stationary version of this MRP, which is stochastically equivalent to the MRP corresponding to an MAP with a stationary environment process X. This last property is well known (*e.g.* see [19]) but the details of the derivation are never reported. We give here, the main steps of the derivation to emphasize the interest in Markov modeling to get an analytic model that is tractable. Intensity λ given in (1), also appears in the moments of the stationary version of the point process.

The MRP associated with Littlewood's failure model is defined as follows.[17] Let us consider the marked point process $(T_n, J_n)_{n \geq 0}$ where J_n is the active module just after the nth failure time T_n. It is easily seen from the assumptions on Littlewood's model, that $(T_n, J_n)_{n\geq 0}$ is a recurrent finite MRP with semi-Markov kernel $(Q_t)_{t\geq 0}$ on $\mathcal{M}$

$$Q_t = (Q_t(i,j))_{i,j\in\mathcal{M}} = \int_0^t \exp(D_0 s)ds\, D_1 \tag{2}$$
$$= (I - \exp(D_0 t))(-D_0)^{-1}D_1.$$

That is, we have $\mathbb{P}\{J_{n+1} = j, T_{n+1} - T_n \leq t \mid J_n = i\} = Q_t(i,j)$ for any $n \geq 0$. We refer the reader to ^{13} for the basic properties of Markov renewal processes. The Markov chain $(J_n)_{n\geq 0}$, with the state space $\mathcal{M}$, has the following transition probability matrix

$$Q_\infty = (Q_\infty(i,j))_{i,j\in\mathcal{M}} = \int_0^\infty \exp(D_0 s)ds\, D_1 = (-D_0)^{-1}D_1.$$

We note that

$$Q_t = Q_\infty - \exp(D_0 t)Q_\infty. \tag{3}$$

The distribution function of the sojourn time in state $i \in \mathcal{M}$ is given by

$$H_t(i) = \sum_{j\in\mathcal{M}} Q_t(i,j) = 1 - \left(\exp(D_0 t)\mathbb{1}^{t}\right)(i).$$

Then, the mean sojourn time in state i is

$$m(i) = -(D_0^{-1}\mathbb{1}^{t})(i). \tag{4}$$

We associate with the semi-kernel $(Q_t)_{t\geq 0}$, its semi-Markov process $(J_{N_t})_{t\geq 0}$:

$$t \geq 0, \quad J_{N_t} = J_n \quad \text{if } T_n \leq t < T_{n+1}.$$

We need the *forward recurrence time* process $(V_t)_{t\geq 0}$ defined by

$$V_t = T_{N_t+1} - t$$

(*i.e.* time until the next failure after t) and the process $(J_{N_t+1})_{t\geq 0}$ (the active module just after the next failure after time t).

Now, we are interested in the stationary version of the MRP $(T_n, J_n)_{n\geq 0}$. In the definition of an MRP, the origin of the observation of the process is assumed to be an epoch of failure ($T_0 = 0$). Intuitively, considering the stationary version corresponds to start the observation at an arbitrary time t (with large t). Thus, the MRP is assumed to be in progress for a long time at the beginning of its observation. Formally, we get from the initial MRP $(T_n, J_n)_{n\geq 0}$, a delayed MRP $(\widetilde{T}_n, \widetilde{J}_n)_{n\geq 0}$, for which the probability distribution of $(\widetilde{J}_{N_t}, \widetilde{J}_{N_t+1}, \widetilde{V}_t)$ (and of $(\widetilde{J}_{N_t}, \widetilde{V}_t)$) does not depend on time

$t.^{20,4}$ The first transition of the new MRP is governed by the semi-Markov matrix

$$\widetilde{Q}_t(i,j) = \frac{1}{m(i)} \int_0^t \big(Q_\infty(i,j) - Q_s(i,j)\big) ds.$$

The others jumps are governed by the initial semi-Markov kernel $(Q_t)_{t\geq 0}$ (see (2)). The probability distribution $\widetilde{\pi}$ of $\widetilde{J}_0$ is

$$\forall i \in \mathcal{M}, \qquad \widetilde{\pi}(i) = \lim_{t\to\infty} \mathbb{P}\{J_{N_t} = i\} = \frac{\nu(i)m(i)}{\nu m^t}. \qquad (5)$$

where ν is the stationary distribution of the Markov chain $(J_n)_{n\geq 0}$. It is easily checked by a direct computation that

$$\nu = \frac{-\pi D_0}{-\pi D_0 \mathbb{1}^t} = \frac{\pi D_1}{\pi D_1 \mathbb{1}^t} \qquad (6)$$

with the stationary distribution π of X $(\pi(D_0 + D_1) = 0)$. Note that[4]

$$\widetilde{\pi}(i)\widetilde{Q}_t(i,j) = \lim_{x\to\infty} \mathbb{P}\{J_{N_x} = i, J_{N_x+1} = j, V_x \leq t\}$$

$$\widetilde{\pi}\widetilde{Q}_t = \big(\lim_{x\to\infty} \mathbb{P}\{J_{N_x+1} = j, V_x \leq t\}\big)_{j\in\mathcal{M}}$$

We check that the stationary MRP $(\widetilde{T}_n, \widetilde{J}_n)_{n\geq 0}$ is stochastically equivalent to the MRP $(T_n, J_n)_{n\geq 0}$ with π as initial distribution for the Markovian environment process X. Let us identify the (unconditional) probability distribution of $(\widetilde{T}_1, \widetilde{J}_1)$:

$$\widetilde{\pi}\widetilde{Q}_t = \frac{1}{\nu m^t} \int_0^t (\nu Q_\infty - \nu Q_s) ds \quad \text{from (5)}$$

$$= \frac{1}{\nu m^t} \int_0^t \frac{-\pi D_0}{-\pi D_0 \mathbb{1}^t} \exp(D_0 s)(-D_0)^{-1} D_1 ds \quad \text{from (2), (3), (6)}$$

$$= \frac{1}{\nu m^t} \frac{1}{\pi D_1 \mathbb{1}^t} \int_0^t \pi \exp(D_0 s) D_1 ds.$$

Now, it follows from (4) and (6) that $\nu m^t = 1/\pi D_1 \mathbb{1}^t$. As a result of the previous statements, we have

$$\widetilde{\pi}\widetilde{Q}_t = \pi Q_t.$$

Therefore, if the probability distribution of X_0 is π, then (T_1, J_1) and $(\widetilde{T}_1, \widetilde{J}_1)$ have the same distribution. Since the next failure times are governed by the same semi-Markov kernel $(Q_t)_{t\geq 0}$, we have the stochastic equivalence between $(\widetilde{T}_n, \widetilde{J}_n)_{n\geq 0}$ and the MRP $(T_n, J_n)_{n\geq 0}$ with stationary Markovian environment process X.

We provide now, some results concerning the moments associated with the stationary Littlewood model. The details of the derivation may be found in [2,19] and are not reported here (see[18] for the non-stationary case). We assume until the end of this section that the execution process X is stationary. The mean value function of the counting process N is given by

$$E_\pi[N_t] = \lambda\, t \quad \forall t \geq 0.$$

We see that the mean function $E_\pi[N_t]$ is identical to that of a homogeneous Poisson process with intensity λ. The scalar λ may be interpreted as the mean number of failures per time unit for the stationary failure model. The variance function $\sigma_\pi^2(t)$ for our counting process is

$$\begin{aligned}
\sigma_\pi^2(t) = {}& \left[\lambda + 2\pi D_0(\mathbb{I}^t\pi - Q)^{-1}D_1\,\mathbb{I}^t - 2\lambda^2\right] t \\
& - 2\pi D_1(I - \mathbb{I}^t\pi)(\mathbb{I}^t\pi - Q)^{-2}D_1\,\mathbb{I}^t \\
& - 2\pi D_1(\mathbb{I}^t\pi - \exp(Qt))(\mathbb{I}^t\pi - Q)^{-2}D_1\,\mathbb{I}^t \quad \forall t \geq 0.
\end{aligned}$$

The first two terms give the linear asymptote of the variance function when t growths to infinity (the last term converges to zero). Note that the variance $\sigma_\pi^2(t)$ cannot be identified to that of a Poisson process with intensity λ.

The last formula concerns the covariance function C_π of the stationary failure point process. If $C_\pi(t, t'; t_0) \stackrel{\text{def}}{=} E_\pi\big[N_t(N_{t'+t+t_0} - N_{t+t_0})\big] - \lambda^2 tt'$ with $t, t' > 0$ and $t_0 \geq 0$, then

$$C_\pi(t, t'; t_0) = \pi D_1\big(I - \exp(Qt)\big)\exp(Qt_0)\big(I - \exp(Qt')\big)(\mathbb{I}^t\pi - Q)^{-2}D_1\,\mathbb{I}^t.$$

When t_0 tends to infinity, this covariance is negligible, *i.e.* $C_\pi(t, t'; t_0) = o(1)$. The number of observed failures in the two disjoint intervals $]0, t]$ and $]t + t_0, t + t_0 + t']$ are asymptotically non-correlated with t_0.

4. Asymptotic Property in case of Reliability Growth

We turn to the main contribution of this chapter. For software systems, it can be expected that a phenomenon of reliability growth takes place with the corrective actions. Thus, we are interested in the behavior of the failure point process when the failure parameters become small. According to Littlewood, the point process tends to be an HPP with intensity λ given in (1). A basic way to represent *reliability growth* is to introduce the perturbated failure parameters

$$\varepsilon\mu(i), \varepsilon\mu(i, j), \quad i, j \in \mathcal{M} \tag{7}$$

where $\varepsilon > 0$ is a small parameter. Then, we investigate the convergence of the failure point process as ε tends to 0.

First of all, we note that the Poisson approximation provided by Littlewood cannot hold as it is stated in [14]. Indeed, consider, as in [16], the time to the first failure T_1 for the model with the failure parameters in (7). Its probability distribution cannot be asymptotically exponential with parameter λ as $\epsilon \to 0$. With the simple model of a Poisson process modulated by a two-states Markov process, specified by

$$Q = \begin{pmatrix} -1 & 1 \\ 1 & -1 \end{pmatrix} \quad D_1^{(\varepsilon)} = \begin{pmatrix} \mu(1)\varepsilon & 0 \\ 0 & \mu(2)\varepsilon \end{pmatrix} \quad D_0^{(\varepsilon)} = Q - D_1^{(\varepsilon)} \quad (\mu(1) \neq \mu(2))$$

and a stationary environment X $(\mathcal{L}(X_0) = \pi = (1/2, 1/2))$, we have

$$\lim_{\varepsilon \to 0} \mathbb{P}\{T_1 > t\} = 1.$$

This agrees with intuition. But, if we investigate the asymptotic distribution of T_1 at time scale t/ε, then it is easily checked that

$$\lim_{\varepsilon \to 0} \mathbb{P}\{T_1 > t/\varepsilon\} = \exp\left(-\pi D_1 \mathbb{1}^t\, t\right).$$

Therefore, this suggests to deal with the counting process $N^{(\varepsilon)} = (N_t^{(\varepsilon)})_{t\geq 0}$ defined by

$$t \geq 0 \qquad N_t^{(\varepsilon)} = N_{t/\varepsilon}$$

where N_t denotes, until the end of the section, the number of failures in the interval $]0, t]$ for Littlewood's model with system (7) of failure parameters.

We heavily use the so-called martingale approach to analyze the counting process $N^{(\varepsilon)}$. The corresponding basic results used in this chapter are reported in Bremaud's book.[3]

4.1. *Compensator and Stochastic Intensity of the Counting Process N*

Considering the bivariate process Z to analyze the counting process N, allows one to deal with a Markov process with discrete state space. One more time, we can take advantage of the powerful analytic theory for such a class of processes. $\mathcal{F}^Z = (\mathcal{F}_t^Z)_{t\geq 0}$ denotes the internal history of the bivariate Markov process $Z = (N, X)$. Due to the special structure of generator G of Z, N may be interpreted as the following counter of specific transitions in Z

$$N_t = \sum_{(x,y)\in T} N_t(x, y) \quad \text{where } T = \left\{ ((n, i); (n+1, j)), \quad i, j \in \mathcal{M}, n \geq 0 \right\}$$

and $N_t(x, y)$ is the number of transitions from state x to state y at time t. It is well known[3] that, for the counting process $N(x, y) = (N_t(x, y))_{t \geq 0}$,

$$N_t(x, y) - \int_0^t \mathbf{1}_{\{Z_{s-} = x\}} G(x, y) ds$$

is a $\mathcal{F}^Z$-*martingale*. We denote the left limit at t of Z by Z_{t-}. The random functions

$$\lambda_t(x, y) = \mathbf{1}_{\{Z_{t-} = x\}} G(x, y), \quad A_t(x, y) = \int_0^t \lambda_s(x, y) ds$$

are the $\mathcal{F}^Z$-intensity and the $\mathcal{F}^Z$-*compensator* ($\mathcal{F}^Z$-dual predictable-projection) of $N(x, y)$, respectively. Then it is easily seen that $(N_t - A_t)_{t \geq 0}$ where

$$A_t = \sum_{(x, y) \in T} A_t(x, y),$$

is a $\mathcal{F}^Z$-martingale. The $\mathcal{F}^Z$-intensity of N is

$$\lambda_t = \sum_{(x, y) \in T} \mathbf{1}_{\{Z_{t-} = x\}} G(x, y) = \sum_{j \in \mathcal{M}} D_1(X_{t-}, j)$$

$$= \varepsilon \big(\mu(X_{t-}) + \sum_{j \neq X_{t-}} Q(X_{t-}, j) \mu(X_{t-}, j) \big).$$

4.2. *Convergence to a Poisson Process*

The $\mathcal{F}^{N^{(\varepsilon)}, (X_{t/\varepsilon})_t}$-compensator of $N^{(\varepsilon)}$ is from the previous subsection

$$A_t^{(\varepsilon)} = \sum_{i \in \mathcal{M}} \left(\mu(i) + \sum_{j \neq i} Q(i, j) \mu(i, j) \right) \varepsilon \int_0^{t/\varepsilon} \mathbf{1}_{\{X_{s-} = i\}} ds \tag{8}$$

with intensity $\lambda_t^{(\varepsilon)} = \mu(X_{t/\varepsilon-}) + \sum_{j \neq X_{t/\varepsilon-}} Q(X_{t/\varepsilon-}, j) \mu(X_{t/\varepsilon-}, j)$. We are now in position to state the convergence of the Littlewood point process to a Poisson process as the small parameter ε tends to zero.

Theorem 1: The probability vector π is such that $\pi Q = 0$. As ε tends to 0, the counting process $N^{(\varepsilon)}$ converges in distribution to the counting process $(P_t)_{t \geq 0}$ of an HPP with an intensity λ defined by

$$\lambda = \sum_i \pi(i) \left(\sum_{j \neq i} Q(i, j) \mu(i, j) + \mu(i) \right).$$

530   *J. Ledoux*

Proof: From the well known time-average properties of the cumulative process $\int_0^t f(X_s)ds$ for an irreducible finite Markov process X, we have

$$\lim_{\varepsilon \to 0} \frac{\varepsilon}{t} \int_0^{t/\varepsilon} \mathbf{1}_{\{X_s = i\}} ds = \pi(i) \quad \text{a.s.}$$

where π satisfies $\pi Q = 0$. Thus, we derive from (8)

$$\lim_{\varepsilon \to 0} A_t^{(\varepsilon)} = t \sum_{i \in \mathcal{M}} \pi(i) \left(\mu(i) + \sum_{j \neq i} Q(i,j)\mu(i,j) \right) = \lambda t \quad \text{a.s.} \ .$$

In particular, this implies the convergence in probability of $A_t^{(\varepsilon)}$ to the compensator of an HPP with an intensity λ as in (1). The theorem follows from Th 1[8].   □

Remark 2: Littlewood extended his model[15] to consider sojourn times in each module with a general probability distribution. Thus, X was assumed to be an irreducible semi-Markov process. He also identified[15] the asymptotic distribution of its counting process to that of an HPP with the intensity

$$\lambda^* = \frac{1}{\nu m^t} \sum_{i \in \mathcal{M}} \nu(i) \left(\mu(i) \sum_j Q_\infty(i,j)m(i,j) + \sum_{j \neq i} Q_\infty(i,j)\mu(i,j) \right)$$

where Q_∞ is the transition probability matrix of the jump Markov chain associated with the semi-Markov process X, ν is such that $\nu Q_\infty = Q_\infty$ and $m(i,j)$ is the mean sojourn time in state i, given that the next state entered is j. We emphasize that a compensator approach leads to the above limiting Poisson process. Indeed, using the concept of compensator for marked point processes, it can be shown that the $\mathcal{F}^{N^{(\varepsilon)},(X_{t/\varepsilon})_t}$-compensator of the counting process $N^{(\varepsilon)}$ is given by

$$\varepsilon \Big[\int_0^{t/\varepsilon} \mu(X_{s-})ds + \sum_i \sum_{j \neq i} \mu(i,j)A_{t/\varepsilon}(i,j) \Big]$$

where $A_t(i,j)$ is the compensator of the counting process $N_t(i,j)$ of transitions from state i to state j for X. For the asymptotic behavior of the first term, we just need time-average properties of cumulative process $\int_0^t f(X_s)ds$ for an irreducible finite semi-Markov process X, which are similar to those for the Markovian case.[13] Using strong laws of large numbers for the second term, we get the convergence in distribution of $N^{(\varepsilon)}$ to an HPP with the intensity λ^*. Note that Anisimov and Korolyuk[1] dealt with the case of a Poisson process with an intensity governed by an irreducible semi-Markov process X. This is the particular case where the $\mathcal{F}^{N^{(\varepsilon)},(X_{t/\varepsilon})_t}$-compensator of $N^{(\varepsilon)}$ is given by

$$\varepsilon \int_0^{t/\varepsilon} \mu(X_{s-})ds$$

(parameters $\mu(\cdot,\cdot)$ are assumed to be 0).

Note that λ in (1) is the scalar product $\langle \pi, D_1 \mathbb{1}^t \rangle$. Moreover, if $Y_s^{(\varepsilon)}$ is vector $\left(1_{\{X_{s/\varepsilon}=i\}}\right)_{i\in\mathcal{M}}$, then the compensator of $N^{(\varepsilon)}$ is from (8)

$$A_t^{(\varepsilon)} = \int_0^t \langle Y_{s-}^{(\varepsilon)}, D_1 \mathbb{1}^t \rangle ds. \tag{9}$$

4.3. *Convergence Rate with Distance in Variation*

Let T be any positive scalar. $\mathbb{T}$ is a finite subdivision $\{t_0, t_1, \ldots, t_n\}$ of the interval $[0,T]$ $(0 = t_0 < t_1 < \cdots < t_n = T)$. To evaluate the proximity between the respective probability distributions $\mathcal{L}(N_{\mathbb{T}}^{(\varepsilon)})$ and $\mathcal{L}(P_{\mathbb{T}})$ of $N_{\mathbb{T}}^{(\varepsilon)} = (N_{t_1}^{(\varepsilon)}, \ldots, N_{t_n}^{(\varepsilon)})$ and $P_{\mathbb{T}} = (P_{t_1}, \ldots, P_{t_n})$, we use their distance in total variation, denoted by $\mathrm{d}_{\mathrm{TV}}\big(\mathcal{L}(N_{\mathbb{T}}^{(\varepsilon)}), \mathcal{L}(P_{\mathbb{T}})\big)$:

$$\mathrm{d}_{\mathrm{TV}}\big(\mathcal{L}(N_{\mathbb{T}}^{(\varepsilon)}), \mathcal{L}(P_{\mathbb{T}})\big) \overset{\text{def}}{=} \sup_{B\subset\mathbb{N}^n} \big|\mathbb{P}\{N_{\mathbb{T}}^{(\varepsilon)} \in B\} - \mathbb{P}\{P_{\mathbb{T}} \in B\}\big|.$$

For a locally bounded variation function $f(\cdot)$, the total variation in the interval $[0,T]$ is

$$\mathrm{Var}_{[0,T]}(f) \overset{\text{def}}{=} \sup_{\mathbb{T}\in\mathcal{P}([0,T])} \sum_{i=1}^n \big|f(t_i) - f(t_{i-1})\big|$$

where $\mathcal{P}([0,T])$ is the set of all finite subdivisions of the interval $[0,T]$.

The main result of this subsection is based on the following estimate Th 3.1[9]

$$\mathrm{d}_{\mathrm{TV}}\big(\mathcal{L}(N_{\mathbb{T}}^{(\varepsilon)}), \mathcal{L}(P_{\mathbb{T}})\big) \leq E\,\mathrm{Var}_{[0,T]}\big(\widehat{A}^{(\varepsilon)} - A\big) \tag{10}$$

where $\widehat{A}^{(\varepsilon)}$, A are the $\mathcal{F}^{N^{(\varepsilon)}}$-compensator of $N^{(\varepsilon)}$ and the compensator of the HPP in Theorem 1, respectively. $\widehat{A}^{(\varepsilon)}$ is from (9) [3]

$$\widehat{A}_t^{(\varepsilon)} = \int_0^t \langle \widehat{Y}_{s-}^{(\varepsilon)}, D_1 \mathbb{1}^t \rangle ds$$

with $\widehat{Y}_t^{(\varepsilon)} = \left(\mathbb{P}\{X_{t/\varepsilon} = i \mid \mathcal{F}_t^{N^{(\varepsilon)}}\}\right)_{i \in \mathcal{M}}$. Therefore, we have from (10)

$$d_{\mathrm{TV}}\left(\mathcal{L}(N_{\mathbb{T}}^{(\varepsilon)}), \mathcal{L}(P_{\mathbb{T}})\right) \leq E \, \mathrm{Var}_{[0,T]}\left(\int_0^t \langle \widehat{Y}_{s-}^{(\varepsilon)} - \pi, D_1 \mathbb{I}^{\mathrm{t}} \rangle ds\right)$$

$$\leq E \int_0^T |\langle \widehat{Y}_{s-}^{(\varepsilon)} - \pi, D_1 \mathbb{I}^{\mathrm{t}} \rangle| ds.$$

Note that $|\langle \widehat{Y}_s^{(\varepsilon)} - \pi, D_1 \mathbb{I}^{\mathrm{t}} \rangle| \leq \delta \, \|\widehat{Y}_s^{(\varepsilon)} - \pi\|_1$ with $\delta = \max_{i \in \mathcal{M}} \left((D_1 \mathbb{I}^{\mathrm{t}})(i)\right) - \min_{i \in \mathcal{M}} \left((D_1 \mathbb{I}^{\mathrm{t}})(i)\right)$ and $\|\cdot\|_1$ denotes the l_1-norm. Thus, it remains to estimate the convergence rate of $\|\widehat{Y}_s^{(\varepsilon)} - \pi\|_1$ to 0 when $\varepsilon \to 0$. The first step consists in writing a filtering equation for vector $Y_t^{(\varepsilon)}$ (Ch IV^3).

Lemma 3: *Put $\widehat{Y}_t^{(\varepsilon)} = E[Y_t^{(\varepsilon)} \mid \mathcal{F}_t^{N^{(\varepsilon)}}]$. We have for all $t \geq 0$,*

$$\widehat{Y}_t^{(\varepsilon)} = \alpha + \frac{1}{\varepsilon} \int_0^t \widehat{Y}_{s-}^{(\varepsilon)} Q \, ds + \int_0^t v_{s-}^{(\varepsilon)} (dN_s^{(\varepsilon)} - \widehat{\lambda}_s ds) \qquad (11)$$

where $v_{s-}^{(\varepsilon)} = -\widehat{Y}_{s-}^{(\varepsilon)} + \widehat{Y}_{s-}^{(\varepsilon)} D_1 / \widehat{\lambda}_s$, $\widehat{\lambda}_s = \langle \widehat{Y}_{s-}^{(\varepsilon)}, D_1 \mathbb{I}^{\mathrm{t}} \rangle$ is the $\mathcal{F}^{N^{(\varepsilon)}}$-intensity of $N^{(\varepsilon)}$ and α is the probability distribution of X_0.

Proof: The MP $X^{(\varepsilon)} = (X_{t/\varepsilon})_{t \geq 0}$ has the generator $Q^{(\varepsilon)} = Q/\varepsilon$. It follows from Dynkin formula that

$$Y_t^{(\varepsilon)} = Y_0^{(\varepsilon)} + \frac{1}{\varepsilon} \int_0^t Y_s^{(\varepsilon)} Q \, ds + M_t \qquad (12)$$

where $(M_t)_{t \geq 0}$ is a $\mathcal{F}^{X^{(\varepsilon)}}$-martingale. Then applying (Ch IV, Th 1^3) to (12), we get for $\widehat{Y}_t^{(\varepsilon)}$

$$\widehat{Y}_t^{(\varepsilon)} = \alpha + \frac{1}{\varepsilon} \int_0^t \widehat{Y}_s^{(\varepsilon)} Q \, ds + \widehat{M}_t$$

where $\widehat{M} = (\widehat{M}_t)_{t \geq 0}$ is a $\mathcal{F}^{N^{(\varepsilon)}}$-martingale. Now, (Ch III, Th 17^3) gives us the following representation of the $\mathcal{F}^{N^{(\varepsilon)}}$-martingale $\widehat{M}$

$$\int_0^t G_s (dN_s^{(\varepsilon)} - \widehat{\lambda}_s ds)$$

where $(G_s)_{s \geq 0}$ is a $\mathcal{F}^{N^{(\varepsilon)}}$-predictable process, called the innovations process. It remains to determine an explicit expression of G_s. This is similar to the case of a Poisson process governed by a Markov process (see pages

$98,107^3$) else but $X^{(\varepsilon)}$ and $N^{(\varepsilon)}$ may have common jumps. So, we omit the details of the derivation. We have $G_s = G_{1,s} - G_{2,s} + G_{3,s}$ with

$$G_{2,s} = \widehat{Y}_{s-}^{(\varepsilon)} \quad \text{and} \quad G_{1,s} = \left(\frac{\widehat{Y}_{s-}^{(\varepsilon)}(i)\left(D_1\mathbb{I}^t\right)(i)}{\widehat{\lambda}_s} \right)_{i\in\mathcal{M}}$$

$$\text{and} \quad G_{3,s} = \left(\frac{\sum_{k\neq i}\widehat{Y}_{s-}^{(\varepsilon)}(k)D_1(k,i) - \widehat{Y}_{s-}^{(\varepsilon)}(i)\sum_{k\neq i}D_1(i,k)}{\widehat{\lambda}_s} \right)_{i\in\mathcal{M}}$$

and the gain G_s is of the form reported in Lemma 3. $\qquad\square$

Equation (11) has the unique solution

$$\widehat{Y}_t^{(\varepsilon)} = \alpha \exp\left(Qt/\varepsilon\right) + \int_0^t v_{s-}^{(\varepsilon)} \exp\left(Q(t-s)/\varepsilon\right)(dN_s^{(\varepsilon)} - \widehat{\lambda}_s ds); \qquad (13)$$

(check, with an integration by parts, that the right hand side term of (13) is a solution of (11) and verify that the difference of the two terms in (13) is a solution of a (first order) homogeneous linear differential equation with initial condition 0).

Theorem 4: Let $(P_t)_{t\geq 0}$ be the counting process of an HPP with the intensity $\langle\pi, D_1\mathbb{I}^t\rangle$, where π is the probability distribution such that $\pi Q = 0$. For any $T > 0$, there exists a constant C_T such that

$$d_{\mathrm{TV}}\left(\mathcal{L}(N_{\mathbb{T}}^{(\varepsilon)}), \mathcal{L}(P_{\mathbb{T}})\right) \leq C_T\,\varepsilon.$$

Proof: Since $v_{s-}^{(\varepsilon)}\mathbb{I}^t = 0$, we can write from (13)

$$\|\widehat{Y}_t^{(\varepsilon)} - \pi\|_1 \leq \|\alpha\exp\left(Qt/\varepsilon\right) - \pi\|_1$$
$$+ \left\|\int_0^t v_{s-}^{(\varepsilon)}\left[\exp\left(Q(t-s)/\varepsilon\right) - \mathbb{I}^t\pi\right](dN_s^{(\varepsilon)} - \widehat{\lambda}_s ds)\right\|_1.$$

Since Q is irreducible, we have the well known exponential estimate:

$$t \geq 0, \quad \|\exp\left(Qt\right) - \mathbb{I}^t\pi\|_1 \leq C\exp(-\rho t) \qquad (14)$$

where ρ is positive and only depends on matrix Q.

In a first step, it follows from (14)

$$\int_0^T \|\alpha\exp\left(Qt/\varepsilon\right) - \pi\|_1 dt \leq C_1\,\varepsilon.$$

In a second step, we have

$$E \left\| \int_0^t v_{s-}^{(\varepsilon)} \left[\exp\left(Q(t-s)/\varepsilon\right) - \mathbb{1}^t\pi\right] (dN_s^{(\varepsilon)} - \widehat{\lambda}_s ds) \right\|_1$$

$$\leq 2E \int_0^t \left\| v_{s-}^{(\varepsilon)} \left[\exp\left(Q(t-s)/\varepsilon\right) - \mathbb{1}^t\pi\right] \right\|_1 \widehat{\lambda}_s ds$$

since $\widehat{\lambda}_s$ is the $\mathcal{F}^{N^{(\varepsilon)}}$-intensity of $N^{(\varepsilon)}$ and $v_{s-}^{(\varepsilon)}$ is $\mathcal{F}^{N^{(\varepsilon)}}$-predictable. Note that $\|v_{s-}^{(\varepsilon)}\|_1 \widehat{\lambda}_s$ is uniformly bounded in ε, t. Moreover, using the exponential estimate (14) for $\|\exp\left(Q(t-s)/\varepsilon\right) - \mathbb{1}^t\pi\|_1$, we derive that, for all $t \geq 0$,

$$E \int_0^t \left\| v_{s-}^{(\varepsilon)} \left[\exp\left(Q(t-s)/\varepsilon\right) - \mathbb{1}^t\pi\right] \right\|_1 \widehat{\lambda}_s ds \leq C_2 t \varepsilon.$$

We deduce from the previous estimate (and Fubini's theorem) that

$$E \int_0^T \left\| \int_0^t v_{s-}^{(\varepsilon)} \left[\exp\left(Q(t-s)/\varepsilon - \mathbb{1}^t\pi\right] (dN_s^{(\varepsilon)} - \widehat{\lambda}_s ds) \right\|_1 dt \leq C_{2,T}\, \varepsilon. \quad \square$$

Remark 5: With respect to Theorem 1, note that Th 1[8] would give the convergence in distribution of the counting process $N^{(\varepsilon)}$ to that of a Poisson process in the space of all counting processes, equipped with the Skorokhod topology. Moreover, convergence in variation also takes place in this space. Indeed, the distance in total variation over interval $[0,T]$ between the distributions of $N^{(\varepsilon)}$ and $P = (P_t)_{t \geq 0}$ is also bounded from above by $E \operatorname{Var}_{[0,T]}(\widehat{A}^{(\varepsilon)} - A)$ (Th 4.1[9]). Thus, it follows from Theorem 4 that the rate of convergence is in ε.

Remark 6: The order of the convergence rate in Theorem 4 cannot be improved in general. This follows from (Section 5, Example 1)[5], where the authors report a lower bound for the distance in variation, that has order 1 in ε for a Poisson process modulated by a 2-states Markov process. We refer the reader to [5] for details.

5. Conclusion

In this chapter, we report some asymptotic properties of the well known Littlewood reliability model for modular softwares. Firstly, we deal with the stationary version of the failure point process, which is a classic concept in point process theory. Secondly, we prove a Poisson approximation for the point process when reliability growth takes place. This fact is reported in [14] but without proof (see[15] in the semi-Markov context). We also provide a convergence rate for the convergence of the point process to its Poisson limit.

All these results may be extended to a more general setting. For instance, Theorems 1 and 4 have their counterpart[6] for a general architecture-based reliability model proposed by Ledoux.[11] The purpose of this work was to take into account the way the failures may affect the execution process. We can also consider that failures can occur in batch. In such a case, using the martingale approach for marked point process, we get different approximating processes depending on the assumptions made on the batch size distributions. In the usual case of independent and identically distributed variables, we obtain a Compound Poisson process. This supports the use of software reliability models as considered by Sahninoglu[21], when clumping of failures exists. Rate of convergence may also be provided. Details will be reported elsewhere.

Acknowledgments

The author would like to thank J.-B. Gravereaux for helpful discussions.

References

1. V. V. Anisimov and V.V. Korolyuk, Asymptotic reliability analysis of semi-Markov systems, *Cybernetics*, **18**, 559–566 (1983).
2. S. Asmussen, Matrix-analytic models and their analysis, *Scandinavian Journal of Statistics*, **27**, 193–226 (2000).
3. P. Bremaud, *Point Processes and Queues* (Springer, 1981).
4. E. Çinlar, Markov renewal theory, *Advances in Applied Probability*, **1**, 123–187 (1969).
5. G. B. Di Masi and Y. M. Kabanov, The strong convergence of two-scale stochastic systems and singular perturbations of filtering equations, *Journal of Mathematical Systems, Estimation, and Control*, **3**, 207–224 (1993).
6. J.-B. Gravereaux and J. Ledoux, *Poisson approximation for some reliability models*, submitted (2002).
7. K. Goseva-Popstojanova and K. S. Trivedi, Architecture-based approach to reliability assessment of software systems, *Performance Evaluation*, **45**, 179–204 (2001).
8. Y. M. Kabanov, R. S. Liptser and A. N. Shiryayev, Some limit theorems for simple point processes (martingale approach), *Stochastics*, **3**, 203–216 (1980).
9. Y. M. Kabanov, R. S. Liptser and A. N. Shiryayev, Weak and strong convergence of the distributions of counting processes, *Theory of Probability and its Applications*, **28**, 303-336 (1983).
10. J. Ledoux, Software reliability modeling, in *Handbook of Reliability Engineering*, Ed. H. Phoam (Springer, 2003) to appear.

11. J. Ledoux, Availability modeling of modular software, *IEEE Transactions on Reliability*, **29**, 159–168 (1999).

12. J. Ledoux and G. Rubino, Simple formulae for counting processes in reliability models, *Advances in Applied Probability*, **29**, 1018–1038 (1997).

13. N. Limnios and G. Oprişan, *Semi-Markov processes and Reliability* (Series in Statistics for Industry and Technology, Birkhuser, 2001).

14. B. Littlewood, A reliability model for systems with Markov structure, *Applied Statistics*, **24**, 172-177 (1975).

15. B. Littlewood, Software reliability model for modular program structure, *IEEE Transactions on Reliability*, **28**, 241-246 (1979).

16. B. Littlewood, A software reliability model for modular program structure (supplement), *ASIS-NAPS document No 03307* (1979).

17. D. M. Lucantoni, K. Meier and M. F. Neuts, A single-server queue with server vacations and a class of non-renewal arrival processes, *Advances in Applied Probability*, **22**, 676–705 (1990).

18. S. Narayana and M. F. Neuts, The first two moment matrices of the counts for the markovian arrival process, *Stochastic Models*, **8**, 459–477 (1992).

19. M. F. Neuts, *Structured Stochastic Matrices of M/G/1 Type and Their Applications* (Marcel Dekker Inc., New-York and Basel, 1989).

20. R. Pyke, Markov renewal processes with finitely many states, *Annals of Mathematical Statistics*, **32**, 1243–1259 (1961).

21. M. Sahninoglu, Compound-Poisson software reliability model, *IEEE Transactions on Software Engineering*, **18**, 624–630 (1992).

22. N. D. Singpurwalla and S. P. Wilson, *Statistical Methods in Software Engineering: Reliability and Risk* (Springer, 1999).

34

ADAPTIVE METHODS UNDER PREDICTION MODELS FOR SOFTWARE RELIABILITY

Shelemyahu Zacks

Department of Mathematical Sciences,
Binghamton University
Binghamton, NY 13902-6000
shelly@math.binghamton.edu

An adaptive sequential procedure for testing software is developed for a data domain model. The approach is that of predictive finite population modeling, when each modular unit is associated with one or more covariates. For a given link function, expressing the probability that there are faults in a unit as a function of the covariates, and for given costs of testing and failures in the field, a rectifying sampling plan is developed. If the link function is not completely known, the rectifying sampling is adaptive sequential.

1. Introduction

Testing of a particular software is designed to activate certain of its branches (chains of modules) in order to detect, via the system failures, the faults which cause it. This can be done by randomly choosing samples of functional requirements for which the software is designed. The certification (testing) process records the execution times (CPU) for test case samples, which are run against successive increments. The interfailure times of the system are recorded. The most important questions are, which modules should be tested and how long one should continue this testing process? The answer depends on the objectives, cost of testing, cost of failure in the field (after release of the software) etc. The present chapter is devoted to the question of which modules to test, how many modules to test, and in which order to test them. In a sense, this is a sequential (adaptive) design of software testing. The development of a sequential testing scheme depends first of all on the model adopted for software reliability, and the measure-

537

ments, or metrics of software reliability. Most models in the literature, as reflected in the books of Singpurwalla and Wilson[8], Musa[7], Lyn[6] and Xie[10], are time domain models. In time domain models, the reliability of software is defined as the probability that a system driven by a given software will function without failure for a specified length of time. Such models must relate the times between failures to a certain stochastic process, like a filtered non-homogeneous Poisson process, which govern the arrivals of requests for various functional units of the system. In my article on sequential methods in software reliability, Zacks[11], most of the models investigated were time domain models. The papers of Dalal and Mallows[2], Fakhre-Zakeri and Slud[5], Dalal and McIntosh[3] are based on time domain models. van Dorp, Mazzuchi and Soyer[9] developed a sequential decision procedure for a single mission system development.

The adaptive sequential procedure presented in the present chapter is different from the sequential procedures mentioned above. It is based on a data domain rather than time domain model. We consider the software as a static population of modular units. All units have specified functional requirements, and should be free of faults. For simplicity we assume that if all units are tested, then all existing faults are corrected, and the software system is at the end free of errors. This is obviously an ideal situation, while in real cases models should be more complicated. Testing might sometimes not find all faults, or the correction process might introduce new complications. Even under the ideal model, the size of the population is often so large, and the essentiallity of certain modular units is so low, that it is not worthy to test certain units. This complexity is related in the present chapter to known *covariates*, which measure the degree of complexity of the units. The probability that a unit contains faults is related to these covariates by means of a *link function*. An important problem is to determine which covariates should be considered, and what kind of link function to use. In Section 2 we discuss the rectifying sampling problem, and the role of the covariates. When the probability of failure due to faults of a unit is known, one can relate the expected cost of failure in the field (risk) to the cost of testing in the plant and make an optimal decision. Generally, the link function has unknown parameters, which have to be estimated from the observed data. The sequential stopping times developed in Section 3, are based on maximum likelihood estimation of the unknown parameters. A special loss function is introduced to evaluate the goodness of a sequential stopping rule. The expected loss is the risk associated with a given sequential procedure due to incomplete knowledge of the link function. Bayesian sequential procedures are introduced in Section 4. Numerical examples based on simulations are given in Sections 3 and 4 for comparison of the different sequential stopping rules. These examples are based on a simple exponential link function, involving one covariate. The present chapter introduces a new methodology for testing software. More research should be performed to evaluate the effectiveness of this methodology.

2. Rectifying Sampling Problems

Chernoff and Ray[1] studied the problem of optimal stopping times in a rectifying sampling inspection. The classical formulation of the problem is the following one. Let $\mathcal{P}$ be a finite population of N elements, having proportion p, $0 < p < 1$, of defective ones. The loss due to releasing $\mathcal{P}$ with $d = [Np]$ defectives in Kd [\$]. The cost of inspecting (and rectifying) n elements is cn [\$]. All defective elements found during an inspection are replaced by good ones. The risk after inspecting n elements is

$$R(p; n) = (N - n)pK + cn$$
$$= NpK - n(pK - c). \tag{1}$$

If $pK > c$ then it is optimal to inspect all N elements. Otherwise, it is optimal to inspect none. The problem is that the exact value of p is unknown. Chernoff and Ray provided approximations to the optimal Bayes sequential stopping rule, assuming that the unknown value of p has a prior beta (a, b) distribution. In the present paper we discuss the sequential stopping rule problem, when different elements of $\mathcal{P}$ have different penalty values for being released while having a fault, and having different inspection cost. In addition, we assume that a vector $\mathbf{x} = (x_1, \cdots, x_p)'$ of known covariates is available for each element of the population, $\mathcal{P}$.

2.1. *Problem Formulation for Software Reliability*

Let $u_1, u_2, \cdots, u_N$ be the elements of $\mathcal{P}$. With the i-th element ($i = 1, \cdots, N$) we associate the values y_i, $\mathbf{x}_i = (x_{i1}, \cdots, x_{ip})'$, K_i, c_i, where $y_i = 1$ if the element is *with* faults, and $y_i = 0$ otherwise; $\mathbf{x}_i$ is a vector of p known covariates; K_i is the penalty (in the field) due to faults in the i-th element and c_i is the cost of testing and rectifying the i-th elements. The values of y_i ($i = 1, \cdots, N$) are unknown. If an element is tested, the value of y_i is recorded, and the element is rectified. Let $\theta(\mathbf{x}) = P[y = 1 \mid \mathbf{x}]$. The values of the covariates are related to y through a *link model*. Examples are the *logistic regression model*

$$\theta(\mathbf{x}) = \exp(\gamma_0 + \boldsymbol{\gamma}'\mathbf{x})/(1 + \exp(\gamma_0 + \boldsymbol{\gamma}'\mathbf{x})), \tag{2}$$

or the *normit regression model*

$$\theta(\mathbf{x}) = \Phi(\gamma_0 + \boldsymbol{\gamma}'\mathbf{x}), \tag{3}$$

where ($\Phi(\cdot)$ is the standard normal distribution and $\boldsymbol{\gamma} = (\gamma_1, \ldots, \gamma_p)$ is a vector of regression coefficients. Other models could be relevant.

The risk of **not** testing any unit of $\mathcal{P}$ is

$$R_0 = \sum_{i=1}^{N} K_i \theta(\mathbf{x}_i). \tag{4}$$

If, on the other hand, one selects n units for testing, $1 \le n \le N$, having indices in $\mathbf{s} = \{i_1, \ldots, i_n\}$, then the cost of testing plus the risk of not testing the units in $\mathbf{r} = \mathcal{P} - \mathbf{s}$ is

$$R_1 = R_0 - \sum_{i \in \mathbf{s}} (K_i \theta(\mathbf{x}_i) - c_i). \tag{5}$$

Obviously, it is optimal to test **all** units for which $K_i \theta(\mathbf{x}_i) > c_i$. If not all such units can be tested, it is optimal to test units having the largest values of $r_i = K_i \theta(\mathbf{x}_i) - c_i$.

The statistical problem addressed in this paper is three-fold:

(1) Which link model to choose?
(2) Which units should be selected for testing, when the parameters of the model are unknown?
(3) How many units to test?

The answer to question (i) requires to investigate how sensitive are the procedures of choosing the units for testing with respect to the choice of model. This is not done here. We focus attention on the sequential adaptive stopping rules for the simple model of one covariate

$$\theta(x) = e^{-\gamma/x}, \quad 0 < x, \ \gamma < \infty, \tag{6}$$

where γ is an unknown parameter. Since for model (6), $\theta(x)$ is increasing in x, the units to be selected for testing should have maximal x values. The problem is to develop a sequential stopping rule, which is based on sequential estimates of γ.

3. Sequential Stopping Based on the MLE

For simplicity, we start with the following simple problem. $\mathcal{P}$ consists of N units, having x values, $x_n = N + 1 - n$, $n = 1, 2, \ldots, N$. We consider model (6), which in the present case is

$$\theta(x_n) = e^{-\gamma/(N+1-n)}, \tag{7}$$

$0 < \gamma < \infty$. Model (7) is reasonable if the parameter γ is of the same order of magnitude of N. For example, if $N = 100$ and $\gamma = 100$, then the maximal value of $\theta(x)$ is 0.3679. On the other hand, if $N = 1,000$ and $\gamma = 100$ then the maximal value of $\theta(x)$ is 0.9048. We will also assume that $K_i = Kx_i$ and $c_i = cx_i$, $i = 1, \ldots, N$. The proof of the following lemma is immediate from (5).

Lemma 1: *If $K_i = Kx_i$, $c_i = cx_i$, $i = 1, \ldots, N$, then under the link model (7), the optimal number of units to test is*

$$n^0(\gamma) = \left(N - \left[\frac{\gamma}{|\log(R)|}\right]\right)^+, \tag{8}$$

where $R = K/c$, $[a]$ denotes the greatest integer smaller than a, and $(\cdot)^+ = \max(0, (\cdot))$. The sample of units for testing is $s^0 = \{1, 2, \ldots, n^0(\gamma)\}$. □

Notice that if $K < c$ then $n^0(\gamma) = 0$, i.e., no units should be tested. The problem is to determine the units to test and when to stop, when γ is unknown. The sequential adaptive procedure is to select an *initial* sample of size n_0, estimate γ, and substitute the estimate in (8). If after sampling and testing n units, the estimator is $\hat{\gamma}_n$, and $n < n^0(\hat{\gamma}_n)$ then another unit is tested and γ is re-estimated. Also, if there is only one covariate, as in (7), and $\theta(x)$ is an increasing function of x, it is optimal to select for testing, irrespective of γ, the units having the largest value of x.

In the present section we consider $\hat{\gamma}_n$ based on the maximum likelihood estimator of γ. For model (6), given a sample of size n, the likelihood function of γ is

$$L(\gamma; \mathbf{x}_n, \mathbf{y}_n) = \prod_{i=1}^{n} e^{-\gamma y_i/x_i}(1 - e^{-\gamma/x_i})^{1-y_i}. \tag{9}$$

Accordingly, the score function for γ is

$$\begin{aligned}
S(\gamma) &= \frac{\partial}{\partial\gamma} \log L(\gamma; \mathbf{x}_n, \mathbf{y}_n) \\
&= \sum_{i=1}^{n} \frac{e^{-\gamma/x_i}}{x_i(1 - e^{-\gamma/x_i})} - \sum_{i=1}^{n} \frac{y_i}{x_i(1 - e^{-\gamma/x_i})}.
\end{aligned} \tag{10}$$

The MLE of γ is the root of the equation

$$\sum_{i=1}^{n} \frac{e^{-\gamma/x_i}}{x_i(1 - e^{-\gamma/x_i})} = \sum_{i=1}^{n} \frac{y_i}{x_i(1 - e^{-\gamma/x_i})}. \tag{11}$$

The MLE does not exist if $\sum_{i=1}^{n} y_i = 0$, since $e^{-\gamma/x}/(1 - e^{-\gamma/x})$ is a decreasing function of γ with $\lim_{\gamma \to \infty} e^{-\gamma/x}/(1 - e^{-\gamma/x}) = 0$.

When $\sum_{i=1}^{n} y_i > 0$, the MLE of γ is unique and can be obtained numerically. The asymptotic variance of the MLE is the inverse of the Fisher information function which is

$$I_n(\gamma) = \sum_{i=1}^{n} \frac{e^{-\gamma/x_i}}{x_i^2(1 - e^{-\gamma/x_i})}. \tag{12}$$

A lower confidence limit, for γ, in large samples is given by

$$\hat{\gamma}_n^{(L)} = \hat{\gamma}_n - 2\mathrm{ASE}(\hat{\gamma}_n), \tag{13}$$

where

$$\mathrm{ASE}(\hat{\gamma}_n) = \left(\frac{1}{I_n(\hat{\gamma}_n)} \right)^{1/2}. \tag{14}$$

We remark in this connection that, in order to establish the usual asymptotic properties of the MLE one needs to assume that all x_i values are in a closed interval $[x', x'']$, where $0 < x' < x'' < \infty$. This is not satisfied in the present model as $N \to \infty$. As stated earlier, we use the model (7) only for illustration purposes of sequential stopping rule.

A sequential stopping time based on the MLE of γ, $\hat{\gamma}_n$, is then

$$n_{\mathrm{MLE}} = \textbf{least} \;\; n, n \geq n_0, \;\; \textbf{such that} \;\; n = n^0(\hat{\gamma}_n), \tag{15}$$

where n_0 is an initial sample size. Since we expect $\hat{\gamma}_n$ to be approximately normally distributed, in large samples, $n_s^{(1)}$ might yield values below the optimal $n^0(\gamma)$ with probability close to 0.5. We therefore consider also the stopping time

$$n_{\mathrm{MLEL}} = \textbf{least} \;\; n \geq n_0, \;\; \textbf{such that} \;\; n = n^0(\hat{\gamma}_n - 2\mathrm{ASE}(\hat{\gamma}_n)), \tag{16}$$

where $\mathrm{ASE}(\hat{\gamma}_n)$ denotes the asymptotic standard error of the MLE, $\hat{\gamma}_n$.

In order to evaluate the goodness of a sequential stopping rule, n_s, we introduce the following loss function

$$
W(\gamma, n_s) = I\{n_s < n^0(\gamma)\} K \sum_{i=n_s+1}^{n^0(\gamma)} (N+1-i) \cdot
$$
$$
\cdot \, e^{-\gamma/(N+1-i)} + I\{n_s > n^0(\gamma)\} c \sum_{i=n^0(\gamma)+1}^{n_s} (N+1-i). \tag{17}
$$

Here $I\{\cdot\}$ denotes the indicator variable, assuming the value 1 if the relation in brackets is true, and the value 0 otherwise. For a given value of γ, a stopping rule having a smaller expected loss (risk) is preferred. In Table 1 we present the results of $M = 500$ simulation runs, in which $\gamma = 100$, $c = 1[\$]$ and $K = 10[\$]$ and $N = 100$. If the value of γ is known it is optimal to stop testing after $n^0(100) = 57$ units. In all the simulation runs, the initial number of units tested is $n_0 = 20$. We compare several characteristics of the empirical distributions of n_{MLE} and n_{MLEL}, namely the .1, .25, .5, .75 and .9 quantiles, the means and standard deviations.

Table 1. Characteristics of The Empirical Distribution of $M = 500$ Simulation Runs: $n_0 = 20$, $\gamma = 100$, $N = 100$, $R = 10$.

Variable	Quantiles					Mean	Standard Deviation
	0.1	0.25	0.5	0.75	0.90		
n_{MLE}	41	50	56	60.25	65	54.3	8.785
$\hat{\gamma}_{\mathrm{MLE}}$	80.70	90.31	101.38	114.57	135.22	104.43	20.166
$W(100, n_s)$	42	87.36	237	407	1,133.6	406.67	505.172
n_{MLEL}	57	59	67	75	81.1	67.79	8.688
$\hat{\gamma}_{\mathrm{MLEL}}$	59.42	82.1	108.61	139.45	150.0	108.04	32.240
$W(100, n_s)$	0	8.3	375	603	733.8	362.4	264.828

As designed n_{MLEL} is stochastically greater than n_{MLE}. However, since it is ten times cheaper to test in the plant rather than in the field, the average loss of n_{MLEL} is smaller than that of n_{MLE}. The average values of the MLE estimates $\hat{\gamma}_{n_s}$ are not significantly different. It is interesting that the median loss of n_{MLE} is significantly smaller than that of the n_{MLEL}, and so is the third quartile. The average loss of n_{MLE} is greater than that of n_{MLEL} due to the effect of 10% of the largest values.

4. Bayesian Sequential Stopping Rules

4.1. *Bayesian Analysis*

For a Bayesian analysis of model (7) one should specify a prior distribution of γ. We focus attention here on the exponential prior distribution with parameter λ (expected value $1/\lambda$). For this prior, and the likelihood function (9), the posterior density of γ, given the data $\mathcal{D}_n = ((x_i, y_i), i = 1, \ldots, n)$ is

$$h(\gamma \mid (\mathbf{x}_n, \mathbf{y}_n)) = \frac{e^{-\lambda\gamma} L(\gamma; \mathbf{x}_n, \mathbf{y}_n)}{\displaystyle\int_0^\infty e^{-\lambda\gamma'} L(\gamma'; \mathbf{x}_n, \mathbf{y}_n) d\gamma'}, \tag{18}$$

where $L(\gamma; \mathbf{x}_n, \mathbf{y}_n)$ is the likelihood function of γ given $\mathcal{D}_n$. Thus, the Bayesian estimator of γ, for the squared-error loss, is the posterior expectation

$$\hat{\gamma}_{B,n} = \frac{\displaystyle\int_0^\infty \gamma e^{-\lambda\gamma} L(\gamma; \mathbf{x}_n, \mathbf{y}_n) d\gamma}{\displaystyle\int_0^\infty e^{-\lambda\gamma} L(\gamma; \mathbf{x}_n, \mathbf{y}_n) d\gamma}. \tag{19}$$

We computed $\hat{\gamma}_{B,n}$ numerically, after the transformation $u = e^{-\lambda\gamma}$, which yields the formula

$$\hat{\gamma}_{B,n} = \frac{\displaystyle\int_0^1 (-\log u) L\left(-\frac{\log u}{\lambda}; \mathbf{x}_n, \mathbf{y}_n\right) du}{\displaystyle\lambda \int_0^1 L\left(-\frac{\log u}{\lambda}; \mathbf{x}_n, \mathbf{y}_n\right) du}. \tag{20}$$

The numerical integration in (20) can be done by the Gauss quadrature or another similar method (see Evans and Swartz[4]).

4.2. *Bayesian Sequential Stopping Rules*

We first consider a stopping rule analogous to n_{MLE}. This is

$$n_{\mathrm{BA}} = \mathbf{least} \ \ n \ge n_0 \ \mathbf{such \ that} \ \ n = n^0(\hat{\gamma}_{B,n}). \tag{21}$$

A sequence of $M = 500$ simulation runs gave characteristics of the empirical distributions, which are presented in Table 2. Comparison of Tables 1 and 2 shows that generally, the stopping rules n_{MLE} and n_{BA} yield similar results.

However, the .9th quantile of $W(100, n_{\mathrm{BA}})$ is about a half of that of $W(100, n_{\mathrm{MLE}})$. This is apparently due to the tendency of n_{MLE} to stop somewhat sooner than the n_{BA}.

Table 2. Characteristics of n_{BA} and Associated Statistics; $n_0 = 20$, $\gamma = 100$, $N = 100$, $R = 10$, $\lambda = 1/100$.

Variable	Quantiles					Mean	Standard Deviation
	0.1	0.25	0.5	0.75	0.90		
n_{BA}	48	54	59.5	64	67	57.63	9.46
$\hat{\gamma}_{\text{BA}}$	77.80	85.38	94.74	106.50	122.73	99.29	23.14
$W(100, n_{\text{BA}})$	40.02	123	200	375	525	389.49	729.04

We derive another Bayesian stopping variable, which minimizes the **posterior risk**. The posterior risk is the expected value of $W(\gamma, n_s)$, with respect to the posterior distribution of γ, with p.d.f. given by (18). We assume that $R > 1$. According to (8)

$$n^0(\gamma) = \sum_{j=0}^{N}(N - j)I\left\{j \le \frac{\gamma}{\log R} < j + 1\right\}. \tag{22}$$

Let $R_n(n_s) = E\{W(\gamma, n_s) \mid \mathcal{D}_n\}$ denote the posterior risk after n tests, if stopping is at $n_s \ge n$. According to (17)

$$R_n(n_s) = \sum_{j=0}^{N}\int_{j\log R}^{(j+1)\log R}\left[cI\{n_s > N - j\}\sum_{n=N-j+1}^{n_s}(N + 1 - n)\right.$$
$$\left. + KI\{n_s < N - j\}\sum_{n=n_s+1}^{N-j}(N + 1 - n)e^{-\gamma/(N+1-n)}\right]. \tag{23}$$
$$\cdot\, h(\gamma \mid \mathbf{x}_n, \mathbf{y}_n)d\gamma.$$

Also, in (23), if $j = N$ the integral is from $N\log R$ to ∞. After some algebraic manipulations we get

Lemma 2: *The posterior risk corresponding to the loss function (17) is*

$$R_n(n_s) = c\sum_{n=1}^{n_s}(N + 1 - n)\int_{(N+1-n)\log R}^{\infty}h(\gamma \mid \mathbf{x}_n, \mathbf{y}_n)d\gamma$$
$$+ K\sum_{n=n_s+1}^{N}(N + 1 - n)\int_{0}^{(N+1-n)\log R}e^{-\gamma/(N+1-n)}. \tag{24}$$
$$\cdot\, h(\gamma \mid \mathbf{x}_n, \mathbf{y}_n)d\gamma.$$

$\square$

Lemma 3: *The stopping time which minimizes the posterior risk is*

$$n_{MPR} = \textbf{\textit{least }} n, n \geq n_0, \textbf{\textit{ such that }} \int_{(N+1-n)\log R}^{\infty} h(\gamma \mid \mathbf{x}_n, \mathbf{y}_n)d\gamma$$

$$\geq R \int_0^{(N+1-n)\log R} \exp\left(-\frac{\gamma}{N+1-n}\right) h(\gamma \mid \mathbf{x}_n, \mathbf{y}_n)d\gamma. \tag{25}$$

Proof: The optimal stopping time is the first n, $n \geq n_0$, such that $\Delta R_n(n_s) = R_n(n_s) - R_n(n_s - 1) \geq 0$. Moreover,

$$\Delta R_n(n_s) = c(N+1-n) \int_{(N+1-n_s)\log R}^{\infty} h(\gamma \mid \mathbf{x}_n, \mathbf{y}_n)d\gamma$$

$$- K(N+1-n) \int_0^{(N+1-n_s)\log R} e^{-\gamma/(N+1-n_s)} \tag{26}$$

$$\cdot\ h(\gamma \mid \mathbf{x}_n, \mathbf{y}_n)d\gamma.$$

This proves (25). $\qquad\qquad\square$

The integrals in (25) are computed numerically after the change of variable $u = e^{-\lambda\gamma}$. In Table 3 we present the characteristics of the empirical distributions associated with n_{MPR}, obtained by $M = 500$ simulation runs. We see in Table 3 that the mean loss of n_{MPR} is smaller than that of n_{BA}, although its median is about 50% larger (compare with Table 2).

Table 3. Characteristics of n_{MPR} and Associated Statistics; $n_0 = 20$, $\gamma = 100$, $N = 100$, $R = 10$, $\lambda = 1/100$.

Variable	Quantiles					Mean	Standard Deviation
	0.1	0.25	0.5	0.75	0.90		
n_{MPR}	56	60	62.5	67	71	62.72	7.685
$\hat{\gamma}_n$	78.53	84.92	95.8	108.3	119.6	99.06	19.637
$W(100, n_{MPR})$	40.02	200	308	443.64	497	337.46	298.43

5. Discussion

Sequential stopping rules were examined, for testing units from a finite population. The test is a rectifying one, and the cost of testing is smaller than the cost of rectifying in the field. In the context of software reliability, the probability that a unit has faults is a function of one or more **known** covariates. An important component of the solution is the choice of a model $\theta(\mathbf{x})$, which correctly links the covariates with the probability of faults. In the present paper we considered the problem with one covariate. In order to simplify the problem to that of an adaptive estimation of one parameter, we considered the link model (6). The logistic regression model (2) should generally be better. Assuming that the faults probability $\theta(\mathbf{x})$ is monotonically increasing function of $\gamma'\mathbf{x}$, units with larger values of $\gamma'\mathbf{x}$ should be tested first. When γ is unknown, an adaptive procedure to consider is that of estimating γ, after n tests, by $\hat{\gamma}_n$, and selecting for the $(n+1)$-st test from the set of $N - n$ untested ones the unit with maximal value of $\hat{\gamma}'_n\mathbf{x}$. A good sequential stopping rule should be chosen, for terminating the testing. For model (6) we examined the characteristics of four sequential stopping rules, based on the MLE of γ, and on the Bayesian estimation of γ. The Bayesian stopping rules depend on the prior distribution and its prior parameter. We illustrated the stopping variables by simulations, with exponential prior, with $\gamma = 100$ and $\lambda = 1/100$. By changing the value of the prior parameter λ the characteristics of the stopping variables would change. We performed $M = 500$ simulation runs for each one of the stopping variables, and compared them with respect to the quantiles, means and standard deviations of the number of tests, n_s, the estimates of γ at stopping, $\hat{\gamma}_{n_s}$, and the loss, $W(\gamma, n_s)$, due to stopping before the optimal $n^0(\gamma)$, or after it. In two out of the four stopping rules, namely n_{MLE} and n_{BA} the medians of the loss W were found to be considerably smaller than the means. This reflects strong positive skewness. The mean loss of n_{MPR} was found to be smallest. The mean loss of n_{MLE} was found to be the largest. The numerical values are given in Tables 1, 2, and 3.

A new approach is demonstrated in the present chapter for rectifying by sampling modular units. A more comprehensive study of this approach should be done for an actual problem of software testing.

References

1. H. Chernoff and S.N. Ray, A Bayes sequential sampling inspection plan, *Annals of Mathematical Statistics*, **36**, 1387-1407 (1965).
2. S.R. Dalal and C.L. Mallows, When should one stop testing software? *Jour. of the Amer. Statist. Assoc.*, **83**, 872-879, (1988).
3. S.R. Dallal and A.A. McIntosh, When to stop testing for large software systems with changing code. *IEEE T Software Eng.*, **20**, 318-323, (1994).
4. M. Evans and T. Swartz, *Approximating Integrals via Monte Carlo and Deterministic Methods* (Oxford University Press, Oxford, 2000).
5. I. Fakhre-Zakeri and E. Slud, Optimal stopping of sequential size dependent search, Technical Report, Department of Mathematics, University of Maryland, College Park, (1994).

6. M.R. Lyn, *Handbook of Software Engineering* (IEEE Computer Science Press, McGraw-Hill, New York, 1996).

7. D. Musa, *Software Reliability Engineering* (McGraw-Hill, N.J., 1998).

8. N.D. Singpurwalla and S.P. Wilson, *Statistical Methods in Software Engineering* (Springer, New York, 1999).

9. J.R. van Dorp, T.A. Mazzuchi and R. Soyer, Sequential inference and decision making for single mission system development, *Jour. Statist. Planning and Inf.*, **62**, 207-219, (1997).

10. M. Xie, *Software Reliability Modelling* (World Scientific, Singapore, 2002).

11. S. Zacks, Sequential Procedures in Software Reliability Testing, in *Recent Advances in Life-Testing and Reliability*, Ed. N. Balakrishnan (CRC Press, Boca Raton, 1995).